WHAT EVERY CHEMICAL TECHNOLOGIST WANTS TO KNOW ABOUT...

Volume VI

POLYMERS AND PLASTICS

Compiled by

Michael and Irene Ash

Chemical Publishing Co., Inc.
New York, N.Y.

Polymers and Plastics Volume 6

ISBN: 978-0-8206-0054-3

Chemical Publishing Company:
www.chemical-publishing.com
www.chemicalpublishing.net

First Edition:

© **Chemical Publishing Company, Inc.** - New York 1990

Second Impression:

Chemical Publishing Company, Inc. - 2011

Printed in the United States of America

PREFACE

This reference book is the sixth volume in the set of books entitled WHAT EVERY CHEMICAL TECHNOLOGIST WANTS TO KNOW ... SERIES. This compendium serves a unique function for those involved in the chemical industry—it provides the necessary information for making the decision as to which trademark chemical product is most suitable for a particular application.

The chemicals included in this sixth book of the series are polymers and plastic materials, however, complete cross-referencing is provided for the multiple functions or classification of all the chemicals.

The first section which is the major portion of each volume contains the most common generic name of the chemicals as the main entry. All these generic entries are in alphabetical order. Synonyms for these chemicals are then listed. The CTFA name appears alongside the appropriate generic name. The structural and/or molecular formula of the chemical is listed whenever possible. The generic chemical is sold under various tradenames and these are listed here in alphabetical order for ease of reference along with their manufacturer in parentheses. *Modifications/Specialty Grades* lists those tradenames that contain or omit additives so that they are used for a variety of specialty purposes. The *Category* subheading classifies the polymer by thermoplastic and/or thermoset types, synthetic rubber, or by functionality, as appropriate. The *Processing* category lists the tradenames that have been formulated to be used for specific or generalized plastics processing. Because of differences in form, activity, etc., individual tradenames of the generic chemical are used in particular applications more frequently. These are delineated in the *Applications* section. The differences in properties, toxicity/handling, storage/handling, and standard packaging are specified in the subsequent sections wherever distinguishing characteristics are known.

The second section of the volume TRADENAME PRODUCTS AND GENERIC EQUIVALENTS helps the user who only knows a chemical by one tradename to locate its main entry in section 1. The user can look up the tradename in this section of the book and be referred to the appropriate, main-entry, generic chemical name.

The third section GENERIC CHEMICAL SYNONYMS AND CROSS REFERENCES provides a way of locating the main entries by knowing only one of the synonyms.

NOTE

The information contained in this series is accurate to the best of our knowledge; however, no liability will be assumed by the publisher for the correctness or comprehensiveness of such information. The determination of the suitability of any of the products for prospective use is the responsibility of the user. It is herewith recommended that those who plan to use any of the products referenced seek the manufacturer's instructions for the handling of that particular chemical.

OTHER BOOKS BY MICHAEL AND IRENE ASH

ABBREVIATIONS

@	at
anhyd.	anhydrous
APHA	American Public Health Association
approx.	approximately
aq.	aqueous
ASTM	American Society for Testing and Materials
avg.	average
B.P.	boiling point
Btu	British thermal unit
C	degrees Centigrade
CAS	Chemical Abstracts Service
cc	cubic centimeter(s)
CC	closed cup
cm	centimeter(s)
cm^3	cubic centimeter(s)
COC	Cleveland Open Cup
compd.	compound, compounded
conc.	concentrated, concentration
cP, cps	centipoise
cs, cSt	centistokes
CTFA	Cosmetic, Toiletry and Fragrance Association
DEA	diethanolamine
disp	dispersible, dispersion
dist	distilled
DOT	Department of Transportation
DW	distilled water
EO	ethylene oxide
equiv.	equivalent
F	degrees Fahrenheit
F.P.	freezing point
FDA	Food and Drug Administration
ft^3	cubic foot, cubic feet
g	gram(s)
gal	gallon(s)
HLB	hydrophile-lipophile balance
insol.	insoluble
IPA	isopropyl alcohol
kg	kilogram(s)
l, L	liter(s)
lb	pound(s)
max	maximum
M.D.	mold direction
MEA	monoethanolamine
MEK	methyl ethyl ketone
mfg.	manufacture
MIBK	methyl isobutyl ketone
min	minute(s)
min.	mineral, minimum
MIPA	monoisopropanolamine
misc.	miscible

ml	milliliter(s)
mm	millimeter(s)
M.P.	melting point
M.W.	molecular weight
NF	National Formulary
no.	number
o/w	oil-in-water
OC	open crucible
PEG	polyethylene glycol
pH	hydrogen-ion concentration
pkgs	packages
PMCC	Pensky Marten closed cup
POE	polyoxyethylene, polyoxyethylated
POP	polyoxypropylene
PPG	polypropylene glycol
pt.	point
R&B	Ring & Ball
RD	Recognized Disclosure
ref.	refractive
rpm	revolutions per minute
R.T.	room temperature
s	second(s)
sol.	soluble, solubility
sol'n.	solution
sp.gr.	specific gravity
SS	stainless steel
std.	standard
SUS	Saybolt Universal seconds
TCC	Taggart closed cup
T.D.	transverse direction
TEA	triethanolamine
tech.	technical
temp.	temperature
theoret.	theoretical
TLV	threshold limit value
TOC	Taggart open cup
UL	Underwriter's Laboratory
USP	United States Pharmacopoeia
uv, UV	ultraviolet
veg	vegetable
visc.	viscosity, viscous
w/o	water-in-oil
wt	weight
≈	approximately equal to
<	less than
>	greater than
≤	less than or equal to
≥	greater than or equal to

TABLE OF CONTENTS

Acrylamides copolymer (CTFA)

TRADENAME EQUIVALENTS:
Reten 521 [Hercules]
Rheolate 1 [NL Treating]

CATEGORY:
Gelling agent, thickener, flocculating agent, retention aid, slip agent, antistat, film former, suspending agent, crosslinking agent

APPLICATIONS:
Industrial applications: adhesives (Rheolate 1; Reten 521); coatings (Rheolate 1); latexes (Rheolate 1); paints (Rheolate 1); paper and pulp industry (Reten 521)

PROPERTIES:
Form:
Emulsion (Rheolate 1)
Powder (Reten 521)
Color:
White (Reten 521)
Milky white (Rheolate 1)
Composition:
30% nonvolatiles in water (Rheolate 1)

GENERAL PROPERTIES:
Ionic Nature:
Anionic (Reten 521)
Solubility:
Sol. in alkali (Rheolate 1)
Dissolves readily in water (Reten 521)
Density:
1.07 g/cm³ (Rheolate 1)
705 kg/m³ (Reten 521)
Visc.:
3300 cps (1%) (Reten 521)

Acrylamide/sodium acrylate copolymer (CTFA)

SYNONYMS:
2-Propenamide, polymer with 2-propenoic acid, sodium salt
2-Propenoic acid, sodium salt, polymer with 2-propenamide

EMPIRICAL FORMULA:
$(C_3H_5NO \cdot C_3H_4O_2 \cdot Na)_x$

CAS No.:
25085-02-3

TRADENAME EQUIVALENTS:
Reten 421, 423, 425 [Hercules]

CATEGORY:
Thickener, suspending agent, flocculant, slip agent, antistat, film-former, adhesive, crosslinking agent

PROPERTIES:
Form:
Powder (Reten 421, 423, 425)

GENERAL PROPERTIES:
Ionic Nature:
Anionic
Solubility:
Sol. in warm or cold water (Reten 421, 423, 425)

Acrylates/steareth-20 methacrylate copolymer (CTFA)

TRADENAME EQUIVALENTS:
Acrysol ICS-1 [Rohm & Haas]

CATEGORY:
Thickener

APPLICATIONS:
Consumer products: cosmetics/toiletries (Acrysol ICS-1); household cleaners (Acrysol ICS-1)
Industrial applications: industrial cleaners (Acrysol ICS-1)

PROPERTIES:
Form:
Liquid emulsion (Acrysol ICS-1)

GENERAL PROPERTIES:
Solubility:
Sol. in alkali (Acrysol ICS-1)

Acrylonitrile-butadiene copolymer

SYNONYMS:
Acrylonitrile rubber
Acrylonitrile-butadiene rubber
Butadiene-acrylonitrile copolymer
NBR
Nitrile-butadiene rubber
Nitrile rubber

EMPIRICAL FORMULA:
—$CH_2CH=CHCH_2CH_2CH(CN)$—

STRUCTURE:

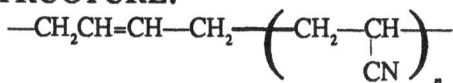

TRADENAME EQUIVALENTS:
Chemigum NBR Polymers [Goodyear]
Darex 110L [W.R. Grace]
Hycar 1312, 1422, 1422X8, 1452P50, Butadiene-Acrylonitrile Rubber, Nitrile, Nitrile Latices [Goodrich]
Krynac 19.65, 21.65, 25.65, 27.50, 29.60, 32.55, 34.35. 34.50, 34.80, 34.140, 34.E50, 34.E80, 38.50, 40.65, 45,55, 50.75, 800, 803, 810, 822, 825, 826, 826, 843 [Polysar]
Nobestos/D-7280 [Rogers]
Perbunan N 1807 NS, N 2802 NS, N 2807 NS, N 2810, N 2818 NS, N 3302 NS, N 3307 NS, N 3310, N 3312 NS, N 3807 NS, N 3810 [Bayer AG]
Perbunan N Latex 1590, 3090, 2818, 3310, 3310 HD, 3810, 3415 M, T, VT [Bayer AG]

MODIFICATIONS/SPECIALTY GRADES:
Carboxy-modified:
Hycar Nitrile
Cross-linked:
Krynac 810
DOP-modified:
Krynac 843 (50 phr DOP)

CATEGORY:
Synthetic rubber

PROCESSING:
Calendering:
Krynac 803, 810
Compression molding:
Krynac 800
Extrusion:
Krynac 34.E50, 34.E80, 810

Acrylonitrile-butadiene copolymer *(cont'd.)*

Injection molding:
Krynac 27.50, 29.60, 34.35, 822, 825, 826, 827

APPLICATIONS:

Automotive applications: brake/clutch linings (Perbunan N 1807 NS, N 2802 NS, N 2807 NS, N 2810, N 2818 NS, N 3302 NS, N 3307 NS, N 3310, N 3312 NS, N 3807 NS, N 3810; Perbunan N Latex 1590, 2818, 3090, 3310, 3310 HD, 3415 M, 3810, T, VT); door panels (Chemigum NBR Polymers); fuel diaphragms (Krynac 50.75); timing chain bumper (Hycar Nitrile)

Consumer products: footwear (Chemigum NBR Polymers; Hycar Butadiene-Acrylonitrile Rubber; Krynac 34.35, 34.50, 803, 826; Perbunan N 1807 NS, N 2802 NS, N 2807 NS, N 2810, N 2818 NS, N 3302 NS, N 3307 NS, N 3310, N 3312 NS, N 3807 NS, N 3810)

Electrical/electronic industry: cable and wire jacketing (Chemigum NBR Polymers; Perbunan N 1807 NS, N 2802 NS, N 2807 NS, N 2810, N 2818 NS, N 3302 NS, N 3307 NS, N 3310, N 3312 NS, N 3807 NS, N 3810)

Functional additives: binder (Perbunan N Latex 1590, 2818, 3090, 3310, 3310 HD, 3415 M, 3810, T, VT); impact modifier (Krynac 810); plasticizer (Hycar 1312, 1422, 1422X8, 1452P50, Butadiene-Acrylonitrile Rubber; Krynac 34.140, 810, 843); viscosity modifier (Krynac 34.140)

Industrial applications: adhesives (Hycar Nitrile Latices; Krynac 38.50; Perbunan N 1807 NS, N 2802 NS, N 2807 NS, N 2810, N 2818 NS, N 3302 NS, N 3307 NS, N 3310, N 3312 NS, N 3807 NS, N 3810); belting/belt covers (Krynac 19.65, 21.65, 32.55, 34.50, 38.50, 800, 803, 825; Perbunan N 1807 NS, N 2802 NS, N 2807 NS, N 2810, N 2818 NS, N 3302 NS, N 3307 NS, N 3310, N 3312 NS, N 3807 NS, N 3810); coatings (Hycar Nitrile Latices); conveyor belts (Hycar Butadiene-Acrylonitrile Rubber); diaphragms (Perbunan N 1807 NS, N 2802 NS, N 2807 NS, N 2810, N 2818 NS, N 3302 NS, N 3307 NS, N 3310, N 3312 NS, N 3807 NS, N 3810); dipped goods (Perbunan N Latex 1590, 2818, 3090, 3310, 3310 HD, 3415 M, 3810, T, VT); drive wheels (Hycar Nitrile); fabric proofings (Perbunan N 1807 NS, N 2802 NS, N 2807 NS, N 2810, N 2818 NS, N 3302 NS, N 3307 NS, N 3310, N 3312 NS, N 3807 NS, N 3810; Perbunan N Latex 1590, 2818, 3090, 3310, 3310 HD, 3415 M, 3810, T, VT); friction stocks (Krynac 34.35); gaskets/seals (Krynac 25.65, 32.55, 40.65, 800, 822; Nobestos/D-7280; Perbunan N 1807 NS, N 2802 NS, N 2807 NS, N 2810, N 2818 NS, N 3302 NS, N 3307 NS, N 3310, N 3312 NS, N 3807 NS, N 3810; Perbunan N Latex 1590, 2818, 3090, 3310, 3310 HD, 3415 M, 3810, T, VT); inks (Hycar Nitrile Latices); insulation (Chemigum NBR Polymers); mechanical goods (Krynac 843); molded goods (Krynac 27.50, 29.60, 34.35, 826, 287); oil seals (Hycar Butadiene-Acrylonitrile Rubber); packings (Krynac 34.50, 34.80, 38.50, 40.65); paper and felt industries (Darex 110L; Hycar Nitrile Latices); paper coatings (Perbunan N Latex 1590, 2818, 3090, 3310, 3310 HD, 3415 M, 3810, T, VT); petroleum industry (Krynac 19.65, 21.65, 40.65, 825); plastics modification (Hycar 1422, 1422X8, 1452P50, Butadiene-Acrylonitrile Rubber; Krynac 810); plastisols (Hycar 1312); printer rolls/blankets (Chemigum NBR Polymers; Krynac

Acrylonitrile-butadiene copolymer *(cont'd.)*

32.55, 34.50, 34.80, 50.75, 800; Perbunan N 1807 NS, N 2802 NS, N 2807 NS, N 2810, N 2818 NS, N 3302 NS, N 3307 NS, N 3310, N 3312 NS, N 3807 NS, N 3810); pumps (Krynac 822, 826); rollers/roll covers (Hycar Butadiene-Acrylonitrile Rubber, Nitrile; Krynac 19.65, 21.65, 25.65, 34.35, 34.50, 34.E50, 34.E80, 38.50, 803, 843; Perbunan N 1807 NS, N 2802 NS, N 2807 NS, N 2810, N 2818 NS, N 3302 NS, N 3307 NS, N 3310, N 3312 NS, N 3807 NS, N 3810); rubber (Hycar 1312); sponge rubber (Perbunan N 1807 NS, N 2802 NS, N 2807 NS, N 2810, N 2818 NS, N 3302 NS, N 3307 NS, N 3310, N 3312 NS, N 3807 NS, N 3810); textile applications (Hycar Nitrile Latices); textile coatings (Perbunan N Latex 1590, 2818, 3090, 3310, 3310 HD, 3415 M, 3810, T, VT); tubing/hoses/fuel hoses (Hycar Butadiene-Acrylonitrile Rubber; Krynac 25.65, 27.50, 29.60, 34.50, 34.80, 34.E50, 34.E80, 40.65, 45.55, 800, 827; Perbunan N 1807 NS, N 2802 NS, N 2807 NS, N 2810, N 2818 NS, N 3302 NS, N 3307 NS, N 3310, N 3312 NS, N 3807 NS, N 3810); valves (Perbunan N 1807 NS, N 2802 NS, N 2807 NS, N 2810, N 2818 NS, N 3302 NS, N 3307 NS, N 3310, N 3312 NS, N 3807 NS, N 3810); vibration damping (Perbunan N 1807 NS, N 2802 NS, N 2807 NS, N 2810, N 2818 NS, N 3302 NS, N 3307 NS, N 3310, N 3312 NS, N 3807 NS, N 3810)

Marine equipment: fuel hose (Krynac 45.55, 50.75)

PROPERTIES:

Form:

Liquid (Chemigum NBR Polymers; Hycar 1312)

Powder (Hycar 1422, 1422X8, 1452P50, Butadiene-Acrylonitrile Rubber)

Solid (Hycar Butadiene-Acrylonitrile Rubber, Nitrile)

Bales (Perbunan N 2818 NS, N 3302 NS, N 3307 NS, N 3310, N 3312 NS, N 3807 NS, N 3810)

Crumb (Perbunan N 2818 NS, N 3810)

Composition:

15% ACN (Perbunan N Latex 1590)

18 ± 1% ACN (Perbunan N 1807 NS)

19% bound ACN (Krynac 19.65)

21% bound ACN (Krynac 21.65)

25% bound ACN (Krynac 25.65)

27% ACN (Perbunan N Latex 2818); 27% bound ACN (Krynac 27.50)

28 ± 1% ACN (Perbunan N 2802 NS, N 2807 NS, N 2810, N 2818 NS)

29% bound ACN (Krynac 29.60, 827)

30% ACN (Perbunan N Latex 3090, VT); 30% bound ACN (Krynac 810)

32% ACN (Perbunan N Latex 3310, 3310 HD); 32% bound ACN (Krynac 32.55, 822)

33% ACN (Perbunan N Latex 3415 M)

34% bound ACN (Krynac 34.35, 34.50, 34.80, 34.140, 34.E50, 34.E80, 800, 803, 843)

34 ± 1% ACN (Perbunan N 3302 NS, N 3307 NS, N 3310, N 3312 NS)

35% ACN (Perbunan N Latex T)

36% bound ACN (Krynac 825, 826)

37% ACN (Perbunan N Latex 3810)

Acrylonitrile-butadiene copolymer (cont'd.)

38% bound ACN (Krynac 38.50)
39 ± 1% ACN (Perbunan N 3807 NS, N 3810)
40% bound ACN (Krynac 40.65)
40% solids (Perbunan N Latex 1590, 3090, 3310 HD)
45% bound ACN (Krynac 45.55)
45% solids (Perbunan N Latex 2818, 3310, 3810, VT)
47% solids (Darex 110L)
47.5% solids (Perbunan N Latex 3415 M)
50% bound ACN (Krynac 50.75)
50% solids (Perbunan N Latex T)

GENERAL PROPERTIES:

Solubility:
Sol. in acetone (Hycar 1312, 1452P50)
Sol. in MEK (Hycar 1312, 1452P50)

Sp. Gr.:
0.96 (Krynac 19.65, 21.65, 25.65, 27.50; Perbunan N 1807 NS)
0.97 (Krynac 29.60, 827)
0.98 (Krynac 32.55, 34.35, 34.50, 34.80, 34.140, 34.E50, 34.E80, 38.50, 800, 803, 810, 822, 825, 826; Perbunan N 2802 NS, N 2807 NS, N 2810, N 2818 NS; Perbunan N Latex 2818, 3810, 3415 M, T)
0.99 (Krynac 40.65; Perbunan N 3302 NS, N 3307 NS, N 3310, N 3312 NS; Perbunan N Latex 1590, 3310, 3310 HD)
0.998 (Perbunan N Latex VT)
0.999 (Perbunan N Latex 3090)
1.00 (Krynac 45.55, 50.75; Perbunan N 3807 NS, N 3810)

Density:
8.4 lb/gal (Darex 110L)

Visc.:
30 cps (Darex 110L)
Mooney 30 ± 5 (ML4, 100 C) (Perbunan N 2802 NS, N 3302 NS)
Mooney 35 (100 C) (Krynac 34.35)
Mooney 38 (100 C) (Krynac 843)
Mooney 40 (100 C) (Krynac 826)
Mooney 45 ± 5 (ML4, 100 C) (Perbunan N 1807 NS, N 2807 NS, N 3307 NS, N 3807 NS)
Mooney 47 (100 C) (Krynac 803, 822, 825)
Mooney 48 (100 C) (Krynac 827)
Mooney 50 (100 C) (Krynac 27.50, 34.50, 34.E50, 38.50)
Mooney 55 (100 C) (Krynac 32.55, 45.55)
Mooney 60 (100 C) (Krynac 29.60, 810)
Mooney 65 (100 C) (Krynac 19.65, 21.65, 25.65, 40.65)
Mooney 65 ± 7 (ML4, 100 C) (Perbuanan N 2810, N 3310, N 3810)
Mooney 75 (100 C) (Krynac 50.75)

Mooney 79 (100 C) (Krynac 800)
Mooney 80 (100 C) (Krynac 34.80, 34.E80)
Mooney 80 ± 5 (ML4, 100 C) (Perbunan 3312 NS)
Mooney 95 ± 7 (ML4, 100 C) (Perbunan N 2818 NS)
Mooney 140 (100 C) (Krynac 34.140)
pH:
4.5–5.5 (Perbunan N Latex 3090)
5.5–6.5 (Perbunan N Latex 3415 M, T)
7.0 (Darex 110L)
7.0–8.0 (Perbunan N Latex 1590)
8.0–9.0 (Perbunan N Latex VT)
10–12 (Perbunan N Latex 2818, 3310, 3310 HD, 3810)
Surface Tension:
40 dynes/cm (Darex 110L)
MECHANICAL PROPERTIES:
Tens. Str.:
13.8 MPa (Nobestos/D-7280)

Adipic acid/dimethylaminohydroxypropyl diethylenetriamine copolymer (CTFA)

STRUCTURE:

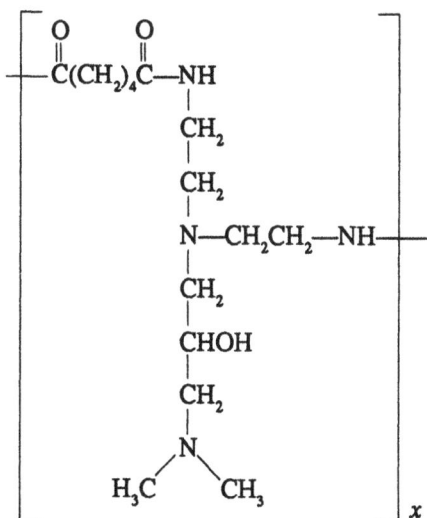

Adipic acid/dimethylaminohydroxypropyl diethylenetriamine copolymer *(cont'd.)*

CAS No.:
61840-27-5
RD No. 977026-96-2

TRADENAME EQUIVALENTS:
Cartaretin F-4, F-8 [Sandoz]

CATEGORY:
Conditioner, wetting agent

APPLICATIONS:
Cosmetic industry preparations: hair products (Cartaretin F-4, F-8); skin care products (Cartaretin F-4, F-8)
Industrial applications: sizing (Cartaretin F-4, F-8)

PROPERTIES:
Form:
Liquid (Cartaretin F-4, F-8)
Color:
Light yellow (Cartaretin F-4)
Amber (Cartaretin F-8)
Odor:
Natural (Cartaretin F-4, F-8)
Composition:
30% active in water (Cartaretin F-4, F-8)

GENERAL PROPERTIES:
Ionic Nature:
Cationic (Cartaretin F-4, F-8)
Solubility:
Sol. in alcohol (Cartaretin F-4, F-8)
Sol. in water (Cartaretin F-4, F-8)
Sp. Gr.:
1.07 (Cartaretin F-4, F-8)
Visc.:
400 cps (Cartaretin F-4)
800 cps (Cartaretin F-8)
F.P.:
–4 C (Cartaretin F-4, F-8)
B.P.:
220 F (Cartaretin F-4, F-8)
Stability:
Very stable (Cartaretin F-4, F-8)

STD. PKGS.:
500 lb fiber drums (Cartaretin F-4, F-8)

Ammonium polyacrylate

TRADENAME EQUIVALENTS:
Acrysol G-100 [Rohm & Haas]
Alcogum 9639 [Alco]
Alcosperse 249 [Alco]
Colloid 102, 118 [Colloids]
Daxad 37LA7 [W.R. Grace]
Dispersol 130 [Aquatec Quimica]
Serpol QPA 150 [Servo]

CATEGORY:
Dispersant, thickener, stabilizer

APPLICATIONS:
Industrial applications: adhesives (Acrysol G-110); carpet backing (Acrysol G-110); cast goods (Acrysol G-110; Alcogum 9639); cements (Acrysol G-110); ceramics (Alcosperse 249; Colloid 102); clay slurries (Colloid 102); coatings (Acrysol G-110; Daxad 37LA7); dipped goods (Acrysol G-110; Alcogum 9639); extrusion compounds (Acrysol G-110); latexes (Acrysol G-110; Alcogum 9639; Daxad 37LA7); molded goods (Acrysol G-110; Alcogum 9639); paints (Daxad 37LA7; Serpol QPA 150); paper coatings (Alcosperse 249; Colloid 102); pigment dispersions (Alcosperse 249; Daxad 37LA7; Serpol QPA 150); spraying, spreading, brushing compounds (Acrysol G-110); textile applications (Alcosperse 249); water treatment compounds (Colloid 118)

PROPERTIES:
Form:
Liquid (Alcosperse 249; Colloid 118; Dispersol 130)
Slightly hazy liquid (Acrysol G-110; Daxad 37LA7)
Aq. solution (Alcogum 9639)
Color:
Colorless (Acrysol G-110)
Slight amber (Daxad 37LA7)
Composition:
40% total solids (Daxad 37LA7)

GENERAL PROPERTIES:
Ionic Nature:
Anionic (Alcosperse 249; Daxad 37LA7)
Solubility:
Sol. in glycols (Daxad 37LA7)
Sol. in water (Alcopserse 249; Colloid 102; Daxad 37LA7; Serpol QPA 150)
Sp. Gr.:
1.06 (Acrysol G-110)
1.17 (Daxad 37LA7)
Density:
9.8 lb/gal (Daxad 37LA7)

Ammonium polyacrylate *(cont'd.)*

Visc.:
   80 cps (Daxad 37LA7)
   90–170 cps (5% solids in water) (Acrysol G-110)
Stability:
   Good (Acrysol G-110)
pH:
   7.0 (Daxad 37LA7)
   8.5–9.5 (Acrysol G-110)
Surface Tension:
   66 dynes/cm (1%) (Daxad 37LA7)

Carbomer 940 (CTFA)

CAS No.:
9003-01-4 (TSCA); 9007-17-4
TRADENAME EQUIVALENTS:
Acritamer 940 [RITA]
Carbopol 940 [B.F. Goodrich]
CATEGORY:
Thickener, viscosity agent, emulsifier, stabilizer, suspending agent
APPLICATIONS:
Cosmetic industry preparations: water and hydroalcoholic gels (Carbopol 940)
Industrial applications: lubricants (Carbopol 940); solvents (Carbopol 940); thixotropic paints (Carbopol 940)
PROPERTIES:
Form:
Powder (Carbopol 940)
Compositon:
100% conc. (Carbopol 940)
GENERAL PROPERTIES:
Ionic Nature:
Anionic (Carbopol 940)
Solubility:
Sol. in many nonpolar solvent blends (Carbopol 940)
Sol. in polar solvents (Carbopol 940)
Sol. in water (Carbopol 940); dispersible (Acritamer 940)

Carbomer 941 (CTFA)

CAS No.:
9003-01-4 (TSCA)
RD No.: 977056-22-6
TRADENAME EQUIVALENTS:
Acritamer 941 [RITA]
Carbopol 941 [B.F. Goodrich]

Carbomer 941 *(cont'd.)*

CATEGORY:
Thickener, viscosity agent, emulsifier, stabilizer

APPLICATIONS:
Cosmetic industry preparations: gels (Carbopol 941); lotions (Carbopol 941); shampoos (Carbopol 941)

PROPERTIES:
Form:
Powder (Carbopol 941)
Compositon:
100% conc. (Carbopol 941)

GENERAL PROPERTIES:
Ionic Nature:
Anionic (Carbopol 941)
Solubility:
Sol. in many nonpolar solvent blends (Carbopol 941)
Sol. in polar solvents (Carbopol 941)
Sol. in water (Carbopol 941); dispersible (Acritamer 941)

Cyclomethicone *(CTFA)*

SYNONYMS:
Cyclic dimethylsiloxane
Cyclic dimethylpolysiloxane

STRUCTURE:

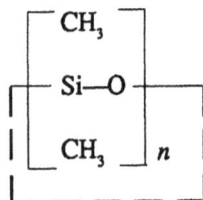

where avg. $n = 3\text{--}6$

CAS No.:
69430-24-6

TRADENAME EQUIVALENTS:
SF 1173, 1202, 1204 [General Electric]
SWS-03314 [Wacker Silicones]
Volatile Silicones 7158, 7207, 7349 [Union Carbide]

CATEGORY:
Lubricant, emollient
APPLICATIONS:
Bath products: bath oils (Volatile Silicones 7158, 7207, 7349)

Cosmetic industry preparations: (SWS-03314); conditioners (SF 1173, 1202, 1204); creams and lotions (Volatile Silicones 7158, 7207, 7349); hair preparations (SF 1173, 1202, 1204; Volatile Silicones 7158, 7207, 7349); makeup (SF 1173, 1202, 1204); perfumery (Volatile Silicones 7158, 7207, 7349); shaving preparations (Volatile Silicones 7158, 7207, 7349); skin preparations (SF 1173, 1202, 1204; Volatile Silicones 7158, 7207, 7349)

Industrial applications: particle treatment (SF 1173, 1202, 1204)

Pharmaceutical applications: antiperspirant/deodorant (SF 1173, 1202, 1204; Volatile Silicones 7158, 7207, 7349); sunscreens (SF 1173, 1202, 1204; Volatile Silicones 7158, 7207, 7349)

PROPERTIES:
Form:

Liquid (SF 1173, 1202, 1204; Volatile Silicones 7158, 7207, 7349)
GENERAL PROPERTIES:
Solubility:

Sol. in aerosol propellants (fluorocarbons, hydrocarbons) (Volatile Silicones 7158, 7207, 7349)

Sol. in alcohols (SWS-03314; Volatile Silicones 7158, 7207, 7349); sol. in lower alcohols (SF 1173, 1202, 1204)

Sol. in aliphatic hydrocarbons (SF 1173, 1202, 1204; SWS-03314)

Sol. in aromatic hydrocarbons (SF 1173, 1202, 1204; SWS-03314)

Sol. in chlorinated hydrocarbons (SF 1202, 1204; SWS-03314)

Sol. in fatty esters (Volatile Silicones 7158, 7207, 7349)

Sol. in ethanol (Volatile Silicones 7158, 7207, 7349)

Sol. in halogenated hydrocarbons (SF 1173, 1202, 1204)

Sol. in mineral oil (Volatile Silicones 7158, 7207, 7349)

Insol. in water (SF 1173, 1202, 1204; Volatile Silicones 7158, 7207, 7349)

Sp. Gr.:

0.954 (SF 1173, 1204)

0.955 (SF 1202)

F.P.:

−40 C (SF 1202)

11 C (SF 1204)

17 C (SF 1173)

B.P.:

175 C (SF 1173)

175–210 C (SF 1204)

190–210 C (SF 1202)

Cyclomethicone *(cont'd.)*

Flash Pt.:
    55 C (CC) (SF 1173, 1204)
    82 C (CC) (SF 1202)
Storage Stability:
    Excellent (SF 1173, 1202, 1204)
Ref. Index:
    1.394 (SF 1173, 1204)
    1.395 (SF 1202)

TOXICITY/HANDLING:
Will cause minor, temporary eye discomfort on direct contact (SF 1173, 1202, 1204)

STORAGE/HANDLING:
Keep containers tightly closed, free from moisture, and away from heat, sparks, and open flame (SF 1173, 1202, 1204)

Dimethicone (CTFA)

SYNONYMS:
Dimethylpolysiloxane
Dimethyl silicone
Dimethylsiloxane
PDMS
Polydimethylsiloxane
Poly [oxy (dimethylsilylene)], α-(trimethylsilyl)-ω-methyl-

EMPIRICAL FORMULA:
$(C_2H_6OSi)_xC_4H_{12}Si$

STRUCTURE:

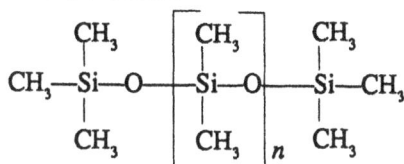

CAS No.:
9006-65-9; 9016-00-6; 63148-62-9

TRADENAME EQUIVALENTS:
Abil 10–10,000 Series [Th. Goldschmidt AG]
AF 60, 66, 72, 75, 9020 [General Electric]
Akrochem Silicone Emulsion 350, 350 Conc., 1M, 10M, 60M [Akron Chem.]
Akrochem Silicone Fluids 350, 1M, 5M, 10M, 30M, 60M [Akron Chem.]
Akrochem SWS-201 [Akron Chem.]
Antifoam Compound SWS-201, SWS-202, SWS-203 [Wacker Silicones]
Dow Corning 200 Fluid, 1500 Silicone Antifoam, 1520 Silicone Antifoam, FI-1630 [Dow Corning]
F-751 [Wacker Silicones]
Foamkill 810F, 830F [Crucible]
Masil EM-14, -62, -100, -100 Conc., -100D, -100P, -250 Conc., -350, -350X, -350X Conc., -1000, -1000 Conc., -1000P, -10,000, -10,000 Conc., -60,000, -100,000, -N [PPG-Mazer]
Masil SF 5, 20, 50, 100, 200, 350, 500, 1000, 5000, 10,000, 12,500, 30,000, 60,000, 100,000, 300,000, 600,000 [PPG-Mazer]
Sag Silicone Antifoam 10, 30, 100, 720 [Union Carbide]
SF 18, 69, 81, 96, 97, 1093 [General Electric]

Dimethicone (cont'd.)

Silicone AF-10FG, AF-10 IND, AF-30 FG, AF-30 IND, AF-600M [Harcros]
Silicone C111 [ICI Specialty]
Silicone Fluid 200, 203, 225, 230 [Dow Corning]
Silicone Mold Release SEM-35 [Harcros]
Silicone Release Agent #5038 [Polymer Research Corp. of Amer.]
SM 2061, 2140, 2155, 2162 [General Electric]
SWS-101, -230 Series, -231, -232, -235 [Wacker Silicones]
Union Carbide Dimethyl Fluid L-45 [Union Carbide]
Union Carbide LE-45, LE-46, LE-420, LE-453HS, LE-458HS, LE-461, LE-462, LE-462HS, LE-467, LE-467HS [Union Carbide]
Viscasil 5M, 10M, 60M [General Electric]

MODIFICATIONS/SPECIALTY GRADES:

Emulsion:

AF 60 (aq.), 72, 75, 9020 (aq.); Akrochem Silicone Emulsion 350, 350 Conc., 1M, 10M, 60M; Dow Corning 1520 Silicone Antifoam; Masil EM-14 (aq.), -62 (aq.), -100 (aq.), -100 Conc. (aq.), -100D (aq.), -100P (aq.), -250 Conc. (aq.), -350 (aq.), 350X (aq.), -350X Conc. (aq.), -1000 (aq.), -1000 Conc. (aq.), -1000P (aq.), -10,000 (aq.), -10,000 Conc. (aq.), -60,000 (aq.), -100,000 (aq.), EM-N (aq.); Sag Silicone Antifoam 10, 30, 720; Silicone AF-10 FG, AF-10 IND, AF-30 FG, AF-30 IND, AF-600M; Silicone Mold Release SEM-35 (aq.); Silicone Release Agent #5038; SM 2061 (o/w), 2162; SWS-230 Series, -231, -232, -235; Union Carbide LE-45 (o/w), LE-46 (o/w), -420 (modified aq.), -453HS, -458HS, -461, -462, -462HS, -467, -467HS

Filled:

AF 66; Antifoam Compound SWS-201 (silica), SWS-203; Dow Corning 1500 Silicone Antifoam (silica), 1520 Silicone Antifoam (silica); Sag Silicone Antifoam 100 (silica)

Amine functional:

F-751; SF 1705

Food grade:

AF 72, 75; Sag Silicone Antifoam 720; Silicone AF-10 FG, AF-30 FG

CATEGORY:

Defoamer, antifoamer, mold release agent, slip agent, lubricant, emollient, water repellent, film modifier, softener, surfactant, leveling agent, plasticizer

APPLICATIONS:

Automobile applications: auto polish (F-751; Masil EM-100P, EM-1000P; Masil SF 5, SF 20, SF 50, SF 100, SF 200, SF 350, SF 500, SF 1000, SF 5000, SF 10,000, SF 12,500, SF 30,000, SF 60,000, SF 100,000, SF 30,000, SF 600,000; Viscasil 5M, 10M, 60M)
Cleansers: hand cleanser (Masil SF Series (lower visc.))
Cosmetic industry preparations: (Foamkill 810F, 830F; Masil SF Series (lower visc.); SF 69, 81, 97; Silicone C111; Silicone Mold Release SEM-35; SWS-101; Union

Carbide LE-45; Viscasil 5M, 10M, 60M); creams and lotions (Masil SF Series (lower visc.)); hair preparations (Abil 10–10,000 Series; Masil SF Series (lower visc.); Viscasil 5M, 10M, 60M); hand creams and lotions (SM 2140, 2155, 2162; Union Carbide Dimethyl Fluid L-45); nail polish (Masil SF Series (lower visc.)); personal care products (Masil SF 5, SF 20, SF 50, SF 100, SF 200, SF 350, SF 500, SF 1000, SF 5000, SF 10,000, SF 12,500, SF 30,000, SF 60,000, SF 100,000, SF 30,000, SF 600,000); shaving preparations (Abil 10–10,000 Series; Masil SF Series (lower visc.)); skin preparations (Abil 10–10,000 Series; Viscasil 5M, 10M, 60M)

Farm products: insecticides/pesticides (Silicone AF-10 IND, AF-30 IND, AF-600M)

FDA-approved applications: (AF 72, 75, 9020; Antifoam Compound SWS-202, SWS-203; Masil EM-350)

Food applications: (AF 72, 75, 9020; Foamkill 810F, 830F; Masil EM-350; Masil SF 5, SF 20, SF 50, SF 100, SF 200, SF 350, SF 500, SF 1000, SF 5000, SF 10,000, SF 12,500, SF 30,000, SF 60,000, SF 100,000, SF 30,000, SF 600,000); Sag Silicone Antifoam 720; SF 18; Silicone AF-10 FG, AF-30 FG; SWS-101); beverage processing (AF 72, 75); direct food additive (AF 75); food packaging (AF 75; Foamkill 810F, 830F); indirect food contact use (Antifoam Compound SWS-202, SWS-203)

Household products: (Masil SF 5, SF 20, SF 50, SF 100, SF 200, SF 350, SF 500, SF 1000, SF 5000, SF 10,000, SF 12,500, SF 30,000, SF 60,000, SF 100,000, SF 30,000, SF 600,000); aerosol oven sprays (SF 1093); aerosol polish (SWS-231, –232); aerosol spray starch (Masil EM-100,000; SM 2061; Union Carbide LE-467, LE-467HS); carpet cleaners (Silicone AF-600M); detergents (AF 72; Sag Silicone Antifoam 100; Silicone AF-10 IND, AF-600M); furniture polish (Masil EM-100P, EM-1000P; Masil SF 5, SF 20, SF 50, SF 100, SF 200, SF 350, SF 500, SF 1000, SF 5000, SF 10,000, SF 12,500, SF 30,000, SF 60,000, SF 100,000, SF 30,000, SF 600,000); leather/vinyl cleaners (Masil EM-14, EM-62, EM-100, EM-100 Conc., EM-100D, EM-100P, EM-1000P, EM-250 Conc., EM-350, EM-350X, EM-350X Conc., EM-1000, EM-1000 Conc., EM-10,000, EM-10,000 Conc., EM-60,000, EM-100,000, EM-N); soap mfg. (AF 60, 72, 75, 9020; Silicone AF-10 FG, AF-30 FG); window cleaners (Masil EM-14, EM-62, EM-100, EM-100 Conc., EM-100D, EM-100P, EM-1000P, EM-250 Conc., EM-350, EM-350X, EM-350X Conc., EM-1000, EM-1000 Conc., EM-10,000, EM-10,000 Conc., EM-60,000, EM-100,000, EM-N; Masil SF 5, SF 20, SF 50, SF 100, SF 200, SF 350, SF 500, SF 1000, SF 5000, SF 10,000, SF 12,500, SF 30,000, SF 60,000, SF 100,000, SF 30,000, SF 600,000; SM 2140, 2155, 2162)

Industrial applications: (AF 9020; Antifoam Compound SWS-201); adhesives (AF 60, 66, 72, 75, 9020; Foamkill 810F, 830F; Silicone AF-10 FG, AF-10 IND, AF-30 FG, AF-10 IND, AF-30 IND, AF-600M); asphalt (Dow Corning 200 Fluid); coatings (Sag Silicone Antifoam 10, 30); corrosion protection (SF 1705); distillation (Dow Corning 200 Fluid); dyes and pigments (AF 72, 75; Silicone AF-30 FG; Union Carbide Dimethyl Fluid L-45); electrical applications (SF 69, 81, 97; Viscasil 5M, 10M, 60M); foundry operations (SM 2140, 2155, 2162); glass molding (Silicone

Dimethicone (cont'd.)

Mold Release SEM-35; Union Carbide LE-45); greases (SF 18); heat transfer fluids (SF 96, SF 1093; Union Carbide Dimethyl Fluid L-45); high temperature applications (SF 1093); hydraulic fluids (Union Carbide Dimethyl Fluid L-45); insect repellents (Sag Silicone Antifoam 100); laboratory use (SF 1093); latex processing (AF 60, 72, 75, 9020; Sag Silicone Antifoam 10, 30; Silicone AF-10 IND, AF-30 FG, AF-30 IND); lubricating/cutting oils (Silicone AF-10 IND, AF-30 IND, AF-600M); mechanical damping (SF 1093); mechanical goods (Akrochem Silicone Emulsion 350, 1M, 10M, 60M; Akrochem Silicone Fluids 350, 1M, 5M, 10M, 30M, 60M; SF 69, 81, 97; Union Carbide LE-46, LE-467, LE-467HS; Viscasil 5M, 10M, 60M); metalworking (AF 75; Masil EM-14, EM-62, EM-100, EM-100 Conc., EM-100D, EM-100P, EM-1000P, EM-250 Conc., EM-350, EM-350X, EM-350X Conc., EM-1000, EM-1000 Conc., EM-10,000, EM-10,000 Conc., EM-60,000, EM-100,000, EM-N; Masil SF 5, SF 20, SF 50, SF 100, SF 200, SF 350, SF 500, SF 1000, SF 5000, SF 10,000, SF 12,500, SF 30,000, SF 60,000, SF 100,000, SF 30,000, SF 600,000; Silicone AF-10FG, AF-30 IND); molded flexible urethane foam (Dow Corning F1-1630); paint mfg. (AF 60, 66, 72, 9020; SF 69, 81, 96, 97; Union Carbide Dimethyl Fluid L-45; Viscasil 5M, 10M, 60M); paper coating (AF 75); paper mfg. (AF 75; Foamkill 810F, 830F; Silicone AF-10FG, AF-30 IND); petroleum industry (AF 66; Dow Corning 200 Fluid; Masil SF 5, SF 20, SF 50, SF 100, SF 200, SF 350, SF 500, SF 1000, SF 5000, SF 10,000, SF 12,500, SF 30,000, SF 60,000, SF 100,000, SF 30,000, SF 600,000; Union Carbide Dimethyl Fluid L-45; Viscasil 5M, 10M, 60M); plastics (Akrochem Silicone Emulsion 350, 1M, 10M, 60M; Akrochem Silicone Fluids 350, 1M, 5M, 10M, 30M, 60M; Masil EM-14, EM-62, EM-100, EM-100 Conc., EM-100D, EM-100P, EM-1000P, EM-250 Conc., EM-350, EM-350X, EM-350X Conc., EM-1000, EM-1000 Conc., EM-10,000, EM-10,000 Conc., EM-60,000, EM-100,000, EM-N; Masil SF 5, SF 20, SF 50, SF 100, SF 200, SF 350, SF 500, SF 1000, SF 5000, SF 10,000, SF 12,500, SF 30,000, SF 60,000, SF 100,000, SF 30,000, SF 600,000; SF 18, 69, 81, 96, 97; Silicone Fluid 200, 203, 225, 230; Silicone Mold Release SEM-35; Silicone Release Agent #5038; SM 2140, 2155, 2162; SWS-230 Series; Union Carbide LE-45, LE-458HS; Viscasil 5M, 10M, 60M); polishes and waxes (SF 69, 81, 97, 1705; SWS-101, -235; Union Carbide LE-453HS, LE-461, LE-462, LE-462HS; Viscasil 5M, 10M, 60M); polymers/polymerization (AF 66, 72, 75, 9020; Sag Silicone Antifoam 10, 30); printing inks (AF 60, 66, 72, 75, 9020; Masil SF 5, SF 20, SF 50, SF 100, SF 200, SF 350, SF 500, SF 1000, SF 5000, SF 10,000, SF 12,500, SF 30,000, SF 60,000, SF 100,000, SF 30,000, SF 600,000; Silicone AF-10 FG, AF-10 IND, AF-30 FG, AF-30 IND, AF-600M; Silicone Mold Release SEM-35); resin mfg. (Dow Corning 200 Fluid); rubber (Akrochem Silicone Emulsion 350, 1M, 10M, 60M; Akrochem Silicone Fluids 350, 1M, 5M, 10M, 30M, 60M; Masil EM-14, EM-62, EM-100, EM-100 Conc., EM-100D, EM-100P, EM-1000P, EM-250 Conc., EM-350, EM-350X, EM-350X Conc., EM-1000, EM-1000 Conc., EM-10,000, EM-10,000 Conc., EM-60,000, EM-100,000, EM-N; Masil SF 5, SF 20, SF 50, SF 100, SF 200, SF 350, SF 500, SF 1000, SF 5000, SF 10,000, SF 12,500, SF 30,000, SF 60,000,

SF 100,000, SF 30,000, SF 600,000; Sag Silicone Antifoam 10, 30; SF 18, 69, 81, 96, 97; Silicone Release Agent #5038; SM 2140, 2155, 2162; SWS-230 Series; Union Carbide LE-45, LE-420, LE-458HS, LE-467, LE-467HS; Viscasil 5M, 10M, 60M); sizing (AF 75); starch processing (AF 60, 72, 75, 9020; Sag Silicone Antifoam 100; Silicone AF-10 IND, AF-30 IND); stripping processes (Sag Silicone Antifoam 10, 30); textile/leather processing (AF 72, 75; Masil EM-14, EM-62, EM-100, EM-100 Conc., EM-100D, EM-100P, EM-1000P, EM-250 Conc., EM-350, EM-350X, EM-350X Conc., EM-1000, EM-1000 Conc., EM-10,000, EM-10,000 Conc., EM-60,000, EM-100,000, EM-N; Masil SF 5, SF 20, SF 50, SF 100, SF 200, SF 350, SF 500, SF 1000, SF 5000, SF 10,000, SF 12,500, SF 30,000, SF 60,000, SF 100,000, SF 30,000, SF 600,000; SF 96, 1705; Silicone AF-10FG, AF-30 IND; Silicone Mold Release SEM-35; SM 2140, 2155, 2162; Union Carbide LE-45, LE-467, LE-467HS; Viscasil 5M, 10M, 60M); waste treatment (AF 9020); water treatment (AF 72; Silicone AF-30 FG); wire and cable (Akrochem Silicone Emulsion 350, 1M, 10M, 60M; Akrochem Silicone Fluids 350, 1M, 5M, 10M, 30M, 60M); wood pulping (AF 75; Silicone AF-10 IND)

Industrial cleaners: hard surface cleaners (Union Carbide LE-453HS); soaps (Silicone AF-30 IND)

Military applications: high-temperature use (SF 1093)

Pharmaceutical applications: (Foamkill 810F, 830F; Silicone C111); antiperspirant/deodorant (Masil SF Series (lower visc.)); drug extraction/processing (Foamkill 810F, 830F; Sag Silicone Antifoam 100; Union Carbide Dimethyl Fluid L-45); suntan/sunscreen preparations (Abil 10–10,000 Series; Masil SF Series (lower visc.)); vitamins (Foamkill 830F)

PROPERTIES:

Form:

Liquid (Abil 10–10,000 Series; AF 66; Dow Corning 1500, 1520 Silicone Antifoam; F-751; SF 96, SF 1093, SF 1705; Silicone C111; Silicone Fluid 200, 203, 225, 230; SWS-101, -230 Series; Union Carbide Dimethyl Fluid L-45)

Clear liquid (Akrochem Silicone Fluids 350, 1M, 5M, 10M, 30M, 60M; Dow Corning 200 Fluid)

Translucent fluid (Akrochem SWS-201; Antifoam Compound SWS-201, SWS-203)

Viscous liquid (Antifoam Compound SWS-202)

Clear oily liquid (Masil SF 5, SF 20, SF 50, SF 100, SF 200, SF 350, SF 500, SF 1000, SF 5000, SF 10,000, SF 12,500, SF 30,000, SF 60,000, SF 100,000, SF 300,000, SF 600,000; SF 18, 69, 81, 97; Viscasil 5M, 10M, 60M)

Emulsion (AF 60, 72, 75, 9020; Akrochem Silicone Emulsion 350, 350 Conc., 1M, 10M, 60M; Masil EM-14, EM-62, EM-100, EM-100 Conc., EM-100D, EM-100P, EM-1000P, EM-250 Conc., EM-350, EM-350X, EM-350X Conc., EM-1000, EM-1000 Conc., EM-10,000, EM-10,000 Conc., EM-60,000, EM-100,000, EM-N; Sag Silicone Antifoam 10, 30, 720; Silicone AF-10FG, AF-10 IND, AF-30 FG, AF-30 IND, AF-600M; Silicone Mold Release SEM-35; Silicone Release Agent #5038; SM 2061, 2140, 2155, 2162; Union Carbide LE-45, LE-46, LE-420, LE-458HS,

Dimethicone *(cont'd.)*

LE-467, LE-467HS)

Color:

Water-white (Akrochem Silicone Fluids 350, 1M, 5M, 10M, 30M, 60M; Masil SF 5, SF 20, SF 50, SF 100, SF 200, SF 350, SF 500, SF 1000, SF 5000, SF 10,000, SF 12,500, SF 30,000, SF 60,000, SF 100,000, SF 300,000, SF 600,000; SF 18, 69, 81, 96, 97; Viscasil 5M, 10M, 60M)

White (AF 60, 72, 75, 9020; Akrochem Silicone Emulsion 350, 350 Conc., 1M, 10M, 60M; Silicone AF-10FG, AF-10 IND, AF-30 FG, AF-30 IND, AF-600M; SM 2140, 2155, 2162; Union Carbide LE-420)

Milky white (Sag Silicone Antifoam 10; Silicone Mold Release SEM-35; Silicone Release Agent #5038; Union Carbide LE-467, LE-467HS)

Opaque white (Sag Silicone Antifoam 30)

Cream (Antifoam Compound SWS-202)

Grayish-white (AF 66)

Gray (Akrochem SWS-201; Antifoam Compound SWS-201, SWS-203)

Amber (SF 1093)

Odor:

None (AF 66; Antifoam Compound SWS-201, SWS-202, SWS-203; Masil SF 5, SF 20, SF 50, SF 100, SF 200, SF 350, SF 500, SF 1000, SF 5000, SF 10,000, SF 12,500, SF 30,000, SF 60,000, SF 100,000, SF 300,000, SF 600,000; SF 69, 81, 97; Viscasil 5M, 10M, 60M)

Essentially odorless (SF 1093; SM 2140)

Mild (SM 2162)

Slight sorbic acid (AF 60)

Taste:

None (Masil SF 5, SF 20, SF 50, SF 100, SF 200, SF 350, SF 500, SF 1000, SF 5000, SF 10,000, SF 12,500, SF 30,000, SF 60,000, SF 100,000, SF 300,000, SF 600,000; SF 69, 81, 97; Viscasil 5M, 10M, 60M)

Composition:

10% active (Silicone AF-10FG, AF-10 IND, AF-600M)

10% silicone content (AF 75)

10% silicone emulsion (Foamkill 810F; Sag Silicone Antifoam 10)

14% active (Masil EM-14)

20% active (Sag Silicone Antifoam 720)

20% silicone (AF 9020)

25% active (Masil EM-62, EM-N)

30% active (AF 60; Masil EM-100D; Silicone AF-30 FG, AF-30 IND)

30% silicone content (AF 72)

30% silicone emulsion (Sag Silicone Antifoam 30)

35% active (Masil EM-100, EM-350, EM-350X, EM-1000, EM-10,000, EM-60,000, EM-100,000; Silicone Mold Release SEM-35; Silicone Release Agent #5038; SM 2061; SWS-230 Series, -231, -232, -235)

35% silicone content (Akrochem Silicone Emulsion 350, 1M, 10M, 60M; Union

Carbide LE-420; Union Carbide LE-467)

44.2 ± 1% total solids (AF 72)

50% silicone content (SM 2140, 2155, 2162; Union Carbide LE-467HS)

50% silicone in mineral spirits/IPA (SF 1705)

60% active (Masil EM-100 Conc., EM-100P, EM-250 Conc., EM-350X Conc., EM-1000 Conc., EM-1000P, EM-10,000 Conc.)

60% silicone content (Akrochem Silicone Emulsion 350 Conc.)

100% active (Abil 10–10,000 Series; AF 66; Akrochem SWS-201; Antifoam Compound SWS-201, SWS-202, SWS-203; Dow Corning 200 Fluid; SF 96; Viscasil 5M, 60M)

Solubility:

Sol. in higher and lower alcohols (SF 96 (5 cs grade))

Sol. in aliphatic solvents (AF 66; SF 96 (5 cs grade); Sag Silicone Antifoam 100; SF 1705; SWS-101; Union Carbide Dimethyl Fluid L-45)

Sol. in aromatic solvents (AF 66; Sag Silicone Antifoam 100; SF 96 (5 cs grade), SF 1705; SWS-101; Union Carbide Dimethyl Fluid L-45)

Sol. in chlorinated solvents (AF 66; F-751; SF 96 (5 cs grade), SF 1705; SWS-101)

Sol. in hydrocarbons (F-751); sol. in some higher hydrocarbons (SF 96 (5 cs grade))

Sol. in higher ketones (SF 96 (5 cs grade))

Miscible with nonpolar liquids (hydrocarbons, ethers, etc.) (Akrochem Silicone Fluids 350, 1M, 5M, 10M, 30M, 60M)

Disp. in solvents (Akrochem SWS-201, SWS-203; Antifoam Compound SWS-201)

Appreciable sol. in water (SM 2140, 2162); readily disp. in warm or cold water with mild agitation (AF 60, 72, 9020; Silicone AF-10FG, AF-10 IND, AF-30 FG, AF-30 IND, AF-600M); readily disp. in water (Silicone Mold Release SEM-35); disp. in water (AF 75; Antifoam Compound SWS-202; Sag Silicone Antifoam 720; Union Carbide LE-45, LE-453HS, LE-458HS, LE-461, LE-462, LE-462HS, LE-467, LE-467HS); readily dilutable with water (Sag Silicone Antifoam 10; Silicone Release Agent #5038); dilutable in water (SWS-231, -232, -235); negligible sol. in water (AF 66; Antifoam Compound SWS-201, SWS-203; Viscasil 5M, 10M, 60M); immiscible (Akrochem Silicone Fluids 350, 1M, 5M, 10M, 30M, 60M); insol. (Sag Silicone Antifoam 100; Union Carbide Dimethyl Fluid L-45)

Ionic Nature:

Nonionic (AF 60, 72, 75, 9020; Akrochem Silicone Emulsion 350, 1M, 10M, 60M; SF 96; Masil EM-14, EM-100, EM-100 Conc., EM-100D, EM-100P, EM-1000P, EM-250 Conc., EM-350, EM-350X, EM-350X Conc., EM-1000, EM-1000 Conc., EM-10,000, EM-10,000 Conc., EM-60,000, EM-100,000; Silicone AF-10FG, AF-10 IND, AF-30 FG, AF-30 IND; Silicone Mold Release SEM-35; SM 2061, 2140, 2155, 2162; Union Carbide LE-45, LE-46, LE-453HS; Viscasil 5M)

Nonionic/anionic (Masil EM-62)

Anionic (Masil EM-N; Union Carbide LE-458HS, LE-467, LE-467HS)

Sp.gr.:

0.82 (SF 1705)

Dimethicone (cont'd.)

0.916 (Masil SF 5)

0.916–0.974 (SF 96)

0.95–0.98 (Akrochem Silicone Fluids 350, 1M, 5M, 10M, 30M, 60M)

0.953 (Masil SF 20)

0.953–0.974 (SF 97)

0.96–1.00 (Akrochem Silicone Emulsion 350 Conc.)

0.963 (Masil SF 50; SF 1093 (50 cs))

0.965 (SF 69)

0.968 (Masil SF 100; SF 1093 (100 cs))

0.970 (Masil EM-N)

0.972 (Masil SF 200; SF 81)

0.973 (Masil SF 350, SF 500; SF 18)

0.974 (Masil SF 1000)

0.975 (Masil SF 5000, SF 10,000, SF 12,500)

0.976 (Masil SF 30,000)

0.977 (Masil SF 60,000)

0.978 (Masil SF 100,000, SF 300,000)

0.979 (Masil SF 600,000)

0.98 (Viscasil 5M, 10M, 60M)

0.98–1.02 (Akrochem Silicone Emulsion 350, 1M, 10M, 60M)

0.988 (Silicone AF-600M)

0.99 (Masil EM-62, EM-100, EM-100 Conc., EM-100D, EM-100P, EM-1000P, EM-250 Conc., EM-350, EM-350X, EM-350X Conc., EM-1000, EM-1000 Conc., EM-10,000 Conc., EM-60,000; Silicone Mold Release SEM-35; Union Carbide LE-45, LE-46, LE-420, LE-467, LE-467HS)

1.00 (Masil EM-14, EM-100,000; Sag Silicone Antifoam 10; Silicone AF-10FG, AF-10 IND)

1.004 (Sag Silicone Antifoam 30)

1.01 (AF 60, 66, 72, 9020; Antifoam Compound SWS-201, SWS-202, SWS-203; SM 2140; Silicone AF-30 FG, AF-30 IND)

1.02 (AF 75)

1.04 (SM 2162)

1.09 (Masil EM-10,000)

Density:

7.8–8.0 lb/gal (Akrochem Silicone Fluids 350, 1M, 5M, 10M, 30M, 60M)

8.0 lb/gal (Silicone Release Agent #5038)

8.25 lb/gal (Akrochem Silicone Emulsion 350, 1M, 10M, 60M; SM 2140, 2155, 2162)

8.3 lb/gal (Akrochem Silicone Emulsion 350 Conc.; Sag Silicone Antifoam 30)

8.31 lb/gal (Sag Silicone Antifoam 10)

8.4 lb/gal (AF 60, 66, 72, 75, 9020; Akrochem SWS-201; Antifoam Compound SWS-201, SWS-202, SWS-203)

Visc.:

50–500 cps (SF 1705)

22

230 cps (Akrochem SWS-201; Antifoam Compound SWS-201)
500 cps max. (AF 60)
1000 cps (SM 2162)
1000 cps max. (AF 72)
1500 cps (SM 2140, 2155)
2400 cps (Antifoam Compound SWS-203)
2500 cps max. (AF 75, 9020)
10,000 cps (Antifoam Compound SWS-202)
0.65–1000 cstk (Silicone C111)
5 cstk (Masil SF 5)
10 cstk (SF 69)
10–60,000 cstk (SWS-101)
15 cstk (Masil EM-N)
20 cstk (Masil SF 20)
25 cstk (Masil EM-62)
50 cstk (Masil SF 50; SF 81; SF 1093 (50 cs grade))
100 cstk (Masil EM-100, EM-100 Conc., EM-100D, EM-100P; Masil SF 100; SF 1093 (100 cs grade))
200 cstk (Masil SF 200)
250 cstk (Masil EM-250 Conc.)
350 cstk (Akrochem Silicone Emulsion 350, 350 Conc.; Akrochem Silicone Fluid 350; Masil EM-14, EM-350, EM-350X, EM-350X Conc.; Masil SF 350; SF 18; SWS-231)
350–60,000 cstk (SWS-230 Series)
500 cstk (Masil SF 350); 500 cstk max. (AF 66)
1000 cstk (Akrochem Silicone Emulsion 1M; Akrochem Silicone Fluid 1M; Masil EM-1000, EM-1000 Conc., EM-1000P; Masil SF 1000; SWS-232)
1000–10,000 cstk (Dow Corning 200 Fluid)
5000 cstk (Akrochem Silicone Fluid 5M; Masil SF 5000)
10,000 cstk (Akrochem Silicone Emulsion 10M; Akrochem Silicone Fluid 10M; Masil EM-10,000, EM-10,000 Conc.; Masil SF 10,000; SWS-235)
12,500 cstk (Masil SF 12,500)
12,500–60,000 cstk (Dow Corning 200 Fluid)
30,000 cstk (Akrochem Silicone Fluid 30M; Masil SF 30,000)
60,000 cstk (Akrochem Silicone Emulsion 60M; Akrochem Silicone Fluid 60M; Masil EM-60,000; Masil SF 60,000)
100,000 cstk (Masil EM-100,000; Masil SF 100,000)
300,000 cstk (Masil SF 300,000)
600,000 cstk (Masil SF 600,000)
5–1000 cs grades (SF 96)
Avail. in 20, 50, 100, 350, 500, and 1000 cstk grades (SF 97)
Avail. in 5000, 10,000, 12,5000, 30,000, 60,000, 100,000, 300,000, and 600,000 cstk grades (Viscasil)

Dimethicone *(cont'd.)*

F.P.:
30 F (Silicone Mold Release SEM-35)
B.P.;
212 F (SM 2140, 2162)
Pour Pt.:
–84 C (Masil SF 5)
–65 C (Masil SF 20)
–55 C (Masil SF 50, SF 100)
–50 C (Masil SF 200, SF 350, SF 500, SF 1000)
–49 C (Masil SF 5000)
–47 C (Masil SF 10,000, SF 12,500)
–46 C (Masil SF 30,000)
–44 C (Masil SF 60,000)
–40 C (Masil SF 100,000, SF 300,000)
–34 C (Masil SF 600,000)
–120 F (SF 81)
–120 to –58 F (SF 96)
–100 F (SF 1093 (50 cs))
–90 F (SF 1093 (100 cs))
–85 to –58 F (SF 97)
–58 F (SF 18)
–58 to –25 F (Viscasil)
–40 F (SF 69)
Flash Pt.:
28 C (CC) (SF 1705)
138 C (Masil SF 5)
202 C (CC) (Masil SF 20)
238 C (CC) (Masil SF 50, SF 100, SF 200)
260 C (CC) (Masil SF 350, SF 500, SF 1000, SF 5000, SF 10,000, SF 12,500, SF 30,000, SF 60,000, SF 100,000, SF 300,000, SF 600,000)
145 F (SF 69)
280–500 F (CC) (SF 96)
395–500 F (SF 97)
425 F (CC) (SF 1093)
460 F (SF 81)
500 F (CC) (SF 18)
> 500 F (TOC) (Viscasil 5M, 10M)\
> 575 F (SM 2162)
600 F (COC) (Akrochem SWS-201; Antifoam Compound SWS-201, SWS-203)
> 600 F (OC) (AF 66); (TOC) (Viscasil 60M)
Stability:
Excellent stability (Union Carbide LE-45, LE-46, LE-420, LE-467, LE-467HS)
Oxidation stability: 300 F @ 200 h (Akrochem Silicone Fluids)

24

Good stability; excellent oxidation resistance (Silicone Mold Release SEM-35)

Heat stable to 43 C; dilution stability < 2% creaming and no settling after 24 h @ 10% silicone content (AF 60, 9020)

Heat stable to 45 C (AF 75)

Heat stable to 45 C; dilution stability < 2% creaming and no settling after 24 h @ 10% silicone content (AF 72)

Heat stable to 110 F (Silicone AF-10FG, AF-10 IND, AF-30 FG, AF-30 IND)

Heat stable to 126 F; freeze-thaw stable (Silicone AF-600M)

Heat and freeze-thaw stable (Akrochem Silicone Emulsion 350 Conc.; Antifoam Compound SWS-201, SWS-202, SWS-203)

Good heat aging and freeze-thaw stability (Akrochem Silicone Emulsion 350, 1M, 10M, 60M)

Good heat and oxidative stability (Akrochem Silicone Fluids 350, 1M, 5M, 10M, 30M, 60M)

High temperature stability; generally stable in hard water (SM 2140, 2155, 2162)

Good thermal, oxidative, and shear stability (SF 18, 96, 1093)

Generally stable; thermal decomposition at temps. > 500 F will produce CO_2 and SiO_2 (Viscasil 5M, 60M)

Generally stable; combustion will produce CO, CO_2, and SiO_2 (Viscasil 10M)

Excellent thermal and oxidative stability; stable to mechanical shear stresses (Masil SF 5, SF 20, SF 50, SF 100, SF 200, SF 350, SF 500, SF 1000, SF 5000, SF 10,000, SF 12,500, SF 30,000, SF 60,000, SF 100,000, SF 300,000, SF 600,000)

Freeze-thaw stable; excellent dilution stability; protected for max. stability with bactericides and a small amount of rust inhibitor (Masil EM-14, EM-62, EM-100, EM-100 Conc., EM-100D, EM-100P, EM-1000P, EM-250 Conc., EM-350, EM-350X, EM-350X Conc., EM-1000, EM-1000 Conc., EM-10,000, EM-10,000 Conc., EM-60,000, EM-100,000, EM-N)

Storage Stability:

Excellent storage stability with proper handling (Silicone AF-10FG, AF-10 IND, AF-30 FG, AF-30 IND, AF-600M; Silicone Mold Release SEM-35)

No visible settling in 60 days; up to 6 mos shelf life (AF 60, 9020)

6 mos. shelf life (Akrochem Silicone Emulsion 350 Conc.; Antifoam Compound SWS-201, SWS-202, SWS-203; Silicone AF-10FG, AF-10 IND, AF-30 FG, AF-30 IND, AF-600M)

Up to 6 mos shelf life (AF 75)

6 mos @ 70 F (Akrochem Silicone Emulsion 350, 1M, 10M, 60M)

Ref. Index:

1.3970 (Masil SF 5)

1.3970–1.4035 (SF 96)

1.4010 (Masil SF 20)

1.401–1.4035 (SF 97)

1.4020 (Masil SF 50; SF 69)

1.4030 (Masil SF 100; SF 18, 81)

Dimethicone *(cont'd.)*

    1.4031 (Masil SF 200)
    1.4032 (Masil SF 350)
    1.4033 (Masil SF 500)
    1.4035 (Masil SF 1000, SF 5000, SF 10,000, SF 12,500, SF 30,000, SF 60,000, SF 100,000, SF 300,000, SF 600,000; Viscasil)
    1.4040 (SF 1093)
pH:
    4.0–5.0 (Silicone AF-10 FG, Silicone AF-30 FG, AF-30 IND)
    5.0–6.0 (Silicone AF-600M)
    7.0 (Akrochem 350, 1M, 10M, 60M)
Surface Tension:
    19.7–21.1 dynes/cm (SF 96)
    20.5 dynes/cm (SF 69)
    20.8–21.1 dynes/cm (SF 97)
    21.0 dynes/cm (SF 81)
    21.1–21.3 dynes/cm (Viscasil)

THERMAL PROPERTIES:
Conductivity:
    0.0082–0.092 Btu/h/ft^{2}F/ft (SF 97)
    0.087 Btu/h/ft^2/F/ft (SF 81)
    0.090–0.92 Btu/h/ft^2/F/ft (Viscasil)
Coeff. of Linear Exp.:
    0.0095 cc/cc/C (SF 81)
    0.000925 cc/cc/C (25–150 C) (Viscasil)
    0.000925–0.00107 cc/cc/C (25–150 C) (SF 97)
Sp. Heat:
    0.36 Btu/lb/F (SF 69, 81, 97; Viscasil)

ELECTRICAL PROPERTIES:
Dissip. Factor:
    0.0001 (SF 81, 97; Viscasil)
Dielec. Str.:
    35.0 kV (SF 81, 97; Viscasil)
Dielec. Constant:
    2.68–2.75 (SF 97)
    2.74 (SF 81)
    2.75 (Viscasil)
Vol. Resist.:
    1×10^{14} ohm-cm (SF 81, 97; Viscasil)

TOXICITY/HANDLING:
    Considered to be nontoxic and physiologically inert (Antifoam Compound SWS-201, SWS-202, SWS-203)

Dimethicone (cont'd.)

Physiologically inert; observe normal safety precautions (Akrochem Silicone Emulsion 350, 1M, 10M, 60M)

Extremely low oral toxicity (Sag Silicone Antifoam 30)

Low toxicity (Union Carbide LE-45, LE-46)

Low toxicity; observe normal precautions (Akrochem Silicone Fluids 350, 1M, 5M, 10M, 30M, 60M)

Low toxicity; eye contact will cause temporary irritation (AF 75; SF 69, 81, 97; Viscasil)

May cause eye irritation (AF 60, 66)

May cause minor eye irritation on eye contact (Viscasil 5M, 60M)

May cause slight transitory eye irritation; safety glasses recommended (SM 2140)

Transitory eye irritant (SM 2155, 2162)

STORAGE/HANDLING:

Store in tightly sealed containers to prevent free water and particulate contamination, in a dry area at temps. of 20–30 C (SF 69, 81, 97; Viscasil)

Store in tightly closed containers; although it is freeze-thaw stable, storage in a heated area is recommended (Silicone Mold Release SEM-35)

Keep from freezing (SM 2140, 2155, 2162)

Store in cool place in tight closed containers; protect from freezing (Silicone AF-10FG, AF-10 IND, AF-30 FG, AF-30 IND)

Store in tightly closed containers in a cool place; although freeze-thaw stable, protection from repeated freezing is recommended (Silicone AF-600M)

Store in cool place (16–24 C) in tightly closed containers; protect from freezing (AF 60, 72, 75, 9020)

Store in well-ventilated areas at temps. above freezing to 130 F max. (Masil EM-14, EM-62, EM-100, EM-100 Conc., EM-100D, EM-100P, EM-1000P, EM-250 Conc., EM-350, EM-350X, EM-350X Conc., EM-1000, EM-1000 Conc., EM-10,000, EM-10,000 Conc., EM-60,000, EM-100,000, EM-N)

STD. PKGS.:

5-gal pails, 55-gal drums, bulk (Masil SF 5, SF 20, SF 50, SF 100, SF 200, SF 350, SF 500, SF 1000, SF 5000, SF 10,000, SF 12,500, SF 30,000, SF 60,000, SF 100,000, SF 300,000, SF 600,000)

5-gal pails, 55-gal (440 lb net) coated drums (Akrochem Silicone Fluids 350, 1M, 5M, 10M, 30M, 60M)

5-gal pails, 55-gal drums, bulk tank car or wagon (Masil EM-14, EM-62, EM-100, EM-100 Conc., EM-100D, EM-100P, EM-1000P, EM-250 Conc., EM-350, EM-350X, EM-350X Conc., EM-1000, EM-1000 Conc., EM-10,000, EM-10,000 Conc., EM-60,000, EM-100,000, EM-N)

5-gal (40 lb net) container, 50-gal (400 lb net) drums (Antifoam Compound SWS-201)

5-gal, 55-gal containers (AF 60, 72, 75, 9020)

5-gal containers, 55-gal fiberpak drums (AF 66)

30/55-gal drums (Silicone Release Agent #5038)

27

Dimethicone copolyol (CTFA)

SYNONYMS:
Dimethylsiloxane-glycol copolymer

CAS No.:
64365-23-7

RD No.: 977058-72-2

TRADENAME EQUIVALENTS:
Dow Corning 190, 193, 196, 198, 1315, 5043, 5098 Surfactant [Dow Corning]

Dow Corning 470A Fluid [Dow Corning]

Dow Corning 3225C Formulation Aid [Dow Corning]

SF 1188 [General Electric]

CATEGORY:
Surfactant, lubricant, release agent, emollient, emulsifier, wetting agent, surface tension depressant, profoamer

APPLICATIONS:
Cosmetic industry preparations: (Dow Corning 470A Fluid; SF 1188); conditioners (SF 1188); hair preparations (SF 1188); hand lotions (SF 1188); personal care products (Dow Corning 3225C Formulation Aid); shampoos (SF 1188); shaving preparations (Dow Corning 470A Fluid; SF 1188); toiletries (SF 1188)

Household detergents: (Dow Corning 470A Fluid); soaps (Dow Corning 470A Fluid)

Industrial applications: dyes and pigments (Dow Corning 470A Fluid); glass polish (Dow Corning 470A Fluid); mold release (SF 1188); paints (SF 1188); polishes (Dow Corning 470A Fluid); polyurethane foam (Dow Corning 190, 193, 196, 198, 1315, 5043, 5098 Surfactants); powders (Dow Corning 470A Fluid); rubber lubricants (SF 1188); textile applications (SF 1188)

Pharmaceutical applications: antiperspirants (SF 1188)

PROPERTIES:
Form:

Liquid (Dow Corning 470A Fluid, 3225C Formulation Aid)

Clear liquid (SF 1188)

Clear, low-viscosity liquid (Dow Corning 1315, 5043, 5098 Surfactant)

Clear viscous liquid (Dow Corning 193 Surfactant)

Low-viscosity liquid (Dow Corning 190, 196, 198 Surfactant)

Color:

Amber (SF 1188)

Light straw (Dow Corning 5043, 5098 Surfactant)

Medium straw (Dow Corning 470A Fluid)

Gardner 2 (Dow Corning 190, 193 Surfactant)

Gardner 2–4 (Dow Corning 1315 Surfactant)

Gardner < 4 (Dow Corning 196, 198 Surfactant)

Composition:

10% conc. in volatile silicone (Dow Corning 3225C Formulation Aid)

100% silicone (SF 1188); 100% solids (Dow Corning 470A Fluid)

Dimethicone copolyol *(cont'd.)*

GENERAL PROPERTIES:
Ionic Nature:
Nonionic (Dow Corning 3225C Formulation Aid)
Solubility:
Sol. in acetone (SF 1188)
Sol. in lower alcohols (SF 1188)
Sol. in some aromatic hydrocarbons (SF 1188)
Sol. in some chlorinated hydrocarbons (SF 1188)
Sol. in fluorocarbon (Dow Corning 196, 198 Surfactant)
Sol. in polyol (Dow Corning 196, 198 Surfactant)
Sol. in most polyol/water/amine premixes (Dow Corning 5043 Surfactant)
Sol. in toluene (SF 1188)
Sol. in water (Dow Corning 196, 198 Surfactant); sol. in water below 43 C (SF 1188);
    sol. in water below 160 F (Dow Corning 470A Fluid)
Sol. in water-amine streams (Dow Corning 196, 198 Surfactant)
Sp. Gr.:
1.00 (Dow Corning 5043 Surfactant)
1.03 (Dow Corning 1315 Surfactant)
1.035 (Dow Corning 190, 196, 198 Surfactant)
1.04 (SF 1188)
1.07 (Dow Corning 193 Surfactant)
1.075 (Dow Corning 470A Fluid)
1.08 (Dow Corning 5098 Surfactant)
Density:
8.65 lb/gal (SF 1188)
Visc.:
1000 cps (SF 1188)
465 cs (Dow Corning 193 Surfactant)
1000 cs (Dow Corning 1315 Surfactant)
250 cSt (Dow Corning 5098 Surfactant)
275 cSt (Dow Corning 470A Fluid)
300 cSt (Dow Corning 5043 Surfactant)
1500 cSt (Dow Corning 190, 198 Surfactant)
2000 cSt (Dow Corning 196 Surfactant)
F.P.:
12 C (Dow Corning 5098 Surfactant)
50 F (Dow Corning 193 Surfactant)
Pour Pt.:
52 F (Dow Corning 193 Surfactant)
Flash Pt.:
56 C (CC) (Dow Corning 5098 Surfactant)
60 C (CC) (Dow Corning 5043 Surfactant)
60.6 C (CC) (Dow Corning 1315 Surfactant)

Dimethicone copolyol *(cont'd.)*

82 C (PM) (SF 1188)
> 150 F (OC) (Dow Corning 198 Surfactant)
225 F (OC) (Dow Corning 196 Surfactant)
250 F (CCC) (Dow Corning 190 Surfactant)
300 F (COC) (Dow Corning 193 Surfactant)
500 F (Dow Corning 470A Fluid)
HLB:
4.0 (Dow Corning 3225C Formulation Aid)
Hydroxyl No.:
< 10 (Dow Corning 5098 Surfactant)
Stability:
Nonhydrolyzable (Dow Corning 5043 Surfactant)
Good stability in water and amine premixes (Dow Corning 190 Surfactant)
Good premix stability; becomes hazy < 77 F; solidifies to soft wax < 50 F (Dow Corning 193 Surfactant)
Storage Stability:
12 mo shelf life (Dow Corning 5098 Surfactant; SF 1188)
Ref. Index:
1.45 (Dow Corning 193 Surfactant, 470A Fluid)
Surface Tension:
25.5 dynes/cm (SF 1188)

TOXICITY/HANDLING:

Mild eye irritant; moderate skin irritant on prolonged contact (Dow Corning 196, 198 Surfactant)

Avoid skin contat; may cause slight skin irritation or very slight, transient conjunctival irritation (Dow Corning 1315 Surfactant)

Repeated/prolonged skin exposure may cause moderate irritation and possibly a superficial burn; may cause at most a very slight transient conjunctival irritation (Dow Corning 5043, 5098 Surfactant)

STORAGE/HANDLING:

Store at R.T. in properly sealed container to avoid contamination and evaporation (SF 1188)

Store ≤ 32 C to obtain 6 mos. shelf life (Dow Corning 1315 Surfactant)

Store ≤ 32 C to obtain 12 mos. shelf life (Dow Corning 5043 Surfactant)

Combustible—keep away from heat, sparks, open flames; becomes hazy when stored below 25 C; solidifies to a soft wax below 15 C (easily reliquified by warming) (Dow Corning 5098 Surfactant)

STD. PKGS.:

1 gal pails, 5 and 55-gal drums (Dow Corning 470A Fluid)
40 and 441 lb containers net wt. (Dow Corning 190, 193 Surfactant)
441 lb drums; 2000 and 4000 gal tank trucks (Dow Corning 196, 198 Surfactant)
18- and 200-kg containers (Dow Corning 5098 Surfactant)
200-kg drums and 6938 or 14,515-kg tank trucks (Dow Corning 1315 Surfactant)

Ethylene/acrylate copolymer (CTFA)

SYNONYMS:
Ethylene/acrylic acid copolymer (EAA)
Ethylene/methacrylate copolymer
Ethylene/methacrylic acid copolymer (EMAA)
CAS No.:
RD No.: 977064-63-3
TRADENAME EQUIVALENTS:
A-C Copolymer 540A [Allied-Signal]
Primacor 4990 Dispersion [Dow]
CATEGORY:
Thermoplastic resin
APPLICATIONS:
FDA-approved applications: (Primacor 4990)
Functional additives: binder (Primacor 4990); dispersant (A-C Copolymer 540A);
    lubricant (A-C Copolymer 540A); processing aid (A-C Copolymer 540A)
Industrial applications: color concentrates (A-C Copolymer 540A); pigments (A-C
    Copolymer 540A); plastics (A-C Copolymer 540A); textile applications (Primacor
    4990)

PROPERTIES:
Form:
Dispersion (Primacor 4990)
Powder (A-C Copolymer 540A)
Color:
Milky white (Primacor 4990)
Composition:
35% solids in ammonia water (Primacor 4990)

GENERAL PROPERTIES:
Solubility:
Sol. in water (Primacor 4990)
Sp. Gr.:
0.960 (Primacor 4990, solids)
0.985 (Primacor 4990, liquid)
Density:
8.22 lb/gal (Primacor 4990)

Ethylane/acrylate copolymer (cont'd.)

Visc.:
> 500 cps (Primacor 4990)

B.P.:
> 100 C (Primacor 4990)

Stability:
> Excellent mechanical stability W pH > 7; good freeze-thaw stability (Primacor 4990)

Storage Stability:
> Excellent @ pH > 7 (Primacor 4990)

Surface Tension:
> 44–46 dynes/cm (Primacor 4990)

Biodegradable:
> Nonbiodegradable (Primacor 4990)

TOXICITY/HANDLING:
> Contains ammonia which may be irritating; prolonged/repeated skin contact may cause moderate irritation; eye contact may cause mild irritation; low toxicity via skin absorption (Primacor 4990)

Ethylene/maleic anhydride copolymer (CTFA)

SYNONYMS:
> 2,5-Furandione, polymer with ethene

CAS No.:
> 9006-26-2

TRADENAME EQUIVALENTS:
Linear and cross-linked grades:
> EMA [Monsanto]

CATEGORY:
> Thickener (cross-linked grades)
> Dispersant, opacifier, anti-redeposition agent, sequestrant (linear grades)

GENERAL PROPERTIES:
Solubility:
> Dispersible in mineral oil (EMA, cross-linked grades)
> Sol. in water (EMA, linear grades); dispersible in water (EMA, cross-linked grades)

Ethylene/methyl acrylate copolymer

SYNONYMS:
EMA

TRADENAME EQUIVALENTS:
PE 2205, 2207, 2255, 2260 [Chevron]

MODIFICATIONS/SPECIALTY GRADES:
High slip additive:
PE 2255
High antiblock additive:
PE 2255

CATEGORY:
Thermoplastic resin

PROCESSING:
Coextrusion:
PE 2205, 2255
Lamination:
PE 2207, 2260

APPLICATIONS:
Industrial applications: coatings (PE 2207, 2260); films (PE 2205, 2255); gloves (PE 2255); laminates (PE 2207, 2260)
Medical applications: (PE 2255)

GENERAL PROPERTIES:
Melt Flow:
2.0 g/10 min (PE 2260)
2.4 g/10 min (PE 2205, 2255)
6.0 g/10 min (PE 2207)
Sp. Gr.:
0.942 (PE 2205, 2207, 2255, 2260)

MECHANICAL PROPERTIES:
Tens. Str.:
1650 psi (break) (PE 2205, 2207, 2255, 2260)
Tens. Elong.:
740% (break) (PE 2205, 2207, 2255, 2260)
Tens. Mod.:
12,000 psi (PE 2205, 2207, 2255, 2260)
Water Absorp.:

Nylon

EMPIRICAL FORMULA:

$(C_6H_{11}NO)_n$

CAS No.:

63428-83-1

Nylon 6:

DERIVATION:

Condensation of caprolactam

SYNONYMS:

Polycaprolactam

STRUCTURE:

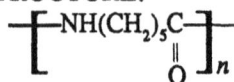

Nylon 6/6:

DERIVATION:

Condensation product of adipic acid and hexamethylenediamine

SYNONYMS:

Poly(hexamethylene adipamide)

STRUCTURE:

$H[NH(CH_2)_6NHCO(CH_2)_4CO]_nOH + (2n - 1)H_2O$

Nylon 6/10:

DERIVATION:

Condensation product of sebacic acid and hexamethylenediamine

STRUCTURE:

$COOH(CH_2)_8COOH$

Nylon 9:

DERIVATION: 9-Aminononanoic acid

Nylon 11:

DERIVATION:

From castor bean oil

SYNONYMS:
 Poly (11-aminoundecanoic acid)
STRUCTURE:
 H[NH(CH$_2$)$_{10}$CO]$_n$OH + (n — 1)H$_2$O

Nylon 12:
DERIVATION:
 Polymerization of dodecyl lactam (laurilactam)
STRUCTURE:

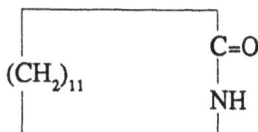

TRADENAME EQUIVALENTS:
Nylon Terpolymer:
 BCI Nylon LX 3249, LX 3250, No. 651, No. 653 [Belding]
Nylon Resin:
 Herox 0200HA, 0200MA, 0200SA [DuPont]
 Monocast MC901 [Polymer Corp.]
 Nylatron GSM, NSB [Polymer Corp.]
 RTP 200H-FR, 204H-FR, 205H-FR [Fiberite]
 Ultramid A3, A3K, A3KN, A3 Pellets 24, A3R, A3SK, A3W, A4, A4H, A4K, A5 B3,
  B3K, B3L, B3S, B4, B4K, B5, B5W, B35, B35M, B35MF01, B35SK, B35W, KR
  4205, KR 4405, KR 4406, KR 4407, KR 4409, KR 4411, KR 4412, S3, S3K, S4
  [Badische]
 Vydyne M-340, M-344, M-345 [Monsanto]
 Zytel ST FR-80 [DuPont]
Carbon-reinforced nylon:
  RTP 203H-CF15 (Pan carbon fiber), 204H-CF10 (Pan carbon fiber) [Fiberite]
  Stat-Kon VC-1003 (15% carbon fiber) [LNP]
Glass-reinforced nylon:
  RTP 201H (10% glass fiber), 203H (20% glass fiber), 205H (30% glass fiber), 207H
   (40% glass fiber) [Fiberite]
  STX Nylon (glass mat) [Allied-Signal]
  Ultramid A3EG5 (25% glass fiber), A3EG6 (30% glass fiber), A3EG7 (35% glass
   fiber), A3EG10 (50% glass fiber), A3G5 (25% glass fiber), A3G6 (30% glass
   fiber), A3G7 (35% glass fiber), A3HG5 (25% glass fiber), A3WG5 (25% glass
   fiber), A3WG6 (30% glass fiber), A3WG7 (35% glass fiber), A3WG10 (50%
   glass fiber), A3X1G7 (35% glass fiber), A3X1G10 (50% glass fiber), A3XG5
   (25% glass fiber), A3XG7 (35% glass fiber), B3EG5 (25% glass fiber), B3EG6
   (30% glass fiber), B3EG7 (35% glass fiber), B3EG10 (50% glass fiber), B3G5

Nylon *(cont'd.)*

(25% glass fiber), B3G5HS (25% glass fiber), B3G6 (30% glass fiber), B3G6HS (30% glass fiber), B3G7 (35% glass fiber), B3G7HS (35% glass fiber), B3G10 (50% glass fiber), B3WG5 (25% glass fiber), B3WG6 (30% glass fiber), B3WG7 (35% glass fiber), B3WG10 (50% glass fiber), B35G3 (15% glass fiber), KR 4445/1 (15% glass fiber), KR 4447/1 (30% glass fiber) [Badische]
Vydyne 909 [Monsanto]

Glass-reinforced amorphous nylon:
Thermocomp XF-1004 (20% glass fiber), XF-1006 (30% glass fiber), XF-1008 (40% glass fiber), XFL-4036 (30% glass fiber) [LNP]

Glass-reinforced partially aromatic nylon:
PDX-85384 (20% glass), -85382 (30% glass), -85473 (20% glass) [LNP]

Glass/ballotini-reinforced nylon:
Ultramid KR 4449 (30% glass fiber/ballotini) [Badische]

Glass/mineral-reinforced nylon:
Ultramid KR 4448 (40% glass fiber/mineral), KR 4450 (30% glass fiber/mineral) [Badische]

Mineral- or mineral/glass-reinforced nylon:
Minlon 11C, 12T, 20B, 21C [DuPont]

Mineral-reinforced nylon:
Ultramid B3WM601 (30% mineral), B3WM602 (30% mineral), KR 4250 (40% mineral), KR 4446 (30% mineral), KR-B3X2V6 (30% mineral) [Badische]
Vydyne R-100, R-200, R-220, R-250 [Monsanto]

Nylon 6:
Capron 8200, 8202, 8202C, 8202CL, 8202L, 8203C, 8207F, 8209F, 8220 HS, 8221 HS, 8253, 8254, 8255, 8259, 8270 HS, 8280 HS, 8350, 8352 [Allied-Signal]
Ertalon 6SAU [Chemplast]
Firestone Nylon 200-001, 210-001L, 213-001, 228-001 [Firestone]
Fosta Nylon 446, 471, 512, 523, 567, 589, 714, 870, 1047, 1210, 1379, 1417, 1525, 1693, 1722 [Hoechst Celanese]
Grilon A23, A23GM, A23G natural 6165, A28, A28GM, A28NX, A28NY, A28NZ, A28VO, A28W10, BT40, F40, F47, R40, R40GM, R40N38, R47, R47HW, R50, W3941, W5744 [Emser Industries]
Nydur B 31 SK, B 40 SK [Mobay]
Nylaflow LP [Polymer Corp.]
Nylatron 2001, 2005, 2010, 2011, 2015, 2029, 2038, 2039, GS-60, GS-63 [Polymer Corp.]
Texalon 600A, 600A HS, 600A NU, 600A PL-2, 600A Zip 1, 600A Zip 3, 670A, 680A, 1106, 1108, 1110A PL, 1110A PL HS, 1203D, XP-1296 [Texapol]
Vekton 6PA, 6PAG, 6PAM, 6PB, 6XAU [Chemplast]
Wellamid 42LH-N, 42L-N, 42LN2-N [Wellman]

Nylon 6/elastomer:
Grilon ELX23NZ [Emser Industries]
Nyrim P-1 1000 (10% rubber), P-1 2000 (20% rubber), P-1 3000 (30% rubber), P-

1 4000 (40% rubber) [Custom Urethane Elastomers]

Carbon-reinforced nylon 6:

EMI-X PC-1008 (40% carbon fiber), PC-100-10 (50% carbon fiber) [LNP]

NY-30CF (30% carbon fiber) [Compounding Technology]

Stat-Kon PC-1006 (30% carbon fiber) [LNP]

Thermocomp PC-1006 (30% PAN carbon fiber) [LNP]

Glass-reinforced nylon 6:

Capron 8230G HS (6% glass fiber), 8231G HS (14% glass fiber), 8232G HS FR (25% glass fiber), 8233G HS (33% glass fiber), 8234G HS (44% glass fiber) [Allied-Signal]

Firestone Nylon 415-001HS (15% short glass fibers), 430-001HS (30% short glass fibers) [Firestone]

Grilon PV-3H (30% glass fiber), PV-5H (50% glass bead), PV-15H (15% glass fiber), PVN-3H (30% glass fiber), PVN-15H (15% glass fiber), PVZ-3H (15% glass fiber) [Emser Industries]

Nydur BKV 15 H (15% glass fiber), BKV 30 H (30% glass fiber), BKV 35 H (35% glass fiber), BKV 35 Z (35% glass fiber), BKV 40 H (40% glass fiber), BKV 50 H (50% glass fiber), GKV 115 (15% glass fiber), BKV 120 (20% glass fiber), BKV 125 (25% glass fiber), BKV 130 (30% glass fiber), BKV 135 (35% glass fiber), RM KU 2-22501/30 (30% glass fiber), RM KU 2-2521/20 (20% glass fiber), RM KU 2-25121/25 (25% glass fiber), RM KU 2-2521/30 (30% glass fiber), RM KU 2-2521/35 (35% glass fiber) [Mobay]

Nylatron 2008 (30% fiber glass), 2040 (40% fiber glass), GS-60-9 (30% glass), GS-60-13 (40% glass) [Polymer Corp.]

RTP 201A (10% glass fibers), 203A (20% glass fiber), 205A (30% glass fiber), 207A (40% glass fiber), 209A (50% glass fiber), 299x27567 [Fiberite]

Texalon GF 600 A(6-33), GF 604 (15 and 35) [Texapol]

Thermocomp PF-1002 (10% fiber glass), PF-1002HI (10% fiber glass), PF-1004 (20% fiber glass), PF-1004HI (20% fiber glass), PF-1006 (30% fiber glass), PF-1006FR (30% glass fiber), PF-1006HI (30% fiber glass), PF-1008 (40% glass fiber), PF-1008HI (40% glass fiber), PF-100-10 (50% glass fiber), PF-100-12 (60% glass fiber), PFL-4216 (30% glass fiber), PFL-4218 (40% glass fiber) [LNP]

Wellamid GS40-60 (40% solid glass spheres) [Wellman]

Glass- and mineral-reinforced nylon 6:

Capron 8266G HS (40% mineral/glass), 8267G HS (40% mineral/glass) [Allied-Signal]

Firestone Nylon 640-001HS (25% mineral, 15% short glass fibers) [Firestone]

Wellamid MRG30/10 (30% mineral, 10% short glass fibers) [Wellman]

Mineral-reinforced nylon 6:

Capron 8360 (34% mineral) [Allied-Signal]

Firestone Nylon 540-001HS (40% mineral) [Firestone]

Nylatron 2007 (35% mineral), 2033 (30% mineral) [Polymer Corp.]

Nylon (cont'd.)

Nylon 6/6:

BCI Nylon #808-809, #818-819, #819-S-10%, #829 [Belding]

Celanese Nylon 1000-1, 1000-2, 1000-4, 1003-1, 1003-2, 1200-1, 1310-1, 1310-2, 1310-4 [Hoechst Celanese]

Ertalon 66SA, 66SAM [Chemplast]

Minlon 10B [DuPont]

Nylaflow H, T [Polymer Corp.]

Nylatron 1022, 1024, GS, GS-21, GS-HS, NSB-90 [Polymer Corp.]

Polypenco Nylon 101 [Polymer Corp.]

RTP 200FR [Fiberite]

Texalon 1200A, 1200A Black 11, 1200A HS, 1200AXL, 1200A ZIP-1, 1200A ZIP-6, 1308A, 1310, XP-1317 [Texapol]

Thermocomp R-1000, R-1000FR-HS, RL-4040, RL-4310, RL-4540 (Migralube Resin) [LNP]

Vydyne 21, 21SP, 21X, 22H, 22HSP, 24 NSL, 24NSP, 25W, 25X, 65B, 66B [Monsanto]

Vylor 7264, ME-1791 [DuPont]

Wellamid 2BRH-NW, 21LN2-NNT, 22LH-N, 22LH13-N, 22L-N, 22LN2-N, 220-N, FR22F-N [Wellman]

Nylon 6/6 alloy:

Nylatron 1028 [Polymer Corp.]

Aluminum-filled nylon 6/6:

EMI-X RA-30 (30% aluminum flake), RA-35 (35% aluminum flake), RA-40 (40% aluminum flake) [LNP]

Carbon-reinforced nylon 6/6:

EMI-X RC-1008 (40% carbon fiber), RC-100-10 (50% carbon fiber), RC-100-12 (60% carbon fiber) [LNP]

RTP 283 (20% PAN carbon fiber), 285 (30% PAN carbon fiber), 285P (pitch carbon fiber), 287 (40% PAN carbon fiber), 287P (pitch carbon fiber) [Fiberite]

Stat-Kon R (carbon powder grade), RC-1002 (10% carbon fiber), RC-1004 (20% carbon fiber), RC-1004 FR (20% carbon fiber), RC-1006 (30% carbon fiber), RC-1006 HI (30% carbon fiber), RCL-4036 (30% carbon fiber), RCL-4042 (10% carbon fiber), RCL-4536 (30% carbon fiber) [LNP]

Thermocomp RC-1002 (10% carbon fiber), RC-1004 (20% carbon fiber), RC-1006 (30% carbon fiber), RC-1006HI (30% carbon fiber), RC-1006PC (30% Pitch carbon fiber), RC-1008 (40% carbon fiber), RCL-4036 (30% carbon fiber), RCL-4536 (30% carbon fiber) [LNP]

Glass/carbon-reinforced nylon 6/6:

RTP 201P25 (glass/pitch carbon fibers), 203P25 (glass/pitch carbon fibers) [Fiberite]

Stat-Kon RCF-1006 (10% carbon fiber/20% glass fiber), RF-15 (carbon powder grade/15% glass fiber) [LNP]

Glass-reinforced nylon 6/6:

Celanese Nylon 1500-1 (33% glass fiber), 1500-2 (33% glass fiber), 1503-1 (33% glass fiber), 1503-2 (33% glass fiber), 1600-1 (40% glass fiber), 1600-2 (40% glass fiber), 1603-1 (40% glass fiber), 1603-2 (40% glass fiber) [Hoechst Celanese]

Nylatron 1018 (33% fiber glass), 1027 (30% fiber glass), GS-51 (30% glass fiber), GS-51-13 (40% glass fiber) [Polymer Corp.]

RTP 201 (10% glass fiber), 201FR (10% glass fiber), 202FR (15% glass fiber), 203 (20% glass fiber), 203FR (20% glass fiber), 204FR (25% glass fiber), 204GB FR, 205 (30% glass fiber), 205FR (30% glass fiber), 205TFE5 (30% glass fiber), 205TFE15 (30% glass fiber), 205TFE20 (30% glass fiber), 207 (40% glass fiber), 209 (50% glass fiber) [Fiberite]

Stat-Kon R-15 (15% fiber glass) [LNP]

Texalon GF 1200A (13-40), GF 1308A (13-33), GF 1310 (13 and 33) [Texapol]

Thermocomp RF-1002 (10% glass fiber), RF-1002HI (10% glass fiber), RF-1004 (20% glass fiber), RF-1004FR-HS (20% glass fiber), RF-1004HI (20% glass fiber), RF-1006 (30% glass fiber), RF-1006FR-HS (30% glass fiber), RF-1006HI (30% glass fiber), RF-1008 (40% glass fiber), RF-1008HI (40% glass fiber), RF-100-10 (50% glass fiber), RF-100-12 (60% glass fiber), RFL-4216 (30% glass fiber), RFL-4218 (40% glass fiber) [LNP]

Vydyne R-513 (13% glass fiber), R-513H (13% glass fiber), R-533 (33% glass fiber), R-533H (33% glass fiber), R-538H-02 (33% glass), R-543 (43% glass fiber), R-543H (43% glass fiber) [Monsanto]

Wellamid FRGF25-66 (25% short glass fiber), FRGS25-66 (25% solid glass spheres), GS25-66 (25% solid glass spheres), GS40-66 (40% solid glass spheres), GSF25/15-66 (25% solid glass spheres, 15% short glass fiber) [Wellman]

Zytel FR-50 (25% glass) [DuPont]

Glass/mineral-reinforced nylon 6/6:

Nylatron 1026 (15% fiber glass, 25% mineral) [Polymer Corp.]

RTP 202M (glass/mineral/beads), 203GB20 (glass/mineral/beads), 203M GB20 (glass/mineral/beads) [Fiberite]

Texalon GMF 600-40 (15% short glass fiber, 25% mineral) [Texapol]

Vydyne R-400G [Monsanto]

Wellamid MRG30/10 (30% mineral/10% short glass fiber) [Wellman]

Mineral-reinforced nylon 6/6:

Nylatron 1025 (40% mineral) [Polymer Corp.]

RTP 227FR [Fiberite]

Texalon MF 1200A-40 (40% mineral), MF 1200A1-40 (40% mineral) [Texapol]

Wellamid MR329HS (32% mineral), MR409HS (40% mineral), MR-410HS (40% mineral) [Wellman]

Mineral/talc-reinforced nylon 6/6:

RTP 225 (20% filler), 227 (40% filler) [Fiberite]

Nylon *(cont'd.)*

Nickel-reinforced nylon 6/6:
　PDX-83392 [LNP]
Nylon 6/9:
　Ultramid KR 4609 [Badische]
　Vydyne 60H [Monsanto]
Nylon 6/10:
　Migralube Q-1000 [LNP]
　Carbon-reinforced nylon 6/10
　　EXI-X QC-1008 (40% carbon fiber), QC-100-10 (50% carbon fiber) [LNP]
　　Stat-Kon QC-1002 (10% carbon fiber), QC-1006 (30% carbon fiber) [LNP]
　　Thermocomp QC-1006 (30% PAN carbon fiber) [LNP]
　Glass-reinforced nylon 6/10:
　　Migralube QFL-4036 (30% fiber glass), QFL-4536 (30% fiber glass) [LNP]
　　RTP 201B (10% glass fibers), 203B (20% glass fiber), 205B (30% glass fibers),
　　　207B (40% glass fiber) [Fiberite]
　　Thermocomp QF-1002 (10% glass fiber), QF-1004 (20% glass fiber), QF-1006
　　　(30% glass fiber), QF-1006FR (30% glass fiber), QF-1008 (40% glass fiber),
　　　QF-100-10 (50% glass fiber), QF-100-12 (60% glass fiber) [LNP]
　Nickel-reinforced:
　　PDX-82428 (40% nickel) [LNP]
Nylon 6/12:
　Grilon CA6, CA6EH, CF35, CR9, CR9 natural 6361 [Emser Industries]
　Nylatron GS-71, GS-73 [Polymer Corp.]
　Carbon-reinforced nylon 6/12:
　　Stat-Kon IC-1006 (15% carbon fiber) [LNP]
　Glass-reinforced nylon 6/12:
　　Nylatron GS-71-9 (30% glass fiber) [Polymer Corp.]
　　RTP 201D (10% glass fiber), 203D (20% glass fiber), 205D (30% glass fiber), 207D
　　　(40% glass fiber) [Fiberite]
　　Thermocomp IF-1002 (10% fiber glass), IF-1004 (20% fiber glass), IF-1006 (30%
　　　fiber glass), IF-1008 (40% fiber glass), IF-100-10 (50% fiber glass), IF-100-12
　　　(60% fiber glass) [LNP]
Nylon 6 and 6/6:
　Durethan A30S, B25T, B30P, B30S, B31SK, B31F, B40E, B40F, B40SK, B50E,
　　BKV35, BKV50H [Bayer AG]
　Ultramid KR 4645 [Badische]
　Glass-reinforced nylon 6 and 6/6:
　　Durethan AKV30 (30% glass fiber), AKV30H (30% glass fiber), BG30X (glass
　　　fiber/glass beads), BKV30 (30% glass fiber), BKV30H, BKV30N, BKV30N1,
　　　BKV35 (35% glass fiber), BKV50H (50% glass fiber) [Bayer AG]
　　Ultramid KR 4653 (30% glass fiber) [Badische]
　Glass/mineral-reinforced nylon 6 and 6/6:
　　Ultramid KR 4651 (30% glass fiber/mineral) [Badische]

Nylon (cont'd.)

Mineral-reinforced nylon 6 and 6/6:
Ultramid KR 4650 (40% mineral), KR 4652 (30% mineral) [Badische]
With olefin polymer:
Durethan BC30, BC40 [Bayer AG]
Nylon 11:
Ertalon 11SA [Chemplast]
Rilsan BECNO, BECVO P40 TL, BESHVO, BESN Black T, BESN F15, BESNO, BESNO P20, BESNO P40, BESNO P40 TL, BESNO TL, BESVO, BMN F15, BMN F25, BMNO, BMNO P20, BMNO P40, BMNY BZ TL, BMV Black T, KMVOTL, RDP15-10 Natural, RDP-17-1 Dispersion Coating, RDP Pigmented [Rilsan]
Carbon-reinforced nylon 11:
Stat-Kon H (carbon powder grade) [LNP]
Carbon and/or carbon/glass-reinforced nylon 11:
SR 80, SR 100, SR 120 [Rilsan]
Glass-reinforced nylon 11:
Rilsan BZM300 (30% glass) [Rilsan]
RTP 201C (10% glass fiber), 203C (20% glass fiber), 205C (30% glass fiber), 207C (40% glass fiber) [Fiberite]
Thermocomp HF-1006 (30% glass) [LNP]
Glass/graphite-reinforced nylon 11:
Rilsan BZM23G9 (23% glass/graphite), BZM43G9 (43% glass/graphite) [Rilsan]
Nylon 12:
Grilamid L16G, L16GM, L20FR, L20G, L20W20, L20W40, L25, L25 natural 6112, L25W10, L25W20, L25W40, L25W40NZ, TR55LX, TR55UV [Emser Industries]
Nylon 12 elastomer:
Grilamid ELY20NZ, ELY60 [Emser Industries]
Glass-reinforced nylon 12:
Grilamid LKN-5H (50% glass bead), LV-3H (30% glass fiber), LV-43H (43% glass fiber) [Emser Industries]
Thermocomp SF-1006 (30% glass fiber), SF-100-10 (50% glass fiber) [LNP]

MODIFICATIONS/SPECIALTY GRADES:

Conductive grade (EMI shielding):
EMI-X PC-1008, PC-100-10, QC-1008, QC-100-10, RA-30, RA-35, RA-40, RC-1008, RC-100-10, RC-100-12; PDX-82428, -83392; RTP 203H-CF15, 204H-CF10, 283, 285, 285P, 287P, 299x27567
Statically dissipative:
Stat-Kon H, IC-1006, PC-1006, QC-1002, QC-1006, R, R-15, RC-1002, RC-1004, RC-1004FR, RC-1006, RC-1006 HI, RCF-1006, RCL-4036, RCL-4042, RCL-4536, RF-15, VC-1003
Flexural-modified:
Grilon W3941, W5744

Nylon (cont'd.)

Super rigid:

SR 80, SR 100, SR 120

Super tough:

RTP 200H-FR, 203H-CF15, 204H-CF10, 204H-FR, 205H-FR; Stat-Kon VC-1003

Impact-modified:

Nydur B 31 SK, B 40 SK, BKV 35 Z, BKV 115, BKV 120, BKV 120, BKV 125, BKV 130, BKV 135; Fosta Nylon 1379, 1417, 1722; Grilon A28NX, A28NY, A28NZ, BT40, PVN-3H, PVN-15H, PVZ-3H; Nylatron 1028, 2029, 2038, 2039; Texalon 1308A, XP-1296; Thermocomp PF-1002HI, PF-1004HI, PF-1006HI, PF-1008HI, RC-1006HI, RF-1002HI, RF-1004HI, RF-1006HI, RF-1008HI; Ultramid B3L, B3WM601, KR 4409, KR 4411, KR 4447/1, KR 4645, KR 4652; Wellamid 22LH13-N

High-impact:

Durethan B40SK; Fosta Nylon 1693; RTP 201H, 203H, 205H, 207H; Stat-Kon RC-1006 HI; Texalon MF1200A-40; Ultramid KR 4412, KR 4609

Controlled crystallization grade:

Celanese Nylon 1310-1, 1310-2, 1310-4

High-flow:

Capron 8202L; Grilamid L16G, L16GM, TR55LX, TR55UV; Grilon A23, A23GM, A23G natural 6165; Ultramid B3L, B3S, KR 4250, KR 4405, KR 4411, KR 4446

Flame-retardant:

Capron 8232G FR; Durethan BKV30N, BKV30N1; Grilamid L20FR; Grilon A28VO; Nylatron 2033; RTP 200FR, 200H-FR, 201FR, 202FR, 203FR, 204FR, 204GB FR, 204H-FR, 205H-FR, 205FR, 227FR; Stat-Kon RC-1004 FR; Thermocomp PF-1006FR, QF-1006FR, R-1000FR-HS, RF-1004FR-HS, RF-1006FR-HS; Ultramid A3X1G5, A3X1G7, A3X1G10, A3XG5, A3XG7, KR 4205 (halogen-free), KR 4406, KR-B3X2V6; Vydyne 909, M-340, M-344, M-345; Wellamid FR22F-N, FRGF25-66, FRGS25-66; Zytel FR-50, ST FR-80

Wear-resistant:

Ertalon 62AU; Nylatron 2039

Hydrolysis-stabilized:

Nylatron GS-51; Rilsan KMVOTL

Hydrolysis-resistant:

Vydyne R-538H-02

Reduced moisture absorption:

Nydur RM KU 2-2501/30, 2-2521/20, 2-2521/25, 2-2521/30, 2-2521/35

Stabilized:

Ultramid A3EG5, A3EG6, A3DG7, A3EG10, A3K, A3KN, A3R, A3SK, A3W, A3WG5, A3WG6, A3WG7, A3WG10, A4H, A4K, B3EG5, B3EG6, B3EG7, B3EG10, B3K, B3L, B3WG5, B3WG6, B3WG7, B3WG10, B3WM602, B4K, B5W, B35SK, B35W, KR 4205, KR 4250, KR 4406, KR 4409, KR 4411, KR 4412, KR 4446, KR 4609, KR 4645, KR 4650, KF 4651, KR 4653, S3K

Nylon *(cont'd.)*

Heat-stabilized:

Capron 8220 HS, 8221 HS, 8230G HS, 8231G HS, 8232 G HS FR, 8233G HS, 8234G HS, 8266G HS, 8267G HS, 8270 HS, 8280 HS; Celanese 1003-1, 1503-1, 1503-2, 1603-1, 1603-2; Durethan AKV30H, BKV30H, BKV50H; Firestone Nylon 210-001L, 213-001, 415-001HS, 430-001HS, 540-001HS, 640-001HS; Fosta Nylon 870, 1047, 1379, 1525; Grilamid LKN-5H, LV-3H, LV-43H, TR55UV; Grilon A28NX, A28NY, A28NZ, A28VO, BT40, CA6EH, PV-3H, PV-5H, PV-15H, PVN-3H, PVN-15H, PVZ-3H, R40GM, W3941, W5744; Minlon 20B; Monocast MC901; Nydur BKV 15H, BKV 30H, BKV 35H, BKV 40H, BKV 50H; Nylaflow LP; Nylatron 1018, 1024, 1026, 1027, 2008, 2010, 2011, GS-51, GS-71, GS-73, GS-HS; Rilsan BECVO P40 TL, BESN Black T, BESNO P40 TL, BESNO TL, BMNY BZ TL, BMV Black T, KMVOTL; Texalon 600A HS, 1110A PL HS, 1200A HS; Thermocomp R-1000FR-HS, RF-1004FR-HS, RF-1006FR-HS; Vydyne 22H, 22H, 22HSP, 60H, R-200, R-220, R-250, R-513H, R-533H, R-543H, R-400G; Wellamid 2BRH-NW, 22LH-N, 22LH13-N, 42LH-N, GS40-60, MR329HS, MR409HS, MR-410HS

UV-stabilized:

Firestone Nylon 213-001, 540-001HS, 640-001HS; Grilamid LKN-5H, LV-3H, LV-43H, TR55UV; Grilon A28NX, A28NY, A28NZ, A28VO, BT40, CA6EH, PV-3H, PV-5H, PV-15H, PVN-3H, PVN-15H, PVZ-3H, R40GM, W3941, W5744; Nylaflow LP; Nylatron 2038, GS-21, GS-63, GS-73; Rilsan BECVO P40 TL, BESNO P40 TL, BESNO TL, BMNY BZ TL, BMV Black T, KMVOTL

UV-resistant:

Texalon 1200A Black 11 (2% carbon black) [Texapol]

Weather-resistant:

Vydyne 25W (with carbon), 25X (with carbon) [Monsanto]

Nucleated:

Firestone Nylon 213-001; Fosta Nylon 446; Texalon 600A NU, 600A Zip 1, 1200A ZIP-1; Vydyne 24NSL, 24NSP; Wellamid 21LN2-NNT, 22LN2-N, 42LN2-N

Plasticized:

Durethan B30P; Fosta Nylon 567, 714, 1210, 1379, 1525; Grilamid L20W20, L20W40, L25W10, L25W20, L25W40, L25W40NZ; Grilon A28W10, R47HW; Rilsan BESNO P20, BESNO P40, BMNO P20, BMNO P40; Texalon 600A PL-2, 1110A PL

Lubricated:

Texalon 600A Zip 1, 600A Zip 3, 1200AXL, 1200A ZIP-1; Vydyne 21SP, 21X, 22H, 22HSP, 60H, 24NSL, 24NSP

Self-lubricating:

Rilsan BMNY BZ TL

Solid lubricant:

Nylatron 1028, 2011, NSB, NSB-90

Internally lubricated:

Grilamid L16G, L16GM, L20G; Grilon A23GM, A23G natural 6165, A28GM,

Nylon (cont'd.)

R40GM; Nylatron 1024, 1025, 1026

Externally lubricated:
Ultramid A3 Pellets 24

Process lubricated:
Wellamid 21LN2-NNT, 22LH-N, 22LH13-N, 22L-N, 22LN2-N, 42LH-N, 42L-N, 42LN2-N

Graphite-lubricated:
Thermocomp RL-4310 (5% graphite); Vekton 6PAG

MoS$_2$-lubricated:
Ertalon 66SAM; Nylatron GS, GSM, GS-21, GS-51, GS-51-13, GS-60, GS-60-9, GS-60-13, GS-63, GS-71, GS-71-9, GS-73, GS-HS, GSM; Thermocomp PFL-4216, PFL-4218, RFL-4216, RFL-4218; Vekton 6PAM

TFE/PTFE-lubricated:
Migralube QFL-4036 (15% TFE); RTP 205TFE5 (5% TFE), 205TFE15 (15% TFE), 205TFE20 (20% TFE); Stat-Kon RCL-4036 (15% PTFE), RCL-4042 (20% PTFE); Thermocomp RCL-4036 (15% PTFE), RL-4040 (20% TFE), XFL-4036 (15% TFE)

TFE/silicone-lubricated:
Migralube QFL-4536 (15% TFE/silicone); Stat-Kon RCL-4536 (15% PTFE/silicone); Thermocomp RCL-4536 (15% PTFE/silicone), RL-4540 (20% TFE/silicone)

Surface-lubricated:
Celanese Nylon 1000-2, 1003-2, 1310-2, 1500-2, 1503-2, 1600-2, 1603-2

CATEGORY:
Thermoplastic resin

PROCESSING:

Blow molding:
Capron 8270 HS; Durethan B 50 E; Fosta Nylon 589; Grilon R47, R50; Rilsan BESHVO, BESN Black T, BESN F15, BESNO, BESNO P20, BESNO P40, BESNO P40 TL, BESNO TL, BESVO; Ultramid B5, B5W

Blown film:
Grilon BT40, F47, R47

Cast film:
Grilon BT40, R47

Electrostatic deposition:
Rilsan RDP15-10 Natural, RDP-17-1 Dispersion Coating, RDP Pigmented

Extrusion:
BCI Nylon No. 653; Capron 8203C, 8207F, 8209F, 8220 HS, 8221 HS, 8253, 8254, 8270 HS, 8350; Celanese Nylon 1000-1, 1000-2, 1003-1, 1003-2, 1200-1; Durethan B 31 F, B 40 E, B 40 F; Ertalon 6SAU, 11SA, 66SA, 66SAM; Firestone Nylon 200-001, 210-001L; Fosta Nylon 446, 471, 567, 589, 714, 870, 1210, 1379, 1525; Grilamid ELY 20NZ, ELY 60, L25, L25 natural 6112, L25 W10, L25 W20, L25 W40, L25 W40NZ, TR55 LX, TR55 UV; Grilon A23, A28, BT40, CA6, CA6 EH,

CF35, CR9, CR9 natural 6361, ELX 23NZ, F40, F47, R40, R40 GM, R40 N38, R47, R47HW, R50, W3941, W5744; Rilsan BECNO, BECVO P40 TL, BESHVO, BESVO, BESN Black T, BESN F15, BESNO, BESNO P20, BESNO P40, BESNO P40 TL, BESNO TL; Texalon 600A, 600A HS, 600A PL-2, 670A, 1106, 1108, 1110A PL, 1200A; Ultramid A3, A3 Pellets 24, A4, A5, B3, B4, B4K, B5, B5W, B35, B35M, B35MF01, B35W, KR4407, KR4609, S3, S4; Vydyne 60H, 65B, 66B

Fluid bed processing:

Rilsan RDP15-10 Natural, RDP-17-1 Dispersion Coating, RDP Pigmented

Injection molding:

BCI Nylon No. 653; Capron 8200, 8202, 8202C, 8202CL, 8202L, 8230G HS, 8231G HS, 8233G HS, 8234G HS, 8253, 8255, 8259, 8260, 8266G HS, 8267G HS, 8350, 8352, 8360; Celanese Nylon 1000-1, 1000-2, 1000-4, 1003-1, 1003-2, 1200-1, 1310-1, 1310-2, 1310-4, 1500-1, 1500-2, 1503-1, 1503-2, 1600-1, 1600-2, 1603-1, 1603-2; Durethan A 30 S, AKV 30, AKV 30 H, B 25 T, B 30 P, B 30 S, B 31 SK, B 40 E, B 40 SK, BC 30, BC 40, BG 30 X, BKV 30, BKV 30 H, BKV 30 N, BKV 30 N1, BKV 35, BKV 50 H; EMI-X PC-1008, PC-100-10, QC-1008, QC-100-10, RA-30, RA-35, RA-40, RC-1008, RC-100-10, RC-100-12; Firestone Nylon 200-001, 213-001, 228-001, 415-001HS; Fosta Nylon 512, 523, 567, 714, 1047, 1210, 1379, 1417, 1693, 1722; Grilamid ELY 20NZ, ELY 60, L16G, L16 GM, L20 FR, L20 G, L20 W20, L20 W40, LKN-5H, LV-3H, LV-43H, TR55 LX, TR55 UV; Grilon A23 GM, A23 G Natural 6165, A28 GM, A28 NX, A28 NY, A28 NZ, A28 VO, A28 W10, BT40, ELX 23NZ, PV-3H, PV-5H, PV-15H, PVN-3H, PVN-15H, PVZ-3H, R40, R40 GM, R40 N38, R47HW, W3941, W5744; Minlon 10B, 11C, 12T, 20B, 21C; Nydur B 31 SK, B 40 SK, BKV 15 H, BKV 30 H, BKV 35 H, BKV 35 Z, BKV 40 H, BKV 50 H, BKV 115, BKV 120, BKV 125, BKV 130, BKV 135, RM KU 2-2501/30, RM KU 2-2521/20, RM KU 2-2521/25, RM KU 2-2521/30, RM KU 2-2521/35; Nylatron 1018, 1022, 1024, 1025, 1026, 1027, 1028, 2001, 2005, 2007, 2008, 2010, 2011, 2015, 2029, 2033, 2038, 2039, 2040, GS, GS-21, GS-51, GS-51-13, GS-60, GS-60-9, GS-60-13, GS-63, GS-71, GS-71-9, GS-73, GS-HS, GSM, NSB, NSB-90; PDX-82428, -83392, -85384; Rilsan BMN F15, BMN F25, BMNO, BMNO P20, BMNO P40, BMNY BZ TL, BMV Black T, BZM23G9, BZM43G9, BZM300, KMVOTL, SR 80, SR 100, SR 120; RTP 200FR, 200H-FR, 201, 201A, 201B, 201C, 201D, 201FR, 201H, , 201P 25, 202FR, 202M, , 203, 203A, 203B, 203CF, 203D, 203FR, 203 GB 20, 203H, 203H-CF15, 203M GB 20, 203P 25, 204FR, 204GB FR, 204H-CF10, 204H-FR, 205, 205A, 205B, 205C, 205D, 205FR, 205H, 205H-FR, 205TFE5, 205TFE15, 205TFE20, 207, 207A, 207B, 207C, 207D, 207H, 209, 209A, 225, 227, 227FR, 283, 285, 285P, 287, 287P, 299x27567; Texalon 600A, 600A HS, 600A NU, 600A Zip 1, 600A Zip 3, 670A, 1108, 1110A PL, 1200A, 1200A HS, 1200 AXL, 1200A Zip 1, 1200A Zip-6, 1203D, 1308A, 1310, GF600A (6–33), GF 604 (15 and 35), GF1200A (13–40), GF1308A (13–33), GF1310 (13 and 33), GMF 600-40, MF1200A-40, MF1200A1-40, XP-1296, XP-1317; Thermocomp HF-1006, IF-1002, IF-1004, IF-1006, IF-1008, IF-100-10, IF-100-12, PC-1006, PF-1002, PF-1002HI, PF-1004, PF-

Nylon *(cont'd.)*

1004HI, PF-1006, PF-1006FR, PF-1006HI, PF-1008, PF-1008HI, PF-100-10, PF-100-12, PFL-4216, PFL-4218, QC-1006, QF-1002, QF-1004,, QF-1006, QF-1006FR, QF-1008, QF-100-10, QF-100-12, R-1000, R-10000FR-HS, RC-1002, RC-1004, RC-1006, RC-1006HI, RC-1006PC, RC-1008, RCL-4036, RCL-4536, RF-1002, RF-1002HI, RF-1004, RF-1004FR-HS, RF-1004HI, RF-1006, RF-1006FR-HS, RF-1006HI, RF-1008, RF-1008HI, RF-100-10, RF-100-12, RFL-4216, RFL-4218, RL-4040, RL-4310, RL-4540, SF-1006, SF-100-10, XC-1006, XF-1004, XF-1006, XF-1008, XFL-4036; Ultramid A3EG5, A3EG6, A3EG7, A3EG10, A3G5, A3G6, A3G7, A3HG5, A3K, A3KN, A3 Pelets 24, A3R, A3SK, A3W, A3WG5, A3WG6, A3WG7, A3WG10, A3X1G5, A3X1G7, A3X1G10, A3XG5, A3XG7, A4, A4H, A4K, B3EG5, B3EG6, B3EG7, B3EG10, B3G5, B3G5HS, B3G6, B3G6HS, B3G7, B3G7HS, B3G10, B3K, B3L, B3S, B3WG5, B3WG6, B3WG7, B3WG10, B3WM601, B3WM602, B4, B4K, B35G3, B35SK, B35W, KR4205, KR4250, KR4405, KR4406, KR4407, KR4409, KR4411, KR4412, KR4445/1, KR4446, KR4447/1, KR4448, KR4449, KR4450, KR4609, KR4645, KR4650, KR4651, KR4652, KR4653, KR-B3X2V6, S3K; Vekton 6PA, 6PAG, 6PAM, 6PB, 6XAU; Vydyne 21, 21SP, 21X, 22H, 22HSP, 24NSL, 24NSP, 25W, 25X, 909, M-340, M-344, M-345, R-100, R-200, R-220, R-250, R-400G, R-513, R-513H, R-533, R-533H, R-538H-02, R-543, R-543H; Wellamid 2BRH-NW, 21LN2-NNT, 22LH-N, 22LH13-N, 22L-N, 22LN2-N, 42LH-N, 42L-N, 42LN2-N, 220-N, FR22F-N, FRGF25-66, FRGS25-66, GS25-66, GS40-60, GS40-66, GSF25/15-66, MR329HS, MR409HS, MR410HS, MRG30/10; Zytel FR-50, ST FR-80

Reaction-injection molding (RIM):
Nyrim P-1 1000, P-1 2000, P-1 3000, P-1 4000

Rotational molding:
Capron 8280 HS; Rilsan BMN F15, BMN F25, BMNO, BMNO P20, BMNO P40, BMNY BZ TL, BMV Black T, BZM23G9, BZM43G9, BZM300, KMVOTL; Texalon 680A

APPLICATIONS:

Agriculture industry: farm machinery coatings (Rilsan RDP15-10 Natural, RDP-17-1 Dispersion Coating, RDP Pigmented); farm machinery parts (Rilsan BMN F15, BMN F25, BMNO, BMNO P20, BMNO P40, BMNY BZ TL, BMV Black T, BZM23G9, BZM43G9, BZM300, KMVOTL); farm machinery tubing (Nylaflow H)

Architectural applications: construction applications (Durethan A 30 S, AKV 30, AKV 30 H, B 25 T, B 39 P, B 30 S, B 31 SK, B 40 E, B 40 SK, BC 30, BC 40, BG 30 X, BKV 30, BKV 30 H, BKV 30 N, BKV 30 N1, BKV 35, BKV 50 H; Monocast MC901; Nylatron GSM)

Automotive applications: (Durethan A 30 S, AKV 30, AKV 30 H, B 25 T, B 39 P, B 30 S, B 31 SK, B 40 E, B 40 SK, BC 30, BC 40, BG 30 X, BKV 30, BKV 30 H, BKV 30 N, BKV 30 N1, BKV 35, BKV 50 H; Rilsan SR 80, SR 100, SR 120; RTP 201,

Nylon *(cont'd.)*

201A, 203, 205, 205A, 207, 207A, 209, 209A; Texalon 600A, 600A HS, 600A PL-2, 1106, 1108, 1200A, 1200A Zip-6, 1308A, GF1200A (13-40), GF1308A (13–33), GF1310 (13 and 33), MF1200A1-40; Ultramid B5W, KR 4653); body parts (Texalon 1108, GMF600-40); cable coverings (Rilsan BECNO, BECVO P40 TL); door/window hardware (Capron 8230G HS, 8231G HS, 8233G HS, 8234G HS, 8260, 8266G HS, 8267G HS, 8360; Nydur B 31 SK, B 40 SK, BKV 15 H, BKV 30 H, BKV 35 H, BKV 35 Z; Ultramid A3X1G5); electrical housings (Nydur RM KU 2-2501/30, RM KU 2-2521/20, RM KU 2-2521/25, RM KU 2-2521/30, RM KU 2-2521/35); exterior parts (Capron 8230G HS, 8231G HS, 8233G HS, 8234G HS, 8266G HS, 8267G HS; Nydur BKV 115; Ultramid KR 4409, KR 4446, KR 4448, KR 4652); fuel lines (Rilsan BECHVO, BESN Black T, BESN F15, BESNO, BESNO P20, BESNO P40, BESNO P40 TL, BESNO TL, BESVO); fuel pumps (Minlon 10B, 11C, 12%, 20B, 21C; Nydur BKV 35 H); gas/clutch pedals (Nydur BKV 35 H, BKV 35 Z, BKV 40 H, BKV 50 H, BKV 115); gas tanks (Rilsan BECHVO, BESN Black T, BESN F15, BESNO, BESNO P20, BESNO P40, BESNO P40 TL, BESNO TL, BESVO); hoses (Rilsan BECHVO, BESN Black T, BESN F15, BESNO, BESNO P20, BESNO P40, BESNO P40 TL, BESNO TL, BESVO); interior parts (PDX-85384); interior lighting (Capron 8260, 8360); mechanical parts (Texalon GF600A (6-33)); mirror housings (Capron 8260, 8360; Nydur BKV 15 H, RM KU 2-2501/30, RM KU 2-2521/20, RM KU 2-2521/25, RM KU 2-2521/30, RM KU 2-2521/35; Ultramid B35G3); motorcycle components (Nydur B 40 SK); sunroof frames (Nydur BKV 115); tubing (Capron 8253, 8254, 8350; Grilamid L25 W10, L25 W20, L25 W40, R47HW, W3941); under-the-hood parts (Capron 8230G HS, 8231G HS, 8233G HS, 8234G HS, 8266G HS, 8267G HS; Celanese Nylon 1003-1, 1003-2; Nydur B 31 SK; Texalon 1200A HS; Vydyne R-538H-02, R-400G; Wellamid MR410 HS); vehicle bodies (Minlon 10B, 11C, 12%, 20B, 21C; Wellamid MR329 HS, MR409 HS, MRG30/10); wheel hubs/trim (Capron 8260, 8360; Ultramid B3G5HS)

Aviation industry: aerospace industry (Rilsan SR 80, SR 100, SR 120); aircraft control cable covering (Rilsan BECNO, BECVO P40 TL); avionics housings (EMI-X PC-1008, PC-100-10, QC-1008, QC-100-10, RA-30, RA-35, RA-40, RC-1008, RC-100-10, RC-100-12; PDX-82428, -83392)

Consumer products: (Nydur BKV 115); appliance housings (Wellamid MR410 HS); appliance parts (Capron 8260, 8360); bicycles (Rilsan SR 80, SR 100, SR 120; Ultramid B35G3); bus seats (Nydur BKV 35 Z, RM KU 2-2501/30, RM KU 2-2521/20, RM KU 2-2521/25, RM KU 2-2521/30, RM KU 2-2521/35); chain links (Nydur B 31 SK); coatings (Rilsan RDP15-10 Natural, RDP-17-1 Dispersion Coating, RDP Pigmented); combs (Fosta Nylon 523; Texalon 1203D)); exercise equipment (Nydur RM KU 2-2501/30, RM KU 2-2521/20, RM KU 2-2521/25, RM KU 2-2521/30, RM KU 2-2521/35); eyeglass frames (Grilamid TR55 LX); fire extinguisher parts (Nydur B 31 SK, B 40 SK); fishing line (Fosta Nylon 567, 714, 1210; Ultramid B3, B35); furniture (Durethan A 30 S, AKV 30, AKV 30 H, B 25 T, B 39 P, B 30 S, B 31 SK, B 40 E, B 40 SK, BC 30, BC 40, BG 30 X, BKV 30, BKV 30

Nylon (cont'd.)

H, BKV 30 N, BKV 30 N1, BKV 35, BKV 50 H; Wellamid MR410 HS); furniture casters/parts (Capron 8202C, 8202CL; Nydur B 31 SK, B 40 SK, BKV 40 H, BKV 50 H, RM KU 2-2501/30, RM KU 2-2521/20, RM KU 2-2521/25, RM KU 2-2521/30, RM KU 2-2521/35); hardware (Celanese 1000-1, 1000-2, 1000-4, 1003-1, 1003-2; Fosta Nylon 523; Grilon R40 N38); housewares (Durethan A 30 S, AKV 30, AKV 30 H, B 25 T, B 39 P, B 30 S, B 31 SK, B 40 E, B 40 SK, BC 30, BC 40, BG 30 X, BKV 30, BKV 30 H, BKV 30 N, BKV 30 N1, BKV 35, BKV 50 H); kitchenware (Texalon 1203D, XP-1317); lawn and garden equipment (Capron 8230G HS, 8231G HS, 8233G HS, 8234G HS, 8253, 8255, 8266G HS, 8267G HS, 8350; Minlon 10B, 11C, 12%, 20B, 21C; Nydur B 31 SK, B 40 SK, BKV 15 H, BKV 30 H, RM KU 2-2501/30, RM KU 2-2521/20, RM KU 2-2521/25, RM KU 2-2521/30, RM KU 2-2521/35); office furniture (Capron 8260, 8360); recreational equipment (Grilamid ELY 20NZ; Nydur BKV 35 H, BKV 40 H, BKV 50 H; Rilsan SR 80, SR 100, SR 120; RTP 201, 201H, 203, 203H, 205, 205H, 207, 207H, 209); shoes/boots (Fosta Nylon 567, 714; Grilamid ELY 20NZ; Grilon R40 N38); sporting goods (Ultramid KR 4409; Wellamid MR410 HS); stadium seats (Nydur B 40 SK)

Electrical/electronic industry: (Capron 8202C, 8202CL; Durethan A 30 S, AKV 30, AKV 30 H, B 25 T, B 39 P, B 30 S, B 31 SK, B 40 E, B 40 SK, BC 30, BC 40, BG 30 X, BKV 30, BKV 30 H, BKV 30 N, BKV 30 N1, BKV 35, BKV 50 H; PDX-85384; Polypenco Nylon 101; Rilsan SR 80, SR 100, SR 120; RTP 201D, 203D, 205D, 207D; Vydyne R-400G; Wellamid MR410 HS); business machines and office equipment (EMI-X PC-1008, PC-100-10, QC-1008, QC-100-10, RA-30, RA-35, RA-40, RC-1008, RC-100-10, RC-100-12; PDX-82428, -83392, -85384); cable and wire jacketing (BCI LX-3250, No. 653; Capron 8220 HS, 8221 HS, 8253, 8254, 8350; Firestone Nylon 210-001L; Fosta Nylon 870, 1379, 1525; Grilon CA6 EH; Rilsan BECNO, BECVO P40 TL, BESNO, BESNO P20, BESNO P40; Texalon 600A HS, 600A PL-2, 670A; Ultramid B35, B35W; Vydyne 60H); cable liners (Capron 8203C); coil bobbins (Thermocomp RF-1004FR-HS); connectors (Capron 8202, 8230G HS, 8231G HS, 8233G HS, 8234G HS, 8253, 8255, 8266G HS, 8267G HS, 8350; RTP 201, 203, 205, 207, 209; Thermocomp RF-1004FR-HS; Zytel FR-50, ST FR-80); dielectric parts (Ultramid A3K, KR 4205, KR 4445/1); electrical components (Rilsan BMN F15, BMN F25, BMNO, BMNO P20, BMNO P40, BMNY BZ TL, BMV Black T, BZM23G9, BZM43G9, BZM300, KMVOTL; RTP 201A, 203A, 205A, 207A, 209A; Stat-Kon H, IC-1006, PC-1006, QC-1002, QC-1006, R, RC-1002, RC-1004, RC-1004 FR, RC-1006, RC-1006 HI, RCF-1006, RCL-4036, RCL-4042, RCL-4536, RF-15, VC-1003); electrical enclosures/packaging (Stat-Kon H, IC-1006, PC-1006, QC-1002, QC-1006, R, RC-1002, RC-1004, RC-1004 FR, RC-1006, RC-1006 HI, RCF-1006, RCL-4036, RCL-4042, RCL-4536, RF-15, VC-1003); electronic devices (EMI-X PC-1008, PC-100-10, QC-1008, QC-100-10, RA-30, RA-35, RA-40, RC-1008, RC-100-10, RC-100-12; PDX-82428, -83392); EMI/RFI shielding applications (EMI-X PC-1008, PC-100-10, QC-1008, QC-100-10, RA-30, RA-35, RA-40, RC-1008, RC-100-10, RC-100-

12; PDX-82428, -83392); housings (Capron 8230G HS, 8231G HS, 8233G HS, 8234G HS, 8266G HS, 8267G HS; EMI-X PC-1008, PC-100-10, QC-1008, QC-100-10, RA-30, RA-35, RA-40, RC-1008, RC-100-10, RC-100-12; Nydur BKV 15 H, BKV 30 H, BKV 35, BKV 35 Z, BKV 115, RM KU 2-2501/30, RM KU 2-2521/20, RM KU 2-2521/25, RM KU 2-2521/30, RM KU 2-2521/35; PDX-82428, –83392; Ultramid B3G6HS, B3G7HS, KR 4412, KR 4447/1); power tools (Capron 8230G HS, 8231G HS, 8233G HS, 8234G HS, 8260, 8266G HS, 8267G HS, 8360; Nydur BKV 15 H, BKV 30 H, BKV 35 H, BKV 35 Z, BKV 115, RM KU 2-2501/30, RM KU 2-2521/20, RM KU 2-2521/25, RM KU 2-2521/30, RM KU 2-2521/35; RTP 201A, 203A, 205A, 207A, 209A; Ultramid KR 4447/1; Wellamid MR329 HS, MR409 HS, MRG30/10); radio/TV (Rilsan SR 80, SR 100, SR 120); receptacles (Capron 8202, 8202L, 8253, 8255, 8350); switchgear (Capron 8202L; Ultramid A3X1G5, A3X1G7, A3XG5, A3XG7, KR 4406); wiring devices (Capron 8200, 8202)

FDA-approved applications: (Capron 8207F, 8209F; Fosta Nylon 471; Grilamid L25, TR55 LX, TR55 UV; Grilon A23 GM, A23G natural 6165, A28 GM, CA6, F40, F47; Nylaflow H, T; Polypenco Nylon 101)

Food-contact applications: (Grilon A23 GM, A23G natural 6165, A28 GM, F40, F47); food handling tubing (Nylaflow H, T); food packaging (Capron 8207F, 8209F; Fosta Nylon 471; Grilamid L25; Grilon CA6); food processing (Polypenco Nylon 101); food processing equipment coatings (Rilsan RDP15-10 Natural, RDP-17-1 Dispersion Coating, RDP Pigmented)

Functional additives: extreme pressure lubricant (Migralube QFL-4536; Thermocomp RL-4540); phenolic resin modifier (BCI Nylon No. 651)

Industrial applications: (Vydyne R-400G); adhesives (BCI Nylon No. 651); barrier coatings (BCI Nylon No. 651); bearing applications (Grilamid LKN-5H; Nylatron NSB-90; Ultramid A3EG5, A3HG5, A3WG5, A4H); bearings and bushings (Celanese 1000-1, 1000-2, 1000-4, 1003-1, 1003-2; Ertalon 6SAU, 11SA, 66SA, 66SAM; Fosta Nylon 523; Monocast MC901; Nylatron 2011, GS, GSM, NSB; Polypenco Nylon 101; Rilsan BMN F15, BMN F25, BMNO, BMNO P20, BMNO P40, BMNY BZ TL, BMV Black T, BZM23G9, BZM43G9, BZM300, KMVOTL, SR 80, SR 100, SR 120; Thermocomp PFL-4216, PFL-4218, RFL-4216, RFL-4218, RL-4310; Ultramid A3K, A3KN, A3R, A3W, A4K, KR 4609; Vekton 6PA, 6PAG, 6PAM, 6XAU); bottles and containers (Capron 8270 HS; Durethan B 50 E, BC 40; Fosta Nylon 589); bristles (Grilon A23, A28; Ultramid A3, A3 Pellets 24, A4, B3, B35, S3); brush filaments (Herox 0200HA, 0200MA, 0200SA); buttons/knobs (Texalon 1200A Zip-6); cams (Monocast MC901; Nylatron GSM, NSB; Polypenco Nylon 101; Thermocomp PFL-4216, PFL-4218, RFL-4216, RFL-4218); cast film (Fosta Nylon 471); chemical engineering parts (Ultramid S3K); clamps (Ertalon 66SA, 66SAM); coatings (BCI Nylon No. 651; Fosta Nylon 471); conveyor belt parts (Nydur BKV 40 H, BKV 50 H, BKV 115); couplings (Vekton 6PB); dies (Nylatron GS; Polypenco Nylon 101); dynamically loaded parts (Nylatron 2010); engineering parts (Ultramid A3K, A3KN, A3 Pellets 24, A3SK, A3W,

Nylon (cont'd.)

A4, A4H, A4K, B3L, B4K, B35SK, B35W, KR 4411, KR 4412, KR 4645; Wellamid 22LH-N, 22LH13-N, 22L-N, 22LN2-N, 42LH-N, 42L-N, 42LN2-N, 220-N, FR22F-N, FRGF25-66, FRGS25-66, GS25-66, GS40-66, GSF25/15-66); fasteners (Capron 8202, 8253, 8255, 8259, 8350; Polypenco Nylon 101; Texalon 1108, 1110A PL); films (BCI Nylon LX-3249; Celanese Nylon 1200-1; Durethan B 31 F, B 40 F; Fosta Nylon 446, 589; Grilamid L25, L25 natural 6112; Grilon BT40, CA6, CF35, CR9, CR9 natural 6361, F40, F47; Ultramid B3, B4, KR 4407, KR 4609, S3); fittings (Capron 8200, 8260, 8360; Ultramid B3L, B3S); frictional/ sliding applications (Migralube QFL-4036, WFL-4536; Stat-Kon R-15; Thermo- comp PFL-4216, PFL-4218, RFL-4216, RFL-4218, RL-4310, RL-4540; Ultramid A3R, A4K, KR 4609); fuel filters (Capron 8202; Grilon A23G natural 6165); fuel lines (Grilamid L25 W10, L25 W20, L25 W40); gaskets/seals (BCI LX-3250, No. 653; Ertalon 66SA, 66SAM; Rilsan BMN F15, BMN F25, BMNO, BMNO P20, BMNO P40, BMNY BZ TL, BMV Black T, BZM23G9, BZM43G9, BZM300, KMVOTL); gears (Capron 8200, 8202C, 8202CL, 8230G HS, 8231G HS, 8233G HS, 8234G HS, 8266G HS, 8267G HS; Celanese 1000-1, 1000-2, 1000-4, 1003-1, 1003-2; Ertalon 6SAU, 11SA; Fosta Nylon 523; Monocast MC901; Nydur BKV 30 H; Nylatron 2010, GS, GSM; Polypenco Nylon 101; Rilsan BMN F15, BMN F25, BMNO, BMNO P20, BMNO P40, BMNY BZ TL, BMV Black T, BZM23G9, BZM43G9, BZM300, KMVOTL, SR 80, SR 100, SR 120; Stat-Kon R-15; Thermocomp PFL-4216, PFL-4218, RFL-4216, RFL-4218; Ultramid A3EG7, A3G7, A3HG5, A3K, A3KN, A3W, A3WG7, A4H, A4K; Vekton 6PA, 6PAG, 6PAM, 6PB, 6XAU; Wellamid MR410 HS); handles (Capron 8200, 8230G HS, 8231G HS, 8233G HS, 8234G HS, 8266G HS, 8267G HS; Ultramid B3S); housings (Capron 8253, 8255, 8260, 8350, 8360; Rilsan BMN F15, BMN F25, BMNO, BMNO P20, BMNO P40, BMNY BZ TL, BMV Black T, BZM23G9, BZM43G9, BZM300, KMVOTL; Texalon 1203D, 1308A, GMF600-40; Ultramid A3EG5, A3EG6, A3EG7, A3EG10, A3G5, A3G6, A3G7, A3HG5, A3SK, A3WG5, A3WG6, A3WG7, A3WG10, A3X1G5, A3X1G7, A3XG5, A3XG7, B3EG5, B3EG6, B3EG7, B3EG10, B3G5, B3G5HS, B3G6, B3G6HS, B3G7, B3G7HS, B3G10, B3L, B3S, B3WG5, B3WG6, B3WG7, B3WG10, B3WM601, B3SM602, B35G3, B35SK, KR 4250, KR 4409, KR 4411, KR 4412, KR 4445/1, KR 4446, KR 4449, KR 4450, KR 4645, KR 4650, KR 4653; Zytel ST FR-80); hydraulic hoses (Rilsan BECHVO, BESN Black T, BESN F15, BESNO, BESNO P20, BESNO P40, BESNO P40 TL, BESNO TL, BESVO); hydraulic systems (Nydur RM KU 2- 2501/30, RM KU 2-2521/20, RM KU 2-2521/25, RM KU 2-2521/30, RM KU 2- 2521/35); impact parts (Texalon 1310, GF1310 (13 and 33), XP-1296; Vekton 6PB); insulation/insulators (Ertalon 11SA; Monocast MC901; Nylatron GSM; Polypenco Nylon 101); intricate/hard-to-fill moldings (Grilon PVZ-3H; Thermo- comp PF-1006FR); jacketing (Texalon 1110A PL); laminates (Grilon F40); large diameter pipe (Rilsan BESHVO, BESVO); large panels (Texalon MF1200A-40); large parts (Texalon MF1200A1-40); load-bearing components (Nydur BKV 35 Z, BKV 40 H); long-flow-path parts (Grilamid L16G, L16GM; Grilon A23 GM);

machine parts (Ultramid A3EG5, A3EG6, A3EG7, A3EG10, A3G5, A3G6, A3G7, A3HG5, A3WG5, A3WG6, A3WG7, A3WG10, B3EG10, B3K, B3WG10, KR 4250, KR 4406, KR 4446, KR 4449, KR 4450, KR 4650, KR 4651, KR 4652, KR 4653, KR-B3X2V6; Wellamid MR410 HS); machining (Ultramid A5, B4, B5, B5W, KR 4609, S3, S4); machinery (Polypenco Nylon 101; Rilsan BMN F15, BMN F25, BMNO, BMNO P20, BMNO P40, BMNY BZ TL, BMV Black T, BZM23G9, BZM43G9, BZM300, KMVOTL); mandrels (Grilon R47HW, W5744); material handling (Monocast MC901; Nylatron GSM); material handling equipment coatings (Rilsan RDP15-10 Natural, RDP-17-1 Dispersion Coating, RDP Pigmented); mechanical cable coverings (Rilsan BECNO, BECVO P40 TL); mechanical components (BCI LX-3250, No. 653; Capron 8200; Celanese 1000-1, 1000-2, 1000-4, 1003-1, 1003-2; Durethan A 30 S, AKV 30, AKV 30 H, B 25 T, B 39 P, B 30 S, B 31 SK, B 40 E, B 40 SK, BC 30, BC 40, BG 30 X, BKV 30, BKV 30 H, BKV 30 N, BKV 30 N1, BKV 35, BKV 50 H; Texalon 600A, 600A Zip 1, 670A, 1200A, 1310, GF604 (15 and 35), XP-1296); metal industry (BCI Nylon #808-809, #818-819, #819-S-10%, #829; Monocast MC901; Nylatron GSM); mining industry (Monocast MC901; Nylatron GSM); molded parts (Celanese Nylon 1200-1, 1500-1, 1500-2, 1503-1, 1503-2, 1600-1, 1600-2, 1603-1, 1603-2; RTP 202M, 203 GB 20, 203M GB 20); monofilament (Grilon A23, A28; Ultramid A3, A3 Pellets 24, A4, B4, B5, B35M, B35MF01, S3); outdoor parts (Texalon 1200A Black 11); packaging/displays (Durethan A 30 S, AKV 30, AKV 30 H, B 25 T, B 39 P, B 30 S, B 31 SK, B 40 E, B 40 SK, BC 30, BC 40, BG 30 X, BKV 30, BKV 30 H, BKV 30 N, BKV 30 N1, BKV 35, BKV 50 H; packaging film/liners (Capron 8207F, 8209F; Grilon F40); panels (Ultramid A5, B4, B5); paper industry (Monocast MC901; Nylatron GSM; Vylor 7264, ME-1791); pipes (Ultramid A5); pipe fittings (Nydur BKV 15 H, BKV 115); plumbing fixtures (PDX-85384); potable water pipe fittings (Fosta Nylon 512); powder coating (Rilsan RDP15-10 Natural, RDP-17-1 Dispersion Coating, RDP Pigmented); precision parts/engineering (Durethan A 30 S, AKV 30, AKV 30 H, B 25 T, B 39 P, B 30 S, B 31 SK, B 40 E, B 40 SK, BC 30, BC 40, BG 30 X, BKV 30, BKV 30 H, BKV 30 N, BKV 30 N1, BKV 35, BKV 50 H; Ultramid S3K); profiles (Celanese Nylon 1200-1; Durethan BCF 40; Grilon R47, R47HW, R50, W3941, W5744; Rilsan BESHVO, BESNO, BESNO P20, BESNO P40, BESVO); protective coatings (BCI Nylon #808-809, #818-819, #819-S-10%, #829); pumps (Ertalon 11SA; Nydur BKV 30 H, BKV 115; RTP 201D, 203D, 205D, 207D); rigid applications (Texalon 600A NU); rods (Fosta Nylon 446; Grilon R47, R47HW, R50, W3941, W5744); rollers (Nylatron GS; Polypenco Nylon 101; Ultramid B4, B35SK; Vekton 6PA, 6PAG, 6PAM, 6PB); rubber industry (BCI Nylon #808-809, #818-819, #819-S-10%, #829); sheet (Celanese Nylon 1200-1; Fosta Nylon 446; Grilon R40); shrink wrapping (Grilon CA6); slabs (Grilon R47, R50); sleeves (Polypenco Nylon 101); small parts (Texalon 600A NU, 600A Zip 1, 600A Zip 3, 1200AXL, 1200A Zip-1); sprockets (Nylatron 2010); sterilization packaging film (BCI Nylon LX-3249); tanks (Capron 8270 HS, 8280 HS; Texalon 680A; Ultramid B5); tapes and coatings

Nylon *(cont'd.)*

(Ultramid B3, B4, B5, B35M); textile applications (BCI Nylon #808-809, #818-819, #819-S-10%, #829; Monocast MC901; Nylatron GSM; Rilsan BMN F15, BMN F25, BMNO, BMNO P20, BMNO P40, BMNY BZ TL, BMV Black T, BZM23G9, BZM43G9, BZM300, KMVOTL, SR 80, SR 100, SR 120); thick sections (Durethan B 40 E; Ultramid KR 4407); thin sections (Capron 8202; Durethan B 25 T; Firestone Nylon 228-001; Grilamid L16G, L16GM; Grilon A23 GM, R40; Ultramid A3SK, B3S, KR 4405); tires (Texalon 1200A Black 11); tool housings (RTP 201, 203, 205, 207, 209); tooling fixtures (Nylatron GS, GSM; Polypenco Nylon 101); trays (Texalon 680A); tubing/hoses (Capron 8203C, 8253, 8254, 8350; Fosta Nylon 446, 589; Grilamid L25 W10, L25 W20, L25 W40; Grilon R40, R47, R47HW, R50, W3941, W5744; Rilsan BECHVO, BESN Black T, BESN F15, BESNO, BESNO P20, BESNO P40, BESNO P40 TL, BESNO TL, BESVO; Texalon 1106, 1110A PL; Ultramid B5, B5W, KR 4609); valves/valve seats (Ertalon 11SA; Fosta Nylon 523; Monocast MC901; Nylatron GS, GSM, NSB; Polypenco Nylon 101); wear applications (Ertalon 6SAU; Monocast MC901; Nylatron GS, GS-71, GSM, NSB, NSB-90; Polypenco Nylon 101; RTP 205TFE5, 205TFE15, 205TFE20; Thermocomp RL-4310; Vekton 6PA, 6PAG, 6PAM, 6XAU); wood industry (BCI Nylon #808-809, #818-819, #819-S-10%, #829)

Marine equipment: hardware (Capron 8260, 8360)

Medical applications: coatings (Rilsan RDP15-10 Natural, RDP-17-1 Dispersion Coating, RDP Pigmented); medical and hospital equipment (Grilon A23G natural 6165)

Military applications: (Polypenco Nylon 101)

USDA-approved applications: (Rilsan RDP15-10 Natural, RDP-17-1 Dispersion Coating, RDP Pigmented)

PROPERTIES:

Form:

Sol'n. (BCI Nylon #819-S-10%)

Solid (Celanese Nylon 1500-1, 1500-2, 1503-1, 1503-2, 1600-1, 1600-2, 1603-1, 1603-2; Ertalon 6SAU, 11SA, 66SA, 66SAM; Fosta Nylon 446, 471, 567, 589, 714, 870, 1047, 1210, 1379, 1417, 1525, 1693, 1722; Vekton 6PA, 6PAG, 6PAM, 6PB, 6XAU)

Crystalline (Fosta Nylon 512)

Fine crystalline (Ultramid B3S, KR 4405)

Semicrystalline (Firestone Nylon 200-001, 210-001L, 213-001, 228-001, 415-001HS, 430-001HS, 540-001HS, 640-001HS)

Powder (Rilsan RDP15-10 Natural)

Granular, to be converted to solvent sol'n. (BCI Nylon #808-809, #818-819, #829)

Small uniform spherulites (Fosta Nylon 523)

Cylindrical granules (Durethan A 30 S, AKV 30, AKV 30 H, B 25 T, B 30 P, B 30 S, B 31 F, B 31 SK, B 40 E, B 40 F, B 40 SK, B 50 E, BC 30, BC 40, BG 30 X, BKV 30, BKV 30 H, BKV 30 N, BKV 30N1, BKV 35, BKV 50 H

Pellet (BCI Nylon LX-3249, LX-3250, No. 651, No. 653; Capron 8200, 8202, 8202C, 8202CL, 8202L, 8203C, 8220 HS, 8221 HS, 8230G HS, 8231G HS, 8233G HS, 8234G HS, 8253, 8254, 8255, 8259, 8260, 8266G HS, 8267G HS, 8350, 8360; Vydyne R-513, R-513H, R-533, R-533H, R-543, R-543H, R-400G)
Free-flowing pellets (Grilamid ELY 20NZ, ELY 60, L16G, L16 GM, L20 FR, L20 G, L20 W20, L20 W40, L25, L25 natural 6112, L25 W10, L25 W20, L25 W40, L25 W40 NZ, LKN-5H, LV-3H, LV-43H, TR55 LX, TR55 UV; Grilon A23, A23 GM, A23 G natural 6165, A28, A28 GM, A28 NX, A28 NY, A28 NZ, A28 VO, A28 W10, BT40, CA6, CA6 EH, CF35, CR9, CR9 natural 6361, ELX 23NZ, F40, F47, PV-3H, PV-5H, PV-15H, PVN-3H, PVN-15H, PVZ-3H, R40, R40 GM, R40 N38, R47, R47HW, R50, W3941, W5744)
Cylindrical, free-flowing pellets (Celanese Nylon 1000-1, 1000-2, 1000-4, 1003-1, 1003-2, 1300 Series, 1310-1, 1310-2, 1310-4)
Filaments avail. in even-numbered diameters from 0.006 in. to 0.022 in.; some avail. in crimped form (Herox 0200HA, 0200MA, 0200SA)
Avail. in rod, plate, bar, etc. (Monocast MC 901; Nylatron GSM)
Avail. in rod and plate in various sizes (Nylatron NSB)
Avail. in rod, disc, strip, etc. (Nylatron GS)
Tubing (Nylaflow H, LP, T)
Precut sheet (STX Nylon)
Color:
Transparent (Durethan B 25 T; Grilamid TR55 LX, TR55 UV)
Translucent (of molded specimens) (BCI Nylon No. 653)
Neutral (Wellamid 21LN2-NNT)
Natural (Firestone Nylon 200-001, 210-001L, 213-001, 228-001, 415-001HS, 430-001HS, 540-001HS, 640-001HS; Nylaflow H, T; Nylatron 1025, 1026, 2001, 2008, 2029; Ultramid KR 4407)
Natural or opaque (Durethan A 30 S, AKV 30, AKV 30 H, B 30 P, B 30 S, B 31 F, B 31 SK, B 40 E, B 40 F, B 40 SK, B 50 E, BC 30, BC 40, BG 30 X, BKV 30, BKV 30 H, BKV 30 N, BKV 30N1, BKV 35, BKV 50 H)
Natural and brown (Ultramid A4H)
Natural and black (Celanese Nylon 1000-1, 1000-2, 1000-4, 1003-1, 1003-2, 1300 Series, 1310-1, 1310-2, 1310-4; Herox 0200HA, 0200MA, 0200SA; Nylatron 1018, 1022, 1024, 2040; RTP 200FR, 200H-FR, 201, 201A, 201B, 201C, 201D, 201FR, 201H, 201P25, 202FR, 202M, , 203, 203A, 203B, 203C, 203D, 203FR, 203 GB20, 203H, 203M GB20, 203P 25, 204FR, 204GB FR, 204H-FR, 205, 205A, 205B, 205C, 205D, 205FR, 205H, 205H-FR, 205TFE5, 205TFE15, 205TFE20, 207, 207A, 207B, 207C, 207D, 207H, 209, 209A, 225, 227, 227FR; Ultramid A3W, A3WG5, A3WG6, A3WG7, A3WG10, A4K, B3, B3WG5, B3WG6, B3WG7, B3WG10, B35W, KR 4653; Vydyne R-513, R-513H, R-533, R-533H, R-543, R-543H, R-400G; Wellamid GS40-60)
White (Nylatron 2005; Wellamid 2BRH-NW)
Beige (Nylatron 2033)

Nylon *(cont'd.)*

Blue (Monocast MC 901)
Brown (Nylatron 2039)
Brown, black (Nylatron 2010)
Gray (Nylatron 2007)
Dark gray (Nylatron 1028)
Gray-black (Nylatron GSM)
Black (Nylaflow LP; Nylatron 1027, 2011, 2015, 2038, GS-21, GS-63, GS-73, NSB-90; RTP 203H-CF15, 204H-CF10, 283, 285, 285P, 287, 287P, 299x27567; Vydyne R-538H-02)

Composition:
10% sol'n. in alcohol/water (BCI Nylon #819-S-10%)

GENERAL PROPERTIES:

Solubility:
Sol. in lower aliphatic alcohols (BCI Nylon #808-809, #818-819, #819-S-10%, #829)
Sol. in phenols (BCI Nylon #808-809, #818-819, #819-S-10%, #829)

Melt Flow:
7.0 g/10 min (Grilamid ELY 60)

Sp. Gr.:
1.00 (Grilamid ELY 20NZ)
1.01 (Grilamid ELY 60)
1.03 (Grilon ELX 23NZ)
1.04 (Ertalon 11SA; Rilsan BECNO, BESHVO, BESN Black T, BESNO, BESNO TL, BESVO, BMNO, BMV Black T, KMVOTL)
1.05 (Rilsan BESNO P20, BMNO P20, RDP-17-1 Dispersion Coating)
1.06 (Rilsan BECVO P40 TL, BESN F15, BESNO P40, BESNO P40 TL, BMN F15, BMN F25, BMNO P40)
1.07 (Capron 8350; Ultramid S3, S3K, S4)
1.08 (Capron 8254, 8255; Grilon W5744; Migralube Q-1000; Texalon 1110A PL HS; Ultramid KR 4409, KR 4609, KR 4645; Vydyne 60H)
1.085 (Texalon 1106)
1.09 (Capron 8253, 8259; Grilon W3941; Nylatron GS-71; Texalon 1108); (cast film). (BCI Nylon #808-809, #818-819, #829)
1.094 (Nyrim P-1 4000)
1.10 (Nylatron GS-73; Ultramid A3R, B3L)
1.102 (Nyrim P-1 3000)
1.109 (Texalon XP-1296)
1.11 (Fosta Nylon 1693, 1722; Rilsan RDP15-10 Natural; RTP 201C; Texalon 1308A)
1.12 (Grilon R47HW; Stat-Kon QC-1002; Texalon 600A PL-2, 1310; Wellamid 42LH-N, 42L-N, 42LN2-N)
1.12–1.13 (cast film) (BCI Nylon LX-3249, No. 651)
1.121 (Nyrim P-1 2000)
1.13 (BCI Nylon LX-3250, No. 653; Capron 8200, 8202, 8202C, 8202CL, 8202L, 8203C, 8220 HS, 8221 HS; Firestone Nylon 200-001, 210-001L, 228-001;

Fosta Nylon 446, 471, 567, 870, 1047, 1379, 1525; Nylatron 2001, 2010, 2011, 2029; Ultramid A3, A3K, A3KN, A3 Pellets 24, A3W, A4, A4H, A4K, A5, B3, B3K, B3S, B4, B5, B5W, B35, B35M, B35SK, B35W, KR 4405, KR 4407, KR 4411, KR 4412; Vekton 6PB)

1.13–1.15 (Wellamid 2BRH-NW, 21LN2-NNT, 22LH-N, 22L-N, 22LN2-N, 220-N)

1.134 (Nyrim P-1 1000)

1.14 (Durethan A 30 S, B 25 T, B 30 P, B 30 S, B 31 F, B 31 SK, B 40 E, B 40 F, B 40 SK, B 50 E, BC 30, BC 40; Ertalon 6SAU; Firestone Nylon 213-001; Fosta Nylon 512, 523, 589, 714; Grilon A28 GM; Herox 0200HA, 0200MA, 0200SA; Nydur B 31 SK, B 40 SK; Nylatron 1022, 1024, 1028, 2005, 2015; RTP 201D; Stat-Kon H; Texalon 600A, 600H HS, 600A NU, 600A Zip 1, 600A Zip 3, 670A, 680A, 1200A, 1200A Black 11, 1200A HS, 1200 AXL, 1200A Zip-1, 1200A Zip-6, 1203D, XP-1317; Thermocomp IF-1002, R-1000; Ultramid A3SK; Vydyne 21, 21SP, 21X, 22H, 22HSP, 24NSL, 24NSP, 25W, 25X, 65B, 66B)

1.14–1.15 (Ertalon 66SA; Polypenco Nylon 101)

1.14–1.18 (Nylatron GS; Vekton 6PA, 6PAG, 6PAM, 6XAU)

1.141 (Vylor 7264, ME-1791)

1.15 (Ertalon 66SAM; Fosta Nylon 1210; Nylatron 2038, 2039, GS-60, GS-63; Rilsan SR 80; RTP 201B, 201H; Thermocomp QF-1002)

1.15–1.17 (Monocast MC 901; Nylatron GSM)

1.16 (Capron 8230G HS; Nylatron GS-21, GS-HS; Stat-Kon VC-1003; Thermocomp RL-4310; Ultramid KR 4205, KR 4406)

1.17 (Texalon GF1308A-13, GF1310-13)

1.18 (Nylatron NSB, NSB-90; Rilsan RDP Pigmented; RTP 203C; Stat-Kon R, RC-1002; Thermocomp RC-1002)

1.19 (Thermocomp PF-1002HI)

1.20 (RTP 203D)

1.21 (RTP 201, 201A; Thermocomp IF-1004, PF-1002, RF-1002, RF-1002HI)

1.22 (Capron 8231G HS; Rilsan BZM23G9; RTP 203B; Stat-Kon IC-1006; Thermocomp QF-1004; Vydyne R-513, R-513H)

1.23 (Nydur BKV 15 H, BKV 115; Rilsan SR 100; RTP 203H, 283; Stat-Kon QC-1006, RC-1004; Thermocomp QC-1006, RC-1004; Ultramid B35G3)

1.24 (Thermocomp HF-1006, SF-1006; Vydyne M-340)

1.25 (Thermocomp RL-4540)

1.26 (Rilsan BZM300; Thermocomp RL-4040, XF-1004)

1.27 (NY-30CF; RTP 200H-FR, 203A; Stat-Kon RC-1006HI; Thermocomp PF-1004, RC-1006HI, XC-1006; Vydyne M-344)

1.28 (Nydur BKV 120, RM KU 2-2521/20; RTP 203, 205C, 285; Stat-Kon PC-1006, RC-1006; Thermocomp PC-1006, PF-1004, RC-1006, RF-1004, RF-1004HI)

1.29 (EMI-X QC-1008; RTP 205D; Stat-Kon R-15; Zytel ST FR-80)

1.30 (Rilsan SR 120; RTP 225; Stat-Kon RF-15; Thermocomp IF-1006, QF-1006, RC-1006PC; Ultramid A3EG5, A3HG5, A3WG5, A3G5, A3XG5, B3EG5, B3G5, B3G5HS, B3WG5, KR 4445/1; Wellamid GS25-66)

Nylon *(cont'd.)*

1.31 (RTP 203H-CF15, 205B, 285P; Stat-Kon RCL-4042; Vydyne M-345)

1.32 (Grilon PVN-3H; Nydur BKV 125, RM KU 2-2521/25; Nylatron GS-71-9; RTP 200FR, 205H)

1.33 (RTP 204H-CF10, 287; Ultramid A3X1G5, KR 4447/1; Wellamid FR22F-N)

1.34 (EMI-X PC-1008, QC-100-10, RC-1008; Stat-Kon RCF-1006; Thermocomp RC-1008; Ultramid KR 4449, KR 4450, KR 4651)

1.35 (Durethan AKV 30; PDX-85384; Thermocomp R-1000FR-HS, XF-1006; Ultramid A3EG6, A3G6, A3WG6, B3EG6, B3G6, B3G6HS, B3WG6, B3WM601, KR 4653; Wellamid MR329HS)

1.36 (Durethan BKV 30, BKV 30 H; Grilon PV-3H; Nydur BKV 30 H, BKV 130, RM KU 2-2501/30, RM KU 2-2521/30; RTP 201P 25, 205, 205A; Stat-Kon RCL-4536; Thermocomp RCL-4536; Ultramid B3WM602, KR 4446)

1.36–1.60 (Durethan AKV 30 H, BKV 30 N, BKV 30 N1, BKV 35)

1.37 (Firestone Nylon 430-001HS; Nylatron GS-60-9; Texalon GF600A-33; Thermocomp PF-1006, PF-1006HI, RF-1006, RF-1006HI; Ultramid KR 4652)

1.38 (Capron 8233G HS; EMI-X RA-30; PDX-83392; RTP 207C, 287P; Stat-Kon RCL-4036; Thermocomp RCL-4036)

1.39 (EMI-X PC-100-10, RC-100-10; PDX-82428; RTP 207D; Texalon GF1200A-33)

1.40 (RTP 207B; Thermocomp IF-1008; Ultramid A3EG7, A3G7, A3WG7, A3XG7, B3EG7, B3G7, B3G7HS, B3WG7; Vydyne R-533, R-533H, R-538H-02; Wellamid GS40-60, GS40-66)

1.41 (Nydur BKV 35 H, BKV 35 Z, BKV 135, RM KU 2-2521/35; Nylatron 1018, 1027, GS-51; RTP 205TFE5, 207H; Stat-Kon RC-1004FR; Thermocomp QF-1008)

1.42 (Minlon 12T, 20B; Nylatron 2008; PDX-85382; Rilsan BZM43G9; Vydyne R-400G)

1.43 (Capron 8360; EMI-X RA-35; Migralube QFL-4536; Thermocomp PFL-4216, RFL-4216; Wellamid FRGF25-66, GSF25/15-66, MR410HS)

1.44 (RTP 299x27567; Thermocomp XF-1008)

1.45 (Migralube QFL-4036; Minlon 21C; Nylatron 2040; Thermocomp SF-100-10; Ultramid A3X1G7, KR-B3X2V6)

1.46 (EMI-X RC-100-12; Firestone Nylon 640-001HS; Nydur BKV 40 H; Nylatron 2007, GS-51-13, GS-60-13; RTP 203 GB20, 203M GB20, 203P 25, 207, 207A; Thermocomp PF-1008, PF-1008HI, RF-1008, RF-1008HI; Vydyne R-220, R-250; Wellamid MR409HS)

1.47 (RTP 202M, 227; Thermocomp XFL-4036; Vydyne 909, R-100, R-200; Wellamid MRG30/10)

1.48 (Capron 8266G HS, 8267G HS; EMI-X RA-40; Minlon 11C; Nylatron 1026; Texalon MF1200A-40; Ultramid KR 4448; Wellamid FRGS25-66)

1.49 (Capron 8234G HS, 8260; RTP 205TFE15; Thermocomp IF-100-10)

1.50 (Firestone Nylon 540-001HS; Texalon MF1200A1-40; Thermocomp QF-100-10; Ultramid KR 4250, KR 4650; Vydyne R-543, R-543H)

1.51 (Minlon 10B; Nylatron 1025; Texalon GMF600-40; Thermocomp RF-1004FR-HS)
1.52 (RTP 201FR; Thermocomp PFL-4218, RFL-4218)
1.54 (PDX-85473; RTP 205TFE20)
1.55 (RTP 204GB FR; Ultramid A3EG10, A3WG10, B3EG10, B3G10, B3WG10)
1.56 (RTP 204FR; Zytel ST FR-80)
1.57 (Nydur BKV 50 H; RTP 209, 209A; Thermocomp PF-100-10, RF-100-10)
1.58 (Durethan BKV 50 H; RTP 227FR; Thermocomp PF-1006FR, QF-1006FR)
1.59 (RTP 203FR; Thermocomp RF-1006FR-HS)
1.60 (RTP 205FR; Ultramid A3X1G10)
1.61 (RTP 202FR)
1.64 (Thermocomp IF-100-12)
1.65 (Thermocomp QF-100-12)
1.70 (Thermocomp PF-100-12, RF-100-12)
1.71 (Nylatron 2033)
2.14 (Firestone Nylon 415-001HS)
4.00 (Rilsan BMNY BZ TL)

Sp. Vol.:

16.3 in.^{3}/lb (Thermocomp PF-100-12, RF-100-12)
16.8 in.^{3}/lb (Thermocomp QF-100-12)
16.9 in.^{3}/lb (Thermocomp IF-100-12)
17.6 in.^{3}/lb (Thermocomp PF-100-10, RF-100-10)
17.7 in.^{3}/lb (Thermocomp PFL-4218, RFL-4218)
18.0 in.^{3}/lb (PDX-85473)
18.2 in.^{3}/lb (Thermocomp QF-100-10)
18.3 in.^{3}/lb (Thermocomp IF-100-10)
19.0 in.^{3}/lb (Thermocomp PF-1008, RF-1008)
19.1 in.^{3}/lb (Thermocomp SF-100-10)
19.4 in.^{3}/lb (Thermocomp PFL-4216, RFL-4216)
19.5 in.^{3}/lb (PDX-85382)
19.8 in.^{3}/lb (Thermocomp QF-1008)
19.9 in.^{3}/lb (Thermocomp IF-1008)
20.2 in.^{3}/lb (Thermocomp PF-1006, RF-1006)
20.5 in.^{3}/lb (PDX-85384)
21.3 in.^{3}/lb (Thermocomp QF-1006)
21.5 in.^{3}/lb (Thermocomp IF-1006)
21.7 in.^{3}/lb (Thermocomp PF-1004, RF-1004)
22.3 in.^{3}/lb (Thermocomp HF-1006, SF-1006)
22.7 in.^{3}/lb (Thermocomp QF-1004)
22.9 in.^{3}/lb (Thermocomp IF-1004, RF-1002, RL-4310)
24.1 in.^{3}/lb (Thermocomp QF-1002)
24.3 in.^{3}/lb (Thermocomp IF-1002)

Nylon (cont'd.)

Density:
0.47 kg/l (Rilsan RDP15-10 Natural)
0.49 kg/l (Rilsan RDP Pigmented)
0.0412 lb/in.³ (Celanese Nylon 1000-1, 1000-2, 1000-4, 1003-1, 1003-2, 1310-1, 1310-2, 1310-4)
0.0498 lb/in.³ (Celanese Nylon 1500-1, 1500-2, 1503-1, 1503-2)
0.0574 lb/in.³ (Celanese Nylon 1600-1, 1600-2, 1603-1, 1603-2)

M.P.:
140–148 C (cast film) (BCI Nylon #818-819, #829)
145–153 C (cast film) (BCI Nylon #808-809)
148–160 C (cast film) (BCI Nylon No. 651)
158–170 C (cast film) (BCI Nylon LX-3249)
160 C (Grilamid ELY 60)
170–180 C (Rilsan BMN F15, BMN F25)
171 C (Grilamid ELY 20NZ)
185 C (Fosta Nylon 714)
190 C (Rilsan RDP15-10 Natural, RDP Pigmented)
209 C (Grilon ELX 23NZ)
210–220 C (Wellamid 42LH-N, 42L-N, 42LN2-N, GS40-60)
213 C (Grilon R47HW; Ultramid KR 4609)
213–216 C (Texalon 1106, 1108, XP-1296)
215 C (Capron 8200, 8202, 8202C, 8202CL, 8202L, 8203C, 8220 HS, 8221 HS, 8230G HS, 8231G HS, 8233G HS, 8234G HS, 8253, 8254, 8255, 8259, 8260, 8266G HS, 8267G HS, 8350, 8360; Firestone Nylon 200-001, 210-001L, 430-001HS; Ultramid B35M, S3, S3K, S4)
215–218 C (Texalon 600A, 600A HS, 600A NU, 600A PL-2, 600A Zip 1, 600A Zip 3, 670A, 680A, 1110A PL HS)
216 C (Grilon W5744)
217–221 C (Durethan B 30 S, B 31 SK, B 40 SK, B 50 E, BKV 30, BKV 30 H, BKV 50 H)
218 C (Fosta Nylon 1210)
218–220 C (Texalon GF600A-33, GMF 600-40)
219 C (Firestone Nylon 213-001, 228-001, 415-001HS, 540-001HS, 640-001HS; Grilon W3941)
220 C (Fosta Nylon 446, 471, 512, 523, 567, 589, 870, 1047, 1379, 1525, 1693, 1722; Ultramid B3, B3EG5, B3EG6, B3EG7, B3EG10, B3G5, B3G5HS, B3G6, B3G6HS, B3G7, B3G7HS, B3G10, B3K, B3L, B3S, B3WG5, B3WG6, B3WG7, B3WG10, B3SM601, B3WM602, B4, B4K, B5, B5W, B35, B35G3, B35SK, B35W, KR 4405, KR 4406, KR 4407, KR 4409, KR 4411, KR 4412, KR 4445/1, KR 4446, KR 4447/1, KR 4448, KR-B3X2V6)
222 C (Grilon A28 GM, PV-3H, PVN-3H)
232–249 C (Wellamid 22LH-N, 22L-N, 22LN2-N, 220-N, GS25-66, GS40-66, GSF25/15-66)

238 C (Zytel ST FR-80)

243 C (Ultramid KR 4645, KR 4650, KR 4651, KR 4652, KR 4653)

249–260 C (Wellamid 2BRH-NW, 21LN2-NNT, FR22F-N, FRGF25-66, FRGS25-66, MR410HS)

249–270 C (Vydyne R-513, R-513H, R-543, R-543H)

250–254 C (Texalon 1308A, 1310)

254 C (Texalon 1203D)

255 C (Durethan A 30 S, AKV 30; Ultramid A3, A3EG5, A3EG6, A3EG7, A3EG10, A3G5, A3G6, A3G7, A3HG5, A3K, A3KN, A3 Pellets 24, A3R, A3SK, A3W, A3WG5, A3WG6, A3WG7, A3WG10, A3X1G5, A3X1G7, A3X1G10, A3XG5, A3XG7, A4, A4H, A4K, A5, KR 4205, KR 4250; Zytel FR-50)

255–258 C (Texalon 1200A, 1200A Black 11, 1200A HS, 1200 AXL, 1200A Zip-1, 1200A Zip-6, Texalon GF1308A-13, GF1310-13, XP-1317)

256 C (Minlon 10B, 11C, 12T; Vylor 7264, ME-1791)

256–260 C (Texalon GF1200A-33, MF1200A-40, MF1200A1-40)

258 C (Minlon 20B)

259 C (Minlon 21C)

260 C (Vydyne R-533, R-533H)

309 F (BCI Nylon No. 653)

319 F (BCI Nylon LX-3250)

364 F (Rilsan BESN F15)

365 ± 5 F (Ertalon 11SA)

367 F (Rilsan BECNO, BECVO P40 TL, BESHVO, BESN Black T, BESNO, BESNO P20, BESNO P40, BESNO P40 TL, BESNO TL, BESVO, BMNO, BMNO P20, BMNO P40, BMNY BZ TL, BMV Black T, BZM23G9, BZM43G9, BZM300, KMVOTL)

410 F (Vydyne 60H)

415 F (Nylatron GS-60-9, GS-71, GS-71-9, GS-73)

419 ± 10 F (Vekton 6PB)

420 ± 10 F (Ertalon 6SAU)

425 F (Nylatron GS-60, GS-63)

428 F (Nylatron GS-60-13)

428 ± 10 F (Vekton 6PA, 6PAG, 6PAM, 6XAU)

430 ± 10 F (Monocast MC 901; Nylaflow LP; Nylatron GSM)

450 F (Vydyne R-250)

460 F (Vydyne M-345)

480 F (Vydyne M-340, M-344, R-220)

482 F (Vydyne 909)

482–500 F (Nylatron GS, NSB; Polypenco Nylon 101; Vydyne 24NSL)

485–495 F (Herox 0200HA, 0200MA, 0200SA)

489 ± 9 F (Nylatron GS-51, GS-51-13, GS-HS, NSB-90)

495 F (Celanese Nylon 1000-1, 1000-2, 1000-4, 1003-1, 1003-2, 1310-1, 1310-2, 1310-4)

Nylon (cont'd.)

495–507 F (Celanese Nylon 1500-1, 1500-2, 1503-1, 1503-2, 1600-1, 1600-2, 1603-1, 1603-2)

496 F (Vydyne 25W)

496 ± 9 F (Nylaflow H, T)

498 F (Vydyne 25X)

500 F (Vydyne 21, 21SP, 21X, 22H, 22HSP, 24NSP, 65B, 66B, R-100, R-200, R-538H-02)

500 ± 5 F (Ertalon 66SA, 66SAM)

520–560 F (NY-30CF)

Stability:

Excellent resistance to aq. alkalis, oxygen, ozone, carbon dioxide, hydrogen, nitrogen, aliphatic hydrocarbons, aromatic hydrocarbons, ethers, mineral oils; good resistance to organic sulfides, chlorinated hydrocarbons, esters, amines, amides, and vegetable oils (BCI Nylon #808-809, #818-819, #819-S-10%, #829)

Excellent resistance to aromatic, chlorinated, and ketone solvents, gasolines, oils, greases, animal and vegetable oils; satisfactory resistance to most mild or moderate base sol'ns.; limited resistance to most acids and poor resistance to strong acids, e.g., sulfuric, hydrochloric, formic, nitric, etc.; attacked by phenols, sodium hypochlorite, chlorine water, potassium permanganate (Celanese Nylon 1000-1, 1000-2, 1000-4, 1003-1, 1003-2, 1310-1, 1310-2, 1310-4, 1500-1, 1500-2, 1503-1, 1503-2, 1600-1, 1600-2, 1603-1, 1603-2)

Excellent resistance to aromatic and aliphatic hydrocarbons, ketones, ethers, esters, neutral salts, and freons; good resistance to acids (to pH 5), to alkalis (to pH 11), to alcohols, and to zinc chloride (Nylaflow H)

Excellent resistance to aromatic and aliphatic hydrocarbons, ketones, ethers, esters, neutral salts, and freons; good resistance to acids (to pH 5), to alkalis (to pH 11), to alcohols (Nylaflow H, T)

Excellent resistance to solvents; good resistance to bases; fair resistance to acids (Thermocomp IF-100-10, PC-1006, QC-1006, RC-1002, RC-1004, RC-1006, RC-1006HI, RC-1006PC, RC-1008, RCL-4036, RCL-4536)

Resistant to common solvents, hydrocarbons, esters, ketones, alkalis, and dilute acids (Vekton 6PA, 6PAG, 6PAM, 6PB)

Wear- and temp.-resistant; resistant to common solvents, hydrocarbons, esters, ketones, alkalis, and dilute acids (Vekton 6XAU)

Generally unaffected by organic reagents (except those which act as solvents, i.e., the lower alcohols, and the phenols); little affected by inorganic bases and by neutral salts; inorganic acids and certain strong organic acids dissolve it with degradation (BCI Nylon LX-3249, No. 651)

Good resistance to alkalis and organic chemicals (Polypenco Nylon 101)

Good thermal stability, and abrasion and impact resistance (Firestone Nylon 200-001, 213-001, 228-001, 415-001HS, 430-001HS, 540-001HS, 640-001HS)

Good thermal stability, and abrasion and impact resistance; outstanding resistance to long term oxidative embrittlement (Firestone Nylon 210-001L)

Good thermal stability and resistance to high dynamic loads, abrasion, and wear; resistant to hydrocarbons, oils and greases, esters, ketones, ethers, and to organic and inorganic bases up to med. concs.; largely resistant to many chlorinated or halogenated hydrocarbons; generally stable under outdoor conditions (Durethan A 30 S, AKV 30, B 30 S, B 31 SK, B 40 SK, B 50 E, BKV 30, BKV 30 H, BKV 30 N, BKV 30 N1, BKV 35, BKV 50 H)

Resistant to abrasion, uv light, mold and fungus attack, and aging under conditions of heat, humidity, and time; good chemical resistance to ammonium hydroxide and ammonia, even at elevated temps.; to sodium and potassium hydroxide at ambient temps.; to most inorganic salts; organic acids and bases; most hydrocarbons; butyl, ethyl, and methyl alcohols and glycerin; most aldehydes and ketones at normal temps.; methyl bromide and methyl chloride; most salts, esters, and ethers; all petroleum-based fuels, etc. (Rilsan BECNO, BECVO P40 TL, BESHVO, BESN Black T, BESN F15, BESNO, BESNO P20, BESNO P40, BESNO P40 TL, BESNO TL, BESVO, BMN F15, BMN F25, BMNO, BMNO P20, BMNO P40, BMNY BZ TL, BMV Black T, BZM23G9, BZM43G9, BZM300, KMVOTL)

Good resistance to strong alkalis, oil, grease, commercial solvents, weathering (Rilsan RDP15-10 Natural, RDP-17-1 Dispersion Coating, RDP Pigmented)

High level of inertness towards bases and salt sol'ns., sea water and marine environments, oils, greases, and petroleum products; good resistance to chemicals, heat and cold (Rilsan SR 80, SR 100, SR 120)

Wear- and temperature-resistant; resistant to common solvents, hydrocarbons, esters, ketones, alkalis, and dilute acids (Ertalon 6SAU, 11SA, 66SA, 66SAM)

Outstanding resistance to oils and grease (Fosta Nylon 471)

High chemical resistance and high resistance to aging (Fosta Nylon 589)

Excellent abrasion resistance and heat stability (Fosta Nylon 870)

Not affected by chlorinated solvents, e.g., carbon tetrachloride, chloroform, trichlorethylene; petroleum hydrocarbons, e.g., gasoline, benzene, naphtha, toluene, kerosene; ester solvents, e.g., butyl acetate, amyl acetate; aldehydes; ketones; alkalis; cleaning agents, e.g., soap, trisodium phosphate, trisodium phosphate-soda ash; may become limp on contact with methanol or ethanol, but will regain original stiffness when freed of these solvents; other alcohols have little or no effect; somewhat resistant to attack by acids although some degradation may be caused; attacked by phenols and chemically related compounds; not permanently affected by continuous exposure to temps. up to 150 F; resistant to attack by rodents, insects, and fungi (Herox 0200HA, 0200MA, 0200SA)

High heat resistance; good resistance to hydrocarbons, engine fluids; good dimensional stability (Minlon 10B, 11C, 12T, 20B, 21C)

Superior heat resistance; improved dimensional stability (Nylatron GS)

Excellent wear and corrosion resistance (Nylatron NSB)

Good resistance to wear, impact, vibration and peening, fatigue, and corrosion (Monocast MC 901; Nylatron GSM)

Nylon *(cont'd.)*

MECHANICAL PROPERTIES:

Tens. Str.:

No break (Grilamid ELY 20NZ)

7 MPa (yield) (Rilsan BMN F25)

11 MPa (yield) (Rilsan BMN F15)

20 MPa (yield) (Grilamid ELY 60)

35 MPa (yield) (Capron 8254, 8255)

38 MPa (break) (Grilon ELX 23NZ)

41.3 MPa (Fosta Nylon 714)

42.7 MPa (Zytel ST FR-80)

48 MPa (yield) (Fosta Nylon 567, 1379, 1693)

50 MPa (yield) (Capron 8350); (break) (Durethan B 50 E; Nydur B 31 SK; Ultramid B35M)

54.5 MPa (Wellamid FRGS25-66)

55 MPa (yield) (Capron 8259); (break) (Nydur B 40 SK; Ultramid KR 4409)

58.4 MPa (yield) (Fosta Nylon 1210)

58.7 MPa (Wellamid FR22F-N)

60 MPa (Ultramid B3WM601, KR-B3X2V6); (yield) (Capron 8221 HS); (break) (Durethan B 30 S, B 31 SK, B 40 SK; Ultramid KR 4645)

62 MPa (yield) (Fosta Nylon 1417, 1722)

65 MPa (yield) (Capron 8253); (break) (Ultramid B3L)

69 MPa (Wellamid GS40-60, GS40-66)

70 MPa (break) (Ultramid A3R, KR 4609, S3, S3K, S4)

70.3 MPa (Fosta Nylon 1525)

72.4 MPa (Wellamid 42LH-N, 42L-N, 42LN2-N)

74.5 MPa (Wellamid GS25-66)

75 MPa (Ultramid KR 4652)

75.9 MPa (Wellamid 21LN2-NNT)

76 MPa (yield) (Fosta Nylon 471, 589)

79 MPa (Firestone Nylon 228-001); (yield) (Fosta Nylon 870, 1047)

79.3 MPa (Wellamid 22LH-N, 22L-N, 220-N)

79.4 MPa (Minlon 12T; Wellamid 2BRH-NW)

80 MPa (Ultramid B3WM602); (yield) (Capron 8202, 8202L); (break) (Ultramid A3, A3 Pellets 24, A4, A4H, A4K, A5, B3, B3K, B4, B4K, B5, B5W, B35, B35W, KR 4205, KR 4406, KR 4411, KR 4412)

83 MPa (yield) (Firestone Nylon 200-001; Fosta Nylon 446)

85 MPa (Ultramid KR 4446); (yield) (Capron 8200, 8202CL, 8203C, 8220 HS, 8230G HS); (break) (Ultramid A3K, A3KN, A3W)

86 MPa (yield) (Firestone Nylon 210-001L; Fosta Nylon 512, 523)

86.3 MPa (Wellamid 22LN2-N)

89 MPa (Minlon 11C)

89.7 MPa (Wellamid FRGF25-66, MR410HS)

90 MPa (Ultramid KR 4250, KR 4650); (yield) (Capron 8202C, 8260, 8360); (break)

Nylon *(cont'd.)*

(Durethan A 30 S; Ultramid A3SK, B3S, B35SK, KR 4405, KR 4407)
95 MPa (Ultramid KR 4450, KR 4651); (yield) (Firestone Nylon 540-001HS)
96.6 MPa (Wellamid GSF25/15-66)
97 MPa (yield) (Firestone Nylon 213-001)
97.9 MPa (Minlon 10B)
100 MPa (Ultramid KR 4445/1)
110 MPa (Ultramid KR 4449)
114 MPa (yield) (Vydyne R-400G)
115 MPa (Ultramid B35G3)
117 MPa (Minlon 20B)
120 MPa (Ultramid KR 4448); (yield) (Firestone Nylon 415-001HS); (break) (Nydur
  B 40 SK, BKV 115; Rilsan SR 80)
120.8 MPa (Minlon 21C)
124 MPa (Vydyne R-513, R-513H)
125 MPa (yield) (Capron 8231G HS)
130 MPa (Ultramid KR 4447/1); (yield) (Capron 8266G HS)
140 MPa (yield) (Capron 8267G HS); (break) (Nydur BKV 120, RM KU 2-2521/20;
  Rilsan SR 100)
145 MPa (yield) (Firestone Nylon 640-001HS)
150 MPa (break) (Nydur BKV 125, Rm KU 2-2521/25; Rilsan SR 120)
157 MPa (Zytel FR-50)
160 MPa (Ultramid A3EG5, A3G5, A3HG5, A3WG5, A3X1G5, A3XG5); (break)
  (Durethan BKV 30 H; Nydur BKV 130)
165 MPa (break) (Nydur RM KU 2-2521/30)
169 MPa (yield) (Firestone Nylon 430-001HS)
170 MPa (Ultramid B3EG5, B3G5, B3G5HS, B3WG5); (break) (Nydur BKV 135)
175 MPa (break) (Nydur RM KU 2-2501/30)
180 MPa (Ultramid A3X1G7, B3EG6, B3G6, B3G6HS, B3WG6, KR 4653); (break)
  (Durethan AKV 30, BKV 50 H; Nydur BKV 30 H, RM KU 2-2521/35)
190 MPa (Ultramid A3EG6, A3G6, A3WG6, B3EG7, B3G7, B3G7HS, B3WG7)
193 MPa (Vydyne R-533, R-533H)
200 MPa (yield) (Capron 8233G HS); (break) (Nydur BKV 35 H, BKV 35 Z)
210 MPa (Ultramid A3EG7, A3G7, A3WG7, A3X1G10, A3XG7); (break) (Nydur
  BKV 40 H)
220 MPa (Ultramid B3EG10, B3G10, B3WG10)
221 MPa (Vydyne R-543, R-543H)
230 MPa (Ultramid A3EG10, A3WG10); (yield) (Capron 8234G HS); (break) (Nydur
  BKV 50 H)
490 MPa (Vylor ME-1791)
531 MPa (Vylor 7264)
7.8 psi (Rilsan BMV Black T)
2000 psi (yield) (Nyrim P-1 4000)
2800 psi (cast film) (BCI Nylon #829)

Nylon *(cont'd.)*

3500 psi (cast film) (BCI Nylon #818-819)
3600 psi (yield) (Nyrim P-1 3000)
3900 psi (cast film) (BCI Nylon #808-809)
4800 psi (Rilsan BMNY BZ TL)
5000 psi (Stat-Kon H)
5500–5800 psi (cast film) (BCI Nylon No. 651)
5800 psi (Rilsan BESN F15)
5900 psi (yield) (Grilon W3941)
6000 psi (Nylaflow LP)
6100 psi (yield) (Nyrim P-1 2000)
6500 psi (BCI Nylon No. 653; RTP 200H-FR; Texalon 1110A PL HS); (yield) (Grilon W5744); 6500 psi (cast film) (BCI Nylon LX-3249)
6800 psi (BCI Nylon LX-3250)
7000 psi (Stat-Kon R)
7100 psi (Rilsan BECVO P40 TL, BESNO TL, BMNO P40)
7300 psi (yield) (Grilon R47HW)
7400 psi (Rilsan BESNO P20)
7500 psi (Nylaflow H, T; Rilsan BESNO P40; Texalon 600A PL-2, 1106)
7700 psi (Ertalon 11SA)
7800 psi (Rilsan BECNO, BESN Black T)
8000 psi (Nylatron 2039; Rilsan BMNO P20)
8100 psi (Rilsan BESNO P40 TL)
8250 psi (Rilsan BMNO)
8500 psi (Migralube Q-1000; Rilsan BESHVO, BESNO, BESVO, KMVOTL; Texalon XP-1296); (yield) (Vydyne 60H, M-344)
8600 psi (yield) (Nyrim P-1 1000)
8700 psi (Texalon 1308A)
9000 psi (Ertalon 6SAU; Nylatron 2038, GS-71; Thermocomp RL-4040)
9000–12,000 psi (Polypenco Nylon 101)
9100 psi (yield) (Vydyne M-345)
9200 psi (Texalon 1108, 1310)
9300 psi (Nylatron GS-73)
9300–10,600 psi (Vekton 6PB)
9500 psi (EMI-X RA-40; Nylatron 2029; RTP 200FR, 204GB FR)
9600 psi (Thermocomp RL-4540)
10,000 psi (EMI-X RA-30, RA-35; Nylatron 2011, 2033; RTP 201FR, 227FR; Stat-Kon R-15; Texalon 1203D)
10,000–14,000 psi (Nylatron GS)
10,100 psi (Nylatron NSB-90)
10,500 psi (Nylatron NSB; RTP 225; Thermocomp R-1000FR-HS); (yield) (Vydyne M-340)
11,000 psi (Nylatron 2001, 2005, 2010, 2015; RTP 201H; Wellamid MR329HS)
11,000–14,000 psi (Monocast MC 901; Nylatron GSM)

11,370–14,000 psi (Vekton 6PA, 6PAG, 6PAM, 6XAU)

11,500 psi (Ertalon 66SA; Nylatron 1028, 2007, GS-60, GS-63); (yield) (Vydyne 25X)

11,600 psi (Texalon 670A, 680A)

11,800 psi (Texalon 600A; Thermocomp R-1000)

11,900 psi (yield) (Vydyne 65B, 66B)

12,000 psi (Celanese Nylon 1000-1, 1000-2, 1000-4, 1003-1, 1003-2; Nylatron 1022, 1024; RTP 201C, 203M GB20; Stat-Kon RF-15; Texalon 1200A; Thermocomp IF-1002, QF-1002); (yield) (Vydyne 21, 21SP, 21X, 25W, R-220)

12,200 psi (Texalon 1200A HS)

12,300 psi (Texalon 600A HS)

12,400 psi (yield) (Vydyne 22H, 22HSP)

12,400–13,500 psi (Ertalon 66SAM)

12,500 psi (RTP 201B; Texalon 1200A Black 11, MF1200A-40; Thermocomp RL-4310; Wellamid MR409HS)

12,600 psi (Texalon 600A Zip 3); (yield) (Grilon A28 GM)

12,800 psi (Texalon 1200 AXL, 1200A Zip-6)

12,900 psi (Texalon GF1310-13)

13,000 psi (RTP 201D, 202FR; Texalon 600A NU, XP-1317; Thermocomp PF-1002HI; Wellamid MRG30/10); (yield) (Vydyne R-250)

13,250 psi (Nylatron GS-21)

13,400 psi (Texalon 600A Zip 1)

13,500 psi (Nylatron 1025, GS-HS; Texalon 1200A Zip-1, GF1308A-13, MF1200A1-40; Thermocomp PF-1002)

13,600 psi (Celanese Nylon 1310-1, 1310-2, 1310-4; RTP 227)

13,700 psi (Rilsan BZM23G9); (yield) (Vydyne 24NSL, 24NSP)

14,000 psi (Rilsan BZM300; RTP 201A, 203C; Thermocomp HF-1006, RF-1002, RF-1002HI); (yield) (Vydyne R-100, R-200)

14,500 psi (RTP 203H)

14,800 psi (RTP 201)

15,000 psi (RTP 203FR)

15,600 psi (RTP 285P)

16,000 psi (RTP 205C)

16,500 psi (yield) (Vydyne 909)

17,000 psi (Thermocomp PF-1004HI)

17,200 psi (Thermocomp SF-1006)

17,300 psi (Stat-Kon VC-1003)

17,500 psi (RTP 202M, 203B; Thermocomp RC-1006PC)

18,000 psi (Migralube QFL-4536; Nylatron 1026; RTP 203D, 204FR, 205H; Stat-Kon RCL-4042; Texalon GMF600-40; Thermocomp IF-1004, QF-1004, RF-1004FR-HS, RF-1004HI)

18,500 psi (PDX-83392; Thermocomp PF-1004, PFL-4216, QF-1006FR, XF-1004, XFL-4036)

19,000 psi (RTP 203, 207C; Thermocomp RF-1004)

Nylon (cont'd.)

19,250 psi (RTP 201P 25, 287P)

19,400 psi (Rilsan BZM43G9)

19,500 psi (RTP 204H-CF10; Stat-Kon QC-1002)

20,000 psi (Migralube QFL-4036; Nylatron 2008; RTP 203A, 203 GB20, 203H-CF15, 299x27567; Stat-Kon RC-1002; Thermocomp RC-1002, RFL-4216)

20,600 psi (Nylatron GS-71-9)

21,000 psi (Nylatron GS-60-9; RTP 205B; Thermocomp QF-1006)

21,500 psi (RTP 207H; Thermocomp RF-1006FR-HS, XF-1006)

22,000 psi (Nylatron GS-51; PDX-82428; RTP 205FR; Thermocomp IF-1006, SF-100-10)

22,250 psi (RTP 205D)

22,400 psi (break) (Grilon PVN-3H)

22,500 psi (Nylatron 1018)

22,700 psi (RTP 203P 25)

23,000 psi (RTP 205A, 205TFE20; Thermocomp PF-1006)

24,000 psi (Nylatron 2040; RTP 205TFE15; Thermocomp RF-1006HI)

24,200 psi (break) (Grilon PV-3H)

24,300 psi (Stat-Kon RCF-1006)

24,500 psi (Nylatron GS-51-13)

24,600 psi (Nylatron GS-60-13)

25,000 psi (Nylatron 1027; PDX-85384; RTP 205TFE5, 207A; Thermocomp PF-1008HI, PFL-4218, XF-1008)

26,000 psi (RTP 205, 283; Stat-Kon RC-1004FR; Thermocomp IF-1008, PF-1008, QF-1008, RF-1006, RFL-4218)

26,250 psi (RTP 207D)

26,500 psi (RTP 207B)

27,000 psi (PDX-85382; Stat-Kon RCL-4536; Thermocomp RCL-4536)

28,000 psi (Celanese Nylon 1500-1, 1500-2, 1503-1, 1503-2; Stat-Kon QC-1006, RC-1004; Texalon GF1200A-33; Thermocomp QC-1006, RC-1004)

29,000 psi (PDX-85473; Stat-Kon IC-1006; Texalon GF600A-33; Thermocomp IF-100-10, QF-100-10, RF-1008HI); (yield) (Vydyne R-538H-02)

30,000 psi (RTP 209A; Thermocomp XC-1006)

30,500 psi (Stat-Kon RCL-4036; Thermocomp RCL-4036)

31,000 psi (Thermocomp IF-100-12, PF-100-10, QF-100-12, RF-1008)

32,000 psi (RTP 207; Stat-Kon PC-1006, RC-1006HI; Thermocomp PC-1006, RC-1006HI, RF-100-10)

33,000 psi (NY-30CF; RTP 209; Thermocomp PF-100-12, RF-100-12)

34,000 psi (Celanese Nylon 1600-1, 1600-2, 1603-1, 1603-2; RTP 285)

35,000 psi (EMI-X PC-1008, PC-100-10, QC-1008, QC-100-10, RC-100-12; Stat-Kon RC-1006; Thermocomp RC-1006)

36,000 psi (RTP 287)

40,000 psi (EMI-X RC-1008, RC-100-10; Thermocomp RC-1008)

62,000 psi (Herox 0200HA, 0200MA, 0200SA)

Tens. Elong.:
1.0% (EMI-X RC-100-12)
1–2% (Thermocomp IF-100-12, PF-100-12, QF-100-12, RF-100-12)
1.3% (break) (Rilsan SR 120)
1.5% (EMI-X RC-100-10; RTP 205FR, 287P)
1.9% (RTP 209A)
2% (EMI-X RA-40, RC-1008; RTP 203, 203 GB20, 203H-CF15, 203M GB20, 204FR, 205, 205TFE5, 205TFE15, 205TFE20, 207, 207A, 207D, 209, 227FR, 285P, 287, 299x27567; Stat-Kon R-15); (break) (Capron 8234G HS; Nydur BKV 50 H; Rilsan SR 100; Ultramid A3EG10, A3WG10, A3X1G10, B3EG10, B3G10, B3WG10)
2–3% (NY-30CF; Thermocomp IF-100-10, PF-1008, PF-100-10, QF-1008, QF-100-10, RF-1008, RF-100-10)
2.1% (break) (Durethan BKV 50 H)
2.2% (RTP 201P 25, 203P 25)
2.3% (RTP 202M, 207C)
2.4% (EMI-X PC-100-10)
2.5% (EMI-X QC-100-10, RA-35; RTP 201, 203FR, 204H-CF10, 205D, 207B; Stat-Kon RC-1006, RCL-4036, RCL-4536); (break) (Durethan BKV 30 H; Rilsan SR 80)
2.6% (RTP 205C)
2.7% (RTP 202FR)
2.8% (break) (Firestone Nylon 415-001HS)
3% (Celanese Nylon 1600-1, 1600-2, 1603-1, 1603-2; EMI-X PC-1008, QC-1008, RA-30; Nylatron 1025, 1026, 1027, 2008, 2033, 2040, GS-51-13; PDX-82428, –83392; RTP 201FR, 203A, 203B, 205A, 205B, 227, 285; Stat-Kon IC-1006, RC-1004, RC-1004FR; Thermocomp PFL-4218, RFL-4218; Zytel FR-50); (break) (Capron 8233G HS; Minlon 10B, 20B; Nydur BKV 30 H, BKV 35 H, BKV 35 Z, BKV 40 H, RM KU 2-2501/30; Texalon GF1200A-33; Ultramid A3EG5, A3EG6, A3G5, A3G6, A3HG5, A3WG5, A3WG6, A3EG7, A3G7, A3WG7, A3XG5, A3XG7, B3EG5, B3EG6, B3EG7, B3G5, B3G5HS, B3G6, B3G6HS, B3G7, B3G7HS, B3WG5, B3WG6, B3WG7, B35G3, KR 4445/1, KR 4448, KR 4449, KR 4653; Vydyne R-513, R-513H, R-543, R-543H)
3–4% (Migralube QFL-4036; PDX-85384, -85382, -85473; Thermocomp HF-1006, IF-1008, PF-1002, PF-1004, PF-1006, PF-1008HI, PFL-4218, QF-1002, QF-1004, QF-1006, RC-1004, RC-1008, RF-1002, RF-1004, RF-1004FR-HS, RF-1006, RF-1006FR-HS, RFL-4216, XF-1006, XF-1008, XFL-4036); (break) (Nydur BKV 135, RM KU 2-2521/30, RM KU 2-2521/35)
3.1% (break) (Firestone Nylon 640-001HS)
3.2% (RTP 201A, 201B, 203D)
3.3% (RTP 207H); (break) (Firestone Nylon 430-001HS)
3.5% (RTP 203C, 205H; Stat-Kon PC-1006, QC-1006, RCL-4042); (break) (Capron 8231G HS)
3.6% (RTP 203H)

Nylon *(cont'd.)*

4% (Celanese Nylon 1500-1, 1500-2, 1503-1, 1503-2; Nylatron 1018, 2007; Rilsan BMNY BZ TL, BZM43G9; RTP 201C, 201H, 204GB FR, 283; Stat-Kon RC-1002, RF-15); (break) (Capron 8266G HS, 8267G HS; Durethan AKV 30; Minlon 21C; Nydur BKV 15 H, BKV 125, BKV 130, RM KU 2-2521/20, RM KU 2-2521/25; Texalon GF600A-33, GF1308A-13, GF1310-13, MF1200A1-40; Ultramid A3X1G5, A3X1G7, B3WM602, KR-B3X2V6)

4–5% (Migralube QFL-4536; Thermocomp IF-1006, PF-1006HI, R-1000FR-HS, RF-1006HI, RF-1008HI)

4–6% (Thermocomp SF-1006, SF-100-10, XF-1004)

4.5% (Stat-Kon RCF-1006)

5% (Nylatron GS-51, GS-60-90, GS-60-13, GS-71-9; Rilsan BZM300; RTP 225; Stat-Kon QC-1002, RC-1006HI); (break) (Capron 8230G HS; Grilon PV-3H, PVN-3H; Nydur BKV 115, BKV 120; Texalon MF1200A-40, GMF600-40; Ultramid KR 4250, KR 4447/1, KR 4450, KR 4650, KR 4651; Vydyne R-533, R-533H, R-538H-02; Wellamid FRGF25-66, GS25-66, GS40-66, GSF25/15-66, MR410HS)

5–6% (Thermocomp IF-1004, PF-1004HI, RF-1004HI)

5–10% (break) (Texalon 600A NU)

5–150% (Ertalon 66SAM; Nylatron GS)

5.3% (RTP 200FR)

5.5% (Thermocomp RL-4540)

6% (Rilsan BZM23G9; RTP 201D)

6–7% (Thermocomp IF-1002, PF-1002HI, RF-1002HI)

7% (Stat-Kon VC-1003); (yield and break) (Vydyne 909); (break) (Vydyne R-400G; Wellamid FRGS25-66, MRG30/10)

8% (break) (Firestone Nylon 540-001HS; Ultramid KR 4652; Wellamid FR22F-N)

8–10% (Thermocomp RL-4310)

9% (Thermocomp RL-4040); (yield) (Vydyne 65B, 66B, M-345, R-538H-02); (break) (Vydyne R-100, R-200)

10% (Nylatron 2039, NSB; Stat-Kon H; Thermocomp R-1000); (yield) (Vydyne 21, 21SP, 21X, 22H, 22HSP, 24NSL, 24NSP, 60H, M-340); (break) (Capron 8202CL, 8260; Texalon 600A Zip 1, 1200A Zip-1; Ultramid A3R, KR 4205, KR 4446; Wellamid GS40-60, MR409HS)

10–60% (Monocast MC 901; Nylatron GSM; Vekton 6PA, 6PAG, 6PAM, 6XAU)

10+% (RTP 200H-FR)

12% (Migralube Q-1000; Nylatron GS-21, NSB-90)

13% (break) (Vydyne R-220)

14% (Stat-Kon R)

15% (Nylatron 1028, 2015, GS-HS); (break) (Capron 8202C, 8360; Ultramid KR 4406, KR 4609; Wellamid 2BRH-NW, 21LN2-NNT, 22LN2-N, MR329HS)

15–20% (Texalon XP-1317)

17% (yield) (Rilsan BMN F15, BMN f25); (break) (Minlon 11C)

20% (Nylatron 1022, 2011); (break) (Fosta Nylon 512, 523; Minlon 12T; Ultramid A3SK, B3WM601; Wellamid 42LN2-N)

20–50% (break) (Ultramid B3S, KR 4405)
20–200% (Polypenco Nylon 101)
21% (break) (Vydyne R-250)
23% (break) (Wellamid 22LH-N, 22L-N, 220-N)
25% (Nylatron 1024); (break) (Firestone Nylon 213-001; Texalon 1200A Black 11, 1203D; Vydyne M-340)
27% (break) (Vylor 7264, ME-1791)
30% (break) (Grilon A28 GM; Vydyne 24NSL, 24NSP, 25W)
35% (break) (Wellamid 42LH-N, 42L-N)
35–39% (Herox 0200HA, 0200MA, 0200SA)
35–55% (Celanese Nylon 1310-1, 1310-2, 1310-4)
40% (Nylatron GS-60, GS-63); (break) (Durethan A 30 S)
40–80% (Celanese Nylon 1000-1, 1000-2, 1000-4, 1003-1, 1003-2)
44% (Zytel ST FR-80)
45% (break) (Texalon 1310; Ultramid A3K, A3KN, A3W)
50% (Nylatron 2029, GS-71); (break) (Capron 8202L; Durethan B 30 S, B 31 SK; Nydur B 31 SK; Texalon 600A Zip 3, 1200A HS, 1200 AXL, 1200A Zip-6; Ultramid A3, A3R, A4, A4H, A4K, A5, B35SK, KR 4407; Vydyne M-344)
50–60% (break) (Texalon 1200A)
50–100% (break) (Ultramid B3, B3K, B4, B4K, B5, B5W, B35, B35W, KR 4411, KR 4412)
50–200% (Ertalon 6SAU)
55% (break) (Texalon 600A HS)
60% (break) (Texalon 600A, 1308A; Vydyne 25X)
67% (Nylatron GS-73)
70% (Nylatron 2005); (break) (Capron 8202; Durethan B 40 SK; Nydur B 40 SK; Texalon 670A; Ultramid S3, S3K, S4; Vydyne 22H, 22HSP, M-345)
80% (Nylatron 2001; Nyrim P-1 1000); (break) (Capron 8220 HS; Texalon 680A; Vydyne 21, 21SP, 21X)
90% (break) (Durethan B 50 E; Firestone Nylon 228-001)
100% (break) (Fosta Nylon 1047; Ultramid B3L)
110% (break) (Fosta Nylon 1210)
125% (break) (Texalon XP-1296)
130% (break) (Fosta Nylon 446)
140% (Nylatron 2010)
150% (break) (Capron 8200, 8253, 8259; Fosta Nylon 471, 870, 1417; Texalon 1108; Ultramid KR 4409, KR 4645)
170% (break) (Firestone Nylon 200-001, 210-001L; Vydyne 65B, 66B)
180% (Nylatron 2038)
200% (break) (Capron 8203C, 8221 HS; Fosta Nylon 589, 1693, 1722; Vydyne 60H); (cast film) (BCI Nylon LX-3249)
210% (break) (Grilon W3941)
220% (break) (Grilon R47HW)

Nylon *(cont'd.)*

225% (break) (Texalon 1106)
230% (break) (Capron 8255)
240% (Ertalon 66SA); (break) (Capron 8254)
250% (break) (Grilon W5744; Texalon 600A PL-2)
250–350% (cast film) (BCI Nylon No. 651)
260% (Nyrim P-1 2000); (break) (Capron 8350)
270% (Rilsan BESNO P20)
275% (BCI Nylon LX-3250)
280% (Rilsan BESNO P40 TL)
280–380% (break) (Rilsan BMN F25)
290% (Rilsan BECVO P40 TL)
300% (BCI Nylon No. 653; Ertalon 11SA; Nyrim P-1 3000; Rilsan BESNO P40, BESNO TL, KMVOTL); (break) (Fosta Nylon 567, 1379; Grilamid ELY 60; Ultramid B35M)
300–370% (break) (Rilsan BMN F15)
300+% (break) (Fosta Nylon 1525; Texalon 1110A PL HS)
310% (Rilsan BECNO, BESHVO, BESN Black T, BESNO, BESVO)
329% (Rilsan BMNO, BMV Black T)
330% (Rilsan BMNO P20, BMNO P40)
335% (Rilsan BESN F15)
390% (break) (Grilon ELX 23NZ)
400% (Nyrim P-1 4000)
420% (break) (Grilamid ELY 20NZ)
550% (cast film) (BCI Nylon #808-809, #818-819)
700% (cast film) (BCI Nylon #829)
Tens. Mod.:
250 MPa (Grilamid ELY 60)
3000 MPa (Nydur B 40 SK)
3200 MPa (Nydur B 31 SK)
4027 MPa (Vylor 7264, ME-1791)
6000 MPa (Nydur BKV 115)
6300 MPa (Nydur BKV 15 H)
6660 MPa (Vydyne R-400G)
6800 MPa (Nydur BKV 120)
7000 MPa (Nydur RM KU 2-2521/200
7900 MPa (Nydur BKV 125)
8000 MPA (Nydur RM KU 2-2521/25)
9000 MPa (Nydur BKV 130)
9500 MPA (Nydur RM KU 2-2521/30)
9600 MPa (Nydur BKV 30 H)
10,000 MPa (Nydur BKV 135, RM KU 2-2501/30)
10,500 MPa (Nydur BKV 35 Z, RM KU 2-2521/35)
11,500 MPa (Nydur BKV 35 H)

13,000 MPa (Nydur BKV 40 H)
16,000 MPa (Nydur BKV 50 H)
2.8 GPa (Durethan B 50 E)
3.0 GPa (Durethan B 40 SK)
3.2 GPa (Durethan B 30 S, B 31 SK)
3.4 GPa (Durethan A 30 S)
9.0 GPa (Durethan BKV 30 H)
9.3 GPa (Durethan AKV 30)
15 GPa (Durethan BKV 50 H)
27,000 psi (Nyrim P-1 4000)
35,000 psi (BCI Nylon No. 653)
37,000 psi (Nyrim P-1 3000)
45,000 psi (BCI Nylon LX-3250)
109,000 psi (Nyrim P-1 2000)
150,000 psi (Ertalon 11SA)
208,000 psi (Nyrim P-1 1000)
220,000 psi (Ertalon 6SAU)
250,000–400,000 psi (Polypenco Nylon 101)
300,000 psi (RTP 200H-FR)
304,000 psi (Nylatron GS-73)
335,000 psi (Nylatron GS-71)
350,000–450,000 psi (Monocast MC 901; Nylatron GSM)
350,000–540,000 psi (Vekton 6PA, 6PAG, 6PAM, 6XAU)
370,000–425,000 psi (Vekton 6PB)
408,000 psi (Nylatron NSB, NSB-90)
430,000 psi (RTP 201C; Vydyne 21, 21SP, 21X, 22H, 22HSP, M-340)
435,000 psi (Vydyne 65B, 66B)
450,000–600,000 psi (Nylatron GS)
480,000–540,000 psi (Ertalon 66SA)
500,000–550,000 psi (Ertalon 66SAM)
515,000 psi (Nylatron GS-60, GS-63)
517,000 psi (Vydyne 24NSL, 24NSP)
530,000 psi (RTP 200FR)
550,000 psi (Nylatron GS-HS; RTP 201H)
600,000 psi (Nylatron GS-21)
700,000 psi (RTP 201FR)
710,000 psi (RTP 204GB FR)
750,000 psi (RTP 203C, 203H)
780,000 psi (RTP 201)
800,000 psi (RTP 225)
860,000 psi (Vydyne R-220)
900,000 psi (RTP 201A, 201B, 201D)
980,000 psi (Nylatron GS-71-9; RTP 203A)

Nylon *(cont'd.)*

1.0×10^6 psi (Nylatron GS-60-9; RTP 202FR, 203B, 203D, 205C, 227, 227FR; Vydyne R-100, R-200)

1.1×10^6 psi (RTP 205H)

1.2×10^6 psi (Nylatron GS-51; RTP 203, 203FR, 205B)

1.25×10^6 psi (RTP 205D)

1.3×10^6 psi (Nylatron GS-51-13; RTP 203M GB20, 205A)

1.4×10^6 psi (RTP 202M, 203 GB20, 207A, 207D, 285P)

1.45×10^6 psi (RTP 207C)

1.5×10^6 psi (RTP 204FR, 205, 205TFE5, 207B, 207H, 299x27567)

1.55×10^6 psi (RTP 205TFE15)

1.6×10^6 psi (RTP 205TFE20)

1.68×10^6 psi (Nylatron GS-60-13)

1.7×10^6 psi (RTP 205FR)

1.8×10^6 psi (RTP 201P 25)

1.9×10^6 psi (RTP 207)

1.95×10^6 psi (RTP 204H-CF10)

2.0×10^6 psi (RTP 203H-CF15, 203P 25, 209A)

2.25×10^6 psi (RTP 287P)

2.4×10^6 psi (RTP 209, 283)

3.2×10^6 psi (RTP 285)

3.7×10^6 psi (RTP 287)

Flex. Str.:

9.6 MPa (Grilamid ELY 20NZ)

10 MPa (Grilon ELX 23NZ)

15 MPa (Grilamid ELY 60)

18.6 MPa (yield) (Fosta Nylon 714)

25 MPa (Capron 8255)

30 MPa (Capron 8254)

41.3 MPa (yield) (Fosta Nylon 1525)

48 MPa (yield) (Fosta Nylon 567, 1379)

51 MPa (yield) (Fosta Nylon 1210)

55 MPa (yield) (Fosta Nylon 1693)

65 MPa (Capron 8350)

69 MPa (yield) (Fosta Nylon 1417)

70 MPa (Capron 8221 HS)

71.6 MPa (yield) (Fosta Nylon 1722)

75 MPa (Capron 8259)

85 MPa (Capron 8253)

103 MPa (yield) (Fosta Nylon 471, 589)

104 MPa (Firestone Nylon 200-001, 210-001L)

105 MPa (Capron 8220 HS)

107 MPa (Firestone Nylon 228-001)

110 MPa (Capron 8200, 8202, 8202C, 8202L); (yield) (Fosta Nylon 523, 870, 1047)

114 MPa (Wellamid FR22F-N, FRGS25-66); (yield) (Fosta Nylon 446)
115 MPa (Capron 8202CL, 8203C)
117 MPa (yield) (Fosta Nylon 512)
124 MPa (Wellamid 42LH-N, 42L-N, 42LN2-N)
125 MPa (Capron 8230G HS)
128 MPa (Wellamid GS40-60, GS40-66)
131 MPa (Firestone Nylon 213-001; Wellamid 22LN2-N)
135 MPa (Capron 8360)
138 MPa (Wellamid 22LH-N, 22L-N, 220-N, GS25-66)
140 MPa (Capron 8260; Ultramid KR 4652)
145 MPa (Ultramid KR 4445/1, KR 4446)
150 MPa (Firestone Nylon 540-001HS; Ultramid KR 4450)
159 MPa (Wellamid FRGF25-66, GSF25/15-66)
160 MPa (Ultramid KR 4650, KR 4651)
162 MPa (Wellamid MR410HS)
170 MPa (Capron 8231G HS; Ultramid B35G3)
175 MPa (Ultramid KR 4447/1, KR 4449)
179 MPa (Vydyne R-513, R-513H)
184 MPa (Vydyne R-400G)
186 MPa (Firestone Nylon 415-001HS)
190 MPa (Capron 8266G HS; Nydur BKV 115; Ultramid KR 4448)
200 MPa (Nydur BKV 15 H)
205 MPa (Capron 8267G HS)
207 MPa (Firestone Nylon 640-001HS)
210 MPa (Ultramid B3EG5, B3G5, B3G5HS, B3WG5)
220 MPa (Durethan BKV 30 H; Nydur BKV 120; Ultramid A3EG5, A3G5, A3HG5, A3WG5, A3XG5)
225 MPa (Nydur RM KU 2-2521/20)
240 MPa (Nydur BKV 125, RM KU 2-2521/25; Ultramid B3EG6, B3G6, B3G6HS, B3WG6)
246 MPa (Firestone Nylon 430-001HS)
260 MPa (Nydur BKV 130, RM KU 2-2521/30; Ultramid B3EG7, B3G7, B3G7HS, B3WG7)
262 MPa (Vydyne R-533, R-533H)
270 MPa (Durethan AKV 30; Ultramid A3EG6, A3G6, A3WG6, KR 4653)
275 MPa (Capron 8233G HS)
280 MPa (Durethan BKV 50 H; Nydur BKV 30 H, BKV 135; Ultramid A3EG7, A3G7, A3WG7, A3XG7)
285 MPa (Capron 8234G HS)
290 MPa (Nydur RM KU 2-2501/30, RM KU 2-2521/35)
300 MPa (Nydur BKV 35 Z; Ultramid B3EG10, B3G10, B3WG10)
310 MPa (Nydur BKV 35 H)
315 MPa (Vydyne R-543, R-543H)

Nylon *(cont'd.)*

320 MPa (Ultramid A3EG10, A3WG10)
330 MPa (Nydur BKV 40 H)
370 MPa (Nydur BKV 50 H)
1242 MPa (Wellamid 21LN2-NNT)
1310 MPa (Wellamid 2BRH-NW)
2600 psi (yield) (Grilon W5744)
2900 psi (yield) (Grilon R47HW)
3900 psi (BCI Nylon No. 653)
4100 psi (yield) (Grilon W3941)
6600 psi (Texalon 600A PL-2)
7000 psi (Stat-Kon H)
10,000 psi (Nylatron 2039)
10,500 psi (RTP 200H-FR)
11,000 psi (Vydyne 60H)
11,500 psi (Stat-Kon R)
12,000 psi (Migralube Q-1000)
12,500–14,000 psi (Polypenco Nylon 101)
12,700 psi (yield) (Nylatron GS-73)
12,900 psi (yield) (Nylatron GS-71)
13,000 psi (Nylatron 2029, 2038; Vydyne 21, 21SP, 21X, M-345)
13,800 psi (Thermocomp RL-4040)
14,000 psi (RTP 201C; Texalon 1203D; Thermocomp RL-4540)
14,200 psi (yield) (Nylatron GS-60, GS-63)
14,400 psi (yield) (Nylatron NSB-90)
14,500 psi (Nylatron NSB)
15,000 psi (Nylatron 2011; Thermocomp R-1000)
15,200 psi (Vydyne 22H, 22HSP)
15,500 psi (RTP 200FR; Vydyne M-340)
16,000 psi (Nylatron 1028, 2001, 2005, 2015)
16,000–17,500 psi (Monocast MC 901; Nylatron GSM)
16,000–19,000 psi (Nylatron GS)
16,500 psi (EMI-X RA-40); (yield) (Nylatron GS-21, GS-HS)
16,800 psi (Vydyne 24NSL, 24NSP)
17,000 psi (Celanese Nylon 1000-1, 1000-2, 1000-4, 1003-1, 1003-2; Nylatron 1022; RTP 203C, 204GB FR, 227FR; Texalon 600A, 600A NU, 1200A, GF1310-13, XP-1317; Thermocomp PF-1002, QF-1002, R-1000FR-HS); (yield) (Grilon A28 GM)
17,200 psi (Texalon 670A, 1200A HS)
17,500 psi (EMI-X RA-30, RA-35; Nylatron 1024; RTP 201B, 201FR, 225; Texalon 600A HS, 600A Zip 3, 680A, 1200 AXL, 1200A Zip-6, GF1308A-13; Thermocomp RL-4310)
17,900 psi (Celanese Nylon 1310-1, 1310-2, 1310-4)
18,000 psi (Nylatron 2010; RTP 201A, 203M GB20; Texalon 600A Zip 1, MF1200A-40; Thermocomp IF-1002; Vydyne R-220)

18,500 psi (Texalon 1200A Zip-1)

19,000 psi (Nylatron 2033; RTP 201D; Texalon MF1200A1-40; Vydyne R-250)

19,500 psi (RTP 202FR)

20,000 psi (Nylatron 2007; RTP 205C; Stat-Kon R-15, RF-15; Thermocomp HF-1006, PF-1002HI, RF-1002)

20,500 psi (RTP 201H)

21,000 psi (Nylatron 1025; RTP 201)

21,500 psi (Thermocomp RF-1002HI)

22,000 psi (RTP 203FR, 227; Vydyne R-100, R-200)

23,000 psi (RTP 203A)

23,500 psi (RTP 203H; Thermocomp SF-1006)

24,000 psi (RTP 202M)

24,700 psi (Stat-Kon VC-1003)

24,800 psi (RTP 285P)

25,000 psi (Nylatron 1026; RTP 207C; Thermocomp RF-1004FR-HS, XF-1004, XFL-4036)

25,500 psi (Thermocomp RC-1006PC)

26,000 psi (PDX-83392; RTP 204FR; Thermocomp PF-1004HI, QF-1004)

27,000 psi (Migralube QFL-4536; RTP 203B; Stat-Kon QC-1002); (yield) (Nylatron GS-60-9)

27,500 psi (RTP 205H)

28,000 psi (RTP 205A; Stat-Kon RCL-4042; Thermocomp IF-1004, PFL-4216, RF-1004HI, XF-1006; Vydyne 909)

28,250 psi (RTP 203D)

28,300 psi (yield) (Nylatron GS-71-9)

28,500 psi (PDX-82428; RTP 203, 299x27567; Thermocomp PF-1004)

29,000 psi (Migralube QFL-4036; RTP 203 GB20, 287P; Texalon GMF600-40; Thermocomp RF-1004); (yield) (Nylatron GS-51)

29,500 psi (RTP 201P 25)

30,000 psi (Nylatron 1027; RTP 204H-CF10; Stat-Kon RC-1002; Thermocomp QF-1006FR, RC-1002, RF-1006FR-HS, RFL-4216)

31,000 psi (Nylatron 2008; RTP 207A)

31,750 (yield) (Nylatron GS-51-13)

32,000 psi (Nylatron 1018; RTP 203H-CF15, 205B, 205FR, 205TFE20, 207H; Thermocomp IF-1006, PF-1006HI, QF-1006, SF-100-10, XF-1008)

33,000 psi (RTP 205D; Thermocomp PF-1006FR, PFL-4218)

33,500 psi (PDX-85384; Thermocomp PF-1008HI)

34,000 psi (Nylatron 2040; RTP 205TFE15; Thermocomp PF-1006)

34,500 psi (RTP 203P 25)

35,000 psi (RTP 205TFE5; Thermocomp QF-1008, RF-1006HI); (yield) (Nylatron GS-60-13)

35,250 psi (yield) (Grilon PVN-3H)

36,000 psi (RTP 205, 209A; Stat-Kon RCL-4536; Thermocomp PF-1008, RCL-4536)

Nylon *(cont'd.)*

36,500 psi (PDX-85382)
37,000 psi (Thermocomp RFL-4218)
38,000 psi (RTP 207B; Texalon GF600A-33, GF1200A-33; Thermocomp RF-1006, RF-1008HI; Vydyne R-538H-02)
38,300 psi (Stat-Kon RCF-1006)
39,000 psi (PDX-85473; RTP 207D; Thermocomp IF-1008)
39,700 psi (yield) (Grilon PV-3H)
40,000 psi (RTP 207; Stat-Kon RC-1004FR)
40,500 psi (Stat-Kon QC-1006; Thermocomp QC-1006)
41,000 psi (Celanese Nylon 1500-1, 1500-2, 1503-1, 1503-2; RTP 283)
42,000 psi (Stat-Kon IC-1006, RC-1004; Thermocomp RC-1004, RF-1008)
43,000 psi (Stat-Kon RCL-4036; Thermocomp IF-100-10, QF-100-10, RCL-4036)
45,000 psi (NY-30CF; RTP 285; Thermocomp PF-100-10)
46,000 psi (Stat-Kon PC-1006; Thermocomp IF-100-12, PC-1006, QF-100-12)
46,250 psi (RTP 209)
46,500 psi (Stat-Kon RC-1006HI; Thermocomp RC-1006HI, RF-100-10)
47,000 psi (EMI-X PC-1008, QC-1008)
47,500 psi (Thermocomp XC-1006)
48,000 psi (Celanese Nylon 1600-1, 1600-2, 1603-1, 1603-2)
49,000 psi (EMI-X PC-100-10, QC-100-10; Thermocomp PF-100-12)
50,000 psi (RTP 287; Thermocomp RF-100-12)
51,000 psi (Stat-Kon RC-1006; Thermocomp RC-1006)
55,000 psi (EMI-X RC-100-12)
60,000 psi (EMI-X RC-1008, RC-100-10; Thermocomp RC-1008)

Flex. Mod.:

100 MPa (Durethan B 50 E)
110 MPa (Durethan B 40 SK)
120 MPa (Durethan B 30 S, B 31 SK; Rilsan BMN F25)
130 MPa (Durethan A 30 S)
150 MPa (Rilsan BMN F15)
186 MPa (Grilamid ELY 20NZ; Grilon ELX 23NZ)
215 MPa (Grilamid ELY 60)
760 MPa (Capron 8254)
840 MPa (Capron 8255)
1448 MPa (Zytel ST FR-80)
1800 MPa (Capron 8350)
2000 MPa (Capron 8221 HS; Firestone Nylon 200-001)
2070 MPa (Firestone Nylon 210-001L)
2100 MPa (Capron 8259)
2205 MPa (Capron 8253)
2750 MPa (Firestone Nylon 228-001)
2760 MPa (Wellamid 21LN2-NNT, 22LN2-N)
2825 MPa (Capron 8200, 8202, 8220 HS)

2830 MPa (Wellamid 2BRH-NW, 22LH-N, 22L-N, 220-N)
2860 MPa (Capron 8202L)
2870 MPa (Wellamid 42LH-N, 42L-N, 42LN2-N)
3035 MPa (Capron 8202CL)
3170 MPa (Capron 8202C, 8203C; Firestone Nylon 213-001)
3245 MPa (Wellamid FR22F-N)
3795 MPa (Wellamid FRGS25-66)
3962 MPa (Wellamid GS40-60)
4000 MPa (Capron 8230G HS; Ultramid B3WM601; Wellamid GS25-66)
4300 MPa (Ultramid KR 4652)
4589 MPa (Minlon 12T)
4830 MPa (Wellamid GS40-66)
5000 MPa (Capron 8360)
5200 MPa (Ultramid KR 4446)
5244 MPa (Minlon 11C)
5500 MPa (Ultramid B35G3, KR 4450, KR 4651; Vydyne R-513, R-513H)
5515 MPa (Capron 8231G HS, 8260)
5520 MPa (Wellamid FRGF25-66)
5700 MPa (Firestone Nylon 415-001HS)
6000 MPa (Firestone Nylon 540-001HS)
6210 MPa (Minlon 21C)
6500 MPa (Ultramid KR 4250, KR 4449, KR 4650)
6555 MPa (Wellamid MR410HS)
6760 MPa (Vydyne R-400G)
6797 MPa (Minlon 20B)
6900 MPa (Wellamid GSF25/15-66)
7000 MPa (Ultramid B3EG5, B3G5, B3G5HS, B3WG5)
7239 MPa (Minlon 10B)
7300 MPa (Ultramid KR 4445/1)
7500 MPa (Ultramid B3WM602, KR-B3X2V6)
7585 MPa (Capron 8267G HS)
8000 MPa (Rilsan SR 80; Ultramid A3EG5, A3G5, A3HG5, A3WG5, A3XG5, KR 4447/1, KR 4448)
8205 MPa (Zytel FR-50)
8250 MPa (Firestone Nylon 640-001HS)
8500 MPa (Ultramid B3EG6, B3G6, B3G6HS, B3WG6)
9000 MPa (Firestone Nylon 430-001HS; Ultramid A3X1G5)
9170 MPa (Capron 8266G HS)
9300 MPa (Vydyne R-533, R-533H)
9375 MPa (Capron 8233G HS)
10,000 MPa (Rilsan SR 100; Ultramid A3EG6, A3G6, A3WG6, B3EG7, B3G7, B3G7HS, B3WG7, KR 4653)
11,000 MPa (Ultramid A3EG7, A3G7, A3WG7, A3X1G7, A3XG7; Vydyne R-543,

Nylon (cont'd.)

R-543H)
11,860 MPa (Capron 8234G HS)
12,000 MPa (Rilsan SR 120)
15,000 MPa (Ultramid B3EG10, B3G10, B3WG10)
16,000 MPa (Ultramid A3EG10, A3WG10, A3X1G10)
22,000 psi (Rilsan BESN F15)
28,000 psi (BCI Nylon No. 653)
47,000 psi (Grilon W5744)
50,000 psi (Nyrim P-1 4000; Rilsan BECVO P40 TL, BESNO P40, BESNO P40 TL, BMNO P40)
53,000 psi (BCI Nylon LX-3250)
58,000 psi (Grilon R47HW)
71,000 psi (Rilsan BESNO P20, BMNO P20)
82,000 psi (Grilon W3941)
120,000 psi (Nyrim P-1 3000)
130,000 psi (Nylaflow LP)
140,000 psi (Rilsan KMVOTL; Texalon 1110A PL HS)
142,000 psi (Rilsan BECNO, BESHVO, BESNO, BESVO, BMNO)
143,000 psi (Rilsan BESN Black T, BESNO TL, BMV Black T)
175,000 psi (Nylaflow H, T; Texalon 600A PL-2)
175,000–410,000 psi (Polypenco Nylon 101)
245,000 psi (Nyrim P-1 2000)
250,000 psi (Stat-Kon H; Texalon 1106)
257,000 psi (Rilsan BMNY BZ TL)
280,000 psi (Migralube Q-1000; Texalon 1308A)
290,000 psi (RTP 200H-FR); (–40 F) (Vydyne 60H)
300,000 psi (Nylatron 2039; Thermocomp RL-4040)
310,000 psi (RTP 201C)
315,000 psi (Texalon 1310)
325,000 psi (Texalon 1108, XP-1296)
340,000 psi (Stat-Kon R)
350,000 psi (Nylatron 2029, 2038)
360,000 psi (Texalon 1203D; Thermocomp RL-4540)
372,000 psi (Nylatron GS-73)
377,500 psi (Nylatron GS-71)
395,000 psi (Nyrim P-1 1000)
397,000 psi (Nylatron NSB-90)
400,000 psi (Nylatron NSB; Vydyne 21, 21SP, 21X); (–40 F) (Vydyne M-345)
400,000–500,000 psi (Nylatron GS)
410,000 psi (Texalon 600A, 1200A; Thermocomp R-1000); (–40 F) (Vydyne 65B, 66B)
414,000 psi (Rilsan BZM23G9, BZM300)
415,000 psi (Texalon 670A; Vydyne 22H, 22HSP)

420,000 psi (Celanese Nylon 1000-1, 1000-2, 1000-4, 1003-1, 1003-2; Nylatron GS-60, GS-63; Texalon 600A HS, 1200A Black 11; Vydyne 25X)
430,000 psi (Texalon 600A Zip 3, 680A, 1200A HS); (–40 F) (Vydyne 25W, M-344)
440,000 psi (Grilon A28 GM; Nylatron 2001, 2005, 2015)
450,000 psi (Nylatron 1028; RTP 200FR; Texalon 1200 AXL, 1200A Zip-6; Texalon XP-1317; Vydyne 24NSL, 24NSP); (–40 F) (Vydyne M-340)
460,000 psi (Nylatron 1022, 2011, GS-HS)
470,000 psi (Celanese Nylon 1310-1, 1310-2, 1310-4; Nylatron 1024, 2010; Texalon 600A NU, 1200A Zip-1)
475,000 psi (Nylatron GS-21)
480,000 psi (Texalon 600A Zip 1)
490,000 psi (RTP 201H)
500,000 psi (RTP 201FR; Thermocomp R-1000FR-HS)
510,000 psi (Texalon GF1310-13)
530,000 psi (RTP 203C; Texalon GF1308A-13)
575,000 psi (Thermocomp RL-4310)
586,000 psi (Rilsan BZM43G9)
600,000 psi (RTP 201A, 204GB FR; Thermocomp PF-1002HI)
640,000 psi (Wellamid MR329HS)
650,000 psi (Thermocomp PF-1002, RF-1002, RF-1002HI)
690,000 psi (RTP 201)
700,000 psi (RTP 201B, 203A, 203H, 225)
730,000 psi (Vydyne R-250)
750,000 psi (RTP 205C; Texalon MF1200A-40; Thermocomp IF-1002, QF-1002; Wellamid MR409 HS)
800,000 psi (RTP 201D, 203B, 227FR; Stat-Kon RF-15; Thermocomp PF-1004HI; Vydyne R-220)
850,000 psi (Stat-Kon R-15; Thermocomp PF-1004, RF-1004, RF-1004HI)
875,000 psi (Thermocomp HF-1006)
900,000 psi (RTP 202FR; Texalon MF1200A1-40; Thermocomp IF-1004, QF-1004, XF-1004; Vydyne R-100, R-200)
920,000 psi (RTP 203D)
950,000 psi (RTP 227; Stat-Kon RCL-4042; Thermocomp RF-1004FR-HS)
970,000 psi (Stat-Kon QC-1002)
980,000 psi (Wellamid MRG30/10)
1.0×10^6 psi (EMI-X RA-30; Migralube QFL-4536; Nylatron GS-60-9; RTP 203, 205B, 205H, 207C; Stat-Kon RC-1002, VC-1003; Thermocomp PF-1006HI, RC-1002, SF-1006)
$1.0–1.5 \times 10^6$ psi (STX Nylon)
1.05×10^6 psi (–40 F) (Vydyne 909)
1.1×10^6 psi (EMI-X RA-35; Nylatron 1025, 2007, GS-71-9; RTP 203FR, 205A, 205D, 299x27567; Thermocomp IF-1006, PFL-4216, QF-1006, RF-1006HI)
1.12×10^6 psi (RTP 203M GB20)

Nylon *(cont'd.)*

1.13×10^6 psi (RTP 203 GB20)

1.15×10^6 psi (Migralube QFL-4036; Thermocomp RF-1006FR-HS, XF-1006, XFL-4036)

1.2×10^6 psi (EMI-X RA-40; PDX-85384; RTP 202M, 285P; Thermocomp PF-1006, PF-1008HI, QF-1006FR)

1.24×10^6 psi (Texalon GMF600-40)

1.25×10^6 psi (Grilon PVN-3H; Thermocomp XF-1008)

1.3×10^6 psi (Celanese Nylon 1500-1, 1500-2, 1503-1, 1503-2; Nylatron 1026, 2033, GS-51; RTP 205, 205TFE5, 207A, 207B, 207D, 207H; Texalon GF600A-33; Thermocomp IF-1008, QF-1008, RC-1006PC, RF-1006, SF-100-10)

1.35×10^6 psi (Texalon GF1200A-33; Thermocomp PF-1006FR); (–40 F) (Vydyne R-538H-02)

1.37×10^6 psi (Grilon PV-3H)

1.4×10^6 psi (Nylatron 1018, 1027, 2008; PDX-83392; RTP 201P 25, 204FR, 205TFE15; Thermocomp RF-1008HI, RFL-4216)

1.42×10^6 psi (Nylatron GS-51-13)

1.5×10^6 psi (PDX-82428, -85382; RTP 204H-CF10, 205FR, 205TFE20; Stat-Kon RCF-1006; Thermocomp PF-1008, PFL-4218)

1.6×10^6 psi (Celanese Nylon 1600-1, 1600-2, 1603-1, 1603-2; Nylatron 2040, GS-60-13; RTP 203P 25; Thermocomp RF-1008)

1.7×10^6 psi (RTP 203H-CF15, 207; Thermocomp RFL-4218)

1.75×10^6 psi (Thermocomp RCL-4536)

1.8×10^6 psi (PDX-85473; RTP 287P)

1.9×10^6 psi (RTP 209A; Thermocomp IF-100-10, QF-100-10)

2.0×10^6 psi (RTP 283; Stat-Kon RCL-4536; Thermocomp PF-100-10)

2.2×10^6 psi (RTP 209; Stat-Kon QC-1006; Thermocomp QC-1006, RF-100-10, XC-1006)

2.3×10^6 psi (Stat-Kon IC-1006, RCL-4036; Thermocomp IF-100-12, QF-100-12, RCL-4036)

2.4×10^6 psi (NY-30CF; RTP 285; Stat-Kon PC-1006, RC-1004, RC-1004FR; Thermocomp PC-1006, RC-1004)

2.5×10^6 psi (Stat-Kon RC-1006HI; Thermocomp RC-1006HI)

2.8×10^6 psi (EMI-X PC-1008; RTP 287; Thermocomp PF-100-12, RF-100-12)

2.9×10^6 psi (Stat-Kon RC-1006; Thermocomp RC-1006)

2.95×10^6 psi (EMI-X QC-1008)

3.3×10^6 psi (EMI-X PC-100-10, QC-100-10)

3.4×10^6 psi (EMI-X RC-1008; Thermocomp RC-1008)

3.8×10^6 psi (EMI-X RC-100-10)

4.2×10^6 psi (EMI-X RC-100-12)

Compr. Str.:

80 MPa (Durethan B 50 E)

85 MPa (Durethan B 40 SK)

90 MPa (Durethan 30 S, B 31 SK)

100 MPa (Durethan A 30 S)
172 MPa (Vydyne R-513, R-513H)
180 MPa (Durethan BKV 30 H, BKV 50 H)
190 MPa (Durethan AKV 30)
241 MPa (Vydyne R-533, R-533H, R-543, R-543H)
800 psi (1% deformation) (Ertalon 11SA)
1200 psi (1% deformation) (Ertalon 6SAU)
1400 psi (Nyrim P-1 2000); (1% deformation) (Ertalon 66SA)
5000 psi (Polypenco Nylon 101)
5800 psi (RTP 200H-FR)
6500 psi (RTP 200FR)
7300 psi (Rilsan BECVO P40 TL, BESNO P40, BESNO P40 TL)
7500 psi (Rilsan BESNO P20, BMNO P20, BMNO P40)
7800 psi (Rilsan BECNO, BESHVO, BESN Black T, BESNO, BESNO TL, BESVO, BMNO, BMV Black T)
8000 psi (Rilsan KMVOTL)
8500 psi (RTP 201FR)
8800 psi (RTP 225)
9000 psi (Nylatron GS-60, GS-63; RTP 201H); (1% deformation) (Ertalon 66SAM)
10,000 psi (Nyrim P-1 1000; RTP 202FR, 227)
10,000–15,000 psi (1% deformation) (Vekton 6PA, 6PAG, 6PAM, 6XAU)
10,500 psi (Nylatron GS-73; RTP 201C)
10,700 psi (Nylatron GS-71)
11,500 psi (RTP 204GB FR)
11,500–13,000 psi (1% deformation) (Vekton 6PB)
12,000–13,000 psi (Nylatron GS)
12,500 psi (Nylatron NSB; RTP 203C)
12,750 psi (Nylatron GS-21)
12,800 psi (Nylatron NSB-90)
13,000 psi (Nylatron GS-HS; Rilsan BZM23G9, BZM300; RTP 203FR)
14,000 psi (RTP 201; Thermocomp IF-1002, QF-1002)
14,500 psi (RTP 201A, 201B, 201D, 205C)
15,000 psi (RTP 227FR)
15,500 psi (Nylatron GS-71-9)
16,000 psi (RTP 203H, 204FR, 207C; Thermocomp PF-1002)
17,000 psi (RTP 203; Thermocomp QF-1004)
17,500 psi (RTP 205TFE20)
18,000 psi (Nylatron GS-60-9; Rilsan BZM43G9; Thermocomp RF-1002)
19,000 psi (NY-30CF; RTP 203M GB20, 205FR, 205TFE15; Thermocomp IF-1004)
20,000 psi (RTP 203D, 205H, 205TFE5, 283; Thermocomp QF-1006)
20,500 psi (RTP 285P)
21,000 psi (RTP 202M, 203A; Thermocomp PFL-4216)
21,400 psi (Nylatron GS-60-13)

Nylon *(cont'd.)*

22,000 psi (RTP 203B, 205A, 205D; Thermocomp IF-1006, PF-1004)
22,500 psi (RTP 205)
23,000 psi (Nylatron GS-51; RTP 204H-CF10, 207, 207A; Thermocomp IF-1008, PF-1006, PF-1008, QF-1008, RF-1004)
23,500 psi (RTP 201P 25, 207D)
23,800 psi (RTP 207H)
24,000 psi (Nylatron GS-51-13; RTP 203H-CF15, 205B, 209A, 285, 287P; Thermocomp PF-100-10, PFL-4218, RF-1006)
25,000 psi (RTP 207B, 287; Thermocomp IF-100-10, PF-100-12, QF-100-10, RF-1008, RFL-4216)
26,000 psi (Thermocomp IF-100-12, QF-100-12, RFL-4218)
27,000 psi (RTP 203 GB20, 203P 25, 209; Thermocomp RF-100-10)
28,750 psi (RTP 299x27567)
29,400 psi (Celanese Nylon 1500-1, 1500-2, 1503-1, 1503-2)
30,000 psi (Thermocomp RF-100-12)
35,000 psi (PDX-85384)
40,000 psi (PDX-85382)
45,000 psi (PDX-85473)

Compr. Mod.:
173,000 psi (Nyrim P-1 2000)
320,000 psi (Nyrim P-1 1000)

Tear Str.:
780 psi (Nyrim P-1 4000)
1100 psi (Nyrim P-1 1000, P-1 3000)
1550 psi (Nyrim P-1 2000)

Shear Str.:
6000 psi (Ertalon 11SA)
6500–7500 psi (Vekton 6PB)
6800 psi (Nylatron 2039)
7200 psi (Nylatron 2038)
7350 psi (Nylatron GS-73)
7400 psi (Nylatron 2029, GS-71)
7500 psi (Ertalon 6SAU)
8000–11,000 psi (Vekton 6PA, 6PAG, 6PAM, 6XAU)
8040 psi (Nylatron NSB-90)
8100 psi (Nylatron 2015)
8200 psi (Nylatron 2011)
8400 psi (Nylatron 2005)
8500 psi (Nylatron 2001, 2010)
8600 psi (Thermocomp IF-1002, QF-1002)
8700 psi (Nylatron 1028, 2033, GS-71-9)
8800 psi (Nylatron 2007)
9000 psi (Nylatron 1022, GS-60, GS-60-9, GS-63; Thermocomp RC-1006PC)

9200 psi (Nylatron 1025)

9400 psi (Celanese Nylon 1310-1, 1310-2, 1310-4; Thermocomp IF-1004, QF-1004)

9500 psi (Celanese Nylon 1000-1, 1000-2, 1000-4, 1003-1, 1003-2; Nylatron 1024, 1026)

9500–10,500 psi (Nylatron GS)

9600 psi (Nylatron 2008; Polypenco Nylon 101)

9700 psi (Thermocomp PF-1002)

10,000 psi (Nylatron 2040, GS-21; Thermocomp PF-1004, RF-1002)

10,200 psi (Nylatron 1018)

10,500 psi (Ertalon 66SA; Nylatron GS-HS)

10,500–11,500 psi (Nylatron GSM)

10,500–11,500 psi (Ertalon 66SAM; Monocast MC 901)

10,600 psi (Nylatron GS-60-13; Thermocomp RF-1004)

11,000 psi (Nylatron 1027; Thermocomp IF-1006, QF-1006)

11,100 psi (Nylatron GS-51-13)

11,500 psi (Thermocomp RCL-4536)

11,700 psi (Nylatron GS-51)

12,000 psi (Thermocomp IF-1008, PF-1006, QC-1006, QF-1008, RC-1002, RC-1004)

12,300 psi (Thermocomp RF-1006)

12,500 psi (Thermocomp IF-100-10, PC-1006, PF-1008, QF-100-10, RC-1006HI, RCL-4036, XC-1006)

12,700 psi (Thermocomp RF-1008)

13,000 psi (Celanese Nylon 1500-1, 1500-2, 1503-1, 1503-2, 1600-1, 1600-2, 1603-1, 1603-2; Thermocomp IF-100-12, PF-100-10, QF-100-12, RC-1006)

13,300 psi (PDX-85384; Thermocomp RF-100-10)

13,500 psi (Thermocomp PF-100-12)

13,800 psi (Thermocomp RF-100-12)

14,000 psi (Thermocomp RC-1008)

14,300 psi (PDX-85382)

15,300 psi (PDX-85473)

Impact Str.:

4 kJ/m² notched (Durethan A 30 S, B 40 SK)

6 kJ/m² notched (Durethan B 50 E)

10 kJ/m² notched (Durethan AKV 30)

Impact Str. (Izod):

No break (Fosta Nylon 714; Nyrim P-1 4000); notched (Capron 8350; Grilamid ELY 20NZ, ELY 60; Grilon ELX 23NZ, R47HW, W5744)

6 kg cm/cm notched (Rilsan SR 80)

8.6 kg cm/cm notched (Rilsan SR 100)

0.9 J/m notched (Wellamid 2BRH-NW, 21LN2-NNT)

10 kg cm/cm (Rilsan SR 120)

30 J/m notched (Capron 8230G HS; Ultramid A3SK, A4H)

32 J/m (Minlon 10B); notched (Firestone Nylon 210-001L, 213-001; Wellamid

Nylon (cont'd.)

MR410 HS)

34.6 J/m notched (Wellamid GS25-66)

35 J/m notched (Capron 8260)

36 J/m notched (Ultramid KR 4449)

37 J/m notched (Firestone Nylon 200-001)

37.3 J/m notched (Wellamid FRGS25-66)

40 J/m notched (Firestone Nylon 228-001; Ultramid A3K, A3KN, A3W, A4, A4K, A5, B3, B3K, KR 4250, KR 4411, KR 4445/1, KR 4650)

42.6 J/m notched (Wellamid GS40-60, GS40-66, GSF25/15-66)

42.7 J/m notched (Wellamid FR22F-N)

43 J/m (Fosta Nylon 512, 523)

44 J/m notched (Firestone Nylon 540-001HS)

45 J/m notched (Capron 8202C, 8202CL, 8266G HS; Ultramid A3, A3 Pellets 24, B3WM602)

48 J/m (Fosta Nylon 471, 870, 1047; Minlon 21C); notched (Firestone Nylon 415-001HS; Wellamid FRGF25-66)

50 J/m notched (Capron 8267G HS, 8360; Durethan B 30 S, B 31 SK; Nydur B 31 SK; Vydyne R-400G)

50.6 J/m notched (Wellamid 22LN2-N)

53.3 J/m notched (Wellamid 22LH-N, 22L-N, 42LN2-N, 220-N)

54 J/m (Fosta Nylon 1210); notched (Firestone Nylon 640-001HS)

55 J/m notched (Capron 8202, 8202L, 8203C; Ultramid KR 4448)

58.6 J/m notched (Wellamid 42LH-N, 42L-N)

59 J/m (Fosta Nylon 446; Minlon 20B)

60 J/m notched (Capron 8221 HS, 8231G HS; Ultramid KR 4651)

64 J/m (Fosta Nylon 589)

65 J/m notched (Capron 8200, 8220 HS; Nydur BKV 15 H; Ultramid B3WM601)

69 J/m (Minlon 11C)

70 J/m notched (Nydur B 40 SK; Ultramid B35G3, KR 4446, KR 4652)

80 J/m notched (Ultramid A3EG5, A3G5, A3HG5, A3WG5, A3XG5; Vydyne R-513, R-513H)

85 J/m notched (Ultramid B5, B5W, KR 4450)

96 J/m notched (Firestone Nylon 430-001HS)

100 J/m (Durethan BKV 30 H); notched (Ultramid B3EG5, B3G5, B3WG5)

101 J/m (Zytel FR-50)

107 J/m notched (Vydyne R-533, R-533H)

110 J/m notched (Nydur RM KU 2-2501/30; Ultramid A3EG6, A3G6, A3WG6, KR 4653)

115 J/m notched (Capron 8233G HS)

117 J/m (Fosta Nylon 567, 1379)

120 J/m (Durethan BKV 50 H) notched (Nydur BKV 30 H, BKV 115; Ultramid A3EG7, A3G7, A3WG7, A3XG7, B3EG6, B3G6, B3WG6)

128 J/m (Minlon 12T)

130 J/m notched (Nydur RM KU 2-2521/20; Ultramid B3L)
135 J/m notched (Capron 8234G HS, 8253; Ultramid A3EG10, A3WG10)
140 J/m notched (Nydur BKV 35 H, BKV 120; Ultramid B3G5HS)
150 J/m notched (Nydur RM KU 2-2521/25; Ultramid A3X1G5, B3EG7, B3G6HS, B3G7, B3WG7)
160 J/m (Fosta Nylon 1417); notched (Capron 8259; Nydur BKV 35 Z, BKV 40 H, BKV 125)
162 J/m (Fosta Nylon 1525)
165 J/m notched (Nydur BKV 50 H)
170 J/m notched (Nydur RM KU 2-2521/30)
180 J/m notched (Ultramid B3G7HS)
190 J/m notched (Nydur RM KU 2-2521/35; Ultramid B3EG10, B3G10, B3WG10, KR 4447/1)
200 J/m notched (Nydur BKV 135)
213 J/m notched (Vydyne R-543, R-543H)
270 J/m notched (Capron 8254)
325 J/m notched (Capron 8255)
850 J/m notched (Ultramid KR 4645)
854 J/m notched (Zytel ST FR-80)
870 J/m notched (Ultramid KR 4409)
1080 J/m (Fosta Nylon 1693, 1722)
0.45 ft lb/in. notched (RTP 227FR)
0.5 ft lb/in. notched (RTP 200FR, 204GB FR; Thermocomp R-1000FR-HS, RL-4310)
0.6 ft lb/in. notched (Migralube Q-1000; Nylatron 2011; RTP 201FR, 285P)
0.7 ft lb/in. notched (Nylatron 1025, 2001, 2005, 2007, 2015, 2033; Rilsan KMVOTL; RTP 201P 25, 225, 287P; Thermocomp RL-4040)
0.75 ft lb/in. notched (Rilsan BMNO; Texalon XP-1317)
0.8 ft lb/in. notched (RTP 202FR, 202M, 203M GB20; Stat-Kon H, RCL-4042; Texalon 600A NU, 1203D, MF1200A1-40; Thermocomp RF-1002, RL-4540; Wellamid MRG30/10)
0.9 ft lb/in. (Ertalon 66SA; Nylatron 1026); notched (RTP 201, 227; Texalon 600A Zip 1, 1200A Black 11, 1200A Zip-1, MF1200A-40, GMF600-40; Thermocomp IF-1002, QF-1002, R-1000; Vydyne 22H, 22HSP, 24NSL, 24NSP, M-344)
1.0 ft lb/in. notched (Celanese Nylon 1000-1, 1000-2, 1000-4, 1003-1, 1003-2, 1310-1, 1310-2, 1310-4; EMI-X RA-30, RA-35, RA-40; Nylatron 1022, 1028; PDX-82428, -83392; RTP 201B, 201D, 203FR; Stat-Kon QC-1002, R-15, RC-1002, RC-1004FR; Texalon 600A, 600A HS, 600A Zip 3, 670A, 1200A, 1200A HS, 1200 AXL, 1200A Zip-6; Thermocomp PF-1002, RC-1002, RC-1006PC, XF-1004, XFL-4036; Vydyne 21, 21SP, 21X, 25W, M-340, R-100, R-200)
1.0–6.0 ft lb/in. notched (STX Nylon)
1.1 ft lb/in. notched (Nylatron 1024; RTP 201A, 201C, 203 GB20, 203P 25, 204FR, 283, 299x27567; Texalon 680A; Thermocomp IF-1004, QF-1004, RC-1004; Vydyne M-345, R-220)

Nylon (cont'd.)

1.2 ft lb/in. notched (Grilon A28 GM; Nylatron 2010, 2039; RTP 203D; Stat-Kon RC-1004; Thermocomp RF-1004, XC-1006, XF-1006; Vydyne 25X, 60H, 65B, 66B, R-250)

1.27 ft lb/in. notched (Rilsan BESHVO, BESNO, BESVO, BMV Black T)

1.3 ft lb/in. notched (Nylatron 1018, 2008; RTP 203, 203A, 205FR; Thermocomp RF-1004FR-HS; Wellamid MR409 HS)

1.4 ft lb/in. (Ertalon 6SAU; Vekton 6PA, 6PAg, 6PAM, 6XAU); notched (PDX-85384; Rilsan BZM23G9; RTP 203C, 205TFE20; Stat-Kon RCF-1006, RCL-4036, RCL-4536, RF-15; Texalon GF1310-13; Thermocomp PF-1004, RCL-4036, XF-1008)

1.5 ft lb/in. notched (PDX-85382; RTP 285; Stat-Kon RC-1006; Thermocomp PF-1002HI, PF-1006FR, QF-1006FR, RC-1006, RF-1002HI; Wellamid MR329HS)

1.6 ft lb/in. notched (EMI-X RC-1008, RC-100-10, RC-100-12; PDX-85473; RTP 203B, 205TFE15, 287; Thermocomp RC-1008, RF-1006FR-HS, RFL-4216)

1.6–2.0 ft lb/in. notched (Thermocomp PFL-4216, PFL-4218, RFL-4218)

1.7 ft lb/in. notched (Nylatron 1027, 2040; RTP 205C, 205TFE5; Texalon GF1308A-13)

1.73 ft lb/in. notched (Rilsan BZM300)

1.75 ft lb/in. (BCI Nylon LX-3250)

1.8 ft lb/in. notched (EMI-X PC-1008, PC-100-10; Nylatron 2029; RTP 207C; Stat-Kon IC-1006, PC-1006, QC-1006; Thermocomp PC-1006, QC-1006)

1.9 ft lb/in. notched (Vydyne 909)

2.0 ft lb/in. (BCI Nylon No. 653; EMI-X QC-1008, QC-100-10; Nyrim P-1 1000); notched (NY-30CF; Nylatron 2038; Rilsan BMNO P20; RTP 205A; Texalon 1310, GF1200A-33; Thermocomp RF-1004HI, RF-1006; Vydyne R-538H-02)

2.1 ft lb/in. notched (RTP 205; Stat-Kon RC-1006HI; Thermocomp RC-1006HI, RCL-4536)

2.2 ft lb/in. (Celanese Nylon 1500-1, 1500-2, 1503-1, 1503-2); notched (Migralube QFL-4036, QFL-4536; Thermocomp HF-1006)

2.3 ft lb/in. notched (Texalon 1108, GF600A-33; Thermocomp PF-1006)

2.4 ft lb/in. notched (RTP 205B; Thermocomp IF-1006, QF-1006)

2.5 ft lb/in. notched (RTP 200H-FR, 205D, 207A; Stat-Kon R; Thermocomp PF-1004HI)

2.6 ft lb/in. (Celanese Nylon 1600-1, 1600-2, 1603-1, 1603-2; Vekton 6PB); notched (RTP 207; Thermocomp RF-1008)

2.7 ft lb/in. notched (RTP 209; Stat-Kon VC-1003)

2.8 ft lb/in. notched (Texalon 600A PL-2)

3.0 ft lb/in. notched (Grilon PV-3H; RTP 203H-CF15, 209A; Texalon 1308A, XP-1296; Thermocomp PF-1008, RF-1006HI)

3.1 ft lb/in. notched (RTP 207D)

3.2 ft lb/in. notched (RTP 204H-CF10; Thermocomp IF-1008, PF-1006HI, QF-1008)

3.3 ft lb/in. notched (RTP 207B; Thermocomp RF-100-10, RF-100-12, SF-1006)

3.5 ft lb/in. notched (Texalon 1106)

3.8 ft lb/in. notched (Thermocomp RF-1008HI)
4.0 ft lb/in. notched (Grilon PVN-3H; RTP 203H; Thermocomp PF-1008HI, PF-100-10, PF-100-12)
4.1 ft lb/in. notched (RTP 205H)
4.2 ft lb/in. notched (Grilon W3941; RTP 207H; Thermocomp SF-100-10)
4.4 ft lb/in. notched (Rilsan BMNO P40)
4.5 ft lb/in. notched (Thermocomp IF-100-10, IF-100-12, QF-100-10, QF-100-12)
5.6 ft lb/in. notched (Rilsan BESNO P40)
6.0 ft lb/in. notched (RTP 201H)
8.0 ft lb/in. (Nyrim P-1 2000)
13.7 ft lb/in. (Nyrim P-1 3000)
15 ft lb/in. notched (Texalon 1110A PL HS)
> 40 ft lb/in. notched (Rilsan BESN F15, BMN F15, BMN F25)

Tens. Impact:
50–180 ft lb/in.^{2} (Nylatron GS)
74 ft lb/in.^{2} (Nylatron GS-60-13)
75 ft lb/in.^{2} (Nylatron GS-71; Vydyne 22H, 22HSP)
80 ft lb/in.^{2} (Nylatron GS-51, GS-51-13; Vydyne 21, 21SP, 21X)
80–130 ft lb/in.^{2} (Monocast MC 901; Nylatron GSM)
82 ft lb/in.^{2} (Nylatron GS-73)
87 ft lb/in.^{2} (Nylatron NSB-90)
90 ft lb/in.^{2} (Nylatron GS-21)
90–180 ft lb/in.^{2} (Polypenco Nylon 101)
110 ft lb/in.^{2} (Nylatron GS-HS; Vydyne R-100, R-200)
114 ft lb/in.^{2} (Nylatron GS-60-9)
115 ft lb/in.^{2} (Nylatron GS-60, GS-63)
122 ft lb/in.^{2} (Nylatron NSB)
124 ft lb/in.^{2} (Nylatron GS-71-9)

Hardness:
Ball Indentation 80 MPa (Ultramid B35M)
Ball Indentation 100 MPa (Ultramid KR 4409)
Ball Indentation 120 MPa (Ultramid B3L, S3, S3K, S4)
Ball Indentation 140 MPa (Ultramid A3R)
Ball Indentation 150 MPa (Ultramid B3, B3K, B4, B4K, B5, B5W, B35, B35W, KR 4411, KR 4412)
Ball Indentation 160 MPa (Ultramid A3, A3K, A3KN, A3 Pellets 24, A3W, A4, A4H, A4K, A5, B3S, B35SK, KR 4405, KR 4407)
Ball Indentation 170 MPa (Ultramid A3SK, B3WM601, B3WM602, KR 4446, KR 4449, KR 4450, KR 4651)
Ball Indentation 175 MPa (Ultramid KR 4652)
Ball Indentation 180 MPa (Ultramid B35G3, KR-B3X2V6)
Ball Indentation 190 MPa (Ultramid KR 4447/1)
Ball Indentation 210 MPa (Ultramid B3EG5, B3G5, B3G5HS, B3WG5, KR 4250,

KR4 4448, KR 4650)

Ball Indentation 220 MPa (Ultramid B3EG6, B3G6, B3G6HS, B3WG6)

Ball Indentation 230 MPa (Ultramid B3EG7, B3G7, B3G7HS, B3WG7, KR 4653)

Ball Indentation 235 MPa (Ultramid KR 4445/1)

Ball Indentation 240 MPa (Ultramid A3EG5, A3G5, A3HG5, A3WG5, A3X1G5, A3XG5)

Ball Indentation 260 MPa (Ultramid A3X1G7)

Ball Indentation 270 MPa (Ultramid A3EG6, A3G6, A3WG6)

Ball Indentation 280 MPa (Ultramid A3EG7, A3G7, A3WG7, A3XG7, B3EG10, B3G10, B3WG10)

Ball Indentation 300 MPa (Ultramid A3EG10, A3WG10, A3X1G10)

Rockwell L109 (Durethan AKV 30)

Rockwell L110 (Durethan A 30 S)

Rockwell M79–86 (Celanese Nylon 1000-1, 1000-2, 1000-4, 1003-1, 1003-2)

Rockwell M80 (Wellamid 2BRH-NW, 22LH-N, 22L-N, 22LN2-N, 220-N)

Rockwell M82 (Wellamid 21LN2-NNT)

Rockwell M84 (Wellamid FR22F-N, FRGF25-66, FRGS25-66)

Rockwell M84–88 (Celanese Nylon 1310-1, 1310-2, 1310-4)

Rockwell M87 (Thermocomp IF-1002, QF-1002)

Rockwell M88 (Thermocomp PF-1004; Vydyne 24NSL, 24NSP)

Rockwell M89 (Thermocomp IF-1004, QF-1004)

Rockwell M90 (Wellamid MR410HS)

Rockwell M92 (Thermocomp PF-1006, PF-1008, RF-1002)

Rockwell M93 (Thermocomp IF-1006, IF-1008, QF-1006, QF-1008, RF-1004)

Rockwell M94 (Thermocomp PFL-4216, PFL-4218)

Rockwell M95 (Vydyne R-513, R-513H, R-533, R-533H, R-538H-02)

Rockwell M96 (Thermocomp RF-1006, RF-1008)

Rockwell M97 (Thermocomp RFL-4216, RFL-4218)

Rockwell M98 (Thermocomp PF-100-10; Vydyne R-400G)

Rockwell M99 (Thermocomp IF-100-10, QF-100-10)

Rockwell M100 (Celanese Nylon 1500-1, 1500-2, 1503-1, 1503-2, 1600-1, 1600-2, 1603-1, 1603-2; Thermocomp RF-100-10; Vydyne R-543, R-543H)

Rockwell M101 (Thermocomp PF-100-12)

Rockwell M102 (Thermocomp IF-100-12, QF-100-12)

Rockwell M104 (PDX-85384; Thermocomp RF-100-12)

Rockwell M105 (PDX-85382)

Rockwell M106 (PDX-85473)

Rockwell R40 (Capron 8254)

Rockwell R42 (Rilsan BMN F25)

Rockwell R52 (Capron 8255; Nylatron 2029)

Rockwell R53 (Rilsan BMN F15)

Rockwell R56 (Fosta Nylon 714)

Rockwell R57 (Rilsan BESN F15)

Rockwell R78 (Capron 8350; Rilsan BECVO P40 TL, BESNO P40, BESNO P40 TL, BMNO P40)

Rockwell R82 (Capron 8253)

Rockwell R84 (Capron 8259)

Rockwell R87 (Rilsan BESNO P20, BMNO P20)

Rockwell R90 (Ertalon 6SAU; Fosta Nylon 1525)

Rockwell R100 (Fosta Nylon 1210)

Rockwell R102 (Fosta Nylon 1693)

Rockwell R103 (Nylatron NSB-90)

Rockwell R106 (Vydyne R-250)

Rockwell R108 (Ertalon 11SA, 66SA; Rilsan BECNO, BESHVO, BESN Black T, BESNO, BESNO TL, BESVO, BMNO, BMV Black T, KMVOTL; RTP 201C, 203C)

Rockwell R109 (RTP 205C)

Rockwell R110 (Fosta Nylon 1722; Rilsan SR 80)

Rockwell R110–115 (Vekton 6PB)

Rockwell R110–120 (Polypenco Nylon 101)

Rockwell R110–125 (Nylatron GS)

Rockwell R111 (Rilsan BZM43G9; Vydyne 60H)

Rockwell R112 (Rilsan BZM23G9; RTP 200H-FR, 207C)

Rockwell R112–120 (Monocast MC 901; Nylatron GSM)

Rockwell R113 (Fosta Nylon 1417; Nylatron GS-60, GS-63; RTP 201H)

Rockwell R114 (Nylatron GS-71, GS-71-9, GS-73; RTP 203H, 205H, 207H)

Rockwell R115 (Durethan B 50 E; Fosta Nylon 471; Nylatron 1024, GS-60-13; RTP 200FR, 201FR, 202FR, 203H-CF15)

Rockwell R115–120 (Vekton 6PA, 6PAG, 6PAM, 6XAU)

Rockwell R116 (Fosta Nylon 1047; Nylatron 1028, 2038, NSB; Rilsan BZM300; RTP 203FR)

Rockwell R117 (RTP 201A, 201B, 201D, 205TFE15, 205TFE20; Vydyne R-220; Wellamid GS40-60)

Rockwell R118 (Firestone Nylon 200-001, 210-001L; NY-30CF; Nylatron 2040; RTP 203A, 204FR, 204GB FR, 204H-CF10, 205FR, 205TFE5, 227FR)

Rockwell R119 (Capron 8200, 8202, 8202L, 8220 HS, 8260, 8360; Fosta Nylon 446; Nylatron GS-21, GS-60-9, GS-HS; RTP 201, 203B, 203 GB20, 203M GB20, 205A, 205B, 207B, 225; Vydyne 22H, 22HSP, R-100, R-200; Wellamid 42LH-N, 42L-N)

Rockwell R120 (Capron 8202C, 8202CL, 8203C; Durethan B 30 S, B 31 SK, B 40 SK, BKV 30 H, BKV 50 H; Ertalon 66SAM; Minlon 12T, 20B; Nylatron 1025, 1027, 2033, GS-51; RTP 201P 25, 202M, 203, 203D, 203P 25, 205, 207, 207A, 227, 283, 285, 285P, 287, 287P, 299x27567; Vydyne 21, 21SP, 21X, 65B, 66B, 909, M340; Wellamid 42LN2-N)

Rockwell R121 (Capron 8230G HS, 8231G HS, 8233G HS, 8234G HS, 8266G HS, 8267G HS; Firestone Nylon 430-001HS; Minlon 10B, 11C; Nylatron 1018, 1022, GS-51-13; RTP 205D, 207D, 209, 209A)

Nylon *(cont'd.)*

Rockwell R122 (Firestone Nylon 540-001HS; Minlon 21C)
Rockwell R123 (Firestone Nylon 415-001HS, 640-001HS)
Rockwell R129 (Firestone Nylon 213-001)
Rockwell R130 (Firestone Nylon 228-001)
Shore D50 (Grilamid ELY 20NZ; Grilon ELX 23NZ)
Shore D60 (Nyrim P-1 4000)
Shore D63 (Grilamid ELY 60)
Shore D65 (BCI Nylon No. 653)
Shore D70 (BCI Nylon LX-3250; Nyrim P-1 3000)
Shore D75 (Nyrim P-1 2000)
Shore D78 (Nyrim P-1 1000)
Shore D82 (Grilon A28 GM)
Shore D83 (Grilon PVN-3H)
Shore D85 (Grilon PV-3H)

Mold Shrinkage:

0.7–1.0% (linear) (Grilamid ELY 60)
1.4–1.8% (Grilon ELX 23NZ)
2.0–2.2% (radial) (Grilamid ELY 20NZ)
0.003 mm/mm (Vydyne R-543, R-543H)
0.004 mm/mm (Vydyne R-533, R-533H)
0.005 mm/mm (Vydyne R-513, R-513H)
< 0.001 in/in. (Nylatron GS-60-13)
0.001 in./in. (Nylatron 2040, GS-51-13)
0.0015 in./in. (EMI-X PC-100-10, QC-100-10, RC-100-10, RC-100-12)
0.002 in./in. (Capron 8234G HS; EMI-X PC-1008, QC-1008, RC-1008; Nylatron 1018, 1026, 1027, 2008, GS-51, GS-71-9; PDX-82428, -83392; Stat-Kon IC-1006, PC-1006, QC-1006, RC-1006, RC-1006HI, RCF-1006)
0.002–0.006 in./in. (Vydyne 909)
0.0025 in./in. (Stat-Kon RC-1004, RCL-4036, RCL-4536)
0.0025–0.0035 in./in. (PDX-85473)
0.003 in./in. (Capron 8233G HS; Nydur BKV 15 H, BKV 30 H, BKV 35 H, BKV 35 Z, BKV 40 H, BKV 50 H, BKV 115, BKV 120, BKV 125, BKV 135, RM KU 2-2501/30, RM KU 2-2521/20, RM KU 2-2521/25, RM KU 2-2521/30, RM KU 2-2521/35; Nylatron 1025, 2033, GS-60-9; Stat-Kon RC-1004FR; Thermocomp HF-1006)
0.003–0.004 in./in. (PDX-85382)
0.004 in./in. (Capron 8266G HS, 8267G HS; EMI-X RA-40; Nylatron 2007; Stat-Kon RC-1002, RCL-4042; Vydyne R-538H-02)
0.004–0.005 in./in. (PDX-85384)
0.004–0.006 in./in. (Stat-Kon QC-1002)
0.005 in./in. (Capron 8231G HS; EMI-X RA-30, RA-35; Stat-Kon VC-1003)
0.006 in./in. (Nylatron GS-71)
0.007 in./in. (Nylatron GS-21, GS-HS; Stat-Kon RF-15)

0.008 in./in. (Capron 8230G HS; Nylatron 1022, 2038)

0.008–0.011 in./in. (Capron 8260)

0.008–0.012 in./in. (Vydyne 24NSL, 24NSP)

0.009 in./in. (Capron 8202C, 8202CL, 8203C; Nylatron 1028, 2005, 2010, 2015, 2039, GS-60, GS-63, GS-73)

0.009–0.012 in./in. (Capron 8360)

0.010 in./in. (Nylatron 1024, 2011, NSB-90; Stat-Kon H)

0.010–0.022 in./in. (Vydyne R-250)

0.011 in./in. (Nylatron 2001; Stat-Kon R)

0.011–0.013 in./in. (Vydyne 60H)

0.012 in./in. (Capron 8200, 8202, 8202L, 8253; Nydur B 31 SK, B 40 SK)

0.012–0.022 in./in. (Vydyne R-100, R-200, R-220)

0.013 in./in. (Capron 8254, 8255, 8259; Nylatron 2029; Vydyne 25X)

0.014 in./in. (Capron 8350)

0.014–0.020 in./in. (Vydyne M-345)

0.015–0.020 in./in. (Vydyne 21, 21SP, 21X, 22H, 22HSP, 65B, 66B, M-340, M-344)

0.016–0.020 in./in. (Vydyne 25W)

Water Absorp.:

0.1% (EMI-X QC-100-10)

0.11% (EMI-X QC-1008; PDX-82428)

0.12% (Stat-Kon QC-1006)

0.15% (Stat-Kon IC-1006)

0.19% (Nylatron GS-71-9; PDX-85473)

0.20% (Thermocomp HF-1006)

0.21% (PDX-85382)

0.23% (Stat-Kon QC-1002)

0.24% (Nylatron GS-71; PDX-85384)

0.26% (Nylatron GS-73)

0.3% (EMI-X RC-100-12)

0.4% (EMI-X PC-100-10, RC-1008, RC-100-10; PDX-83392; Stat-Kon H; Texalon MF1200A1-40)

0.45% (Stat-Kon RCL-4536)

0.48% (Stat-Kon RCL-4036; Vydyne 60H)

0.5% (EMI-X PC-1008; Grilamid ELY 60; Nylatron 1025; Stat-Kon RC-1004FR, RC-1006, RCF-1006; Texalon GF1308A-13, GF1310-13, MF1200A-40)

0.52% (Nylatron GS-51-13)

0.6% (Nydur RM KU 2-2521/35; Nylatron 1018, GS-51; Stat-Kon RC-1004; Vydyne R-100, R-543, R-543H)

0.62% (Grilamid ELY 20NZ)

0.65% (Nydur RM KU 2-2521/30)

0.7% (EMI-X RA-40; Nydur RM KU -2-2501-30; Nylatron 2033; Vydyne 909, R-533, R-533H, R-538H-02)

0.75% (Nydur RM KU 2-2521/25)

Nylon (cont'd.)

0.8% (Nydur BKV 50 H; Stat-Kon PC-1006, RC-1002, RC-1006HI, RCL-4042, RF-15; Texalon GMF600-40; Vydyne R-200, R-220)

0.85% (Nydur BKV 40 H, BKV 135, RM KU 2-2521/20)

0.9% (Capron 8234G HS, 8266G HS, 8267G HS; EMI-X RA-35; Nydur BKV 35 H, BKV 35 Z; Nylatron 2040, GS-HS, NSB-90, VC-1003; Texalon GF600A-33; Vydyne M-344, R-250)

1.0% (Capron 8360; EMI-X RA-30; Nydur BKV 30 H; Nylatron 1026, 1028, 2007, 2008, GS-21, GS-60-13; Texalon GF1200A-33; Vydyne R-513, R-513H)

1.05% (Nydur BKV 125)

1.1% (Capron 8233G HS, 8260, 8350; Nylatron 1022, 1024, 1027, GS-60-9; Vydyne 24NSL, 24NSP, M-340)

1.15% (Nydur BKV 120)

1.2% (Capron 8254, 8255; Stat-Kon R; Texalon 1308A, 1310; Vydyne 21, 21SP, 21X, 25W, 65B, 66B)

1.3% (Capron 8259; Nydur BKV 15 H, BKV 115; Texalon 1106; Vydyne M-345)

1.36% (Grilon ELX 23NZ)

1.4% (Capron 8231G HS; Texalon 1108, 1200A Black 11, 1200A Zip-1, XP-1296; Vydyne 22H, 22HSP)

1.5% (Capron 8230G HS, 8253; Nylatron 2039; Texalon 600A NU, 600A Zip 1, 1200A, 1200A HS, 1200 AXL, 1200A Zip-6, 1203D, XP-1317)

1.6% (Capron 8200, 8202, 8202C, 8202CL, 8202L, 8203C, 8220 HS, 8221 HS; Nydur B 31 SK, B 40 SK)

1.7% (Nylatron 2038; Texalon 600A, 600A HS, 600A Zip 3, 670A, 680A)

1.8% (Texalon 1110A PL HS)

1.9% (Nylatron GS-60, GS-63)

2.0% (Nylatron 2001, 2005, 2010, 2015)

2.1% (Nylatron 2011)

2.2% (Grilon PVN-3H)

2.5% (Nylatron 2029; Texalon 600A PL-2)

2.5–2.6% (Grilon PV-3H)

3.2–3.5% (Grilon A28 GM)

3.3 ± 0.3% (Ultramid S3, S3K, S4)

3.6 ± 0.3% (Ultramid KR 4609)

4.0 ± 0.3% (Ultramid A3EG10, A3WG10, A3X1G10)

4.5 ± 0.2% (Ultramid KR 4250, KR 4650)

4.7 ± 0.3% (Ultramid A3X1G7)

4.8 ± 0.3% (Ultramid B3EG10, B3G10, B3WG10)

5.0 ± 0.3% (Ultramid A3EG7, A3G7, A3WG7, A3XG7)

5.5 ± 0.3% (Ultramid A3EG6, A3G6, A3WG6)

5.6 ± 0.2% (Ultramid KR 4651, KR 4652, KR 4653)

6.0 ± 0.3% (Ultramid A3EG5, A3G5, A3HG5, A3WG5, A3X1G5, A3XG5, B3WM601, B3WM602)

6.2 ± 0.3% (Ultramid B3EG7, B3G7, B3G7HS, B3WG7, KR 4446, KR 4447/1)

6.6 ± 0.3% (Ultramid B3EG6, B3G6, B3G6HS, B3WG6, KR 4448, KR 4449, KR 4450, KR-B3X2V6)

7.1 ± 0.3% (Ultramid B3EG5, B3G5, B3G5HS, B3WG5)

8.0 ± 0.3% (Ultramid B35G3, KR 4445/1)

8.5 ± 0.5% (Ultramid A3, A3K, A3KN, A3 Pellets 24, A3SK, A3W, A4, A4H, A4K, A5, KR 4409, KR 4645)

8.5 ± 1.0% (Ultramid A3R)

9.0 ± 0.5% (Ultramid B3L)

9.5 ± 0.5% (Ultramid B3, B3K, B3S, B4, B4K, B5, B5W, B35, B35M, B35SK, B35W, KR 4405, KR 4407, KR 4411, KR 4412)

THERMAL PROPERTIES:

Soften. Pt. (Vicat):

123–127 C (Rilsan BMN F25)

140–145 C (Rilsan BMN F15)

150 C (Grilamid ELY 60)

168 C (Grilon ELX 23NZ)

169 C (Rilsan BMNO P40)

170 C (Grilamid ELY 20NZ)

382 F (Nyrim P-1 4000)

390 F (Nyrim P-1 3000)

401 F (Nyrim P-1 2000)

407 F (Nyrim P-1 1000)

Conduct.:

0.20 W/Km (Durethan AKV 30; Ultramid A3EG5, A3EG6, A3EG7, A3EG10, A3G5, A3G6, A3G7, A3HG5, A3WG5, A3WG6, A3WG7, A3WG10, A3X1G5, A3X1G7, A3X1G10, A3XG5, A3XG7)

0.23 W/Km (Durethan A 30 S; Ultramid A3, A3K, A3KN, A3 Pellets 24, A3R, A3SK, A3W, A4, A4H, A4K, A5, B3, B3EG5, B3EG6, B3EG7, B3EG10, B3G5, B3G5HS, B3G6, B3G6HS, B3G7, B3G7HS, B3G10, B3K, B3L, B3S, B3WG5, B3WG6, B3WG7, B3WG10, B3WM601, B4, B4K, B5, B5W, B35, B35G3, B35M, B35SK, B35W, KR 4205, KR 4405, KR 4406, KR 4407, KR 4409, KR 4411, KR 4412, S3, S3K, S4)

0.25 W/Km (Durethan B 30 S, B 31 SK, B 40 SK, B 50 E; Ultramid KR 4250, KR 4446)

0.33 W/Km (Durethan BKV 30, BKV 30 H)

0.38 W/Km (Durethan BKV 50 H; Vydyne R-400G)

6.9×10^{-4} Cal/cm/s/cm²/C (Rilsan BESNO P40, BMNO P20)

8×10^{-4} Cal/cm/s/cm²/C (Rilsan BESHVO, BESNO, BESVO, BMNO, BMV Black T, KMVOTL)

22.1×10^{-4} Cal/cm/s/cm²/C (Rilsan BMNY BZ TL)

1.5 Btu/h/ft²/F/in. (Celanese Nylon 1500-1, 1500-2, 1503-1, 1503-2; Vydyne 60H)

1.7 Btu/h/ft²/F/in. (Stat-Kon R; Vydyne 21, 21SP, 21X, 22H, 22HSP, 24NSL, 24NSP, 25X)

1.9 Btu/h/ft²/F/in. (RTP 225)

Nylon (cont'd.)

1.92 Btu/h/ft²/F/in. (Ertalon 6SAU)

2.00 Btu/h/ft²/F/in. (Ertalon 66SA, 66SAM; Rilsan BECVO P40 TL, BESNO P20, BESNO P40 TL, BMNO P40; RTP 201FR; Thermocomp IF-1002, QF-1002; Vekton 6PA, 6PAG, 6PAM, 6XAU)

2.01 Btu/h/ft²/F/in. (Ertalon 11SA)

2.1 Btu/h/ft²/F/in. (RTP 201B, 201D, 202FR, 204GB FR; Vydyne M-340)

2.2 Btu/h/ft²/F/in. (RTP 203D, 203FR, 204FR; Thermocomp PF-1002HI; Vydyne M-344)

2.3 Btu/h/ft²/F/in. (Rilsan BECNO, BESN Black T, BESNO TL; RTP 205FR)

2.4 Btu/h/ft²/F/in. (RTP 201A; Stat-Kon RF-15; Vydyne 909)

2.45 Btu/h/ft²/F/in. (RTP 201C)

2.5 Btu/h/ft²/F/in. (Stat-Kon H; Thermocomp PF-1002)

2.6 Btu/h/ft²/F/in. (RTP 203C)

2.7 Btu/h/ft²/F/in. (Thermocomp PF-1004, PF-1004HI, RF-1002, RF-1002HI)

2.8 Btu/h/ft²/F/in. (Rilsan BZM300; RTP 201, 201H, 205C; Vydyne R-250)

2.9 Btu/h/ft²/F/in. (RTP 203B; Thermocomp RF-1004, RF-1004HI)

3.0 Btu/h/ft²/F/in. (PDX-85384; RTP 203, 203A, 203H, 207C; Thermocomp IF-1004, QF-1004, XF-1004, XFL-4036)

3.1 Btu/h/ft²/F/in. (RTP 203H-CF15)

3.12 Btu/h/ft²/F/in. (Vydyne R-100, R-200, R-220)

3.2 Btu/h/ft²/F/in. (RTP 200FR, 203M GB20, 204H-CF10, 205D, 227)

3.3 Btu/h/ft²/F/in. (Thermocomp PF-1006, PF-1006HI)

3.4 Btu/h/ft²/F/in. (PDX-85382; RTP 200H-FR, 205H; Thermocomp PF-1008, PF-1008HI, RF-1006, RF-1006HI, XF-1006)

3.5 Btu/h/ft²/F/in. (RTP 202M, 203 GB20, 205, 205A, 205B, 205TFE5, 207H; Thermocomp IF-1006, PF-100-10, QF-1006)

3.6 Btu/h/ft²/F/in. (RTP 205TFE15, 205TFE20, 207, 207A; Thermocomp PF-100-12, RF-1008, RF-1008HI)

3.65 Btu/h/ft²/F/in. (RTP 209A)

3.7 Btu/h/ft²/F/in. (PDX-85473; RTP 209; Thermocomp IF-1008, QF-1008, XF-1008)

3.8 Btu/h/ft²/F/in. (RTP 207B; Stat-Kon QC-1002, RCL-4042; Thermocomp RF-100-10)

3.9 Btu/h/ft²/F/in. (RTP 207D, 299x27567)

4.0 Btu/h/ft²/F/in. (Stat-Kon RC-1002; Thermocomp IF-100-10, QF-100-10, RC-1002, RF-100-12)

4.2 Btu/h/ft²/F/in. (Thermocomp IF-100-12, QF-100-12)

4.5 Btu/h/ft²/F/in. (Stat-Kon VC-1003)

4.6 Btu/h/ft²/F/in. (Stat-Kon RCF-1006)

5.0 Btu/h/ft²/F/in. (RTP 283)

5.2 Btu/h/ft²/F/in. (Stat-Kon RC-1004FR)

5.3 Btu/h/ft²/F/in. (EMI-X RA-30)

5.4 Btu/h/ft²/F/in. (RTP 285P)

5.5 Btu/h/ft²/F/in. (RTP 201P 25, 287P; Stat-Kon RC-1004; Thermocomp RC-1004)

5.6 Btu/h/ft²/F/in. (EMI-X RA-35)
5.8 Btu/h/ft²/F/in. (EMI-X RA-40)
6.0 Btu/h/ft²/F/in. (RTP 203P 25)
6.1 Btu/h/ft²/F/in. (Thermocomp RC-1006PC)
6.5 Btu/h/ft²/F/in. (RTP 285; Stat-Kon IC-1006, QC-1006; Thermocomp QC-1006)
6.8 Btu/h/ft²/F/in. (Stat-Kon RC-1006HI; Thermocomp RC-1006HI)
7.0 Btu/h/ft²/F/in. (Stat-Kon PC-1006, RC-1006, RCL-4036, RCL-4536; Thermo-comp PC-1006, RC-1006, RCL-4036, RCL-4536, XC-1006)
7.5 Btu/h/ft²/F/in. (EMI-X PC-1008, QC-1008)
7.6 Btu/h/ft²/F/in. (RTP 227FR)
8.0 Btu/h/ft²/F/in. (EMI-X PC-100-10, QC-100-10; RTP 287)
8.5 Btu/h/ft²/F/in. (EMI-X RC-1008; Thermocomp RC-1008)
8.7 Btu/h/ft²/F/in. (EMI-X RC-100-10)
8.9 Btu/h/ft²/F/in. (EMI-X RC-100-12)

Distort. Temp.:

35 C (1820 kPa) (Grilon ELX 23NZ)
40–50 C (1.85 MPa) (Rilsan BMN F15)
45 C (264 psi) (Capron 8255)
50 C (264 psi) (Capron 8254; Fosta Nylon 1693; Texalon 1110A PL HS)
51 C (264 psi) (Texalon 600A PL-2)
51.5 C (264 psi) (Fosta Nylon 1210, 1525)
55 C (264 psi) (Capron 8220 HS, 8221 HS; Texalon 1106; Ultramid B5, B5W)
55–75 C (264 psi) (Ultramid B3, B3K, B4, B4K, B35, B35W, KR 4411, KR 4412)
56 C (264 psi) (Fosta Nylon 1417; Grilamid ELY 60)
57 C (264 psi) (Fosta Nylon 1722; Grilamid ELY 20NZ; Texalon 1108)
58 C (264 psi) (Texalon XP-1296)
58–65 C (1.85 MPa) (Rilsan BMN F25)
60 C (264 psi) (Capron 8253, 8259, 8350; Texalon 600A, 600A Zip 3; Ultramid B3WM601)
63 C (1820 kPa) (Firestone Nylon 200-001; Texalon 670A)
64 C (264 psi) (Capron 8200)
65 C (264 psi) (Capron 8202, 8202L; Grilon A28 GM; Texalon 1310; Ultramid B3L, KR 4409, KR 4609, KR 4645)
65–85 C (264 psi) (Ultramid S3, S3K, S4)
66 C (1820 kPa) (Firestone Nylon 210-001L; Texalon 680A, 1203D; Wellamid 42LH-N, 42L-N, 42LN2-N)
68 C (264 psi) (Texalon 600A HS, XP-1317)
70 C (264 psi) (Texalon 1308A)
71 C (1820 kPa) (Firestone Nylon 228-001; Texalon 600A NU, 600A Zip 1)
74 C (1.82 MPa) (Wellamid GS40-60)
75 C (264 psi) (Capron 8202C, 8202CL, 8203C; Nydur B 31 SK; Ultramid KR 4406)
75–80 C (1.81 MPa) (Durethan B 30 S, B 31 SK, B 40 SK, B 50 E)
75–90 C (264 psi) (Ultramid B3S, B35SK, KR 4405, KR 4407)

Nylon (cont'd.)

76 C (1820 kPa) (Firestone Nylon 213-001)

79 C (264 psi) (Texalon 1200A, 1200 AXL, 1200A Zip-1, 1200A Zip-6)

80 C (264 psi) (Nydur B 40 SK)

85 C (264 psi) (Ultramid A3R)

86 C (264 psi) (Texalon 1200A HS)

90 C (264 psi) (Capron 8360)

100 C (264 psi) (Ultramid A3, A3K, A3KN, A3 Pellets 24, A3W, A4, A4H, A4K, A5, KR 4205)

100–110 C (1.81 MPa) (Durethan A 30 S)

105 C (264 psi) (Texalon 1200A Black 11; Ultramid A3SK; Wellamid 2BRH-NW, 21LN2-NNT, 22LH-N, 22L-N, 22LN2-N, 220-N)

107 C (264 psi) (Wellamid FR22F-N, FRGS25-66)

120 C (264 psi) (Capron 8260; Ultramid KR 4446, KR 4652)

145 C (264 psi) (Ultramid KR 4650)

149 C (1.82 Mpa) (Wellamid GS25-66)

150 C (264 psi) (Capron 8230G HS; Ultramid B3WM602, KR-B3X2V6)

157 C (1.82 MPa) (Wellamid GS40-66)

160 C (1820 kPa) (Firestone Nylon 540-001HS)

164 C (1.85 MPa) (Rilsan SR 80)

165 C (1.85 MPa) (Rilsan SR 100; Texalon MF1200A-40)

168 C (1.85 MPa) (Rilsan SR 120)

175 C (1.8 MPa) (Minlon 12T)

180 C (264 psi) (Ultramid KR 4450)

185 C (1.8 MPa) (Minlon 11C; Ultramid KR 4250)

190 C (264 psi) Nydur RM KU 2-2501/30, RM KU 2-2521/20, RM KU 2-2521/25, RM KU 2-2521/30, RM KU 2-2521/35; Ultramid B35G3)

198 C (66 psi) (Zytel ST FR-80)

200 C (264 psi) (Capron 8231G HS; Durethan BKV 30, BKV 30 H, BKV 50 H; Grilon PVN-3H; Nydur BKV 15 H, BKV 30 H, BKV 35 H, BKV 35 Z, BKV 40 H, BKV 50H, BKV 115, BKV 120, BKV 125, BKV 130, BKV 135; Ultramid KR 4448, KR 4449, KR 4651)

202 C (264 psi) (Capron 8267G HS; Firestone Nylon 415-001HS)

204 C (264 psi) (Texalon GF600A-33)

205 C (264 psi) (Grilon PV-3H; Texalon GMF600-40; Ultramid KR 4445/1)

206 C (264 psi) (Capron 8266G HS)

208 C (1820 kPa) (Firestone Nylon 640-001HS)

210 C (264 psi) (Capron 8233G HS, 8234G HS; Ultramid B3EG5, B3EG6, B3G5, B3G5HS, B3G6, B3G6HS, B3WG5, B3WG6, KR 4447/1)

215 C (264 psi) (Texalon MF1200A1-40; Ultramid B3EG7, B3EG10, B3G7, B3G7HS, B3G10, B3WG7, B3WG10)

218 C (1820 kPa) (Vydyne R-400G)

222 C (1.82 MPa) (Wellamid MR410 HS)

225 C (264 psi) (Texalon GF1310-13; Ultramid KR 4653)

230 C (1.8 MPa) (Minlon 10B; Texalon GF1308A-13)

232 C (1.8 MPa) (Minlon 21C)

241 C (264 psi) (Zytel FR-50)

245 C (1.8 MPa) (Minlon 20B; Vydyne R-513, R-513H; Wellamid GSF25/15-66)

250 C (1.81 MPa) (Durethan AKV 30; Ultramid A3EG5, A3EG6, A3EG7, A3EG10, A3G5, A3G6, A3G7, A3HG5, A3WG5, A3WG6, A3WG7, A3WG10, A3X1G5, A3X1G7, A3X1G10, A3XG5, A3XG7; Vydyne R-533, R-533H)

252 C (264 psi) (Texalon GF1200A-33)

254 C (1.82 MPa) (Wellamid FRGF25-66)

255 C (1820 kPa) (Vydyne R-543, R-543H)

118 F (264 psi) (Rilsan BECVO P40 TL, BESNO P40, BESNO P40 TL, BMNO P40)

129 F (264 psi) (Rilsan BESNO P20, BMNO P20)

131 F (66 psi) (BCI Nylon No. 653); (264 psi) (Ertalon 11SA; Rilsan BMNO, BMV Black T, KMVOTL)

133 F (66 psi) (BCI Nylon LX-3250); (264 psi) (Nyrim P-1 4000)

134 F (264 psi) (Rilsan BECNO, BESHVO, BESNO, BESVO)

135 F (Migralube Q-1000); (264 psi) (Rilsan BESN Black T, BESNO TL)

140 F (264 psi) (Nylatron 2010, 2011; Stat-Kon H)

142 F (264 psi) (Nylatron 2038; Vydyne 60H)

144 F (264 psi) (Vydyne M-345)

145 F (264 psi) (Nyrim P-1 3000)

149 F (264 psi) (Nylatron 2001)

150 F (264 psi) (Nylatron 2005)

158 F (264 psi) (Nyrim P-1 2000)

160 F (264 psi) (Nylatron 2015)

168 F (264 psi) (Ertalon 6SAU; Vekton 6PA, 6PAG, 6PAM, 6XAU)

170 F (264 psi) (Celanese Nylon 1000-1, 1000-2, 1000-4, 1003-1, 1003-2; Vydyne M-340, M-344)

171 F (264 psi) (Celanese Nylon 1310-1, 1310-2, 1310-4)

172 F (264 psi) (Nyrim P-1 1000)

175 F (264 psi) (Nylatron GS-60, GS-63)

180 F (264 psi) (Nylatron GS-71, GS-73; Vydyne 21, 21SP, 21X, 65B, 66B)

185 F (264 psi) (Nylatron 1024; Thermocomp R-1000FR-HS)

190 F (264 psi) (Nylatron 1022; Vydyne 22H, 22HSP, 25W)

198 F (264 psi) (Nylatron NSB-90)

200 F (264 psi) (Nylatron 1028, NSB; RTP 200H-FR; Thermocomp RL-4310)

200–425 F (264 psi) (Monocast MC 901; Nylatron GSM)

200–450 F (264 psi) (Polypenco Nylon 101)

200–470 F (264 psi) (Nylatron GS)

203 F (264 psi) (Ertalon 66SA)

208 F (264 psi) (Vydyne 25X)

210 F (264 psi) (Nylatron GS-21, GS-HS)

212 F (264 psi) (Ertalon 66SAM; Rilsan BMNY BZ TL)

Nylon (cont'd.)

215 F (264 psi) (Stat-Kon R; Thermocomp RL-4540)

220 F (264 psi) (RTP 200FR; Thermocomp R-1000, RL-4040)

221 F (264 psi) (Vydyne 24NSL, 24NSP)

266 F (264 psi) (Vydyne R-250)

280 F (264 psi) (Thermocomp XF-1004, XFL-4036)

285 F (264 psi) (Thermocomp XF-1006)

290 F (264 psi) ((Thermocomp XC-1006, XF-1008)

300 F (264 psi) (RTP 227FR)

310 F (264 psi) (Nylatron 2007, GS-60-9; RTP 201C)

320 F (264 psi) (Nylatron 2033; Vydyne R-220)

325 F (264 psi) (RTP 225)

335 F (264 psi) (Thermocomp HF-1006)

340 F (264 psi) (RTP 203C; Thermocomp SF-1006; Wellamid MR329HS)

343 F (264 psi) (Rilsan BZM300)

348 F (264 psi) (Rilsan BZM23G9)

350 F (264 psi) (RTP 205C; Thermocomp SF-100-10)

360 F (264 psi) (RTP 204GB FR, 207C; Vydyne R-100, R-200)

362 F (264 psi) (Rilsan BZM43G9)

370 F (264 psi) (RTP 201A, 201H; Wellamid MR409 HS)

375 F (264 psi) (Thermocomp PF-1002)

380 F (264 psi) (RTP 203H)

385 F (264 psi) (Nylatron GS-71-9)

390 F (264 psi) (RTP 201B, 201FR, 203A, 203H-CF15, 204H-CF10, 205H, 207H)

395 F (264 psi) (Nylatron 2008; RTP 201D)

400 F (264 psi) (Migralube QFL-4536; RTP 202FR, 205A, 207A, 227, 299x27567; Thermocomp IF-1002, PF-1002HI, QF-1002, QF-1006FR; Wellamid MRG30/10)

400–420 F (264 psi) (STX Nylon)

403 F (264 psi) (Nylatron GS-60-13)

405 F (264 psi) (Stat-Kon QC-1002; Thermocomp PF-1006FR)

410 F (264 psi) (Migralube QFL-4036; Nylatron 2040; PDX-82428; RTP 203B; Thermocomp IF-1004, PF-1004, QF-1004)

415 F (264 psi) (RTP 203D, 203FR, 205B, 209A; Thermocomp IF-1006, PF-1004HI, PF-1006HI, PF-1008HI)

420 F (264 psi) (RTP 204FR, 205D, 207B, 207D; Stat-Kon IC-1006; Thermocomp IF-1008, PF-1006, PF-1008, PF-100-10, PF-100-12, PFL-4216, QF-1006, QF-1008)

425 F (264 psi) (EMI-X PC-1008, PC-100-10, QC-1008, QC-100-10; Stat-Kon PC-1006, QC-1006; Thermocomp IF-100-10, IF-100-12, PC-1006, PFL-4218, QC-1006, QF-100-10, QF-100-12)

430 F (264 psi) (NY-30CF; PDX-85384)

435 F (264 psi) (PDX-85382, -85473)

440 F (264 psi) (Nylatron 1025; RTP 205FR)

450 F (264 psi) (RTP 205TFE5)

460 F (264 psi) (EMI-X RA-30, RA-35; Nylatron 1026; RTP 205TFE15, 205TFE20;

Thermocomp RF-1004FR-HS)

464 F (264 psi) (Vydyne 909)

465 F (264 psi) (EMI-X RA-40)

470 F (264 psi) (Nylatron 1027; RTP 201, 203; Thermocomp RF-1002HI, RF-1006FR-HS)

475 F (264 psi) (Nylatron 1018; Thermocomp RFL-4216)

480 F (264 psi) (RTP 202M, 203 GB20, 203M GB20, 205, 207; Thermocomp RC-1006PC; Vydyne R-538H-02)

485 F (264 psi) (Nylatron GS-51; Stat-Kon R-15, RC-1002, RC-1004FR, RCL-4042, RCL-4536, RF-15, VC-1003; Thermocomp RC-1002, RCL-4536, RF-1002, RF-1004, RF-1004HI, RF-1006HI, RF-1008HI, RFL-4218)

488 F (264 psi) (Celanese Nylon 1600-1, 1600-2, 1603-1, 1603-2)

490 F (264 psi) (Nylatron GS-51-13; RTP 201P 25, 203P 25, 283, 285P, 287P; Stat-Kon RCL-4036; Thermocomp RCL-4036, RF-1006)

495 F (264 psi) (Celanese Nylon 1500-1, 1500-2, 1503-1, 1503-2; PDX-83392; Stat-Kon RC-1004, RC-1006, RC-1006HI, RCF-1006; Thermocomp RC-1004, RC-1006, RC-1006HI)

500 F (264 psi) (EMI-X RC-1008, RC-100-10, RC-100-12; RTP 209, 285, 287; Thermocomp RC-1008, RF-1008, RF-100-10, RF-100-12)

Coeff. of Linear Exp.:

1.0–1.5×10^{-5} 1/K (Ultramid A3EG10, A3WG10, A3X1G10, B3EG10, B3G10, B3WG10)

1.5–2.0×10^{-5} 1/K (Ultramid A3EG6, A3EG7, A3G6, A3G7, A3WG6, A3WG7, A3X1G5, A3X1G7, A3XG7, B3EG7, B3G7, B3G7HS, B3WG7)

2.0–2.5×10^{-5} 1/K (Ultramid B3EG5, B3EG6, B3G5, B3G5HS, B3G6, B3G6HS, B3WG5, B3WG6, KR 4447/1, KR 4653)

2.3–3.5×10^{-5} 1/K (Ultramid KR 4449)

2.5–3.0×10^{-5} 1/K (Ultramid KR 4448)

2.5–3.5×10^{-5} 1/K (Ultramid A3EG5, A3G5, A3HG5, A3WG5, A3XG5)

3.0–3.5×10^{-5} 1/K (Ultramid B35G3)

3.0–4.0×10^{-5} 1/K (Ultramid KR 4445/1)

3.0–5.0×10^{-5} 1/K (Ultramid KR-B3X2V6)

3.5–4.5×10^{-5} 1/K (Ultramid B3WM602)

4.0–4.5×10^{-5} 1/K (Ultramid KR 4450)

5.0×10^{-5} 1/K (Ultramid KR 4651)

5.0–7.0×10^{-5} 1/K (Ultramid KR 4446)

5.0–8.0×10^{-5} 1/K (Ultramid KR 4250, KR 4650, KR 4652)

6.0–7.0×10^{-5} 1/K (Ultramid B3WM601)

6.0–8.5×10^{-5} 1/K (Ultramid B3L, KR 4409)

7–10×10^{-5} 1/K (Ultramid A3, A3K, A3KN, A3 Pellets 24, A3R, A3SK, A3W, A4, A4H, A4K, A5, B3, B3K, B3S, B4, B4K, B5, B5W, B35, B35M, B35SK, B35W, KR 4205, KR 4405, KR 4406, KR 4407, KR 4411, KR 4412)

8.0–10.0×10^{-5} 1/K (Ultramid S3, S3K, S4)

Nylon *(cont'd.)*

1.2×10^{-4} K^{-1} (Grilamid ELY 60)

12×10^{-5} K^{-1} (Grilon ELX 23NZ)

15×10^{-5} K^{-1} (Grilamid ELY 20NZ)

16×10^{-6} K^{-1} (Durethan BKV 50 H)

$20-30 \times 10^{-5}$ K^{-1} (Durethan AKV 30)

25×10^{-6} K^{-1} (Durethan BKV 30, BKV 30 H)

$60-80 \times 10^{-6}$ K^{-1} (Durethan A 30 S)

$80-100 \times 10^{-6}$ K^{-1} (Durethan B 30 S, B 31 SK, B 40 SK, B 50 E)

12×10^{-5}/K (Rilsan BMN F15, BMN F25)

3.0×10^{-5} m/m/C (Vydyne R-400G)

3.6×10^{-5} m/m/C (Minlon 10B, 11C, 20B, 21C)

5.4×10^{-5} m/m/C (Minlon 12T)

1.0×10^{-4} mm/mm/C (Capron 8259, 8350)

1.1×10^{-4} mm/mm/C (Capron 8254)

1.4×10^{-4} mm/mm/C (Capron 8255)

2.2×10^{-5} mm/mm/C (Vydyne R-543, R-543H)

2.3×10^{-5} mm/mm/C (Vydyne R-533, R-533H)

2.7×10^{-5} mm/mm/C (Vydyne R-513, R-513H)

3.1×10^{-5} mm/mm/C (Capron 8234G HS, 8266G HS, 8267G HS)

3.8×10^{-5} mm/mm/C (Capron 8233G HS)

5.0×10^{-5} mm/mm/C (Capron 8231G HS, 8260)

5.2×10^{-5} mm/mm/C (Wellamid MR410 HS)

5.4×10^{-5} mm/mm/C (Capron 8360)

5.9×10^{-5} mm/mm/C (Capron 8230G HS)

7.7×10^{-5} mm/mm/C (Wellamid 2BRH-NW, 21LN2-NNT, 22LH-N, 22L-N)

8.1×10^{-5} mm/mm/C (Capron 8202C, 8202CL, 8203C)

8.3×10^{-5} mm/mm/C (Capron 8200, 8202, 8202L; Wellamid 42LH-N, 42L-N, 42LN2-N)

9.9×10^{-5} mm/mm/C (Capron 8253)

1.1×10^{-5}/C (Rilsan SR 120)

3.0×10^{-5}/C (Rilsan SR 80, SR 100)

7.0×10^{-5}/C (Rilsan BMNY BZ TL)

9.1×10^{-5}/C (Rilsan BMNO, BMNO P20, BMV Black T, KMVOTL, RDP Pigmented)

10.1×10^{-5}/C (Rilsan RDP15-10 Natural)

0.5×10^{-5} in./in./F (EMI-X RC-100-12)

0.6×10^{-5} in./in./F (EMI-X PC-100-10, QC-100-10, RC-100-10)

0.8×10^{-5} in./in./F (EMI-X PC-1008, QC-1008, RC-1008; Thermocomp IF-100-12, PF-100-12, QF-100-12, RC-1008)

0.9×10^{-5} in./in./F (NY-30CF; Stat-Kon IC-1006, QC-1006; Thermocomp IF-100-10, PF-100-10, QC-1006, QF-100-10, RF-100-12)

1.0×10^{-5} in./in./F (Nylatron 2040, GS-60-9, GS-60-13; PDX-82428; Stat-Kon PC-1006; Thermocomp PFL-4218, RF-100-10)

1.1×10^{-5} in./in./F (Nylatron 1018, 2008; Stat-Kon RC-1006, RC-1006HI, RCL-4036,

RCL-4536; Thermocomp RC-1006, RC-1006HI)

1.2×10^{-5} in./in./F (Nylatron GS-51-13; Thermocomp IF-1008, PF-1008, PF-1008HI, QF-1008)

1.23×10^{-5} in./in./F (Nylatron GS-71-9)

1.3×10^{-5} in./in./F (Thermocomp XF-1008, RFL-4218; Vydyne R-538H-02)

1.4×10^{-5} in./in./F (Celanese Nylon 1600-1, 1600-2, 1603-1, 1603-2; PDX-83392; Stat-Kon RC-1004; Thermocomp RC-1004, RF-1008, RF-1008HI)

1.5×10^{-5} in./in./F (Nylatron GS-51; Stat-Kon RC-1004FR; Thermocomp IF-1006, QF-1006, RC-1006PC)

1.6×10^{-5} in./in./F (Celanese Nylon 1500-1, 1500-2, 1503-1, 1503-2; Nylatron 2007; Stat-Kon RCF-1006)

1.7×10^{-5} in./in./F (Thermocomp PF-1006, PF-1006HI, RFL-4216)

1.8×10^{-5} in./in./F (Stat-Kon RCL-4042, VC-1003; Thermocomp PF-1006FR, QF-1006FR, RF-1006, RF-1006FR-HS, RF-1006HI, XF-1006)

1.9×10^{-5} in./in./F (Thermocomp PFL-4216)

2.0×10^{-5} in./in./F (Stat-Kon RC-1002; Thermocomp XFL-4036)

2.2×10^{-5} in./in./F (Stat-Kon QC-1002; Thermocomp IF-1004, PF-1004, PF-1004HI, QF-1004)

2.3×10^{-5} in./in./F (Thermocomp RF-1004, RF-1004HI, XF-1004)

2.5×10^{-5} in./in./F (Thermocomp IF-1002, PF-1002, QF-1002)

2.7×10^{-5} in./in./F (Migralube QFL-4036, QFL-4536; Nylatron GS-21, GS-HS; Thermocomp PF-1002HI, RF-1002, RF-1002HI; Vydyne R-100, R-200, R-220)

2.8×10^{-5} in./in./F (Stat-Kon RF-15; Vydyne R-250)

3.2×10^{-5} in./in./F (Nylatron 2033)

3.44×10^{-5} in./in./F (Nylatron GS-71)

3.5×10^{-5} in./in./F (Nylatron GS)

3.6×10^{-5} in./in./F (Nylatron 2005, 2010)

3.63×10^{-5} in./in./F (Nylatron GS-73)

3.7×10^{-5} in./in./F (Nylatron 2001, GS-60, GS-63)

3.8×10^{-5} in./in./F (Nylatron 2011, 2015)

4.0×10^{-5} in./in./F (Ertalon 66SAM; Nylatron 2029; Stat-Kon R)

$4.0–5.0 \times 10^{-5}$ in./in./F (Vekton 6PA, 6PAG, 6PAM, 6XAU)

4.1×10^{-5} in./in./F (Nylatron 2038; Vydyne 909, M-344)

4.2×10^{-5} in./i./F (Ertalon 66SA; Thermocomp RL-4310)

4.3×10^{-5} in./in./F (Nylatron 2039; Vydyne M-345)

4.4×10^{-5} in./in./F (Nylatron NSB-90)

4.5×10^{-5} in./in./F (Ertalon 6SAU; Thermocomp R-1000, R-1000FR-HS; Vydyne 21, 21SP, 21X, 22H, 22HSP, 24NSL, 24NSP, 25X, 65B, 66B, M-340)

4.6×10^{-5} in./in./F (Thermocomp RL-4040, RL-4540)

4.7×10^{-5} in./in./f (Vekton 6PB)

5.0×10^{-5} in./in./F (Celanese Nylon 1000-1, 1000-2, 1000-4, 1003-1, 1003-2; Migralube Q-1000; Monocast MC 901; Nylatron GSM; Stat-Kon H)

5.5×10^{-5} in./in./F (Polypenco Nylon 101)

Nylon *(cont'd.)*

5.9–7.0 × 10⁻⁵ in./in./F (Nyrim P-1 1000)

6.1–7.0 × 10⁻⁵ in./in./F (Nyrim P-1 2000)

7.6 × 10⁻⁵ in./in./F (Nylatron 1024)

7.7–7.8 × 10⁻⁵ in./in./F (Nyrim P-1 3000, P-1 4000)

8.3 × 10⁻⁵ in./in./F (Ertalon 11SA; Vydyne 60H)

8.8 × 10⁻⁵ in./in./F (Nylatron 1028)

1.7 × 10⁻⁵/F (Rilsan BZM43G9, BZM300)

5.1 × 10⁻⁵/F (Rilsan BECNO, BECVO P40 TL, BESN Black T, BESNO P20, BESNO P40 TL, BESNO TL)

6.7 × 10⁻⁵/F (Rilsan BESN F15)

8.3 × 10⁻⁵/F (Rilsan BMNO P40)

820,000 F (BCI Nylon LX-3250, No. 653)

Sp. Heat:

0.3 (Celanese Nylon 1500-1, 1500-2, 1503-1, 1503-2)

0.4 (Celanese Nylon 1000-1, 1000-2, 1000-4, 1003-1, 1003-2, 1310-1, 1310-2, 1310-4)

0.42 (Rilsan BZM300)

0.47 (Rilsan BECNO, BECVO P40 TL, BESN Black T, BESNO P20, BESNO P40 TL, BESNO TL, BMNO, BMNO P20, KMVOTL)

1.7 (Rilsan BZM43G9)

1.1 kJ/kg K (Durethan BKV 50 H)

1.3 kJ/kg K (Durethan BKV 30, BKV 30 H)

1.5 kJ/kg K (Durethan AKV 30)

1.6 kJ/kg K (Durethan B 30 S, B 31 SK, B 40 SK, B 50 E)

1.7 kJ/kg K (Durethan A 30 S)

1.3 J/gK (Ultramid A3EG10, A3WG10, A3X1G10, B3EG10, B3G10, B3WG10)

1.4 J/gK (Ultramid A3EG7, A3G7, A3WG7, A3X1G7, A3XG7, B3EG7, B3G7, B3G7HS, B3WG7)

1.5 J/gK (Ultramid A3EG5, A3EG6, A3G5, A3G6, A3HG5, A3WG5, A3WG6, A3X1G5, A3XG5, B3EG5, B3EG6, B3G5, B3G5HS, B3G6, B3G6HS, B3L, B3WG5, B3WG6, KR 4409)

1.6 J/gK (Ultramid B35G3)

1.7 J/gK (Ultramid A3, A3K, A3KN, A3 Pellets 24, A3R, A3SK, A3W, A4, A4H, A4K, A5, B3, B3K, B3S, B3WM601, B4, B4K, B5, B5W, B35, B35M, B35SK, B35W, KR 4205, KR 4405, KR 4406, KR 4407, KR 4411, KR 4412, S3, S3K, S4)

0.37 Cal/g/C (Vydyne R-400G)

0.47 Cal/g/C (Rilsan BESHVO, BESVO, BESNO, BESNO P40)

0.40 Cal/g/F (BCI Nylon LX-3250, No. 653); (cast film) (BCI Nylon #808-809, #818-819, #829, LX-3249, No. 651)

0.29 Btu/lb/F (Nylatron 1025, 1026, 2033, 2040)

0.3 Btu/lb/F (Nylatron 1018, 1027, 2007, 2008)

0.33 Btu/lb/F (Vydyne 909)

0.34 Btu/lb/F (Nylatron 2011, 2029, 2038, 2039)

0.39 Btu/lb/F (Vydyne M-344)

0.35 Btu/lb/F (Nylatron 1022, 1024, 1028, 2001, 2005, 2010, 2015; Vydyne R-100, R-200, R-220)

0.37 Btu/lb/F (Vydyne R-250)

0.4 Btu/lb/F (Vydyne 21, 21SP, 21X, 22H, 22HSP, 24NSL, 24NSP, 25X, 60H)

1.0 Btu/lb/F (Vydyne M-345)

Flamm.:

5V/V-0 (Vydyne 909, M-340, M-344, M-345)

V-0 (Capron 8232G HS FR; Grilon A28 VO; Nylatron 2033; RTP 200FR, 200H-FR, 201FR, 202FR, 203FR, 204FR, 204GB FR, 205FR, 227FR; Stat-Kon RC-1004FR; Thermocomp PF-1006FR, QF-1006FR, R-1000FR-HS, RF-1004FR-HS, RF-1006FR-HS; Ultramid KR 4205, KR 4406; Wellamid FRGS25-66; Zytel FR-50, ST FR-80)

V-0/V-1 (Ultramid A3X1G5, A3X1G7; Wellamid FR22F-N, FRGF25-66)

V-1 (Ultramid A3X1G10, KR-B3X2V6)

V-2 (Capron 8202, 8202C, 8202CL, 8202L, 8220 HS; Celanese Nylon 1000-1, 1000-2, 1000-4, 1003-1, 1003-2, 1310-1, 1310-2, 1310-4; Durethan A 30 S, B 30 S, B 31 SK, B 40 SK, B 50 E; Grilamid L20 FR; Grilon A28 GM; Nydur B 31 SK; Nylatron 1022; Ultramid A3, A3K, A3KN, A3 Pellets 24, A3SK, A3W, A4, A4H, A4K, A5, B3, B3K, B3S, B3WM601, B35SK, KR 4405, KR 4407, KR 4411, KR 4609, S3, S3K, S4; Vydyne 21, 21SP, 21X, 22H, 22HSP, 24NSL, 24NSP, 25W, 25X; Wellamid 42LH-N, 42L-N, 220-N)

V-2/HB (Ultramid B4, B4K, B35, B35W, KR 4412; Wellamid 2BRH-NW, 21LN2-NNT, 22LH-N, 22L-N, 22LN2-N)

HB (Capron 8200, 8203C, 8230G HS, 8231G HS, 8233G HS, 8234G HS, 8253, 8254, 8255, 8259, 8260, 8266G HS, 8267G HS, 8350, 8360; Celanese Nylon 1500-1, 1500-2, 1503-1, 1503-2, 1600-1, 1600-2, 1603-1, 1603-2; Durethan AKV 30, BKV 30, BKV 30 H; EMI-X PC-1008, PC-100-10, QC-1008, QC-100-10, RA-30, RA-35, RA-40, RC-1008, RC-100-10, RC-100-12; Grilamid ELY 60; Grilon PV-3H, PVN-3H; Nydur B 40 SK, BKV 15 H, BKV 30 H, BKV 35 H, BKV 35 Z, BKV 40 H, BKV 50 H, BKV 115, BKV 120, BKV 125, BKV 130, BKV 135, RM KU 2-2501/30, RM KU 2-2521/20, RM KU 2-2521/25, RM KU 2-2521/30, RM KU 2-2521/35; Nylatron 1018, 1024, 1026, 2001, 2008, 2010; PDX-82428, -83392; Rilsan BMN F15, BMN F25; RTP 201, 201A, 201B, 201C, 201D, 201H, 201P 25, 202M, 203, 203A, 203B, 203C, 203D, 203 GB20, 203H, 203H-CF15, 203M GB20, 203P 25, 204H-CF10, 205, 205A, 205B, 205CF, 205D, 205H, 205TFE5, 205TFE15, 205TFE20, 207, 207A, 207B, 207C, 207D, 207H, 209, 209A, 225, 227, 283, 285, 285P, 287, 287P, 299x27567; Stat-Kon H, IC-1006, PC-1006, QC-1002, QC-1006, R, RC-1002, RC-1004, RC-1006, RCF-1006HI, RCF-1006, RCL-4036, RCL-4042, RCL-4536, R-15, VC-1003; Thermocomp IF-1006, IF-100-10, PC-1006, PF-1002, PF-1002HI, PF-1004, PF-1004HI, PF-1006, PF-1006HI, PF-1008, PF-1008HI, PFL-4216, PFL-4218, QC-1006, QF-100-10, RC-1002, RC-1004, RC-1006, RC-1006HI, RC-1006PC, RC-1008, RCL-4036, RCL-4536, RF-1002, RF-

Nylon (cont'd.)

1002HI, RF-1004, RF-1004HI, RF-1006, RF-1006HI, RF-1008, RF-1008HI, RF-100-10, RFL-4216, RFL-4218, RL-4310, XC-1006, XF-1004, XF-1006, XF-1008, XFL-4036; Ultramid A3EG5, A3EG6, A3EG7, A3EG10, A3G5, A3G6, A3G7, A3HG5, A3R, A3WG5, A3WG6, A3WG7, A3WG10, A3XG5, A3XG7, B3EG5, B3EG6, B3EG7, B3EG10, B3G5, B3G5HS, B3G6, B3G6HS, B3G7, B3G7HS, B3G10, B3L, B3WG5, B3WG6, B3WG7, B3WG10, B3WM602, B35G3, KR 4250, KR 4409, KR 4445/1, KR 4446, KR 4447/1, KR 4448, KR 4449, KR 4450, KR 4645, KR 4650, KR 44651, KR 4652, KR 4653; Vydyne R-100, R-200, R-220, R-513, R-513H, R-533, R-533H, R-543, R-543H; Wellamid GS25-66, GS40-60, GS40-66, GSF25/15-66, MR410 HS)

Self-extinguishing (Ertalon 6SAU, 11SA, 66SA, 66SAM; Nylatron GS-21, GS-60, GS-63, GS-HS; Rilsan BESHVO, BESNO, BESVO, BMNO, BMNY BZ TL, BMV Black T, KMVOTL; Vekton 6PA, 6PAG, 6PAM, 6PB, 6XAU)

Slow burning (Nylatron GS-51, GS-51-13, GS-60-9)

0.10 in./min (Rilsan BZM300)

0.20 in./min (Rilsan BMNO P20)

0.28 in./min (Rilsan BESNO P20, BESNO P40)

0.36 in./min (Rilsan BECVO P40 TL, BESNO P40 TL, BMNO P40)

0.47 in./min (Nylatron GS-73)

0.65 in./min (Nylatron GS-71-9)

0.66 in./min (Nylatron GS-71)

0.7 in./min (Nylatron GS-60-13)

1.08 in./min (Rilsan BZM23G9, BZM43G9)

1.1 in./min (Celanese Nylon 1500-1, 1500-2, 1503-1, 1503-2)

ELECTRICAL PROPERTIES:

Dissip. Factor:

0.006 (50 Hz) (Durethan BKV 30); (60 Hz) (Celanese Nylon 1500-1, 1500-2, 1503-1, 1503-2); (1 kHz) (Vydyne M-344)

0.007 (50 Hz) (Durethan B 30 S, B 31 SK, B 40 SK, B 50 E, BKV 30 H)

0.009 (50 Hz) (Durethan BKV 50 H); (1 kHz) (Vydyne R-200); (100 Hz) (Celanese Nylon 1600-1, 1600-2, 1603-1, 1603-2)

0.010 (1 kHz) (Vydyne 24NSL, 24NSP, 25W, M-340; Wellamid MR410 HS); (1 MHz) (RTP 201C, 203C)

0.011 (1 kHz) (Vydyne R-100, R-220); (1 MHz) (RTP 201B, 201D, 203B; Ultramid KR 4409)

0.011–0.17 (60 to 10^6 Hz) (Thermocomp PF-1006FR)

0.012 (1 kHz) (Thermocomp RF-1006, RF-1008); (1 MHz) (RTP 203D, 205B, 227)

0.012–0.017 (60 to 10^6 Hz) (Thermocomp QF-1006FR)

0.013 (1 kHz) (Vydyne 22H, 22HSP); (1 MHz) (RTP 205H, 207H)

0.013–0.015 (60 to 10^6 Hz) (Thermocomp IF-1006)

0.014 (Ultramid A3EG5, A3EG6, A3G5, A3G6, A3HG5, A3WG5, A3WG6, A3XG5, B3EG10, B3G10, B3WG10); (1 kHz) (Thermocomp PF-1006); (1 MHz) (RTP 203 GB20, 203H, 205D, 205FR, 209, 209A, 225)

0.015 (Ultramid A3EG10, A3WG10); (60 Hz) (Vekton 6PA, 6XAU); (1 MHz) (RTP 203M GB20, 204FR, 204GB FR, 205C, 205TFE15, 205TFE20, 207, 207A, 207B; Ultramid A3R)

0.016 (1 kHz) (Thermocomp PF-1008, QF-1006); (1 MHz) (RTP 200H-FR, 202FR, 203FR, 205, 205TFE5, 207C, 207D)

0.017 (1 MHz) (RTP 200FR, 201FR, 202M, 205A, 227FR)

0.018 (1 kHz) (Thermocomp QF-1008); (1 MHz) (RTP 203A)

0.019 (1 MHz) (RTP 203)

0.02 (Ultramid A3EG7, A3G7, A3WG7, A3X1G5, A3X1G7, A3XG7, B3WM601, B3WM602, B35G3, KR 4446, KR 4447/1, KR 4448, KR-B3X2V6); (60 Hz) (Ertalon 66SAM; Vekton 6PAG, 6PAM); (1 kHz) (Vydyne 21, 21SP, 21X, 60H; Wellamid 22LH-N, 22L-N, 42LH-N, 42L-N, 42LN2-N, 220-N, FR22F-N, FRGF25-66, FRGS25-66); (100 kHz) (Nylatron 1022, 1025, 1027, 2007, 2011, 2029, 2033); (1 MHz) (Durethan A 30 S, AKV 30; RTP 201, 201A, 201H)

0.021 (Ultramid B3EG7, B3G7, B3G7HS, B3WG7, KR 4250); (1 MHz) (Ultramid KR 4205)

0.023 (Ultramid B3EG6, B3G6, B3G6HS, B3WG6); (1 MHz) (Ultramid B3, B3K, B4, B4K, B35, B35W, KR 4411, KR 4412)

0.024 (100 kHz) (Nylatron 2040); (1 MHz) (Ultramid B3L)

0.025 (Ultramid B3EG5, B3G5, B3G5HS, B3WG5); (1 MHz) (Ultramid A3K, A3KN, A3SK, A3W, A4H)

0.026 (1 MHz) (Ultramid A3, A3 Pellets 24, A4, A4K, A5)

0.027 (1 MHz) (Ultramid S3, S3K, S4)

0.03 (1 kHz) (Rilsan BZM23G9, BZM43G9, BZM300); (100 kHz) (Nylatron 1024, 1028, 2010); (1 MHz) (Ultramid B3S, B35SK, KR 4405, KR 4407)

0.031 (1 MHz) (Ultramid B5, B5W)

0.034 (1 kHz) (Thermocomp R-1000FR-HS)

0.036 (1 kHz) (Thermocomp RF-1004FR-HS)

0.037 (1 kHz) (Thermocomp RF-1006FR-HS); (100 Hz) (Celanese Nylon 1310-1, 1310-2, 1310-4)

0.04 (1 kHz) (Vydyne M-345); (100 kHz) (Nylatron 1018, 2038)

0.045 (60 Hz) (Ertalon 11SA); (1 kHz) (Rilsan BESNO)

0.048 (1 kHz) (Rilsan BEHSVO, BESN Black T, BESNO TL, BESVO, BMNO, BMV Black T, KMVOTL)

0.050 (1 kHz) (Rilsan BECNO)

0.06 (60 Hz) (Ertalon 6SAU, 66SA)

0.07 (100 kHz) (Nylatron 2001, 2008)

0.08 (100 kHz) (Nylatron 2005, 2015)

0.1 (10^5 Hz) (Grilamid ELY 60)

0.15 (1 kHz) (Vydyne 909)

0.192 (1 kHz) (Rilsan BESNO P20, BMNO P20)

0.20 (60 Hz) (BCI Nylon LX-3250, No. 653; Celanese Nylon 1000-1, 1000-2, 1000-4, 1003-1, 1003-2)

Nylon *(cont'd.)*

0.224 (1 kHz) (Rilsan BESNO P40 TL)
0.229 (1 kHz) (Rilsan BESNO P40, BMNO P40)
0.23 (1 kHz) (Rilsan BECVO P40 TL)

Dielec. Str.:

35 kV/mm (Grilon A28 GM)
40 kV/mm (Grilon PV-3H)
45 kV/mm (Grilamid ELY 60; Grilon PVN-3H)
50 kV/mm (Grilon ELX 23NZ)
> 50 kV/mm (Ultramid KR 4446)
60 kV/mm (Ultramid KR 4448)
70 kV/mm (Durethan AKV 30; Ultramid A3X1G7, A3X1G10)
80 kV/mm (Ultramid A3X1G5, B3EG5, B3EG6, B3EG7, B3G5, B3G5HS, B3G6, B3G6HS, B3G7, B3G7HS, B3WG5, B3WG6, B3WG7, KR 4447/1)
> 80 kV/mm (Durethan B 30 S, B 31 SK, B 40 SK, B 50 E, BKV 30, BKV 30 H, BKV 50 H)
85 kV/mm (Ultramid B3WM601, B35G3)
90 kV/mm (Ultramid A3EG5, A3EG6, A3EG7, A3EG10, A3G5, A3G6, A3G7, A3HG5, A3WG5, A3WG6, A3WG7, A3WG10, A3XG5, A3XG7, B3EG10, B3G10, B3WG10)
95 kV/mm (Ultramid KR-B3X2V6)
100 kV/mm (Durethan A 30 S; Ultramid B3, B3K, B3L, B3S, B4, B4K, B5, B5W, B35, B35SK, B35W, KR 4205, KR 4405, KR 4406, KR 4407, KR 4409, KR 4411, KR 4412, S3, S3K, S4)
105 kV/mm (Ultramid B3WM602)
110 kV/mm (Ultramid A4H, KR 4250)
120 kV/mm (Ultramid A3, A3K, A3KN, A3 Pellets 24, A3R, A3SK, A3W, A4, A4K, A5)
16.4 V/mm (Wellamid 22LH-N, 22L-N, 220-N)
16.8 V/mm (Wellamid 42LH-N, 42L-N, 42LN2-N)
17.6 V/mm (Wellamid FR22F-N)
20 V/mm (Wellamid MR410 HS)
22 V/mm (Wellamid FRGF25-66, FRGS25-66)
270 V/mil (Nylatron 2008)
295 V/mil (Ertalon 6SAU, 66SA; Vekton 6PA, 6XAU)
300 V/mil (Nylatron 2005, 2015)
300–400 V/mil (Ertalon 66SAM; Nylatron GS; Polypenco Nylon 101)
315 V/mil (Nylatron 1024)
320 V/mil (Nylatron 1025, 2010)
330 V/mil (Nylatron 1022, 2001, 2033, 2040)
340 V/mil (Nylatron 1018, 2011)
350 V/mil (Nylatron 1027, 1028, 2029, 2038)
400 V/mil (BCI Nylon LX-3250, No. 653; Celanese Nylon 1500-1, 1500-2, 1503-1, 1503-2, 1600-1, 1600-2, 1603-1, 1603-2; Nylatron 2007; RTP 227FR; Thermo-

comp R-1000FR-HS, RF-1008)

415 V/mil (Thermocomp RF-1004FR-HS)

420 V/mil (Thermocomp PF-1006FR, PF-1008, QF-1008, RF-1006FR-HS)

425 V/mil (Ertalon 11SA)

430 V/mil (Vydyne M-345)

440 V/mil (Thermocomp IF-1006, QF-1006, RF-1006)

450 V/mil (Thermocomp PF-1006)

470 V/mil (RTP 203M GB20, 225)

475 V/mil (RTP 203FR, 204FR, 204GB FR, 205FR)

480 V/mil (RTP 203 GB20, 227)

490 V/mil (Vydyne R-100, R-200, R-220)

500 V/mil (RTP 200H-FR, 201, 201A, 201B, 201C, 201D, 201FR, 201H, 202FR, 202M, 203, 203A, 203B, 203C, 203D, 203H, 205, 205A, 205B, 205C, 205D, 205H, 205TFE5, 205TFE15, 205TFE20, 207, 207A, 207B, 207C, 207D, 207H, 209, 209A; Thermocomp QF-1006FR)

500–600 V/mil (Monocast MC 901; Nylatron GSM; Vekton 6PAG, 6PAM)

530 V/mil (Rilsan BZM23G9)

536 V/mil (Vydyne 22H, 22HSP)

540 V/mil (Vydyne 909, M-344)

550 V/mil (Vydyne 24NSL, 24NSP, M-340)

570 V/mil (Vydyne 21, 21SP, 21X, 60H)

600 V/mil (Celanese Nylon 1000-1, 1000-2, 1000-4, 1003-1, 1003-2, 1310-1, 1310-2, 1310-4)

650 V/mil (Rilsan BECVO P40 TL, BESNO P40, BESNO P40 TL, BMNO P40; RTP 200FR)

700 V/mil (Rilsan BESNO P20, BMNO P20, KMVOTL)

750 V/mil (Rilsan BECNO, BESHVO, BESN Black T, BESNO, BESNO TL, BESVO, BMNO, BMV Black T)

1050 V/mil (Rilsan RDP Pigmented)

1100 V/mil (Rilsan BZM300, RDP15-10 Natural)

Dielec. Const.:

3.0 (1 MHz) (Ultramid KR 4645)

3.1 (100 kHz) (Nylatron 2011); (1 MHz) (Ultramid KR 4409)

3.2 (100 kHz) (Nylatron 1022); (10^5 Hz) (Grilon PVN-3H); (1 MHz) (RTP 201C; Ultramid A3K, A3KN, A3SK, A3W, A4, A4H, A4K, A5)

3.3 (Ultramid B3WM601); (100 kHz) (Nylatron 1024, 1028); (1 MHz) (Ultramid A3R, B3S, B35SK, KR 4405, KR 4407, S3, S3K, S4)

3.5 (Ultramid A3EG5, A3EG6, A3EG7, A3G5, A3G6, A3G7, A3HG5, A3WG5, A3WG6, A3WG7, A3XG5, A3XG7, B3WM602, KR 4446); (100 kHz) (Nylatron 2010); (1 MHz) (Ultramid B3, B3K, B3L, B4, B4K, B5, B5W, B35, B35W, KR 4411, KR 4412, KR 4609)

3.55 (1 MHz) (RTP 201H)

3.6 (Ultramid A3X1G7, A3X1G10, KR-B3X2V6); (1 kHz) (Vydyne 21, 21SP, 21X,

Nylon (cont'd.)

22H, 22HSP, 24NSL, 24NSP, 60H, M-344; Wellamid 22LH-N, 22L-N, 220-N); (100 kHz) (Nylatron 2029); (10^5 Hz) (Grilon A28 GM, PV-3H); (1 MHz) (Durethan A 30 S; RTP 201A, 201B, 201D, 203C, 203H; Ultramid A3, A3 Pellets 24, KR 4205, KR 4406)

3.62 (1 kHz) (Wellamid FR22F-N, FRGS25-66)

3.68 (1 MHz) (RTP 205H)

3.7 (Ultramid A3X1G5); (60 Hz) (Ertalon 66SAM; Monocast MC 901; Nylatron GSM; Vekton 6PAG, 6PAM); (1 kHz) (Rilsan BECNO, BESHVO, BESN Black T, BESNO, BESNO TL, BESVO, BMNO, BMV Black T, KMVOTL; Vydyne M-340; Wellamid 42LH-N, 42L-N, 42LN2-N); (1 MHz) (RTP 200H-FR, 201, 203A, 203B, 203D, 205C, 205TFE5)

3.73 (1 MHz) (RTP 200FR)

3.75 (1 MHz) (RTP 203, 207H)

3.8 (Ultramid A3EG10, A3WG10, B3EG5, B3EG6, B3G5, B3G5HS, B3G6, B3G6HS, B3WG5, B3WG6, B35G3, KR 4250, KR 4447/1); (50 Hz) (Durethan B 30 S, B 31 SK, B 40 SK, B 50 E); (1 kHz) (Thermocomp QF-1006; Vydyne 25W, R-100, R-200); (1 MHz) (Durethan AKV 30; RTP 205, 205A, 205B, 205D, 205TFE15, 205TFE20, 207, 207B, 207C, 207D)

3.81 (Ultramid KR 4448); (1 kHz) (Thermocomp R-1000FR-HS)

3.83 (1 kHz) (Wellamid FRGF25-66)

3.85 (1 MHz) (RTP 209)

3.9 (Ultramid B3EG7, B3G7, B3G7HS, B3WG7); (1 kHz) (Thermocomp PF-1006, QF-1008, RF-1006; Vydyne 909, R-220; Wellamid MR410 HS); (100 kHz) (Nylatron 1026); (1 MHz) (RTP 202M)

4.0 (60 Hz) (Ertalon 11SA); (1 kHz) (Thermocomp PF-1008); (100 kHz) (Nylatron 1027); (1 MHz) (RTP 201FR, 207A, 225)

4.05 (1 MHz) (RTP 202FR)

4.1 (50 Hz) (Durethan BKV 30); (60 Hz) (Celanese Nylon 1500-1, 1500-2, 1503-1, 1503-2; Polypenco Nylon 101); (1 kHz) (Vydyne M-345); (1 MHz) (RTP 209A)

4.12 (1 kHz) (Thermocomp RF-1004FR-HS); (1 MHz) (RTP 203FR, 227FR)

4.2 (Ultramid B3EG10, B3G10, B3WG10); (50 Hz) (Durethan BKV 30 H); (1 MHz) (RTP 203 GB20, 203M GB20, 204GB FR, 227)

4.20–3.50 (60 to 10^6 Hz) (Thermocomp IF-1006)

4.23 (1 MHz) (RTP 204FR)

4.26 (1 kHz) (Thermocomp RF-1006FR-HS)

4.3 (1 MHz) (RTP 205FR)

4.4 (1 kHz) (Thermocomp RF-1008); (100 Hz) (Celanese Nylon 1600-1, 1600-2, 1603-1, 1603-2)

4.4–4.0 (60 to 10^6 Hz) (Thermocomp QF-1006FR)

4.4–4.1 (60 to 10^6 Hz) (Thermocomp PF-1006HI)

4.6 (50 Hz) (Durethan BKV 50 H)

4.8 (1 kHz) (Rilsan BZM300)

5.0 (100 kHz) (Nylatron 2001, 2040); (10^5 Hz) (Grilamid ELY 60)

5.1 (100 kHz) (Nylatron 2005, 2015)

5.4 (100 Hz) (Celanese Nylon 1310-1, 1310-2, 1310-4); (100 kHz) (Nylatron 1018, 2007)

5.5 (100 Hz) (Celanese Nylon 1000-1, 1000-2, 1000-4, 1003-1, 1003-2); (100 kHz) (Nylatron 2008, 2038)

5.7 (100 kHz) (Nylatron 2033)

5.9 (1 kHz) (Rilsan BESNO P20, BMNO P20)

6.0 (60 Hz) (Vekton 6PA, 6XAU)

6.5 (1 kHz) (Rilsan BZM23G9, BZM43G9)

7.5 (60 Hz) (Ertalon 6SAU)

7.6 (60 Hz) (Ertalon 66SA)

9.7 (1 kHz) (Rilsan BECVO P40 TL, BESNO P40, BESNO P40 TL, BMNO P40)

11.0 (60 Hz) (BCI Nylon LX-3250, No. 653)

Vol. Resist.:

10 ohm-cm (EMI-X PC-100-10, QC-100-10, RC-100-12; PDX-82428, -83392)

35 ohm-cm (RTP 287)

50 ohm-cm (RTP 287P)

75 ohm-cm (RTP 285)

100 ohm-cm (EMI-X PC-1008, QC-1008, RA-35, RA-40, RC-1008; RTP 283; Stat-Kon IC-1006, PC-1006, QC-1006, RC-1006, RC-1006HI, RCL-4036)

200 ohm-cm (RTP 299x27567)

400 ohm-cm (RTP 203H-CF15, 285P)

420 ohm-cm (RTP 201P 25)

480 ohm-cm (RTP 203P 25)

1000 ohm-cm (EMI-X RA-30; Stat-Kon H, RC-1004, RC-1004FR, RCF-1006, RCL-4536)

10^4–10^6 ohm-cm (Ultramid KR 4445/1)

10,000 ohm-cm (Stat-Kon R, RF-15, VC-1003)

22,000 ohm-cm (RTP 204H-CF10)

10^4–10^8 ohm-cm (Rilsan BMNY BZ TL)

100,000 ohm-cm (Stat-Kon QC-1002, RC-1002, RCL-4042)

10^7 ohm-cm (Rilsan SR 100)

3×10^7 ohm-cm (Rilsan SR 80)

10^9 ohm-cm (Rilsan BESN F15)

3×10^9 ohm-cm (Rilsan BMN F25)

5×10^9 ohm-cm (Rilsan BMN F15)

10^{11} ohm-cm (Rilsan BECVO P40 TL, BESNO P20, BESNO P40, BESNO P40 TL, BMNO P20, BMNO P40)

10^{12} ohm-cm (Grilamid ELY 60; RTP 201B, 203B, 205B, 207B)

3.8×10^{12} ohm-cm (Vydyne 909)

0.6×10^{13} ohm-cm (BCI Nylon LX-3250, No. 653)

1.0×10^{13} ohm-cm (Rilsan BZM23G9, GZM43G9; RTP 201C, 201D, 203C, 203D, 205C, 205D, 207C, 207D; Vekton 6PAG, 6PAM)

Nylon *(cont'd.)*

2.5 × 10^{13} ohm-cm (Ertalon 66SAM; Nylatron GS)

3.8 × 10^{13} ohm-cm (Thermocomp RF-1006FR-HS)

4.5 × 10^{13} ohm-cm (Ertalon 66SA; Polypenco Nylon 101)

4.7 × 10^{13} ohm-cm (Ertalon 6SAU; Vekton 6PA, 6XAU)

6.1 × 10^{13} ohm-cm (Thermocomp RF-1004FR-HS)

7.0 × 10^{13} ohm-cm (Thermocomp PF-1008)

7.8 × 10^{13} ohm-cm (Ertalon 11SA)

10^{14} ohm-cm (Celanese Nylon 1000-1, 1000-2, 1000-4, 1003-1, 1003-2, 1310-1, 1310-2, 1310-4, 1500-1, 1500-2, 1503-1, 1503-2, 1600-1, 1600-2, 1603-1, 1603-2; Grilon PV-3H, PVN-3H; Rilsan BECNO, BESHVO, BESN Black T, BESNO, BESNO TL, BESVO, BMNO, BMV Black T, BZM300, KMVOTL; RTP 201, 201A, 201H, 202FR, 202M, 203, 203A, 203FR, 203H, 204FR, 204GB FR, 205, 205A, 205FR, 205H, 205TFE5, 205TFE15, 205TFE20, 207, 207A, 207H, 209, 209A, 225, 227; Vydyne M-344)

1.1 × 10^{14} ohm-cm (Thermocomp R-1000FR-HS)

1.4 × 10^{14} ohm-cm (Vydyne 60H)

1.8 × 10^{14} ohm-cm (Thermocomp RF-1008)

2.8 × 10^{14} ohm-cm (Thermocomp PF-1006)

3.7 × 10^{14} ohm-cm (Wellamid FR22F-N)

4.6 × 10^{14} ohm-cm (Vydyne M-345)

5 × 10^{14} ohm-cm (Durethan BKV 50 H; Thermocomp RF-1006)

7.5 × 10^{14} ohm-cm (Wellamid 42LH-N, 42L-N, 42LN2-N)

7.7 × 10^{14} ohm-cm (Vydyne R-220)

10^{15} ohm-cm (Durethan A 30 S, AKV 30, B 30 S, B 31 SK, B 40 SK, B 50 E, BKV 30 H; Grilon A28 GM, ELX 23NZ; Nylatron 1018, 1022, 1024, 1025, 1026, 2001, 2005, 2007, 2008, 2010, 2011, 2015, 2029, 2033, 2038, 2040; RTP 201FR; Ultramid A3, A3EG5, A3EG6, A3EG7, A3EG10, A3G5, A3G6, A3G7, A3HG5, A3K, A3KN, A3 Pellets 24, A3R, A3SK, A3W, A3WG5, A3WG6, A3WG7, A3WG10, A3X1G5, A3X1G7, A3X1G10, A3XG5, A3XG7, A4, A4H, A4K, A5, B3, B3EG5, B3EG6, B3EG7, B3EG10, B3G5, B3G5HS, B3G6, B3G6HS, B3G7, B3G7HS, B3G10, B3K, B3L, B3S, B3WG5, B3WG6, B3WG7, B3WG10, B3WM601, B3WM602, B4, B4K, B5, B5W, B35, B35G3, B35SK, B35W, KR 4205, KR 4250, KR 4405, KR 4406, KR 4407, KR 4409, KR 4411, KR 4412, KR 4446, KR 4447/1, Kr 4448, KR 4449, KR 4450, KR 4609, KR 4645, KR 4650, KR 4651, KR 4652, KR 4653, KR-B3X2V6, S3, S3K, S4; Vydyne M-340, R-100; Wellamid 22LH-N, 22L-N, 220-N)

1.5 × 10^{15} ohm-cm (Thermocomp QF-1008; Vydyne R-200)

2 × 10^{15} ohm-cm (Durethan BKV 30)

3.6 × 10^{15} ohm-cm (Vydyne 22H, 22HSP)

6 × 10^{15} ohm-cm (Vydyne 21, 21SP, 21X)

7.6 × 10^{15} ohm-cm (Thermocomp QF-1006)

10^{16} ohm-cm (RTP 200FR, 200H-FR, 203 GB20, 203M GB20, 227FR; Wellamid MR410 HS)

2×10^{16} ohm-cm (Vydyne 24NSL, 24NSP)

Surf. Resist.:

0 ohm/sq. (PDX-82428. -83392)

10 ohm/sq. (EMI-X PC-100-10, QC-100-10, RC-100-10, RC-100-12)

10–50 ohm/sq. (NY-30CF)

75 ohm/sq. (Thermocomp RC-1008)

100 ohm/sq. (EMI-X PC-1008, QC-1008, RA-35, RA-40, RC-1008; Stat-Kon IC-1006, PC-1006, QC-1006, RC-1006, RC-1006HI, RCL-4036)

10^2–10^5 ohm (Ultramid KR 4445/1)

150 ohm/sq. (Thermocomp PC-1006, RC-1006, RC-1006HI, XC-1006)

200 ohm/sq. (Thermocomp RCL-4036)

250 ohm/sq. (Thermocomp QC-1006)

1000 ohm/sq. (EMI-X RA-30; Stat-Kon H, R-15, RC-1004, RC-1004FR, RCF-1006, RCL-4536, VC-1003; Thermocomp RCL-4536)

1300 ohm/sq. (Thermocomp RC-1004)

2800 ohm/sq. (Thermocomp RC-1006PC)

10,000 ohm/sq. (Stat-Kon R, RF-15)

100,000 ohm/sq. (Stat-Kon QC-1002, RC-1002, RCL-4042)

10^7 ohm/sq. (Thermocomp RC-1002)

10^{12} ohm (Grilamid ELY 60; Ultramid A3EG5, A3EG6, A3EG7, A3EG10, A3G5, A3G6, A3G7, A3HG5, A3WG5, A3WG6, A3WG7, A3WG10, A3XG5, A3XG7, B3EG5, B3EG6, B3G5, B3G5HS, B3G6, B3G6HS, B3WG5, B3WG6, B3WM601, B3WM602, B35G3, KR 4205, KR 4409, KR 4446, KR 4447/1, KR 4448, KR 4449, KR 4450, KR 4645, KR 4650, KR 4651, Kr 4652, KR 4653, S3, S3K, S4)

10^{12}–10^{13} ohm (Ultramid A3R)

1.69×10^{12} ohm (Thermocomp RF-1006FR-HS)

3.49×10^{12} ohm (Thermocomp RF-1004FR-HS)

5×10^{12} ohm (Durethan B 30 S, B 31 SK, B 40 SK, B 50 E)

7.3×10^{12} ohm (Thermocomp R-1000FR-HS)

10^{13} ohm (Durethan A 30 S, BKV 30 H; Ultramid A3, A3K, A3KN, A3 Pellets 24, A3SK, A3W, A3X1G7, A4, A4H, A4K, A5, B3, B3EG7, B3EG10, B3G7, B3G7HS, B3G10, B3K, B3L, B3S, B3WG7, B3WG10, B4, B4K, B5, B5W, B35, B35SK, B35W, KR 4250, KR 4405, KR 4406, KR 4407, KR 4411, KR 4412, KR 4609, KR-B3X2V6)

$> 10^{13}$ ohm (Ultramid A3X1G5, A3X1G10)

2×10^{13} ohm (Thermocomp PF-1008)

7.8×10^{13} ohm (Thermocomp RF-1008)

9×10^{13} ohm (Thermocomp PF-1006)

10^{14} ohm (Durethan AKV 30, BKV 30, BKV 50 H)

2.9×10^{14} ohm (Thermocomp RF-1006)

6.8×10^{14} ohm (Thermocomp QF-1008)

1.5×10^{15} ohm (Thermocomp QF-1006)

Nylon *(cont'd.)*

Arc Resist.:
70 s (Thermocomp R-1000FR-HS)
74 s (RTP 203M GB20)
77 s (RTP 203 GB20)
82 s (RTP 207H)
83 s (RTP 205H)
85 s (RTP 203H)
90 s (RTP 201H)
91 s (Thermocomp RF-1004FR-HS)
100 s (RTP 201, 201C, 203C; Thermocomp RF-1006FR-HS)
110 s (RTP 203, 203A, 203D, 205C, 225)
112 s (RTP 201A)
115 s (RTP 201D)
120 s (RTP 201B, 203B, 205, 205A, 205D, 205TFE5, 205TFE15, 205TFE20, 207, 207A, 207C, 207D, 209, 209A)
125 s (RTP 202M, 205B; Thermocomp QF-1006)
130 s (RTP 207B; Thermocomp QF-1008, RF-1006)
135 s (Thermocomp PF-1006, PF-1008, RF-1008)
140 s (RTP 227)

TOXICITY/HANDLING:
May cause burns or ignite other flamm. materials @ normal processing temps. ≥ 500 F (Celanese Nylon 1000-1, 1000-2, 1000-4, 1003-1, 1003-2, 1310-1, 1310-2, 1310-4, 1500-1, 1500-2, 1503-1, 1503-2, 1600-1, 1600-2, 1603-1, 1603-2)

STORAGE/HANDLING:
Hygroscopic—avoid moisture (Firestone Nylon 200-001, 210-001L, 213-001, 228-001, 415-001HS, 430-001HS, 540-001HS, 640-001HS)

Keep resin dry (Wellamid 2BRH-NW, 21LN2-NNT, 42LH-N, 42L-N, 42LN2-N, FR22F-N, FRGF25-66, FRGS25-66, GS25-66, GS40-66, GSF25/15-66)

STD. PKGS.:
50 lb net hermetically sealed bags (Firestone Nylon 200-001, 210-001L, 213-001, 228-001, 415-001HS, 430-001HS, 540-001HS, 640-001HS)

25-kg net multiwall, moistureproof bags (Grilamid ELY 20NZ, ELY 60, L16G, L16 GM, L20 FR, L20 G, L20 W20, L20 W40, L25, L25 natural 6112, L25 W10, L25 W20, L25 W40, L25 W40 NZ, LKN-5H, LV-3H, LV-43H, TR55 LX, TR55 UV; Grilon A23, A23 GM, A23 G natural 6165, A28, A28 GM, A28 NX, A28 NY, A28 NZ, A28 VO, A28 W10, BT40, CA6, CA6 EH, CF35, CR9, CR9 natural 6361, ELX 23NZ, F40, F47, PV-3H, PV-5H, PV-15H, PVN-3H, PVN-15H, PVZ-3H, R40, R40 GM, R40 N38, R47, R47HW, R50, W3941, W5744)

46-in. hanks; cut pieces 1 in. to 46 in. inclusive in increments of 1/16 in. (Herox 0200HA)

Multistrand rope on $12^1/_2$-lb spool (Herox 0200MA)

Single strand on $12^1/_2$-lb spool (Herox 0200SA)

PEG-4 (CTFA)

SYNONYMS:
Macrogol 200
PEG 200
POE (4)
Polyethylene glycol 200
Polyoxyethylene (4)

STRUCTURE:
$H(OCH_2CH_2)_nOH$
where avg. $n = 4$

CAS No.:
25322-68-3 (generic); 112-60-7
RD No.: 977007-39-8

TRADENAME EQUIVALENTS:
Alkapol PEG 200 [Alkaril]
Carbowax Sentry PEG 200 [Union Carbide]
Merpoxen PEG 200 [Kempen]
Nopalcol 200 [Henkel/Process]
Pluracol E-200 [BASF; BASF AG]

CATEGORY:
Polymer, lubricant, humectant, plasticizer, softener, intermediate, emollient, conditioner, binder, emulsifier, base, vehicle, extender, coupling agent, solvent, stabilizer, antistat

APPLICATIONS:
Cosmetic industry preparations (Alkapol PEG 200; Merpoxen PEG 200); creams and lotions (Carbowax Sentry PEG 200; Pluracol E-200); hair preparations (Alkapol PEG 200; Pluracol E-200); makeup (Carbowax Sentry PEG 200)

Industrial applications: dyes (Alkapol PEG 200); electropolishing (Pluracol E-200); hydraulic systems (Merpoxen PEG 200); inks (Alkapol PEG 200; Pluracol E-200); metalworking (Pluracol E-200); paper industry (Alkapol PEG 200; Merpoxen PEG 200); rubber industry (Pluracol E-200); sheet (Alkapol PEG 200); sizing (Pluracol E-200); starch paste (Alkapol PEG 200); surfactants processing (Alkapol PEG 200; Pluracol E-200); textile applications (Pluracol E-200); wood treatment (Pluracol E-200)

Pharmaceutical applications: (Carbowax Sentry PEG 200); antiperspirants (Pluracol

PEG-4 *(cont'd.)*

E-200); ointments (Pluracol E-200); suppositories (Pluracol E-200); tablets (Alkapol PEG 200; Pluracol E-200)

PROPERTIES:
Form:
Liquid (Alkapol PEG 200; Nopalcol 200)
Clear liquid (Merpoxen PEG 200; Pluracol E-200)
Color:
Colorless (Pluracol E-200)
Composition:
99% conc. (Nopalcol 200)
100% active (Alkapol PEG 200; Merpoxen PEG 200; Pluracol E-200)

GENERAL PROPERTIES:
Ionic Nature:
Nonionic (Alkapol PEG 200; Merpoxen PEG 200; Pluracol E-200)
Solubility:
Sol. in most organic solvents except aliphatic hydrocarbons (Pluracol E-200)
Sol. in water (Alkapol PEG 200; Carbowax Sentry PEG 200)
M.W.:
190–210 (Alkapol PEG 200)
200 (Carbowax Sentry PEG 200; Merpoxen PEG 200)
Sp. Gr.:
1.12 (Pluracol E-200)
Density:
1.13 g/ml (Alkapol PEG 200)
9.4 lb/gal (Pluracol E-200)
Visc.:
4.36 cs (210 F) (Pluracol E-200)
F.P.:
–48 C (Pluracol E-200)
Flash Pt.:
360 F (Pluracol E-200)
pH:
5–8 (5% DW) (Alkapol PEG 200)
6.5 (5% aq.) (Pluracol E-200)
Surface Tension:
57.2 dynes/cm (1% sol'n.) (Pluracol E-200)

STD. PKGS.:
200 kg net iron drums (Merpoxen PEG 200)
5 and 55-gal drums, tank cars, tank trucks (Pluracol E-200)

PEG-6 (CTFA)

SYNONYMS:
Macrogol 300
PEG 300
POE (6)
Polyethylene glycol 300
Polyoxyethylene (6)

STRUCTURE:
$H(OCH_2CH_2)_nOH$
where avg. $n = 6$

CAS No.:
25322-68-3 (generic)
RD No.: 977007-40-1

TRADENAME EQUIVALENTS:
Alkapol PEG 300 [Alkaril]
Carbowax Sentry PEG 300 [Union Carbide]
Emery 6687 [Henkel/Emery]
Lutrol E300 [BASF AG]
Merpoxen PEG 300 [Kempen]
PGE-300 [Hefti]
Pluracol E-300 [BASF]

CATEGORY:
Polymer, intermediate, binder, lubricant, plasticizer, softener, humectant, solvent, antistat, emollient, conditioner, base, stabilizer, vehicle, extender, coupling agent

APPLICATIONS:
Cosmetic industry preparations: (Alkapol PEG 300; Merpoxen PEG 300; Pluracol E-300); creams and lotions (Carbowax Sentry PEG 300); makeup (Carbowax Sentry PEG 300); toiletries (Pluracol E-300)
Electrical/electronic industry: (Emery 6687)
Food-contact applications: (Carbowax Sentry PEG 300)
Industrial applications: (Merpoxen PEG 300); adhesives (Emery 6687); coatings (Emery 6687); dyes (Alkapol PEG 300); inks (Alkapol PEG 300; PGE 300); lubricants (Emery 6687); metalworking (Emery 6687; Pluracol E-300); paper industry (Alkapol PEG 300); plastics processing (Alkapol PEG 300; PGE 300); printing (Emery 6687; PGE 300); rubber (Emery 6687; Pluracol E-300); sheet (Alkapol PEG 300); sizing (Pluracol E-300); starch pastes (Alkapol PEG 300); surfactants processing (Alkapol PEG 300; Pluracol E-300); textile applications (Alkapol PEG 300; Emery 6687; Pluracol E-300); wood treatment (Pluracol E-300)
Pharmaceutical applications: (Carbowax Sentry PEG 300; PGE 300; Pluracol E-300); tablets (Alkapol PEG 300); tablets (Alkapol PEG 300)

PEG-6 *(cont'd.)*

PROPERTIES:
Form:
> Liquid (Alkapol PEG 300; Lutrol E300; PGE 300)
> Clear liquid (Merpoxen PEG 300; Pluracol E-300)
> Clear viscous liquid (Emery 6687)

Color:
> Colorless (Pluracol E-300)
> Gardner < 1 (Emery 6687)

Composition:
> 100% active (Alkapol PEG 300; Emery 6687; Lutrol E300; Merpoxen PEG 300; PGE 300)

GENERAL PROPERTIES:
Ionic Nature:
> Nonionic (Emery 6687; Lutrol E300)

Solubility:
> Sol. in organic solvents except aliphatic hydrocarbons (Pluracol E-300); partly sol. in some organic solvents (Carbowax Sentry PEG 300)
> Sol. in water (Alkapol PEG 300; Carbowax Sentry PEG 300; Pluracol E-300); (@ 5%) (Emery 6687)
> Sol. in xylene (@ 5%) (Emery 6687)

M.W.:
> 285–315 (Alkapol PEG 300)
> 300 (Carbowax Sentry PEG 300; Merpoxen PEG 300)

Sp. Gr.:
> 1.12 (Pluracol E-300)

Density:
> 1.13 g/ml (Alkapol PEG 300)
> 9.4 lb/gal (Emery 6687; Pluracol E-300)

Visc.:
> 5.75 cs (210 F) (Pluracol E-300)
> 161.2 cSt (100 F) (Emery 6687)

Pour Pt.:
> –10 C (Emery 6687)

Flash Pt.:
> > 400 F (Emery 6687)
> 410 F (Pluracol E-300)

pH:
> 5–8 (5% DW) (Alkapol PEG 300)
> 5.7 (5% aq.) (Pluracol E-300)

Surface Tension:
> 62.9 dynes/cm (1%) (Pluracol E-300)

PEG-6-32 (CTFA)

SYNONYMS:
Macrogol 1500
PEG 1500
POE 1500
Polyethylene glycol 1500
Polyoxyethylene 1500

CAS No.:
25322-68-3 (generic)
RD No.: 977053-63-6

TRADENAME EQUIVALENTS:
Alkapol PEG 1500 [Alkaril]
Carbowax Sentry PEG 1450 [Union Carbide]
Lutrol E-1500 [BASF AG]
Merpoxen PEG 1500S [Kempen]
PGE-1500 [Hefti]
Pluracol E-1500 [BASF]

CATEGORY:
Polymer, lubricant, binder, solubilizer, resorption promoter, base, softener, intermediate, vehicle, extender, coupling agent, stabilizer, solvent, conditioner, antistat

APPLICATIONS:
Cosmetic industry preparations: (Merpoxen PEG 1500S; PGE-1500; Pluracol E-1500); creams and lotions (Carbowax Sentry PEG 1450; Pluracol E-1500); hair preparations (Pluracol E-1500); makeup (Carbowax Sentry PEG 1450)

Food-contact applications: (Carbowax Sentry PEG 1450)

Industrial applications: electropolishing (Pluracol E-1500); hydraulic systems (Merpoxen PEG 1500S); inks (Pluracol E-1500); latexes (Alkapol PEG 1500); metalworking (Pluracol E-1500); paper industry (Merpoxen PEG 1500S); rubber (Alkapol PEG 1500; Pluracol E-1500); sizing (Pluracol E-1500); surfactants processing (Pluracol E-1500); textile applications (Alkapol PEG 1500; Pluracol E-1500); wood treatment (Pluracol E-1500)

Pharmaceutical applications: (Carbowax Sentry PEG 1450; PGE-1500; Pluracol E-1500); antiperspirants (Pluracol E-1500); ointments (Pluracol E-1500); suppositories (Pluracol E-1500); tablets (Pluracol E-1500)

PROPERTIES:
Form:
Solid (Alkapol PEG 1500)
Microbeads (Lutrol E1500)
Flakes (Merpoxen PEG 1500S; PGE-1500)
Waxy solid (Pluracol E-1500)

Color:
White (Merpoxen PEG 1500S; Pluracol E-1500)

117

PEG-6-32 *(cont'd.)*

Composition:
100% active (Alkapol PEG 1500; Lutrol E1500; Merpoxen PEG 1500S; PGE-1500)

GENERAL PROPERTIES:

Ionic Nature:
Nonionic (Alkapol PEG 1500; Lutrol E1500; PGE-1500)

Solubility:
Sol. in most organic solvents except aliphatic hydrocarbons (Pluracol E-1500); partly sol. in some organic solvents (Carbowax Sentry PEG 1450)
Sol. in water (Alkapol PEG 1500; Carbowax Sentry PEG 1450; Pluracol E-1500)

M.W.:
1400–1600 (Alkapol PEG 1500)
1450 (Carbowax Sentry PEG 1450)
1500 (Pluracol E-1500)

Sp. Gr.:
1.20 (Pluracol E-1500)

Density:
10.0 lb/gal (Pluracol E-1500)

M.P.:
46.0–47.5 C (Pluracol E-1500)

Flash Pt.:
> 490 F (Pluracol E-1500)

Hydroxyl No.:
75.5 (Pluracol E-1500)

pH:
5–8 (Alkapol PEG 1500)
6.7 (5% aq.) (Pluracol E-1500)

Surface Tension:
62.8 dynes/cm (1% sol'n.) (Pluracol E-1500)

STD. PKGS.:

50-kg net poly bags (Merpoxen PEG 1500S)
55-gal (510 lb net) drums; bulk (Pluracol E-1500)

PEG-8 (CTFA)

SYNONYMS:

Macrogol 400
PEG 400
POE (8)
Polyethylene glycol 400
Polyoxyethylene (8)

STRUCTURE:

$H(OCH_2CH_2)_nOH$
where avg. n = 8

CAS No.:

25322-68-3 (generic)
RD No.: 977007-41-2

TRADENAME EQUIVALENTS:

Alkapol PEG 400 [Alkaril]
Carbowax Sentry PEG 400 [Union Carbide]
Droxol 400 [Henkel/Process]
Emery 6709 [Henkel/Emery]
Lutrol E 400 [BASF AG]
Merpoxen PEG 400 [Kempen]
PEG 400 [ICI PLC]
PGE-400 [Hefti Ltd.]
Pluracol E-400, E-400 NF [BASF]
Pluriol E400 [BASF AG]
Pogol 400 [Hart Chem. Ltd.]
Poly-G 400 [Olin]

CATEGORY:

Polymer, emollient, lubricant, solvent, base, binder, plasticizer, dispersant, intermediate, extender, coupling agent, stabilizer, conditioning agent, antistat, softener, humectant, viscosity modifier, thickener, mold release agent, defoamer, emulsifier, solubilizer

APPLICATIONS:

Cosmetic industry preparations: (Alkapol PEG 400; Carbowax Sentry PEG 400; Merpoxen PEG 400; PGE-400; Pluracol E-400, E-400 NF; Poly-G 400); aerosols (Poly-G 400); creams and lotions (Alkapol PEG 400; Carbowax Sentry PEG 400; Pluracol E-400); hair preparations (Pluracol E-400; Poly-G 400); makeup (Carbowax Sentry PEG 400; Poly-G 400); perfumery (Poly-G 400); personal care products (Poly-G 400)

Farm products: (Poly-G 400); herbicides/pesticides (Pogol 400)

FDA-approved applications: (Poly-G 400)

Food applications: (Carbowax Sentry PEG 400; Pluracol E-400; Poly-G 400)

Household detergents: (Pluriol E400; Poly-G 400)

Industrial applications: (Merpoxen PEG 400); adhesives (Pluriol E400; Poly-G 400); ceramics (Pluriol E400; Poly-G 400); coatings (Pluriol E400); dyes and pigments (PGE-400; Pluriol E400; Poly-G 400); electrical equipment (Poly-G 400); electropolishing for metals (Pluracol E-400, E-400 NF); EPA applications (Poly-G 400); metalworking (Pluracol E-400, E-400 NF; Pluriol E400; Poly-G 400); paint mfg. (Poly-G 400); paper mfg. (Pluracol E-400 NF; Pluriol E400; Poly-G 400); petroleum industry (Poly-G 400); photography (Poly-G 400); plastics (PGE-400);

119

PEG-8 *(cont'd.)*

printing inks (PGE-400; Pluracol E-400, E-400 NF; Pluriol E400; Poly-G 400); rubber (Pluracol E-400, E-400 NF; Poly-G 400); textile/leather processing (Pluracol E-400, E-400 NF; Pluriol E400; Poly-G 400); wood treatment (Pluracol E-400, E-400 NF; Poly-G 400)

Pharmaceutical applications: (Alkapol PEG 400; Carbowax Sentry PEG 400; PGE-400; Pluracol E-400, E-400 NF; Poly-G 400); antiperspirant/deodorant (Pluracol E-400; Poly-G 400); dental preparations (Poly-G 400); depilatories (Poly-G 400); ointments (Pluracol E-400; Poly-G 400); oral care products (Pluracol E-400 NF); suppositories (Pluracol E-400; Poly-G 400); tablet mfg. (Pluracol E-400; Poly-G 400)

PROPERTIES:

Form:

Liquid (Alkapol PEG 400; Emery 6709; Lutrol E 400; PGE-400; Pluracol E-400 NF; Poly-G 400)

Clear liquid (Merpoxen PEG 400; Pluracol E-400; Pluriol E400; Pogol 400)

Color:

Colorless (Pluracol E-400; Pluriol E400)

White (Alkapol PEG 400)

APHA 25 max. (Poly-G 400)

Composition:

100% active (Emery 6709; Lutrol E 400; Merpoxen PEG 400; PGE-400; Pluracol E-400; Pluriol E400; Pogol 400)

Solubility:

Sol. in acetone (Poly-G 400)

Sol. in ethanol (Poly-G 400)

Sol. in ethyl acetate (Poly-G 400)

Sol. in most organic solvents except aliphatic hydrocarbons (Pluracol E-400); partly sol. in some organic solvents (Carbowax Sentry PEG 400)

Sol. in toluene (Poly-G 400)

Sol. in water (Alkapol PEG 400; Carbowax Sentry PEG 400; Droxol 400; PEG 400; Pluracol E-400; Pluriol E400; Poly-G 400)

Ionic Nature:

Nonionic (Emery 6709; Lutrol E 400; PGE-400; Pluracol E-400)

M.W.:

380–420 (Alkapol PEG 400)

400 (Carbowax Sentry PEG 400; Merpoxen PEG 400; PEG 400; Pluracol E-400 NF; Poly-G 400)

Sp.gr.:

1.12 (Pluracol E-400; Pogol 400)

1.127 (20/20 C) (Poly-G 400)

1.13 (20/15.5 C) (Alkapol PEG 400)

Density:
  1.13 g/cc (20 C) (Pluriol E400)
  9.4 lb/gal (Pluracol E-400; Poly-G 400)
Visc.:
  7.3 cs (99 C) (Poly-G 400)
  7.4 cs (99 C) (Pluracol E-400 NF)
  7.39 cs (210 F) (Pluracol E-400)
  92 cps (Pogol 400)
  105–120 cs (20 C) (Pluriol E400)
F.P.:
  2.5 C (Pluracol E-400)
  4–8 C (Alkapol PEG 400)
  4–10 C (Poly-G 400)
M.P.:
  −10 to 10 C (Pluriol E400)
Pour Pt.:
  3–10 C (Poly-G 400)
  5 C (Pluracol E-400 NF)
Flash Pt.:
  > 150 C (Pluriol E400)
  182 C (COC) (Pluracol E-400 NF)
  224 C (COC) (Poly-G 400)
  460 F (Pluracol E-400)
Hydroxyl No.:
  266–294 (Poly-G 400)
Stability:
  Good physical and chemical stability (Alkapol PEG 400)
  Good heat stability in absence of oxygen; resistant to moderately strong acids, alkalis, and saturated sol'ns. (Pluriol E400)
  Does not hydrolyze or deteriorate; thermally stable (Poly-G 400)
Ref. Index:
  1.465 (Poly-G 400)
pH:
  Neutral (Pogol 400)
  5–8 (5% DW) (Alkapol PEG 400)
  6.0–7.5 (1% aq.) (Pluriol E400)
  6.2 (5% aq.) (Pluracol E-400)
Surface Tension:
  66.6 dynes/cm (1% sol'n.) (Pluracol E-400)

TOXICITY/HANDLING:
  Trace eye irritation, not a primary skin irritant by testing (Poly-G 400)
  Skin and eye irritant; avoid contact (Pogol 400)

PEG-8 *(cont'd.)*

STORAGE/HANDLING:
Store in original drums; tanks must be equipped with heating coils to maintain in fluid state; avoid contact of heated product with hot ignition source (Poly-G 400)

STD. PKGS.:
Drums, T/T (Pogol 400)
5- and 55-gal drums, tank cars, tank trucks (Pluracol E-400)
55-gal (500 lb net) drums, tank cars, tank trucks (Poly-G 400)

PEG-12 (CTFA)

SYNONYMS:
Macrogol 600
PEG 600
POE (12)
Polyethylene glycol 600
Polyoxyethylene (12)

STRUCTURE:
$H(OCH_2CH_2)_nOH$
where avg. $n = 12$

CAS No.:
25322-68-3 (generic)
RD No.: 977053-61-4

TRADENAME EQUIVALENTS:
Alkapol PEG 600 [Alkaril]
Carbowax Sentry PEG 600 [Union Carbide]
Emery 6686 [Henkel/Emery]
Merpoxen PEG 600 [Kempen]
PGE-600 [Hefti]
Pluracol E-600, E-600 NF [BASF; BASF AG]

CATEGORY:
Polymer, lubricant, emollient, conditioner, softener, solvent, humectant, binder, viscosity modifier, stabilizer, intermediate, antistat, base, coupling agent, plasticizer, vehicle, extender, mold release agent, defoamer

APPLICATIONS:
Cleansers: (Pluracol E-600)
Consumer products: footwear (Pluracol E-600)
Cosmetic industry preparations: (Merpoxen PEG 600; PGE-600; Pluracol E-600); creams and lotions (Carbowax Sentry PEG 600; Pluracol E-600); hair preparations (Pluracol E-600); makeup (Carbowax Sentry PEG 600)

Food-contact applications: (Carbowax Sentry PEG 600; Pluracol E-600)

Industrial applications: (Merpoxen PEG 600); alkyd resins (PGE-600); electropolishing (Pluracol E-600); hoses (Pluracol E-600); inks (Pluracol E-600); latex foams (Pluracol E-600); metalworking (Pluracol E-600); paper coatings (PGE-600); petroleum industry (Alkapol PEG 600); pigments (PGE-600); plastics (PGE-600); printing (Pluracol E-600); rubber (Pluracol E-600); sizing (Pluracol E-600); surfactants processing (Alkapol PEG 600; Pluracol E-600); textile applications (Alkapol PEG 600; Pluracol E-600); wood pulping/treatment (Pluracol E-600)

Pharmaceutical applications: (Carbowax Sentry PEG 600; PGE-600; Pluracol E-600); depilatories (Pluracol E-600); ointments (Pluracol E-600); suppositories (Pluracol E-600); tablets (Alkapol PEG 600; Pluracol E-600)

PROPERTIES:

Form:

Liquid (Alkapol PEG 600; Emery 6686; Pluracol E-600 NF)

Clear liquid (Pluracol E-600)

Semisolid (Pluriol E 600)

Clear paste (Merpoxen PEG 600)

Pasty (PGE-600)

Color:

Colorless (Pluracol E-600)

White (Alkapol PEG 600)

Gardner < 1 (Emery 6686)

Odor:

Odorless (Pluracol E-600)

Composition:

100% active (Alkapol PEG 600; Emery 6686; Merpoxen PEG 600; PGE-600; Pluracol E-600)

GENERAL PROPERTIES:

Ionic Nature:

Nonionic (Alkapol PEG 600; Emery 6686; Merpoxen PEG 600; PGE-600; Pluracol E-600)

Solubility:

Sol. in most organic solvents except aliphatic hydrocarbons; partly sol. in some organic solvents (Carbowax Sentry PEG 600)

Sol. in water (Alkapol PEG 600; Carbowax Sentry PEG 600; Pluracol E-600); (@ 5%) (Emery 6686)

Sol. in xylene (@ 5%) (Emery 6686)

M.W.:

570–630 (Alkapol PEG 600)

600 (Carbowax Sentry PEG 600; Merpoxen PEG 600; Pluracol E-600, E-600 NF; Pluriol E 600)

Sp. Gr.:

1.12 (Pluracol E-600)

PEG-12 *(cont'd.)*

Density:
   1.13 g/ml (Alkapol PEG 600)
   9.4 lb/gal (Emery 6686; Pluracol E-600)
Visc.:
   10.8 cs (99 C) (Pluracol E-600 NF)
   10.83 cSt (210 F) (Pluracol E-600)
   63 cSt (100 F) (Emery 6686)
F.P.:
   16.5 C (Pluracol E-600)
   20–25 C (Alkapol PEG 600)
Pour Pt.:
   20 C (Pluracol E-600 NF)
   22 C (Emery 6686)
Flash Pt.:
   249 C (COC) (Pluracol E-600)
   475 F (Emery 6686)
   480 F (Pluracol E-600)
Hydroxyl No.:
   183 (Pluracol E-600)
pH:
   5–8 (5% DW) (Alkapol PEG 600)
   5.3 (5% aq.) (Pluracol E-600)
Surface Tension:
   65.2 dynes/cm (1% sol'n.) (Pluracol E-600)
STD. PKGS.:
   5 and 55-gal drums, tank cars, tank trucks (Pluracol E-600)

PEG-20 (CTFA)

SYNONYMS:
   PEG 1000
   POE (20)
   Polyethylene glycol 1000
   Polyoxyethylene (12)
STRUCTURE:
   $H(OCH_2CH_2)_nOH$
      where avg. $n = 20$
CAS No.:
   25322-68-3 (generic)
   RD No.: 977053-62-5

124

TRADENAME EQUIVALENTS:

Carbowax Sentry PEG 1000 [Union Carbide]
Merpoxen PEG 1000 [Kempen]
PEG 1000 [ICI PLC]
PGE-1000 [Hefti]

CATEGORY:

Polymer, lubricant, solubilizer, coupling agent, hydrotrope, plasticizer, release agent, dispersant, emollient, softener, humectant, solvent, antistat, base

APPLICATIONS:

Agriculture industry: (Carbowax Sentry PEG 1000)

Cosmetic industry preparations: (Merpoxen PEG 1000; PGE-1000); creams and lotions (Carbowax Sentry PEG 1000; PGE-1000); makeup (Carbowax Sentry PEG 1000); shaving preparations (PGE-1000); skin preparations (PGE-1000)

Food-contact applications: (Carbowax Sentry PEG 1000)

Industrial applications: adhesives (Carbowax Sentry PEG 1000); ceramics (Carbowax Sentry PEG 1000); hydraulic systems (Merpoxen PEG 1000); paper coatings (Carbowax Sentry PEG 1000); paper industry (Merpoxen PEG 1000); plastics/ resins (Carbowax Sentry PEG 1000; PGE-1000); textile applications (Carbowax Sentry PEG 1000)

Pharmaceutical applications: (Carbowax Sentry PEG 1000); suppositories (PGE-1000); tablets (Carbowax Sentry PEG 1000); tooth paste (PGE-1000)

PROPERTIES:

Form:
Solid (PGE-1000)
Wax (Merpoxen PEG 1000)
Color:
White (Merpoxen PEG 1000)
Composition:
100% active (Merpoxen PEG 1000; PGE-1000)

GENERAL PROPERTIES:

Ionic Nature:
Nonionic (PGE-1000)
Solubility:
Partly sol. in some organic solvents (Carbowax Sentry PEG 1000)
Sol. in water (Carbowax Sentry PEG 1000)
M.W.:
1000 (Carbowax Sentry PEG 1000; Merpoxen PEG 1000; PEG 1000)

STD. PKGS.:

200 kg net iron drums (Merpoxen PEG 1000)

PEG-75 (CTFA)

SYNONYMS:
Macrogol 4000
PEG 4000
POE (75)
Polyethylene glycol 4000
Polyoxyethylene (75)

STRUCTURE:
$H(OCH_2CH_2)_n OH$
where avg. $n = 75$

CAS No.:
25322-68-3 (generic)
RD No.: 977007-43-4

TRADENAME EQUIVALENTS:
Alkapol PEG 3350 [Alkaril]
Carbowax Sentry PEG 3350 [Union Carbide]
Lutrol E 4000 [BASF AG]
Pluracol E-4000, E-4000 NF [BASF]

CATEGORY:
Polymer, lubricant, mold release agent, binder, base, plasticizer, vehicle, extender, coupling agent, solubilizer, softener, defoamer, conditioner, antistat, intermediate, resorption promoter

APPLICATIONS:
Cleansers: (Pluracol E-4000)

Consumer products: footwear (Pluracol E-4000)

Cosmetic industry preparations: (Pluracol E-4000); creams and lotions (Carbowax Sentry PEG 3350; Pluracol E-4000); hair preparations (Pluracol E-4000); makeup (Carbowax Sentry PEG 3350)

Food-contact applications: (Carbowax Sentry PEG 3350; Pluracol E-4000)

Industrial applications: electropolishing (Pluracol E-4000); hoses (Pluracol E-4000); inks (Pluracol E-4000); latex foams (Pluracol E-4000); metalworking (Pluracol E-4000); printing (Pluracol E-4000); rubber (Alkapol PEG 3350; Pluracol E-4000); sizing (Pluracol E-4000); surfactants processing (Pluracol E-4000); textiles (Pluracol E-4000); tires (Pluracol E-4000); wood pulping/treatment (Pluracol E-4000)

Pharmaceutical applications: (Carbowax Sentry PEG 3350; Pluracol E-4000); depilatories (Pluracol E-4000); ointments (Pluracol E-4000); suppositories (Pluracol E-4000); tablets (Alkapol PEG 3350; Pluracol E-4000)

PROPERTIES:
Form:
Flakes (Alkapol PEG 3350)
Flakes or powder (Carbowax Sentry PEG 3350)
Microbeads (Lutrol E 4000)

Waxy solid (Pluracol E-4000, E-4000 NF)
Color:
White (Pluracol E-4000, E-4000 NF)
Composition:
100% active (Alkapol PEG 3350; Lutrol E 4000)
GENERAL PROPERTIES:
Ionic Nature:
Nonionic (Alkapol PEG 3350; Lutrol E 4000)
Solubility:
Sol. in most organic solvents except aliphatic hydrocarbons (Pluracol E-4000)
Sol. in water (Alkapol PEG 3350; Pluracol E-4000)
M.W.:
3000–3700 (Alkapol PEG 3350)
3350 (Carbowax Sentry PEG 3350)
4000 (Pluracol E-4000, E-4000 NF)
Sp. Gr.:
1.20 (Pluracol E-4000)
Density:
10.0 lb/gal (Pluracol E-4000)
Visc.:
134 cs (99 C) (Pluracol E-4000 NF)
M.P.:
59 C (Pluracol E-4000 NF)
59.5 C (Pluracol E-4000)
Flash Pt.:
> 260 C (COC) (Pluracol E-4000 NF)
> 490 F (Pluracol E-4000)
Hydroxyl No.:
28.6 (Pluracol E-4000)
pH:
5–8 (5% DW) (Alkapol PEG 3350)
6.7 (5% aq.) (Pluracol E-4000)
Surface Tension:
61.9 dynes/cm (1% sol'n.) (Pluracol E-4000)
STD. PKGS.:
55-gal (250 lb net) drums, bulk (Pluracol E-4000)

PEG-150 (CTFA)

SYNONYMS:
Macrogol 6000
PEG 6000
POE (150)
Polyethylene glycol 6000
Polyoxyethylene (150)

STRUCTURE:
$H(OCH_2CH_2)_nOH$
where avg. $n = 150$

CAS No.:
25322-68-3 (generic)
RD No.: 977053-64-7

TRADENAME EQUIVALENTS:
Carbowax Sentry PEG 8000 [Union Carbide]
Lutrol E 6000 [BASF AG]
Merpoxen PEG 6000S [Kempen]
Pluracol E-6000 [BASF]

CATEGORY:
Polymer, lubricant, intermediate, softener, binder, solubilizer, resorption promoter, plasticizer, base, vehicle, extender, coupling agent, release agent, defoamer, conditioner, antistat

APPLICATIONS:
Cleansers: (Pluracol E-6000)
Consumer products: footwear (Pluracol E-6000)
Cosmetic industry preparations: (Merpoxen PEG 6000S; Pluracol E-6000); creams and lotions (Carbowax Sentry PEG 8000; Pluracol E-6000); hair preparations (Pluracol E-6000); makeup (Carbowax Sentry PEG 8000)
Food-contact applications: (Carbowax Sentry PEG 8000 Pluracol E-6000)
Industrial applications: electropolishing (Pluracol E-6000); hoses (Pluracol E-6000); hydraulic systems (Merpoxen PEG 6000S); inks (Pluracol E-6000); latex foams (Pluracol E-6000); metalworking (Pluracol E-6000); paper industry (Merpoxen PEG 6000S); printing (Pluracol E-6000); rubber (Pluracol E-6000); sizing (Pluracol E-6000); surfactants processing (Pluracol E-6000); tires (Pluracol E-6000); wood pulping/treatment (Pluracol E-6000)
Pharmaceutical applications: (Carbowax Sentry PEG 8000; Pluracol E-6000); depilatories (Pluracol E-6000); ointments (Pluracol E-6000); suppositories (Pluracol E-6000); tablets (Pluracol E-6000)

PROPERTIES:
Form:
Microbeads (Lutrol E 6000)
Flakes (Merpoxen PEG 6000S)

128

Flakes or powder (Carbowax Sentry PEG 8000)
Waxy solid (Pluracol E-6000)
Color:
White (Pluracol E-6000)
Composition:
100% active (Lutrol E 6000; Merpoxen PEG 6000S)
GENERAL PROPERTIES:
Ionic Nature:
Nonionic (Lutrol E 6000; Merpoxen PEG 6000S)
Solubility:
Sol. in most organic solvents except aliphatic hydrocarbons (Pluracol E-6000); partly
sol. in some organic solvents (Carbowax Sentry PEG 8000)
Sol. in water (Carbowax Sentry PEG 8000; Pluracol E-6000)
M.W.:
6000 (Merpoxen PEG 6000S)
8000 (Carbowax Sentry PEG 8000)
Sp. Gr.:
1.21 (Pluracol E-6000)
Density:
10.0 lb/gal (Pluracol E-6000)
M.P.:
61 C (Pluracol E-6000)
Flash Pt.:
> 490 F (Pluracol E-6000)
pH:
6.7 (5% aq.) (Pluracol E-6000)
Surface Tension:
62.1 dynes/cm (1% sol'n.) (Pluracol E-6000)
STD. PKGS.:
55 gal drums, tank cars, tank trucks (Pluracol E-6000)

PEG-45/dodecyl glycol copolymer (CTFA)

STRUCTURE:

$$H(OCHCH_2)_xO(CH_2CH_2O)_y(CH_2CHO)_zH$$
$$C_{10}H_{21} \qquad\qquad C_{10}H_{21}$$

where   avg. $x = 11$
avg. $y = 45$
avg. $z = 11$

PEG-45/dodecyl glycol copolymer *(cont'd.)*

TRADENAME EQUIVALENTS:
Elfacos ST9 [Akzo Chemie]
CATEGORY:
Copolymer, stabilizer, emollient
APPLICATIONS:
Cosmetic industry preparations: decorative cosmetics (Elfacos ST9)
PROPERTIES:
Form:
Paste (Elfacos ST9)
Composition:
100% conc. (Elfacos ST9)
Ionic Nature:
Nonionic (Elfacos ST9)
HLB:
7.0 (Elfacos ST9)

Phenol-formaldehyde resin

SYNONYMS:
Phenolic resin
One-step:
A-stage resin
One-stage resin
Resole
Two-step
Novalac resin
Novalak resin
Phenol-formaldehyde novolak resin
Two-stage resin
STRUCTURE:

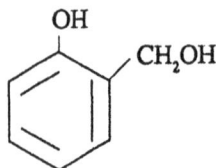

Phenol-formaldehyde resin *(cont'd.)*

TRADENAME EQUIVALENTS:

One-step:
Akrochem P-478 [Akron]
Durez 17702, 30417, 30698, 30806 [Occidental]
PA-52-333, -52-334 [Polymer Applications]
SP-8014 [Schenectady]
Varcum 1417, 1481, 1494, 5145, 5160, 5302, 5485, 6820, 8366 [Reichhold]
Asbestos-filled:
RX 468A [Rogers]
Cellulose-filled:
FM 1389, 14111 [Fiberite]
Cotton flock-filled:
Plenco 368 Black, 417 Black, 476 Black [Plastics Engineering]
Glass-reinforced:
FM 16771 (glass roving), 21288 (glass roving) [Fiberite]
RX 640, 655A, 867A (glass fiber) [Rogers]
Mineral/flock-filled:
Plenco 527 Black, 586 Brown [Plastics Engineering]
Woodflour-filled:
Plenco 369 Black, 480 Black, 557 Black, 567 Black, 4543 Red [Plastics Engineering]
Two-step:
Durez 115, 157, 16038, 21206, 30959 [Occidental]
FM 510 [Fiberite]
Phenolic Molding Compound 868 [Shenango Phenolics]
Varcum 647, 1359, 1364, 4326, 4631, 4727, 9836A, 9874, C-86, DR-406 [Reichhold]
Filled:
Plenco 507 Black [Plastics Engineering]
Aluminum-filled:
Plenco 201, 203 [Plastics Engineering]
Asbestos-filled:
FM 6101 (asbestos yarn), 11547 (asbestos yarn), 17610 (asbestos yarn) [Fiberite]
RX 462, 465, 466, 467, 468, 490, 495 [Rogers]
Asbestos/TFE-filled:
RX 363 [Rogers]
Cellulose-filled:
FM 3000, 3001, 3002, 4201, 7676, 10365 [Fiberite]
RX 429 (cellulose fiber), 448 (cellulose fiber), 525 (cellulose fiber), 950 (cellulose fiber) [Rogers]
Cellulose/glass-filled:
FM 5367 [Fiberite]
Cellulose/graphite-filled:
RX 340, 350, 352 [Rogers]

Phenol-formaldehyde resin (cont'd.)

Cellulose/TFE-filled:
RX 342 [Rogers]
Fabric/fiber-reinforced:
Durez 31735 [Occidental]
FM 1132 (macerated fabric), 1132P (macerated fabric), 3510 (cotton fabric), 7700 (cotton fabric), 9294 (cotton cord) [Fiberite]
Plenco 321 Black (cotton flock), 554 Black (cotton flock) [Plastics Engineering]
RX 431 (rag fiber), 475 (rag fiber) [Rogers]
Glass-reinforced:
Durez 31219 [Occidental]
FM 4004, 4004F, 4005, 4005F, 4007, 4007F, 4030-190 (glass roving), 404464 (glass roving), 8130 (glass roving) [Fiberite]
RX 610N, 611 (short glass), 625, 630 (short glass), 650, 655, 660 (short glass), 842, 850, 862 (glass fiber), 865 (glass fiber), 866D (glass fiber), 867 (glass fiber), 867D (glass fiber) [Rogers]
Mica-filled:
Plenco 343 Natural [Plastics Engineering]
Mineral-filled:
Plenco 349 Black, 4443 (iron powder) [Plastics Engineering]
Mineral/flock-filled:
Plenco 414 Black, 466 Black, 466 Brown, 467 Black, 509 Black, 509 Brown [Plastics Engineering]
Mineral/graphite-filled:
Plenco 118-AF Dark Grey, 118 Dark Grey [Plastics Engineering]
Mineral/woodflour-filled:
Plenco 400 Black, 400 Brown, 485 Black, 485 Brown, 548 Black [Plastics Engineering]
Nylon-reinforced:
FM 1303 [Fiberite]
Woodflour-filled:
Plenco 300 Black, 300 Brown, 300 Red, 308 Black, 308 Brown, 317 Red, 320 Green, 411 Walnut Mottle, 482 Black, 512 Black, 535 Black, 571 Black, 579 Black [Plastics Engineering]
Woodflour/flock-filled:
Plenco 307 Black, 500 Black, 500 Brown, 523 Black [Plastics Engineering]

MODIFICATIONS/SPECIALTY GRADES:

General purpose molding grade:
Durez 115, 16038, 21206; Plenco 300 Black, 300 Brown, 300 Red, 307 Black, 308 Black, 308 Brown, 317 Red, 320 Green, 369 Black, 411 Walnut Mottle, 480 Black, 482 Black, 512 Black, 535 Black, 567 Black, 571 Black, 4543 Red
High performance engineering grade:
RX 340, 342, 350, 352, 363, 429, 431, 448, 462, 465, 466, 467, 468, 468A, 475, 490, 495, 525, 610N, 611, 625, 630, 640, 650, 655, 655A, 660

Phenol-formaldehyde resin *(cont'd.)*

Electrical grade:
Durez 157, 30698, 30959; Plenco 343 Natural, 414 Black, 482 Black, 509 Black, 509 Brown, 548 Black, 557 Black, 586 Brown

RF shielding:
Plenco 4443

Impact grade:
Durez 31735; FM 3510, 4030-190, 5064, 5367, 6101, 8130, 9294, 10365, 16771, 21288; Plenco 321 Black, 368 Black, 417 Black, 467 Black, 476 Black, 500 Black, 500 Brown, 507 Black, 523 Black, 579 Black; RX 525, 862, 865, 866D, 867, 867A, 867D, 950

Flame-retardant:
Durez 115, 157, 16038, 30698, 30959; Plenco 586 Brown; RX 462, 465, 466, 467, 468, 468A, 490, 495, 525, 610N, 611, 625, 630, 640, 650, 660, 842, 850, 862, 865, 866D, 867, 950

Heat-resistant:
Durez 157, 30698, 30959; Phenolic Molding Compound 868; Plenco 201, 203, 349 Black, 400 Black, 400 Brown, 414 Black, 466 Black, 466 Brown, 485 Black, 485 Brown, 509 Black, 509 Brown, 527 Black, 548 Black; Varcum 5145

Shock resistant:
FM 1132; Plenco 368 Black, 467 Black; RX 429, 448, 525, 950

Lubricated:
FM 1389; Plenco 554 Black (Teflon); RX 340 (graphite), 342 (TFE), 350 (graphite), 352 (TFE), 363 (TFE), 842

CATEGORY:
Thermosetting resin

PROCESSING:

Compression molding:
Durez 115, 157, 16038, 17702, 21206, 30417, 30698, 30806, 30959, 31219, 31735; FM 1132, 1132P, 4004F, 4005, 4005F, 4007F; Plenco 118-AF Dark Grey, 118 Dark Grey, 201, 203, 300 Black, 300 Brown, 300 Red, 307 Black, 308 Black, 308 Brown, 317 Red, 320 Green, 321 Black, 343 Natural, 349 Black, 368 Black, 369 Black, 400 Black, 411 Walnut Mottle, 414 Black, 417 Black, 466 Black, 466 Brow, 467 Black, 476 Black, 480 Black, 482 Black, 485 Black, 485 Brown, 500 Black, 500 Brown, 507 Black, 509 Black, 509 Brown, 512 Black, 523 Black, 527 Black, 535 Black, 548 Black, 554 Black, 557 Black, 567 Black, 571 Black, 579 Black, 586 Brown, 4443, 4543 Red; RX 429, 431, 448, 475, 525, 610N, 611, 625, 630, 640, 650, 655, 655A, 660, 862, 865, 950

Injection molding:
Durez 115, 157, 30959, 31219; FM 4004, 4007; Phenolic Molding Compound 868; Plenco 300 Black, 308 Black, 317 Red, 349 Black, 369 Black, 400 Brown, 414 Black, 466 Black, 466 Brown, 480 Black, 482 Black, 485 Black, 485 Brown, 500 Black, 507 Black, 509 Black, 512 Black, 527 Black, 535 Black, 548 Black, 571 Black, 579 Black; RX 429, 431, 448, 475, 525, 862, 865, 950

Phenol-formaldehyde resin (cont'd.)

Molding:
FM 510, 1303, 1389, 3000, 3001, 3002, 3510, 4030-190, 4201, 5064, 5367, 6101, 7676, 7700, 8130, 9294, 10365, 11547, 14111, 17610, 16771, 21288; RX 842, 850

Transfer molding:
Durez 115, 157, 16038, 21206, 30698, 30806, 30959, 31219, 31735; FM 1132, 1132P, 4004F, 4005, 4005F, 4007F; Phenolic Molding Compound 868; Plenco 118-AF Dark Grey, 118 Dark Grey, 300 Black, 300 Brown, 300 Red, 307 Black, 308 Black, 308 Brown, 343 Natural, 349 Black, 368 Black, 369 Black, 400 Black, 400 Brown, 414 Black, 417 Black, 466 Black, 466 Brown, 467 Black, 476 Black, 480 Black, 482 Black, 485 Black, 485 Brown, 500 Black, 500 Brown, 507 Black, 509 Black, 509 Brown, 512 Black, 523 Black, 527 Black, 535 Black, 548 Black, 535 Black, 548 Black, 571 Black, 579 Black, 586 Brown; RX 429, 431, 448, 475, 525, 862, 865, 950

APPLICATIONS:

Automotive applications: (Phenolic Molding Compound 868; RX 610N, 611, 625, 630, 640, 650, 655, 655A, 660); braking systems (Durez 115, 16038, 21206); ignition parts (Plenco 557 Black); transmission parts (Durez 115, 16038, 21206); wiring (Plenco 523 Black)

Consumer products: appliances (Phenolic Molding Compound 868); appliance parts (RX 429, 431, 448, 475, 525); rifle parts (FM 3000, 4030-190, 4201, 5064, 7676, 7700, 8130, 9294)

Electrical/electronic industry: (Plenco 414 Black; RX 462, 465, 466, 467, 490, 495, 525, 610N, 611, 625, 630, 640, 650, 655, 655A, 660; Varcum 1359); circuit breakers (FM 6101, 11547, 17610; RX 462, 465, 466, 467, 490, 495, 525, 610N, 611, 625, 630, 640, 650, 655, 655A, 660); commutators (Durez 31219; RX 462, 465, 466, 467, 490, 495, 525, 866D); computer frames (FM 4004, 4005, 4007; RX 610N, 611, 625, 630, 640, 650, 655, 655A, 660); connectors (FM 4004, 4005, 4007; RX 610N, 611, 625, 630, 640, 650, 655, 655A, 660); control devices (Plenco 509 Black, 509 Brown); housings (FM 4004, 4005, 4007; RX 462, 465, 466, 467, 490, 495, 525, 610N, 611, 625, 630, 640, 650, 655, 655A, 660); insulation (FM 4004, 4005, 4007); power tools (FM 4030-190, 5064, 8130); switchgear (FM 3000, 3001, 3002, 4030-190, 4201, 5064, 6101, 7676, 7700, 8130, 9294, 11547, 17610; Plenco 369 Black); wiring (Plenco 509 Black, 509 Brown)

Food-contact applications: food packaging (Varcum 1417, 1481, 1494, 8366); USDA-approved (Varcum 6820)

Functional additives: plasticizer (Akrochem P-478; PA-52-333, -52-334); processing aid (Varcum DR-406); reinforcer (PA-52-333, -52-334; Varcum C-86); softener (PA-52-333, -52-334); tackifier (Varcum 647, 4727, DR-406)

Industrial applications: (Plenco 523 Black, 579 Black); adhesives (Akrochem P-478; SP-8014; Varcum 1364, 5145, 5302, 5485); agitators (Plenco 476 Black); bearings and bushings (FM 1303, 1389, 3000, 3001, 6101, 7676, 7700, 9294, 11547, 17610, 14111); binders (Varcum 5160); closures (Durez 30417; Plenco 567 Black, 4543 Red); conveyor wheels (FM 3000, 3001, 3002, 4030-190, 4201, 5064, 7676, 7700, 8130, 9294); die cast parts (RX 862, 865); epoxy novolac mfg. (Varcum 4326);

Phenol-formaldehyde resin *(cont'd.)*

frictional/sliding applications (Plenco 554 Black; RX 340, 342, 350, 352, 363); gears (FM 1303, 1389, 3000, 3001, 7676, 7700, 9294, 14111); grinding wheels (Plenco 201, 203); hardware finishes (Varcum 1494); heavy-duty machine parts (FM 13389, 14111); housings (FM 1389, 4030-190, 5064, 8130, 14111); insulators (FM 3000, 3001, 3002, 4201, 7676. 7700, 9294; Phenolic Molding Compound 868); instrument housings (FM 3000, 3001, 3002, 4201, 7676, 7700, 9294); intricate/hard-to-fill moldings (FM 4005); mechanical parts (FM 4004, 4005, 4007); metal coatings (Varcum 1417, 1481, 1494, 6820, 8366); motor starters (FM 6101, 11547, 17610); pulleys (Durez 31735; FM 1303, 1389, 3000, 3001, 7676, 7700, 9294, 14111); pump housings (FM 1303; Plenco 417 Black, 476 Black); pump impellers (Plenco 467 Black); rubber (Akrochem P-478; SP-8014; Varcum 647, 1364, 4727, 5145, 5160, 5302, 5485, 9874, C-86, DR-406); sealing rings (Plenco 118-AF Dark Grey, 118 Dark Grey); terminal strips (FM 4004, 6101, 11547, 17610); textile applications (FM 1389, 3000, 3001, 7676, 7700, 9294, 14111); trays (FM 1389, 14111); vaporizers (Plenco 417 Black, 476 Black); wear applications (FM 1389, 3000, 3001, 7676, 7700, 9294; Plenco 554 Black); welding gun tips (FM 6101, 11547, 17610); wet/dry applications (Durez 17702, 30806; Plenco 417 Black, 476 Black, 480 Black, 507 Black, 527 Black)

Military applications: missile parts (FM 4030-190, 5064, 8130)

PROPERTIES:

Form:

Liquid (Varcum 1481, 1494, 5160, 5302)

Powder (Akrochem P-478; PA-52-334; SP-8014; Varcum 1359, 1364, 4631, 4727, 9874, C-86)

Fine powder (Plenco 201)

Granular (Durez 115, 157, 16038, 17702, 21206, 30417, 30698, 30806, 30959, 31735; FM 1303, 4004, 4005, 4007; Phenolic Molding Compound 868)

Coarse granular (RX-850)

Nodular (Durez 31219; FM 510, 1132P, 3000, 3001, 3002, 4201, 7676, 7700; RX 866D, 867, 867A, 867D)

Crushed (Varcum 1417, 5145, 5485)

Lumps (PA-52-333; Varcum 4326, 6820, 8366, 9836A)

Flakes (Varcum 647, DR-406)

Coarse flakes (RX 862, 865)

Macerated (FM 1132, 1389, 3510, 5367, 10365, 14111)

Chopped fiber (FM 4030-190, 6101)

Chopped fiber avail. in lengths from $1/_4$–$1 1/_2$ in. (FM 16771, 21288)

Chopped fiber avail. in lengths from $1/_2$–$1 1/_2$ in. (FM 5064, 8130)

Chopped strands (FM 9294, 11547, 17610)

Color:

Off-white (SP-8014)

Natural (FM 17610; Plenco 343 Natural; RX 866D, 867A, 867D)

Natural, black (FM 4030-190, 6101, 8130, 9294)

Phenol-formaldehyde resin (cont'd.)

Natural, black, brown (FM 1132, 3510)
Natural, black, green (FM 1132P)
Yellow, black (FM 5064)
Green (FM 4007; Plenco 320 Green)
Green, black (FM 1389, 4004, 4004F, 4005, 4005F, 4007F)
Red (Plenco 300 Red, 317 Red, 4543 Red)
Tan (FM 14111; PA-52-334)
Tan, black (FM 10365)
Brown (Durez 16038; Plenco 300 Brown, 308 Brown, 400 Brown, 466 Brown, 485 Brown, 500 Brown, 509 Brown, 586 Brown)
Brown to dark brown (PA-52-333)
Brown, black (Durez 21206; FM 1303)
Walnut mottle (Plenco 411 Walnut Mottle)
Dark grey (Plenco 118-AF Dark Grey, 118 Dark Grey)
Black (Durez 115, 157, 17702, 30417, 30698, 30806, 30959, 31219, 31735; FM 510, 3000, 3001, 3002, 4201, 5367, 7676, 7700, 11547; Plenco 300 Black, 307 Black, 308 Black, 321 Black, 349 Black, 368 Black, 369 Black, 400 Black, 414 Black, 417 Black, 466 Black, 467 Black, 476 Black, 480 Black, 482 Black, 485 Black, 500 Black, 507 Black, 509 Black, 512 Black, 523 Black, 527 Black, 535 Black, 548 Black, 554 Black, 557 Black, 567 Black, 571 Black, 579 Black; RX 862, 865, 867)

Odor:
Phenolic (PA-52-333)

Composition:
50–54% nonvolatiles in ethanol (Varcum 1481)
52–56% nonvolatiles in ethanol (Varcum 1494)
63–66% nonvolatiles in MEK (Varcum 5302)
65–69% nonvolatiles in MEK (Varcum 5160)

GENERAL PROPERTIES:

Solubility:
Sol. in alcohols (Akrochem P-478; Varcum 1364, 1481, 1494, 4727, 5145, 5160, 5302, 8366, C-86); partly sol. (Varcum 647, 1417, 5485, 6820, 9836A, 9874, DR-406)
Sol. in aliphatics (Varcum 647, 4727, DR-406); insol. (Varcum 1364, 1417, 1481, 1494, 5145, 5160, 5302, 5485, 6820, 8366, 9836A, 9874, C-86)
Sol. in aromatics (Varcum 647, 4727, DR-406); partly sol. (Varcum 5160, 5302); insol. (Varcum 1364, 1417, 1481, 1494, 5145, 5485, 6820, 8366, 9836A, 9874, C-86)
Sol. in esters (Akrochem P-478; Varcum 647, 1364, 1417, 1481, 1494, 4727, 5145, 5160, 5302, 5485, 6820, 8366, 9836A, 9874, C-86, DR-406)
Sol. in ketones (Akrochem P-478; Varcum 647, 1364, 1417, 1481, 1494, 4727, 5145, 5160, 5302, 5485, 8620, 8366, 9836A, 9874, C-86, DR-406)

Sp. Gr.:
1.02–1.07 (Varcum 1481)
1.04–1.06 (Varcum 1494)
1.14–1.18 (PA-52-333, -52-334)

Phenol-formaldehyde resin *(cont'd.)*

1.190–1.210 (Varcum 1417)
1.210–1.220 (Varcum 8366)
1.225–1.235 (Varcum 6820)
1.23–1.25 (FM 1303 black)
1.25 (Akrochem P-478; SP-8014)
1.27–1.29 (FM 1303 brown)
1.35–1.37 (FM 14111)
1.36–1.38 (FM 1132, 1132P, 3510)
1.37 (Plenco 500 Black)
1.37–1.39 (FM 510, 4201, 7676, 7700, 9294, 10365)
1.38 (Durez 21206 compr. molded; Plenco 300 Black, 307 Black, 369 Black, 480 Black, 500 Brown, 512 Black, 523 Black, 579 Black; RX 340, 350, 475, 950)
1.38–1.40 (FM 3002)
1.39 (Durez 115 compr. molded; Plenco 321 Black, 368 Black, 535 Black; RX 525)
1.40 (Plenco 317 Red, 320 Green, 571 Black)
1.40–1.42 (FM 3001)
1.41 (Durez 16038 compr. molded, 17702 compr. molded, 30417 compr. molded, Plenco 300 Brown, 411 Walnut Mottle, 476 Black, 557 Black, 567 Black, 4543 Red; RX 431)
1.42 (Durez 30806 compr. molded; Plenco 308 Black, 417 Black, 467 Black, 482 Black, 554 Black; RX 352, 448)
1.43 (Durez 31735 compr. molded; Plenco 300 Red, 308 Brown; RX 342)
1.44 (RX 429)
1.44–1.46 (FM 3000, 5367)
1.45–1.47 (FM 1389)
1.46 (Plenco 485 Black)
1.48 (Plenco 485 Brown, 548 Black)
1.51 (Plenco 507 Black)
1.52 (Phenolic Molding Compound 868)
1.53 (Plenco 349 Black)
1.54 (Plenco 400 Black)
1.55 (Plenco 400 Brown, 414 Black)
1.57 (Plenco 527 Black)
1.58 (Durez 30959 compr. molded; Plenco 466 Black, 509 Black)
1.60 (Durez 157 compr. molded, 30698 compr. molded; Plenco 466 Brown)
1.61 (Plenco 509 Brown)
1.62 (Plenco 586 Brown)
1.64–1.68 (FM 11547, 17610)
1.68 (RX 465, 466)
1.70 (Durez 31219 compr. molded; RX 468A)
1.71 (RX 867)
1.72 (RX 468)
1.72–1.75 (FM 5064)

Phenol-formaldehyde resin *(cont'd.)*

1.72–1.76 (FM 4030-190)
1.73 (RX 650)
1.74 (RX 611)
1.75 (Plenco 343 Natural; RX 610N, 630, 640)
1.76–1.80 (FM 8130)
1.77 (RX 660)
1.78 (RX 867D)
1.78–1.81 (FM 4004F, 4005, 4005F, 4007F)
1.80 (RX 462, 467, 625)
1.81–1.83 (FM 4007)
1.82 (RX 363, 490, 850)
1.82–1.85 (FM 6101)
1.83 (RX 842)
1.83–1.87 (FM 16771)
1.84 (Plenco 118-AF Dark Grey, 118 Dark Grey)
1.85 (Plenco 203)
1.85–1.87 (FM 21288)
1.86 (RX 867A)
1.88 (RX 862, 865)
1.90–1.95 (FM 4004)
1.94 (RX 866D)
1.95 (RX 495)
2.08 (RX 655, 655A)
2.22 (Plenco 201)
3.42 (Plenco 4443)

Density:

0.32 g/cm^3 (apparent) (Plenco 321 Black)
0.46 g/cm^3 (apparent) (Plenco 417 Black)
0.48 g/cm^3 (apparent) (Durez 30806 compr. molded; Plenco 476 Black, 500 Black)
0.50 g/cm^3 (apparent) (Durez 17702 compr. molded; Plenco 467 Black, 500 Brown, 523 Black, 554 Black)
0.52 g/cm^3 (apparent) (Plenco 307 Black, 579 Black)
0.54 g/cm^3 (apparent) (Plenco 308 Brown, 414 Black)
0.55 g/cm^3 (apparent) (Durez 31735 compr. molded; Plenco 368 Black, 480 Black)
0.56 g/cm^3 (apparent) (Plenco 300 Black, 369 Black)
0.57 g/cm^3 (apparent) (Plenco 507 Black, 535 Black)
0.58 g/cm^3 (apparent) (Plenco 308 Black, 411 Walnut Mottle, 512 Black, 527 Black, 567 Black)
0.59 g/cm^3 (apparent) (Plenco 300 Brown, 466 Black, 557 Black)
0.60 g/cm^3 (apparent) (Durez 16038 compr. molded, 21206 compr. molded, 30698 compr. molded, 30959 compr. molded; Plenco 317 Red, 466 Brown, 482 Black, 485 Black, 485 Brown, 571 Black, 586 Brown, 4543 Red)
0.61 g/cm^3 (apparent) (Plenco 548 Black)

Phenol-formaldehyde resin *(cont'd.)*

0.62 g/cm³ (apparent) (Plenco 300 Red, 320 Green, 509 Black, 509 Brown)
0.63 g/cm³ (apparent) (Durez 30417 compr. molded)
0.64 g/cm³ (apparent) (Phenolic Molding Compound 868; Plenco 400 Black)
0.65 g/cm³ (apparent) (Durez 115 compr. molded, 31219 compr. molded; Plenco 349 Black)
0.66 g/cm³ (apparent) (Plenco 400 Brown)
0.70 g/cm³ (apparent) (Durez 157 compr. molded)
0.75 g/cm³ (apparent) (Plenco 118-AF Dark Grey, 118 Dark Grey)
0.80 g/cm³ (apparent) (Plenco 343 Natural)
0.90 g/cm³ (apparent) (Plenco 201)
1.00 g/cm³ (apparent) (Plenco 203)
2.00 g/cm³ (apparent) (Plenco 4443)
0.65 mg/m³ (apparent) (RX 867A, 867D)
0.70 mg/m³ (apparent) (RX 867)
0.725 mg/m³ (apparent) (RX 865)
0.75 mg/m³ (apparent) (RX 862)
0.80 mg/m³ (apparent) (RX 866D)

Bulking Value:
2.1 (Durez 115 compr. molded)
2.2 (Durez 30417 compr. molded)
2.3 (Durez 157 compr. molded, 16038 compr. molded, 21206 compr. molded)
2.4 (RX 850)
2.6 (Durez 30959 compr. molded, 31219 compr. molded, 31735 compr. molded)
2.7 (Durez 30698 compr. molded)
2.8 (Durez 17702 compr. molded)
2.9 (Durez 30806 compr. molded)

Visc.:
100–175 cps (Varcum 8366)
100–250 cps (60% sol'n. in 1:1 ethanol:MIBK) (Varcum 1417)
175–275 cps (50% sol'n. in ethanol) (Varcum 6820)
600–1000 cps (Varcum 1481)
700–1000 cps (Varcum 1494)
1200–1600 cps (Varcum 5160)
6000–9000 cps (Varcum 5302)

M.P.:
55–75 C (SP-8014)
70–80 C (PA-52-334)
70–85 C (PA-52-333)

Softening Pt.:
65–75 C (Cap.) (Varcum 5485, 6820, 9874)
68–78 C (Cap.) (Varcum 1417, 5145)
70–80 C (Cap.) (Varcum C-86)
75–80 C (R&B) (Varcum 4326)

Phenol-formaldehyde resin *(cont'd.)*

80–90 C (Cap.) (Varcum 1359, 1364, 4631)

85–95 C (Cap.) (Varcum 8366)

85–105 C (R&B) (Varcum DR-406)

95–105 C (R&B) (Varcum 647)

96–106 C (R&B) (Varcum 9836A)

105–120 C (R&B) (Varcum 4727)

Stability:

Good resistance to chemicals and moisture (FM 1389)

Resistant to cracking in wet-dry exposures (Plenco 480 Black)

Resistant to cracking in wet-dry exposures, to soap and detergent solutions at elevated temps. (Plenco 417 Black, 476 Black)

Storage Stability:

3 mos storage life (Plenco 368 Black, 369 Black, 417 Black, 476 Black, 480 Black, 507 Black, 527 Black, 557 Black, 567 Black, 586 Brown, 4543 Red)

1–2 yrs storage life (Plenco 118-AF Dark Grey, 118 Dark Grey, 201, 203, 300 Black, 300 Brown, 300 Red, 307 Black, 308 Black, 308 Brown, 317 Red, 320 Green, 321 Black, 343 Natural, 349 Black, 400 Black, 400 Brown, 411 Walnut Mottle, 414 Black, 466 Black, 466 Brown, 467 Black, 482 Black, 485 Black, 485 Brown, 500 Black, 500 Brown, 509 Black, 509 Brown, 512 Black, 523 Black, 535 Black, 548 Black, 554 Black, 571 Black, 579 Black, 4443)

MECHANICAL PROPERTIES:

Tens. Str.:

34 MPa (Durez 157 compr. molded)

41 MPa (Durez 16038 compr. molded, 30417 compr. molded, 30698 compr. molded; RX 866D, 867, 867A)

41.4 MPa (RX 867D)

45 MPa (Durez 17702 compr. molded, 21206 compr. molded; RX 862)

48 MPa (Durez 30806 compr. molded)

52 MPa (Durez 31219 compr. molded; RX 865)

55 MPa (Durez 115 compr. molded, 31735 compr. molded)

59 MPa (Durez 30959 compr. molded)

3500 psi (Plenco 4443 compr. molded)

4000 psi (Plenco 203 compr. molded)

4100 psi (Plenco 118-AF Dark Grey compr. molded)

5000 psi (FM 16771, 21288; Plenco 118 Dark Grey compr. molded, 201 compr. molded, 343 Natural compr. molded; RX 363)

5500 psi (FM 3000; RX 342, 468, 468A)

5700 psi (RX 352)

5800 psi (FM 3001)

6000 psi (FM 3002, 14111; Plenco 321 Black compr. molded, 349 Black compr. molded, 417 Black compr. molded, 509 Black compr. molded; RX 340, 350, 431, 448, 462, 467, 475, 490, 655, 842, 850)

6100 psi (FM 7700)

Phenol-formaldehyde resin *(cont'd.)*

6300 psi (FM 4201)

6500 psi (FM 510, 1132P, 1389, 7676; Plenco 466 Black compr. molded, 466 Brown compr. molded, 509 Brown compr. molded; RX 429, 465, 466, 525, 950)

6600 psi (RX 495)

6700 psi (Plenco 485 Black compr. molded)

6800 psi (Plenco 557 Black compr. molded)

6900 psi (Plenco 476 Black compr. molded, 527 Black compr. molded)

7000 psi (FM 1132, 3510; Plenco 300 Black compr. molded, 400 Black compr. molded, 467 Black compr. molded, 480 Black compr. molded, 485 Brown compr. molded, 500 Black compr. molded, 500 Brown compr. molded, 507 Black compr. molded, 512 Black compr. molded, 523 Black compr. molded, 571 Black compr. molded, 579 Black compr. molded)

7200 psi (FM 8130; Plenco 300 Red compr. molded, 414 Black compr. molded)

7300 psi (Plenco 320 Green compr. molded, 368 Black compr. molded, 369 Black compr. molded, 411 Walnut Mottle compr. molded, 535 Black compr. molded)

7500 psi (FM 1303, 4030-190, 5064; Plenco 308 Black compr. molded, 554 Black compr. molded, 567 Black compr. molded, 4543 Red compr. molded; RX 625)

7600 psi (Plenco 317 Red compr. molded)

7800 psi (Plenco 482 Black compr. molded)

8000 psi (FM 4004, 6101, 10365; Phenolic Molding Compound 868; Plenco 300 Brown compr. molded, 307 Black compr. molded, 308 Brown compr. molded)

8500 psi (Plenco 548 Black compr. molded, 586 Brown compr. molded)

8600 psi (Plenco 400 Brown inj. molded)

9000 psi (FM 4007, 9294, 11547, 17610; RX 640, 655A, 660)

10,000 psi (FM 4005, 5367; RX 610N, 611, 650)

12,000 psi (FM 4004F)

13,500 psi (RX 630)

14,000 psi (FM 4007F)

16,000 psi (FM 4005F)

Tens. Mod.:

6.9 GPa (Durez 17702 compr. molded, 21206 compr. molded, 30806 compr. molded, 31735 compr. molded)

8.3 GPa (Durez 16038 compr. molded, 30417 compr. molded, 30698 compr. molded)

9.6 GPa (Durez 115 compr. molded)

10.3 GPa (Durez 157 compr. molded, 30959 compr. molded)

13.0 GPa (Durez 31219 compr. molded)

Flex. Str.:

55 MPa (Durez 157 compr. molded)

59 MPa (Durez 17702 compr. molded)

62 MPa (Durez 30698 compr. molded)

66 MPa (Durez 21206 compr. molded)

68 MPa (Durez 30806 compr. molded)

69 MPa (Durez 16038 compr. molded, 30417 compr. molded)

Phenol-formaldehyde resin *(cont'd.)*

72 MPa (Durez 115 compr. molded)

79 MPa (Durez 31219 compr. molded; RX 866D, 867, 867A)

79.4 MPa (RX 867D)

83 MPa (Durez 30959 compr. molded, 31735 compr. molded; RX 862)

117 MPa (RX 865)

5500 psi (Plenco 4443 compr. molded)

7500 psi (Plenco 118-AF Dark Grey compr. molded, 203 compr. molded)

8000 psi (FM 3000; Plenco 343 Natural compr. molded, 466 Brown compr. molded)

8100 psi (Plenco 349 Black compr. molded)

8300 psi (Plenco 118 Dark Grey compr. molded)

8500 psi (FM 3001, 3002)

8900 psi (Plenco 466 Black compr. molded, 527 Black compr. molded)

9000 psi (Plenco 321 Black compr. molded, 480 Black compr. molded, 509 Black compr. molded, 579 Black compr. molded; RX 342)

9200 psi (Plenco 554 Black compr. molded)

9400 psi (Plenco 500 Black compr. molded)

9500 psi (Plenco 308 Black compr. molded, 308 Brown compr. molded, 369 Black compr. molded, 500 Brown compr. molded, 509 Brown compr. molded, , 4543 Red compr. molded)

9800 psi (Plenco 482 Black compr. molded, 507 Black compr. molded)

9900 psi (Plenco 414 Black compr. molded)

10,000 psi (FM 510, 1132P, 4201; Plenco 300 Brown compr. molded, 300 Red compr. molded, 368 Black compr. molded, 400 Black compr. molded, 485 Black compr. molded, 557 Black compr. molded; RX 340, 350, 352, 363, 431, 448, 468A, 525, 950)

10,200 psi (Plenco 411 Walnut Mottle compr. molded)

10,400 psi (Plenco 467 Black compr. molded, 476 Black compr. molded, 523 Black compr. molded)

10,500 psi (Plenco 317 Red compr. molded, 320 Green compr. molded, 485 Brown compr. molded)

10,600 psi (Plenco 300 Black compr. molded, 417 Black compr. molded)

10,700 psi (Plenco 567 Black compr. molded)

10,800 psi (Plenco 512 Black compr. molded)

11,000 psi (FM 1389, 7676, 14111; Plenco 201 compr. molded, 535 Black compr. molded, 548 Black compr. molded, 571 Black compr. molded; RX 429, 466, 468, 495)

11,200 psi (Plenco 307 Black compr. molded)

11,500 psi (RX 490)

12,000 psi (FM 1303, 4004; Phenolic Molding Compound 868; RX 462, 465, 467, 842)

12,500 psi (FM 1132, 7700; Plenco 586 Brown compr. molded; RX 475, 850)

13,000 psi (RX 625, 655)

13,800 psi (Plenco 400 Brown inj. molded)

14,000 psi (FM 3510, 4007, 10365; RX 640)

Phenol-formaldehyde resin *(cont'd.)*

14,500 psi (RX 655A)
15,000 psi (FM 4005, 6101, 21288)
15,800 psi (FM 8130)
16,000 psi (FM 4004F, 4030-190, 9294, 11547, 17610)
17,000 psi (FM 5064; RX 660)
18,000 psi (FM 16771; RX 650)
20,000 psi (FM 4007F; RX 610N, 611)
21,000 psi (FM 5367; RX 630)
22,000 psi (FM 4005F)

Flex. Mod.:

12,400 MPa (RX 867)
13,700 MPa (RX 867A, 867D)
15,800 MPa (RX 862)
17,300 MPa (RX 865, 866D)
500,000 psi (FM 1303)
760,000 psi (Plenco 554 Black compr. molded)
780,000 psi (Plenco 4543 Red compr. molded)
870,000 psi (Plenco 308 Brown compr. molded)
900,000 psi (FM 14111)
940,000 psi (Plenco 417 Black compr. molded, 557 Black compr. molded)
950,000 psi (Plenco 300 Brown compr. molded)
960,000 psi (FM 3000; Plenco 476 Black compr. molded, 480 Black compr. molded, 500 Brown compr. molded, 579 Black compr. molded)
970,000 psi (FM 3002; Plenco 369 Black compr. molded)
980,000 psi (Plenco 368 Black compr. molded, 500 Black compr. molded, 567 Black compr. molded)
1.0×10^6 psi (FM 1132, 7676, 10365; Plenco 308 Black compr. molded, 321 Black compr. molded, 400 Brown inj. molded, 411 Walnut Mottle compr. molded, 512 Black compr. molded, 523 Black compr. molded)
1.1×10^6 psi (FM 510, 1389, 3001, 4201; Plenco 300 Red compr. molded, 307 Black compr. molded, 317 Red compr. molded, 482 Black compr. molded, 485 Brown compr. molded, 535 Black compr. molded, 548 Black compr. molded, 571 Black compr. molded, 586 Brown compr. molded)
1.2×10^6 psi (FM 1132P, 3510; Plenco 300 Black compr. molded, 320 Green compr. molded, 414 Black compr. molded, 467 Black compr. molded, 485 Black compr. molded; RX 350, 352)
1.3×10^6 psi (FM 7700; Plenco 466 Brown compr. molded, 507 Black compr. molded; RX 340, 342, 429, 431, 448, 475, 525, 950)
1.4×10^6 psi (FM 9294; Plenco 400 Black compr. molded, 466 Black compr. molded, 509 Brown compr. molded, 527 Black compr. molded)
1.5×10^6 psi (FM 5367; Plenco 349 Black compr. molded, 509 Black compr. molded)
1.7×10^6 psi (Plenco 343 Natural compr. molded; RX 842)
1.8×10^6 psi (RX 465, 466, 468, 468A, 640)

Phenol-formaldehyde resin *(cont'd.)*

1.9×10^6 psi (Plenco 118-AF Dark Grey compr. molded, 118 Dark Grey compr. molded; RX 363)

2.0×10^6 psi (FM 4005, 4007, 5064, 11547, 17610; RX 462, 467, 610N, 611, 625, 650, 660)

2.2×10^6 psi (FM 6101; RX 850)

2.3×10^6 psi (FM 4004, 8130; RX 490, 495)

2.4×10^6 psi (FM 4030-190)

2.6×10^6 psi (RX 630)

2.8×10^6 psi (RX 655)

3.2×10^6 psi (RX 655A)

Compr. Str.:

152 MPa (Durez 30698 compr. molded)

158 MPa (Durez 31219 compr. molded)

172 MPa (Durez 17702 compr. molded)

186 MPa (Durez 157 compr. molded, 30806 compr. molded; RX 867)

193 MPa (Durez 30417 compr. molded, 30959 compr. molded; RX 862, 866D, 867D)

207 MPa (Durez 16038 compr. molded, 21206 compr. molded, 31735 compr. molded; RX 867A)

221 MPa (Durez 115 compr. molded)

228 MPa (RX 865)

15,000 psi (FM 3000; Plenco 203 compr. molded)

16,000 psi (FM 16771, 21288)

18,000 psi (Plenco 554 Black compr. molded)

20,000 psi (FM 510, 10365; Plenco 4443 compr. molded)

20,800 psi (Plenco 343 Natural compr. molded)

21,000 psi (FM 1389)

22,000 psi (FM 3002, 9294; Plenco 466 Black compr. molded; RX 363)

22,500 psi (Plenco 118 Dark Grey compr. molded, 509 Brown compr. molded)

22,700 psi (Plenco 527 Black compr. molded)

22,800 psi (FM 3001)

23,000 psi (FM 1303, 11547, 17610)

24,000 psi (FM 6101, 7700; Plenco 201 compr. molded, 509 Black compr. molded; RX 462, 465, 475, 495, 842)

24,300 psi (Plenco 586 Brown compr. molded)

24,800 psi (Plenco 466 Brown compr. molded, 507 Black compr. molded)

25,000 psi (FM 14111; Plenco 480 Black compr. molded, 557 Black compr. molded; RX 467)

25,800 psi (Plenco 369 Black compr. molded)

26,000 psi (FM 3510; Plenco 321 Black compr. molded, 417 Black compr. molded; RX 342, 431)

27,000 psi (FM 4030-190, 8130; Plenco 476 Black compr. molded, 567 Black compr. molded, 579 Black compr. molded; RX 466, 468, 468A, 490, 850)

28,000 psi (FM 1132, 1132P, 5064, 7676; Phenolic Molding Compound 868; Plenco

400 Brown inj. molded; RX 340, 525)

28,500 psi (Plenco 523 Black compr. molded, 548 Black compr. molded)

28,600 psi (Plenco 368 Black compr. molded)

28,700 psi (Plenco 349 Black compr. molded)

28,800 psi (Plenco 414 Black compr. molded, 571 Black compr. molded)

29,000 (Plenco 308 Black compr. molded, 485 Black compr. molded, 500 Brown compr. molded, 512 Black compr. molded, 4543 Red compr. molded)

29,200 psi (Plenco 317 Red compr. molded)

29,300 psi (Plenco 411 Walnut Mottle compr. molded)

29,500 psi (Plenco 300 Black compr. molded, 308 Brown compr. molded, 500 Black compr. molded)

29,700 psi (Plenco 467 Black compr. molded)

29,900 psi (Plenco 320 Green compr. molded)

30,000 psi (FM 5367; Plenco 300 Red compr. molded, 482 Black compr. molded, 485 Brown compr. molded; RX 352, 448, 655)

31,000 psi (FM 4201; Plenco 535 Black compr. molded)

31,200 psi (Plenco 400 Black compr. molded)

31,500 psi (Plenco 307 Black compr. molded)

32,000 psi (RX 640, 950)

35,000 psi (RX 350, 429)

37,000 psi (RX 655A)

38,000 psi (FM 4004; RX 625)

40,000 psi (FM 4005, 4007; RX 610N, 611, 630, 650, 660)

45,000 psi (FM 4004F, 4005F, 4007F)

Shear Str.:

14,000 psi (FM 4005)

Impact Str. (Izod):

14 J/m (Durez 21206 compr. molded)

15 J/m (Durez 30417 compr. molded)

16 J/m (Durez 157 compr. molded, 16038 compr. molded)

18 J/m (Durez 115 compr. molded)

20 J/m (Durez 30698 compr. molded)

25 J/m (Durez 30806 compr. molded)

27 J/m (Durez 17702 compr. molded)

28 J/m (Durez 30959 compr. molded, 31735 compr. molded)

32 J/m (RX 867)

37.4 J/m (RX 866D, 867A, 867D)

40 J/m (Durez 31219 compr. molded)

48 J/m (RX 862)

69.4 J/m (RX 865)

0.26 ft lb/in. (Plenco 118-AF Dark Grey compr. molded, 203 compr. molded, 320 Green compr. molded, 349 Black compr. molded)

0.27 ft lb/in. (Plenco 300 Red compr. molded, 482 Black compr. molded, 567 Black

Phenol-formaldehyde resin *(cont'd.)*

compr. molded)

0.28 ft lb/in. (Plenco 308 Brown compr. molded, 317 Red compr. molded, 411 Walnut Mottle compr. molded, 480 Black compr. molded, 512 Black compr. molded)

0.29 ft lb/in. (Plenco 118 Dark Grey compr. molded, 308 Black compr. molded, 343 Natural compr. molded, 400 Brown inj. molded, 535 Black compr. molded, 548 Black compr. molded, 557 Black compr. molded, 571 Black compr. molded, 586 Brown compr. molded)

0.30 ft lb/in. (Plenco 300 Black compr. molded, 300 Brown compr. molded, 400 Black compr. molded, 485 Black compr. molded, 509 Black compr. molded, 509 Brown compr. molded, 4543 Red compr. molded)

0.31 ft lb/in. (Phenolic Molding Compound 868; Plenco 369 Black compr. molded, 485 Brown compr. molded, 527 Black compr. molded)

0.32 ft lb/in. (Plenco 307 Black compr. molded)

0.33 ft lb/in. (Plenco 414 Black compr. molded, 466 Black compr. molded, 466 Brown compr. molded, 507 Black compr. molded, 523 Black compr. molded)

0.34 ft lb/in. (Plenco 554 Black compr. molded)

0.36 ft lb/in. (Plenco 467 Black compr. molded)

0.37 ft lb/in. (Plenco 368 Black compr. molded, 500 Black compr. molded, 500 Brown compr. molded)

0.38 ft lb/in. (Plenco 476 Black compr. molded)

0.42 ft lb/in. notched (RX 655A)

0.45 ft lb/in. notched (FM 4004)

0.46 ft lb/in. (Plenco 579 Black compr. molded)

0.49 ft lb/in. (Plenco 417 Black compr. molded)

0.50 ft lb/in. (Plenco 201 compr. molded); notched (FM 3000, 4007; RX 340, 342, 525, 625, 655, 950)

0.55 ft lb/in. (Plenco 321 Black compr. molded); notched (FM 1303, 4004F)

0.6 ft lb/in. notched (FM 510, 3001, 4005, 4005F; RX 468A)

0.64 ft lb/in. notched (RX 466)

0.7 ft lb/in. (Plenco 4443 compr. molded); notched (FM 3002, 4007F, 4201; RX 468, 660, 842, 850)

0.75 ft lb/in. notched (RX 640)

0.8 ft lb/in. notched (RX 350, 352, 429, 448, 650)

0.85 ft lb/in. notched (RX 610N, 611)

0.9 ft lb/in. notched (RX 465)

1.0 ft lb/in. notched (RX 363)

1.1 ft lb/in. notched (FM 1132P, 7676; RX 467, 630)

1.2 ft lb/in. notched (RX 431)

1.3 ft lb/in. notched (FM 1389)

1.4 ft lb/in. notched (RX 462, 490)

1.5 ft lb/in. notched (FM 14111)

1.6 ft lb/in. notched (FM 1132)

1.75 ft lb/in. notched (RX 475)

Phenol-formaldehyde resin *(cont'd.)*

1.9 ft lb/in. notched (FM 7700)
2.0 ft lb/in. notched (RX 495)
2.2 ft lb/in. notched (FM 3510)
3.5 ft lb/in. notched (FM 6101, 10365, 11547, 17610)
4.5 ft lb/in. notched (FM 5367, 9294)
9 ft lb/in. notched (FM 16771, 21288)
11 ft lb/in. notched (FM 5064)
12 ft lb/in. notched (FM 4030-190, 8130)

Hardness:

Rockwell E54 (FM 16771)
Rockwell E64 (FM 21288)
Rockwell M95 (RX 862, 865)
Rockwell M105 (FM 1389, 7676, 7700, 14111)
Rockwell M110 (FM 510, 1303, 4004, 6101, 9294, 11547, 17610)
Rockwell M113 (FM 3001)
Rockwell M114 (FM 3000)
Rockwell M115 (FM 1132, 1132P, 3002, 3510, 4030-190, 4201, 8130)
Rockwell M120 (FM 4005, 4007, 5064)

Mold Shrinkage:

0.0012 in./in. (RX 842)
0.0015 in./in. (RX 850)
0.003 in./in. (Durez 31219 compr. molded)
0.004 in./in. (Durez 30698 compr. molded)
0.005 in./in. (Durez 157 compr. molded, 30806 compr. molded, 30959 compr. molded)
0.006 in./in. (Durez 115 compr. molded, 17702 compr. molded, 30417 compr. molded)
0.0065 in./in. (Durez 21206 compr. molded, 31735 compr. molded)
0.0075 in./in. (Durez 16038 compr. molded)

Water Absorp.:

0.05% (RX 842)
0.07% (RX 850)
0.20% (Durez 157 compr. molded, 31219 compr. molded)
0.40% (Durez 115 compr. molded, 30959 compr. molded)
0.50% (Durez 16038 compr. molded, 17702 compr. molded, 21206 compr. molded, 30698 compr. molded, 31735 compr. molded)
0.80% (Durez 30417 compr. molded, 30806 compr. molded)

THERMAL PROPERTIES:

Conduct.:

7×10^{-4} Cal/s/cm²/C/cm (FM 1303, 3002)
8×10^{-4} Cal/s/cm²/C/cm (FM 3000, 3510)
9×10^{-4} Cal/s/cm²/C/cm (FM 1132, 1132P, 3001, 4201, 5064, 7676, 7700, 9294)
10×10^{-4} Cal/s/cm²/C/cm (FM 4005, 4030-190, 8130)
11×10^{-4} Cal/s/cm²/C/cm (FM 4007)
12×10^{-4} Cal/s/cm²/C/cm (FM 1389, 11547, 17610)

Phenol-formaldehyde resin *(cont'd.)*

13×10^{-4} Cal/s/cm²/C/cm (FM 4004)

8×10^{-5} Cal/s/cm²/C/cm (FM 14111)

15×10^{-5} Cal/s/cm²/C/cm (FM 6101)

Distort. Temp.:

505 K (RX 867)

530 K (RX 862)

533 K (RX 867D)

561 K (RX 865, 866D, 867A)

124 C (264 psi) (FM 1303)

149 C (264 psi) (FM 1389, 9294)

160 C (264 psi) (FM 1132P, 7676)

163 C (264 psi) (FM 1132)

166 C (264 psi) (Durez 21206 compr. molded; FM 3510, 7700)

171 C (264 psi) (Durez 16038 compr. molded)

174 C (264 psi) (FM 4201)

177 C (264 psi) (Durez 30417 compr. molded, 31735 compr. molded; FM 4004, 11547, 17610)

182 C (264 psi) (Durez 30959 compr. molded)

183 C (264 psi) (FM 6101)

191 C (264 psi) (Durez 115 compr. molded, 157 compr. molded, 17702 compr. molded; FM 4005, 4007)

200 C (264 psi) (FM 14111)

204 C (264 psi) (Durez 30806 compr. molded)

218 C (264 psi) (Durez 30698 compr. molded, 31219 compr. molded)

250 C (264 psi) (FM 5064)

260 C (264 psi) (FM 4030-190, 8130)

295 F (264 psi) (Plenco 554 Black compr. molded)

300 F (264 psi) (FM 10365; Plenco 300 Black compr. molded, 300 Brown compr. molded, 308 Brown compr. molded, 317 Red compr. molded, 321 Black compr. molded, 411 Walnut Mottle compr. molded, 535 Black compr. molded, 579 Black compr. molded)

305 F (264 psi)) (Plenco 308 Black compr. molded, 500 Brown compr. molded, 512 Black compr. molded)

310 F (264 psi) (Plenco 300 Red compr. molded, 320 Green compr. molded, 500 Black compr. molded, 523 Black compr. molded)

315 F (264 psi) (RX 340, 342, 350, 352)

320 F (264 psi) (FM 510; Plenco 203 compr. molded, 307 Black compr. molded, 467 Black compr. molded, 480 Black compr. molded, 482 Black compr. molded, 485 Black compr. molded, 485 Brown compr. molded, 509 Black compr. molded, 509 Brown compr. molded, 571 Black compr. molded, 586 Brown compr. molded; RX 431)

325 F (264 psi) (Plenco 548 Black compr. molded; RX 475, 525, 950)

330 F (264 psi) (Plenco 368 Black compr. molded, 4443 compr. molded, 4543 Red

Phenol-formaldehyde resin (cont'd.)

compr. molded)

335 F (264 psi) (Plenco 417 Black compr. molded, 507 Black compr. molded)

340 F (264 psi) (Plenco 343 Natural compr. molded, 369 Black compr. molded, 414 Black compr. molded, 466 Black compr. molded, 476 Black compr. molded, 567 Black compr. molded; RX 429, 448)

345 F (264 psi) (Plenco 557 Black compr. molded)

350 F (264 psi) (FM 3000, 3002; Plenco 349 Black compr. molded, 400 Black compr. molded, 400 Brown inj. molded, 466 Brown compr. molded, 527 Black compr. molded; RX 625)

360 F (264 psi) (Plenco 201 compr. molded)

365 F (264 psi) (FM 3001)

375 F (Phenolic Molding Compound 868)

380 F (264 psi) (Plenco 118 Dark Grey compr. molded)

395 F (264 psi) (RX 660)

400 F (264 psi) (RX 363)

410 F (264 psi) (FM 5367; RX 630)

425 F (264 psi) (RX 650)

430 F (264 psi) (RX 610N, 611, 842)

445 F (264 psi) (FM 4004F, 4005F, 4007F)

450 F (264 psi) (RX 467, 468, 640)

475 F (264 psi) (RX 468A, 490, 495)

480 F (264 psi) (RX 850)

500 F (264 psi) (RX 462, 465, 466)

550+ F (264 psi) (RX 655, 655A)

600 F (264 psi) (FM 16771, 21288)

Coeff. of Linear Exp.:

1.3×10^{-3} m/m/C (RX 865)

1.4×10^{-5} m/m/C (RX 862)

1.6×10^{-5} m/m/C (RX 866D)

1.1×10^{-5} in./in./C (FM 11547, 17610)

1.2×10^{-5} in./in./C (FM 8130)

1.3×10^{-5} in./in./C (FM 9294)

1.4×10^{-5} in./in./C (FM 4004)

1.5×10^{-5} in./in./C (FM 3510, 4030-190, 6101)

1.8×10^{-5} in./in./C (FM 1132, 5064)

1.9×10^{-5} in./in./C (FM 4005, 4007)

2.0×10^{-5} in./in./C (FM 7700)

2.1×10^{-5} in./in./C (FM 14111)

2.2×10^{-5} in./in./C (FM 1389)

2.4×10^{-5} in./in./C (FM 1132P, 3001, 7676)

2.6×10^{-5} in./in./C (FM 3000, 4201)

5.2×10^{-5} in./in./C (FM 1303)

26×10^{-5} in./in./C (FM 3002)

Phenol-formaldehyde resin *(cont'd.)*

Flamm.:

V-0 (Durez 115 compr. molded, 157 compr. molded, 16038 compr. molded; Plenco 586 Brown compr. molded; RX 462, 466, 468, 468A, 842, 850, 862, 865, 866D, 867)

V-0/V-1 (Durez 30698 compr. molded, 30959 compr. molded; RX 525, 610N, 611, 630, 650, 660, 950)

HB (Durez 21206 compr. molded; RX 429, 431, 448)

ELECTRICAL PROPERTIES:

Dissip. Factor:

0.011 (1 MHz) (FM 4007)

0.012 (1 MHz) (FM 4005, 4005F, 4007F)

0.018 (1 MHz) (RX 611)

0.020 (1 MHz) (FM 4004, 4030-190, 8130, 16771; RX 610N, 650, 660)

0.023 (1 MHz) (RX 865)

0.024 (1 MHz) (RX 867)

0.025 (1 MHz) (FM 5064; RX 640)

0.027 (1 MHz) (RX 862)

0.028 (1 MHz) (RX 630)

0.03 (1 MHz) (FM 1303)

0.032 (1 MHz) (RX 625)

0.035 (1 MHz) (FM 3000)

0.04 (1 kHz) (Durez 16038 compr. molded, 21206 compr. molded, 30959 compr. molded); (1 MHz) (FM 3001, 3002, 21288)

0.047 (1 MHz) (RX 867A)

0.05 (1 kHz) (Durez 31219 compr. molded); (1 MHz) (FM 510, 1132P, 3510, 4201, 7676, 7700, 9294)

0.06 (1 kHz) (Durez 30417 compr. molded); (1 MHz) (FM 1132, 14111)

0.08 (1 kHz) (Durez 115 compr. molded, 157 compr. molded)

0.09 (1 kHz) (Durez 31735 compr. molded)

0.10 (1 kHz) (Durez 30698 compr. molded, 30806 compr. molded)

0.12 (60 Hz) (Phenolic Molding Compound 868)

0.13 (1 kHz) (Durez 17702 compr. molded)

0.3 (1 MHz) (FM 6101, 11547, 17610)

2.9 (1 kHz) (Plenco 343 Natural compr. molded)

4.8 (1 kHz) (Plenco 512 Black compr. molded)

4.9 (1 kHz) (Plenco 482 Black compr. molded)

5.1 (1 kHz) (Plenco 557 Black compr. molded)

6.0 (1 kHz) (Plenco 369 Black compr. molded, 548 Black compr. molded)

6.4 (1 kHz) (Plenco 571 Black compr. molded)

6.5 (1 kHz) (Plenco 300 Red compr. molded, 480 Black compr. molded)

6.7 (1 kHz) (Plenco 579 Black compr. molded)

7.0 (1 kHz) (Plenco 300 Brown compr. molded, 485 Black compr. molded)

7.6 (1 kHz) (Plenco 400 Black compr. molded)

Phenol-formaldehyde resin *(cont' d.)*

7.7 (1 kHz) (Plenco 307 Black compr. molded)
7.9 (1 kHz) (Plenco 308 Black compr. molded)
8.5 (1 kHz) (Plenco 308 Brown compr. molded, 500 Black compr. molded)
8.6 (1 kHz) (Plenco 500 Brown compr. molded)
8.7 (1 kHz) (Plenco 485 Brown compr. molded)
8.8 (1 kHz) (Plenco 586 Brown compr. molded)
9.0 (1 kHz) (Plenco 411 Walnut Mottle compr. molded)
9.1 (1 kHz) (Plenco 4543 Red compr. molded)
9.2 (1 kHz) (Plenco 535 Black compr. molded)
9.3 (1 kHz) (Plenco 466 Brown compr. molded)
9.5 (1 kHz) (Plenco 317 Red compr. molded)
10.0 (1 kHz) (Plenco 320 Green compr. molded)
10.4 (1 kHz) (Plenco 368 Black compr. molded)
10.5 (1 kHz) (Plenco 300 Black compr. molded)
10.6 (1 kHz) (Plenco 466 Black compr. molded)
10.9 (1 kHz) (Plenco 414 Black compr. molded)
12.4 (1 kHz) (Plenco 523 Black compr. molded)
12.5 (1 kHz) (Plenco 509 Black compr. molded)
14.0 (1 kHz) (Plenco 509 Brown compr. molded)
14.6 (1 kHz) (Plenco 321 Black compr. molded)
16.5 (1 kHz) (Plenco 349 Black compr. molded)
18.1 (1 kHz) (Plenco 507 Black compr. molded)
18.5 (1 kHz) (Plenco 400 Brown inj. molded)
18.7 (1 kHz) (Plenco 476 Black compr. molded)
19.0 (1 kHz) (Plenco 467 Black compr. molded)
20.7 (1 kHz) (Plenco 527 Black compr. molded)
24.3 (1 kHz) (Plenco 417 Black compr. molded)
Dielec. Str.:
6.9 MV/m (Durez 30806 compr. molded)
7.9 MV/m (Durez 30417 compr. molded)
9.0 MV/m (RX 867A)
9.1 MV/m (RX 867)
9.8 MV/m (Durez 17702 compr. molded)
10.8 MV/m (Durez 31735 compr. molded)
11.4 MV/m (RX 866D)
11.8 MV/m (Durez 157 compr. molded, 16038 compr. molded; RX 862, 865, 867D)
12.8 MV/m (Durez 115 compr. molded)
13.8 MV/m (Durez 30698 compr. molded)
14.8 MV/m (Durez 21206 compr. molded)
16.7 MV/m (Durez 31219 compr. molded)
60 V/mil (FM 11547, 17610)
80 V/mil (FM 6101)
100 V/mil (RX 490, 495)

Phenol-formaldehyde resin *(cont'd.)*

125 V/mil (RX 462)

150 V/mil (RX 465, 466, 467, 468, 468A)

200 V/mil (FM 14111; RX 429)

250 V/mil (Plenco 349 Black compr. molded, 417 Black compr. molded, 467 Black compr. molded, 507 Black compr. molded; RX 431)

260 V/mil (Plenco 476 Black compr. molded)

270 V/mil (RX 448)

275 V/mil (FM 21288; Plenco 527 Black compr. molded, 4543 Red compr. molded)

280 V/mil (Plenco 411 Walnut Mottle compr. molded, 480 Black compr. molded; RX 475)

290 V/mil (Plenco 400 Brown inj. molded)

300 V/mil (FM 510, 1132, 1132P, 3001, 3510, 7676, 7700, 10365; Plenco 300 Brown compr. molded, 300 Red compr. molded, 317 Red compr. molded, 321 Black compr. molded, 368 Black compr. molded, 466 Black compr. molded, 571 Black compr. molded, 579 Black compr. molded, 586 Brown compr. molded)

305 V/mil (Plenco 308 Black compr. molded)

310 V/mil (Plenco 369 Black compr. molded, 414 Black compr. molded, 482 Black compr. molded, 485 Black compr. molded, 500 Brown compr. molded, 512 Black compr. molded; RX 525, 950)

315 V/mil (Phenolic Molding Compound 868; Plenco 500 Black compr. molded)

320 V/mil (FM 4004, 5064, 9294; Plenco 308 Brown compr. molded)

325 V/mil (Plenco 300 Black compr. molded, 307 Black compr. molded, 400 Black compr. molded, 485 Brown compr. molded)

330 V/mil (Plenco 320 Green compr. molded, 466 Brown compr. molded, 509 Black compr. molded, 523 Black compr. molded, 535 Black compr. molded)

340 V/mil (FM 1303, 4201)

350 V/mil (Plenco 509 Brown compr. molded, 557 Black compr. molded)

375 V/mil (Plenco 548 Black compr. molded)

380 V/mil (FM 3000, 3002, 4007, 4030-190; Plenco 343 Natural compr. molded)

400 V/mil (FM 4005, 8130, 16771)

400+ V/mil (RX 610N, 611, 625, 630, 640, 650, 660)

430 V/mil (FM 4004F, 4005F, 4007F)

Dielec. Const.:

3.6 (1 MHz) (FM 1303, 3000)

3.8 (1 MHz) (FM 3002)

4.0 (1 MHz) (RX 630, 660)

4.1 (1 MHz) (FM 4007)

4.2 (1 MHz) (FM 510, 1132P, 7700)

4.3 (1 MHz) (FM 4201; RX 610N, 611)

4.4 (1 MHz) (FM 1132, 3001)

4.5 (1 MHz) (FM 3510, 4004, 4005F, 7676, 9294)

4.52 (1 kHz) (Plenco 343 Natural compr. molded)

4.53 (1 kHz) (Plenco 320 Green compr. molded)

Phenol-formaldehyde resin *(cont'd.)*

4.6 (1 MHz) (FM 4004F, 4005, 4007F, 11547, 17610)

4.78 (1 kHz) (Plenco 512 Black compr. molded)

4.8 (1 kHz) (Plenco 482 Black compr. molded); (1 MHz) (FM 14111)

4.87 (1 kHz) (Plenco 308 Brown compr. molded)

4.9 (1 kHz) (Plenco 548 Black compr. molded)

5.0 (1 MHz) (FM 4030-190, 8130; RX 625, 650, 865)

5.04 (1 kHz) (Plenco 300 Brown compr. molded)

5.14 (1 kHz) (Plenco 317 Red compr. molded)

5.2 (1 MHz) (RX 640, 862)

5.23 (1 kHz) (Plenco 300 Red compr. molded)

5.4 (1 kHz) (Durez 21206 compr. molded)

5.49 (1 kHz) (Plenco 500 Brown compr. molded, 579 Black compr. molded)

5.5 (1 kHz) (Durez 30959 compr. molded); (1 MHz) (FM 6101)

5.67 (1 kHz) (Plenco 307 Black compr. molded)

5.7 (1 kHz) (Durez 16038 compr. molded, 31219 compr. molded)

5.76 (1 kHz) (Plenco 411 Walnut Mottle compr. molded)

5.8 (1 MHz) (RX 867A)

5.85 (1 kHz) (Plenco 300 Black compr. molded, 523 Black compr. molded, 571 Black compr. molded)

6.0 (60 Hz) (Phenolic Molding Compound 868); (1 kHz) (Plenco 400 Black compr. molded); (1 MHz) (FM 5064, 16771)

6.02 (1 kHz) (Plenco 557 Black compr. molded)

6.04 (1 kHz) (Plenco 308 Black compr. molded)

6.20 (1 kHz) (Plenco 485 Brown compr. molded, 535 Black compr. molded)

6.3 (1 MHz) (FM 21288)

6.36 (1 kHz) (Plenco 369 Black compr. molded)

6.4 (1 kHz) (Plenco 480 Black compr. molded)

6.44 (1 kHz) (Plenco 500 Black compr. molded)

6.5 (1 kHz) (Durez 115 compr. molded)

6.56 (1 kHz) (Plenco 466 Brown compr. molded)

6.64 (1 kHz) (Plenco 321 Black compr. molded)

6.7 (1 kHz) (Durez 157 compr. molded, 30417 compr. molded)

6.73 (1 kHz) (Plenco 368 Black compr. molded, 586 Brown compr. molded)

6.9 (1 kHz) (Plenco 414 Black compr. molded)

7.0 (1 kHz) (Durez 30698 compr. molded)

7.09 (1 kHz) (Plenco 400 Brown inj. molded, 4543 Red compr. molded)

7.45 (1 kHz) (Plenco 509 Black compr. molded)

7.5 (1 kHz) (Durez 31735 compr. molded)

7.79 (1 kHz) (Plenco 485 Black compr. molded)

7.84 (1 kHz) (Plenco 417 Black compr. molded)

7.97 (1 kHz) (Plenco 509 Brown compr. molded)

8.0 (1 kHz) (Durez 17702 compr. molded, 30806 compr. molded)

8.50 (1 kHz) (Plenco 349 Black compr. molded)

Phenol-formaldehyde resin *(cont'd.)*

9.39 (1 kHz) (Plenco 466 Black compr. molded)

9.92 (1 kHz) (Plenco 476 Black compr. molded, 507 Black compr. molded)

10.90 (1 kHz) (Plenco 467 Black compr. molded)

12.14 (1 kHz) (Plenco 527 Black compr. molded)

Vol. Resist.:

10^9 ohm-cm (RX 625, 640, 650)

10^{10} ohm-cm (Durez 115 compr. molded, 16038 compr. molded, 17702 compr. molded, 21206 compr. molded, 30417 compr. molded, 30698 compr. molded, 30806 compr. molded, 30959 compr. molded, 31219 compr. molded, 31735 compr. molded; RX 475, 490, 495)

3×10^{10} ohm-cm (RX 660)

9×10^{10} ohm-cm (Phenolic Molding Compound 868)

1×10^{11} ohm-cm (Durez 157 compr. molded; RX 429, 431, 448, 465, 466, 467, 468, 525, 950)

3×10^{11} ohm-cm (RX 611, 630)

1×10^{12} ohm-cm (FM 16771; RX 468A, 610N)

2×10^{12} ohm-cm (FM 14111)

4.5×10^{12} ohm-cm (Plenco 343 Natural compr. molded)

6×10^{12} ohm-cm (FM 7676)

7×10^{12} ohm-cm (FM 1132)

8×10^{12} ohm-cm (FM 6101, 11547, 17610)

1×10^{13} ohm-cm (FM 1303, 4030-190, 5064, 21288)

2×10^{13} ohm-cm (FM 510, 4004, 4005, 4007, 4201, 7700, 8130, 9294)

3×10^{13} ohm-cm (FM 1132P)

10^{14} ohm-cm (Plenco 557 Black compr. molded)

10^{13} ohm-m (RX 862, 865, 866D, 867, 867A, 867D)

Surf. Resist.:

2.2×10^{12} ohm (Plenco 343 Natural compr. molded)

Arc Resist.:

75 s (FM 510)

115 s (RX 495)

120 s (RX 431, 448, 475, 490, 525, 950)

125 s (RX 429, 462, 467)

140 s (RX 465, 468A)

150 s (RX 466, 468, 610N, 625, 630, 640, 650, 660)

155 s (RX 611)

180 s (FM 4004, 4005, 4007; RX 867)

180+ s (FM 6101)

181 s (Durez 157 compr. molded)

182 s (Durez 30959 compr. molded)

183 s (Plenco 586 Brown compr. molded; RX 850, 862)

184 s (RX 867A, 867D)

185 s (Durez 31219 compr. molded; RX 865, 866D)

Phenol-formaldehyde resin (cont'd.)

187 s (Durez 30698 compr. molded)

CURING CHARACTERISTICS:

Hot plate cure 13–25 s @ 150 C (Varcum 1481)
Hot plate cure 15–30 s @ 150 C (Varcum 1494, 5145)
Hot plate cure 18–25 s @ 150 C (Varcum 5302)
Hot plate cure 25–35 s @ 150 C (Varcum 9874)
Hot plate cure 30–50 s @ 150 C (Varcum 8366)
Hot plate cure 70–100 s @ 150 C (Varcum C-86)
Hot plate cure 70–110 s @ 150 C (Varcum 5485, 6820)
Hot plate cure 80–100 s @ 150 C (Varcum 1359, 1364)
Hot plate cure 120–180 s @ 150 C (Varcum 5160)
Hot plate cure > 4 min @ 185 C (Varcum 1417)
Stroke cure 25–45 s @ 160 C (SP-8014)
Stroke cure 20–40 s @ 165 C (PA-52-334)

STORAGE/HANDLING:

Store in a cool, dry area (Plenco 118-AF Dark Grey, 118 Dark Grey, 201, 203, 300 Black, 300 Brown, 300 Red, 307 Black, 308 Black, 308 Brown, 317 Red, 320 Green, 321 Black, 343 Natural, 349 Black, 400 Brown, 411 Walnut Mottle, 414 Black, 466 Black, 466 Brown, 467 Black, 482 Black, 485 Black, 485 Brown, 500 Black, 500 Brown, 509 Black, 509 Brown, 512 Black, 523 Black, 535 Black, 548 Black, 554 Black, 571 Black, 579 Black, 4443, 4543 Red)

Most one-stage compounds will polymerize slowly in storage—store in cool, dry area no longer than 3 mos. (Plenco 368 Black, 369 Black, 400 Black, 417 Black, 476 Black, 480 Black, 507 Black, 527 Black, 557 Black, 567 Black, 586 Brown)

STD. PKGS.:

50-lb bags and 200-lb drums (PA-52-333)

Phenylmethyl polysiloxane

CAS No.:

68083-14-7

TRADENAME EQUIVALENTS:

Abil AV 20-1000, 8853 [Goldschmidt AG]

CATEGORY:

Conditioner, emollient

APPLICATIONS:

Cosmetic industry preparations: aerosols (Abil AV 20-1000, 8853); creams and lotions (Abil AV 20-1000, 8853); hair preparations (Abil AV 20-1000, 8853); shaving preparations (Abil AV 20-1000, 8853); skin care products (Abil AV 20-1000, 8853)
Pharmaceutical applications: sunscreens/tanning products (Abil AV 20-1000, 8853)

Phenylmethyl polysiloxane *(cont'd.)*

PROPERTIES:
Form:
Liquid (Abil AV 20-1000, 8853)
Composition:
100% conc. (Abil AV 20-1000, 8853)
GENERAL PROPERTIES:
Ionic Nature:
Nonionic (Abil AV 20-1000)

Polyacrylamide (CTFA)

SYNONYMS:
2-Propenamide, homopolymer
EMPIRICAL FORMULA:
$(C_3H_5NO)_x$
STRUCTURE:

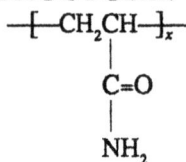

CAS No.:
9003-05-8
TRADENAME EQUIVALENTS:
Cyanamer A-370, P-26, P-35, P-35 Solution, P-70, P-250 [Amer. Cyanamid; Cyanamid BV]
Reten 420 [Hercules]
CATEGORY:
Polymer, dispersant, antiprecipitant, solubilizer, binder, flocculant, suspending agent, lubricant, thickener, stabilizer, film former, cross-linking agent, filtering aid, hydrotrope, slip agent, antistat, antiscalant
APPLICATIONS:
Agriculture industry: (Cyanamer A-370, P-26, P-35, P-250)
Cosmetic industry preparations: (Cyanamer P-26, P-250)
Electric/electronic industry: phospher processing for TV tubes (Cyanamer P-35)
Industrial applications: adhesives (Cyanamer A-370, P-26, P-35, P-250; Reten 420); cement/concrete (Cyanamer A-370, P-26, P-35, P-250); ceramics (Cyanamer A-370, P-35, P-250); coatings (Cyanamer A-370, P-250); detergents (Cyanamer A-370, P-26, P-35, P-250); dyes and pigments (Cyanamer A-370, P-35, P-250); fire

Polyacrylamide *(cont'd.)*

fighting (Cyanamer P-250); foundry molds (Cyanamer P-26); graphic arts (Cyanamer P-35, P-250); insulation (Cyanamer P-35, P-250); intricate/hard-to-fill moldings; latexes (Cyanamer P-26, P-35, P-250); metal processing (Cyanamer P-26, P-35); paints (Cyanamer P-250); plaster (Cyanamer P-26, P-250); printing inks (Cyanamer A-370, P-26, P-35, P-250); processing (Cyanamer A-370, P-26, P-250); sizing (Cyanamer P-250); textile/leather (Cyanamer A-370, P-26, P-35, P-250)

PROPERTIES:
Form:
    Liquid (Cyanamer P-35 Solution)
    Powder (Cyanamer P-35, P-250; Reten 420)
    Free-flowing powder (Cyanamer A-370, P-26)
Color:
    White (Cyanamer P-26, P-35, P-250)
    Off-white to medium tan (Cyanamer A-370)
    Amber (Cyanamer P-35 Solution)
Composition:
    0.20% max. acrylamide monomer (Cyanamer P-26)
    ≤ 7% moisture (Cyanamer A-370)
    50% active (Cyanamer P-35 Solution)
    93% min. solids (Cyanamer P-35)

GENERAL PROPERTIES:
Ionic Nature:
    Essentially nonionic (Cyanamer P-250)
Solubility:
    Limited sol. in organic solvents (Cyanamer A-370, P-26, P-35, P-250)
    Sol. in water (Cyanamer A-370, P-26, P-35, P-70, P-250; Reten 420)
M.W.:
    ≈ 200,000 (Cyanamer P-26)
    ≈ 5–6 × 10^6 (Cyanamer P-250)
Visc.:
    1.8–2.2 cps (0.1% sol'n.) (Cyanamer P-250)
    6.5–10.5 cps (15% sol'n.) (Cyanamer P-35)
    300–800 cps (10% sol'n.) (Cyanamer P-26)
Stability:
    Stable over wide pH range; excellent tolerance for electrolytes (Cyanamer A-370, P-26, P-35, P-250)
Storage Stability:
    Excellent (Cyanamer A-370, P-26, P-35, P-250)
pH:
    6.0–6.5 (1% sol'n.) (Cyanamer P-250)
    10–12 (15% sol'n.) (Cyanamer P-35)

Polyacrylamide *(cont'd.)*

TOXICITY/HANDLING:
Practically nontoxic (Cyanamer A-370, P-26, P-250)
Innocuous by ingestion or skin contact; however, the alkalinity of the powder and its solutions may cause skin and eye irritation (Cyanamer P-35)

STORAGE/HANDLING:
Protect from moisture in storage (Cyanamer A-370, P-26, P-250)

Polyacrylic acid (CTFA)

SYNONYMS:
2-Propenoic acid, homopolymer

STRUCTURE:

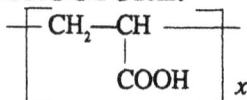

CAS No.:
9003-01-4

TRADENAME EQUIVALENTS:
Acrysol A-1, A-3, A-5, ASE-60, ASE-75, LMW-Series [Rohm & Haas]
Alcogum L [Alco]
Alcosperse 404, 409 [Alco]
Antiprex 461 [Allied Colloids]
Carbopol 907 [BF Goodrich]
Colloid 117/50, 119/50, 204, 274 [Colloids Inc.]
Daxad 37L [W.R. Grace]

CATEGORY:
Polymer, thickener, dispersant, suspending agent, antiscalant, emulsifier, stabilizer

APPLICATIONS:
Agriculture industry: (Carbopol 907)
Consumer products: dishwashing (Colloid 204)
Cosmetics: (Carbopol 907)
Industrial applications: (Carbopol 907); adhesives (Alcogum L; Acrysol A-1, A-3, A-5); binders (Acrysol A-1, A-3, A-5); ceramics (Alcosperse 404); chemical specialties (Carbopol 907); cleaners (Acrysol A-1, A-3, A-5; Colloid 204); detergent additive (Acrysol LMW-Series); dyes and pigments (Acrysol ASE-60, ASE-75; Alcosperse 404; Daxad 37L); latexes (Alcogum L); mining applications (Alcosperse 404); paints (Alcogum L; Acrysol A-1, A-3, A-5; Alcosperse 404; Daxad 37L); paper coatings (Alcogum L; Alcosperse 404); polymers (Acrysol A-1, A-3,

A-5); rubber latexes (Acrysol ASE-60, ASE-75; Alcosperse 404, 409); textile applications (Alcosperse 404; Carbopol 907); water treatment (Antiprex 461; Colloid 117/50, 119/50; Daxad 37L)

Pharmaceuticals: (Carbopol 907)

PROPERTIES:
Form:

Liquid (Acrysol A-1, A-5, LMW-Series; Alcosperse 404, 409; Antiprex 461; Colloid 117/50, 119/50, 204; Daxad 37L)

Clear, visc. sol'n. (Acrysol A-3)

Emulsion (Alcogum L; Acrysol ASE-60, ASE-75)

Solid (Carbopol 907)

Color:

Milky white (Acrysol ASE-60, ASE-75)

White (Carbopol 907)

Light amber (Antiprex 461)

Straw (Daxad 37L)

Composition:

25% active in water (Alcosperse 404)

28% solids in water (Acrysol ASE-60)

40% solids in water (Acrysol ASE-75)

48% active in water (Alcosperse 409)

50% active (Antiprex 461; Colloid 274)

50% solids (Daxad 37L)

100% conc. (Carbopol 907)

GENERAL PROPERTIES:
Solubility:

Sol. in water (Alcosperse 404; Colloid 204)

Ionic Nature:

Anionic (Acrysol ASE-60, ASE-75; Carbopol 907)

Sp. Gr.:

1.054 (25/25 C) (Acrysol ASE-60)

1.08 (Acrysol ASE-75)

1.26 (Antiprex 461)

Visc.:

10 cps (Acrysol ASE-60)

20 cps (Acrysol ASE-75)

250 cps (Brookfield, LVF, #2 spindle, 30 rpm) (Daxad 37L)

270 cps (Alcosperse 404)

400–700 cps (Antiprex 461)

1100 cps (Alcosperse 409)

Stability:

Unlimited stability; unaffected by aging; resistant to bacteria (Carbopol 907)

Excellent chlorine stability (Colloid 204)

Polyacrylic acid (cont'd.)

pH:
1.6 (Alcosperse 409)
2.5 (Acrysol ASE-75)
3.0 (Acrysol ASE-60; Daxad 37L)
TOXICITY/HANDLING:
Keep away from skin and eyes (Antiprex 461)
STD. PKGS.:
55-gal steel drums or bulk (Antiprex 461)

Polychloroprene

SYNONYMS:
Chloroprene rubber
CR
Neoprene
EMPIRICAL FORMULA:
$(CH_3ClC : CHCH_2)_n$
STRUCTURE:

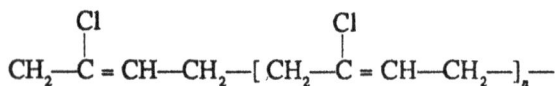

CAS No.:
126-99-8
TRADENAME EQUIVALENTS:
Baypren 110, 112, 115, 124, 130, 210, 211, 213, 214, 215, 220, 230, 233, 235, 236, 243, 320, 320 GR, 321, 321 GR, 330, 331, , 610, 710 [Bayer AG]
Neoprene Series, AC, AD, AF, AG, AH, CG, FB, GN, GNA, GRT, GW, TRT, TW, TW-100, W, WB, WD, WHV, WHV-A, WK, WRT [DuPont]
Nobestos/D-7102 [Rogers]
Polychloroprene latex:
Baypren Latex Series, B, GK, MKB, SK, T [Bayer AG]
Neoprene Latex Series, 101, 102, 950 [DuPont]
   Polychloroprene homopolymer:
   Neoprene Latex 601A, 622, 635, 654, 671, 735-A, 842-A [DuPont]
   Chloroprene-sulfur copolymer:
   Neoprene Latex 571, 572 [DuPont]
   Chloroprene-2,3 dichloro-1,2-butadiene copolymer:
   Neoprene Latex 650, 750 [DuPont]

Polychloroprene *(cont'd.)*

MODIFICATIONS/SPECIALTY GRADES:
Precrosslinked grade:
Baypren 115, 124, 214, 215, 235
Antioxidant:
Neoprene GNA (discoloring)
Carboxyl functionality:
Neoprene Latex 101, 102
Methacrylic acid-modified:
Baypren Latex 4 R (4% methacrylic acid)
Sulfur-modified:
Baypren 610, 710
Thiuram disulfide-stabilized:
Baypren 321, 321 GR, 331

CATEGORY:
Synthetic elastomer

PROCESSING:
Calendering:
Neoprene WB, WK
Coating:
Baypren Latex Series, 4 R, B, GK, MKB, SK, T; Neoprene Latex Series
Dipping:
Baypren Latex Series, 4 R, B, GK, MKB, SK, T; Neoprene Latex Series
Extrusion:
Baypren 110, 112, 115, 124, 130, 210, 211, 213, 214, 215, 220, 230, 233, 235, 236, 243, 320, 320 GR, 321, 321 GR, 330, 331, 610, 710; Neoprene Series, WB, WK
Injection molding:
Baypren 110, 112, 115, 124, 130, 210, 211, 213, 214, 215, 220, 230, 233, 235, 236, 243, 320, 320 GR, 321, 321 GR, 330, 331, 610, 710
Lamination:
Baypren Latex Series, 4 R, B, GK, MKB, SK, T; Neoprene Latex 101
Spraying:
Neoprene Latex Series

APPLICATIONS:
Architectural applications: (Neoprene Series); construction applications (Neoprene Latex 671); road construction (Baypren Latex Series, 4 R, B, GK, MKB, SK, T); roof sheeting (Baypren 110, 112, 115, 124, 130, 210, 211, 213, 214, 215, 220, 230, 233, 235, 236, 243, 320, 320 GR, 321, 321 GR, 330, 331, 610, 710)

Automotive applications: (Neoprene Series); tire sidewalls (Neoprene Series)

Consumer products: footwear (Baypren 110, 112, 115, 124, 130, 210, 211, 213, 214, 215, 220, 230, 233, 235, 236, 243, 320, 320 GR, 321, 321 GR, 330, 331, 610, 710; Neoprene Series); gloves (Neoprene Latex Series); hair treatments (Neoprene Latex 842-A); home crafts (Neoprene Series); sporting goods (Neoprene Series)

Polychloroprene (cont'd.)

Electrical/electronic industry: cable and wire sheathing/insulation (Baypren 110, 112, 115, 124, 130, 210, 211, 213, 214, 215, 220, 230, 233, 235, 236, 243, 320, 320 GR, 321, 321 GR, 330, 331, 610, 710; Neoprene Series)

Food-contact applications: (Baypren 110, 112, 115, 124, 130, 210, 211, 213, 214, 215, 220, 230, 233, 235, 236, 243, 320, 320 GR, 321, 321 GR, 330, 331, 610, 710)

Functional additives: binder (Baypren Latex Series, 4 R, B, GK, MKB, SK, T; Neoprene Latex Series); blending (Neoprene TW-100; Neoprene Latex 635, 735-A); impregnant (Neoprene Latex Series); modifier (Baypren Latex Series, 4 R, B, GK, MKB, SK, T); plasticizer (Neoprene FB); stiffener (Neoprene WHV)

Industrial applications: adhesives/sealants (Baypren 110, 112, 115, 124, 130, 210, 211, 213, 214, 215, 220, 230, 233, 235, 236, 243, 320, 320 GR, 321, 321 GR, 330, 331, 610, 710; Baypren Latex Series, 4 R, B, GK, MKB, SK, T; Neoprene Series, AC, AD, AF, AG, AH, CG, FB, WHV-A; Neoprene Latex Series, 101, 102, 572, 601A, 654, 671, 750, 842-A, 950); belting (Baypren 110, 112, 115, 124, 130, 210, 211, 213, 214, 215, 220, 230, 233, 235, 236, 243, 320, 320 GR, 321, 321 GR, 330, 331, 610, 710; Neoprene Series); bituminous compositions (Baypren Latex Series, 4 R, B, GK, MKB, SK, T); bonded batts (Neoprene Latex 101, 102, 601A, 654); caulks (Neoprene FB); cellular products (Neoprene Series); cements (Neoprene FB, WHV-A; Neoprene Latex 842-A, 950); coated fabrics (Neoprene Latex Series); coatings (Baypren Latex Series, 4 R, B, GK, MKB, SK, T; Neoprene Series, AH; Neoprene Latex 101, 102, 601A, 622, 654, 842-A, 950); concrete (Neoprene Latex Series); dipped goods (Baypren Latex Series, 4 R, B, GK, MKB, SK, T; Neoprene Latex Series, 571, 601A, 622, 650, 671, 750, 842-A); fabric proofings (Baypren 110, 112, 115, 124, 130, 210, 211, 213, 214, 215, 220, 230, 233, 235, 236, 243, 320, 320 GR, 321, 321 GR, 330, 331, 610, 710); foam/sponge rubber (Baypren 110, 112, 115, 124, 130, 210, 211, 213, 214, 215, 220, 230, 233, 235, 236, 243, 320, 320 GR, 321, 321 GR, 330, 331, 610, 710; Neoprene Latex Series, 101, 102, 622, 650); gaskets/seals (Neoprene Series); laminates (Baypren Latex Series, 4 R, B, GK, MKB, SK, T; Neoprene Latex 101, 572, 601A, 654, 842-A); linings (Baypren 110, 112, 115, 124, 130, 210, 211, 213, 214, 215, 220, 230, 233, 235, 236, 243, 320, 320 GR, 321, 321 GR, 330, 331, 610, 710); mastics (Neoprene AG, AH; Neoprene Latex 101); mining industry (Baypren Latex Series, 4 R, B, GK, MKB, SK, T); paper industry (Baypren Latex Series, 4 R, B, GK, MKB, SK, T; Neoprene Latex Series, 101, 102, 654, 735-A, 842-A); proofings (Baypren Latex Series, 4 R, B, GK, MKB, SK, T); roll covers (Baypren 110, 112, 115, 124, 130, 210, 211, 213, 214, 215, 220, 230, 233, 235, 236, 243, 320, 320 GR, 321, 321 GR, 330, 331, 610, 710); sheet (Baypren 110, 112, 115, 124, 130, 210, 211, 213, 214, 215, 220, 230, 233, 235, 236, 243, 320, 320 GR, 321, 321 GR, 330, 331, 610, 710; Neoprene Latex Series; Nobestos/D-7102); textile applications (Baypren 110, 112, 115, 124, 130, 210, 211, 213, 214, 215, 220, 230, 233, 235, 236, 243, 320, 320 GR, 321, 321 GR, 330, 331, 610, 710; Baypren Latex Series, 4 R, B, GK, MKB, SK, T; Neoprene Latex Series); thread (Neoprene Latex Series, 671); tubing/hoses (Baypren 110, 112, 115, 124, 130, 210, 211, 213, 214, 215, 220, 230, 233, 235, 236, 243, 320, 320 GR, 321, 321

Polychloroprene (cont'd.)

GR, 330, 331, 610, 710; Neoprene Series)
PROPERTIES:
Form:
    Liquid (Baypren Latex 4 R, B, GK, MKB, SK, T; Neoprene Latex 101, 102, 571, 572, 601A, 622, 635, 650, 654, 671, 735-A, 750, 842-A, 950)

    Chips (Bapyren 213, 243; Neoprene AC, AD, AF, AG, AH, CG, GN, GNA, GRT, GW, TRT, TW, TW-100, W, WB, WD, WHV, WHV-A, WK, WRT)

    Grains (Neoprene AC, AD, WHV-A)

    Flat chips (Baypren 110, 112, 115, 124, 130, 210, 211, 214, 215, 220, 230, 233, 235, 236, 320, 321, 330, 331, 610, 710)

    Granules (Baypren 320 GR, 321 GR)

    Wax-like solid (Neoprene FB)

Color:
    Milky white (Neoprene Latex 101, 102, 571, 572, 601A, 622, 635, 650, 654, 671, 735-A, 750, 842-A, 950)

    Creamy white (Neoprene AC, AD, TW, TW-100, W, WK)

    Creamy white to light amber (Neoprene WB, WD, WHV, WHV-A, WRT)

    Pale cream (Neoprene AF)

    Whitish to pale brown (Baypren 110, 112, 115, 124, 130, 210, 211, 214, 215, 220, 230, 233, 235, 236, 320, 321, 330, 331, 610, 710)

    Light pink to silver-gray (Neoprene AG)

    Light amber (Neoprene TRT)

    Amber (Neoprene AH, CG, GN, GNA, GRT, GW)

    Dark brown (Neoprene FB)

Composition:
    $32 \pm 1.0\%$ solids (Baypren Latex GK)

    45% solids, 38.5% chlorine (Neoprene Latex 735-A)

    46% solids in water, 36% chlorine (Neoprene Latex 101, 102)

    50% solids, 37.5% chlorine (Neoprene Latex 571, 842-A)

    50% solids, 38% chlorine (Neoprene Latex 572, 950)

    50% solids, 40% chlorine (Neoprene Latex 750)

    $50 \pm 0.5\%$ solids (Baypren Latex 4 R)

    $55 \pm 0.5\%$ solids (Baypren Latex SK)

    $58 \pm 0.5\%$ solids (Baypren Latex B, MKB, T)

    $\approx 59\%$ solids (Neoprene Latex 654)

    59% solids, 38% chlorine (Neoprene Latex 671)

    60% solids, 38% chlorine (Neoprene Latex 601-A, 635)

    60% solids, 40% chlorine (Neoprene Latex 650)

    $\approx 61\%$ solids (Neoprene Latex 622)

GENERAL PROPERTIES:
Ionic Nature:
    Anionic (Neoprene Latex 571, 572, 601-A, 622, 635, 650, 654, 671, 735-A, 750, 842-A)

Polychloroprene *(cont'd.)*

Cationic (Neoprene Latex 950)

Crystallization:

Very slight (Baypren Latex 4 R, T)

Slight (Baypren 110, 115, 124, 130, 610; Baypren Latex GK)

Slight to medium (Baypren 112; Baypren Latex B, MKB)

Medium (Baypren 210, 211, 213, 214, 215, 220, 230, 233, 235, 236, 243, 710)

Strong (Baypren 320, 320 GR, 321, 321 GR, 330, 331)

Very strong (Baypren Latex SK)

Sp.Gr.:

1.08 (Baypren Latex GK)

1.08–1.15 (Neoprene Latex 101, 102, 571, 572, 601-A, 622, 635, 650, 654, 671, 735-A, 750, 842-A, 950)

1.11 (Baypren Latex 4 R)

1.12 (Baypren Latex SK)

1.13 (Baypren Latex B, MKB, T)

1.23 (Baypren 110, 112, 115, 124, 130, 210, 211, , 214, 215, 220, 230, 233, 235, 236, 320, 320 GR, 321, 321 GR, 330, 331, 610, 710; Neoprene AC, AD, AF, AG, AH, CG, FB, GN, GNA, GRT, GW, TRT, TW, TW-100, W, WB, WD, WHV, WHV-A, WK, WRT)

Visc. (Brookfield):

5 cps (Neoprene Latex 735-A)

10 cps (Neoprene Latex 572, 750)

15 cps (Neoprene Latex 571, 842-A, 950)

69 cps (Neoprene Latex 671)

140 cps (Neoprene Latex 601-A)

250 cps (Neoprene Latex 102)

350 cps (Neoprene Latex 101)

450 cps (Neoprene Latex 635)

465 cps (Neoprene Latex 650)

Visc. (Mooney):

35–40 (ML4, 100 C) (Baypren 211)

35–55 (ML4, 100 C) (Baypren 610, 710)

40–50 (ML4, 100 C) (Baypren 110, 112, 210)

45 (Baypren 213)

45–55 (ML4, 100 C) (Baypren 115, 215)

50–60 (ML4, 100 C) (Baypren 214)

65–75 (ML4, 100 C) (Baypren 124)

70–75 (ML4, 100 C) (Baypren 321 GR)

70–80 (ML4, 100 C) (Baypren 220)

70–90 (ML4, 100 C) (Baypren 321)

75–85 (ML4, 100 C) (Baypren 320 GR)

80–95 (ML4, 100 C) (Baypren 320)

90–105 (ML4, 100 C) (Baypren 236, 331)

90–110 (ML4, 100 C) (Baypren 235)
95–105 (ML4, 100 C) (Baypren 330)
95–125 (ML4, 100 C) (Baypren 230, 233)
100–120 (ML4, 100 C) (Baypren 130)
120 (Baypren 236, 243)

Stability:

Good resistance to weathering, ozone, and chemicals; very good resistance to aging and hot air; med. resistance to oil and petrol (Baypren 110, 112, 115, 124, 130, 210, 211, 214, 215, 220, 230, 233, 235, 236, 320, 320 GR, 321, 321 GR, 330, 331, 610, 710)

Good resistance to oils, solvents, chemicals, ozone, weather, oxidation, heat, flame, and abrasion (Neoprene AC, AD, AF, AG, AH, CG, FB, GN, GNA, GRT, GW, TRT, TW, TW-100, W, WB, WD, WHV, WHV-A, WK, WRT)

pH:

≈ 6.5 (Baypren Latex 4 R)
12.5–13.0 (Baypren Latex B, GK, MKB, SK, T)

MECHANICAL PROPERTIES:

Tens. Str.:

15.8 MPa (Nobestos/D-7102)

STD. PKGS.:

25-kg polyethylene-lined paper bags on 1000-kg pallets (Baypren 110, 112, 115, 124, 130, 210, 211, 214, 215, 220, 230, 233, 235, 236, 320, 320 GR, 321, 321 GR, 330, 331, 610, 710)

125-kg or 625-kg plastic drums; 16–23 ton tank wagons (Baypren Latex 4 R, B, BK, MKB, SK, T)

55-lb bags (Neoprene AC, AD, AF, AG, AH, CG, GN, GNA, GRT, GW, TRT, TW, TW-100, W, WB, WD, WHV, WHV-A, WK, WRT)

55-lb fiber drums (Neoprene FB)

55-gal steel drums; 4000-gal tank trucks; 8000-gal tank cars (Neoprene Latex 101, 102, 571 572, 601-A, 622, 635, 650, 654, 671, 735-A, 750, 842-A)

Polyethylene

SYNONYMS:

Ethene, homopolymer
PE

EMPIRICAL FORMULA:

$(H_2C=CH_2)_x$

Polyethylene (cont'd.)

STRUCTURE:
Basic structure for LDPE:

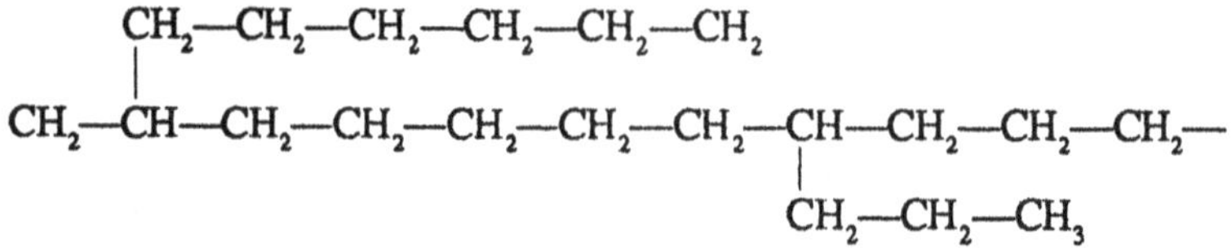

$$CH_2-CH_2-CH_2-CH_2-CH_2-CH_2$$
$$CH_2-CH-CH_2-CH_2-CH_2-CH_2-CH_2-CH-CH_2-CH_2-CH_2-$$
$$CH_2-CH_2-CH_3$$

CAS No.:
9002-88-4

TRADENAME EQUIVALENTS:
Chemplex 1005, 1007, 1008, 1013, 1014, 1015, 1016, 1017 [Chemplex]
Norcast 6368 [R.H. Carlson]
PE005, PE012, PE012/C [M.A. Industries]
RTP 700FR [Fiberite]
Tenite Polyethylene [Eastman]

Polyethylene copolymer:
Arcel Moldable Polyethylene Copolymers [Arco]

Polyethylene emulsion:
Jonwax 26 [S.C. Johnson]
Peem 122 Conc., 397, 410 [GAF]

Biaxially oriented polyethylene film:
Clysar ECL, EH [DuPont]

Crosslinkable polyethylene:
Marlex CL-50-35, CL-100, CL-100B [Phillips]

Branched polyethylene:

LDPE (low density polyethylene) homopolymer:
Bapolene 1030, 1052, 1072, 1082 [Bamberger Polymers]
Chemplex 1005, 1007, 1008, 1013, 1014, 1015, 1016, 1017, 1040, 3401, 3402, 3404, 3405 [Chemplex]
LMA-000 Polyethylene [Mobil]
LBA-133 Polyethylene [Mobil]
LGA-563 Polyethylene [Mobil]
LKA-753 Polyethylene [Mobil]
LLA-533 Polyethylene [Mobil]
Modic L-100F, L-400F, L-400H [Mitsubishi]
Norchem NPE 130, NPE 190, NPE 320, NPE 330, NPE 333, NPE 334, NPE 336, NPE 350, NPE 353, NPE 510, NPE 810, NPE 820, NPE 831, NPE 840, NPE 853, NPE 860, NPE 861, NPE 930, NPE 931, NPE 940, NPE 941, NPE 950, NPE 952, NPE 953, NPE 954 [Norchem]
PE 1017, 1018, 1019, 1028, 1117, 2130, 2151, 4517, 4560, 5325, 5554-H, 5555, 5561, 5565, 5613, 5619, 5622, 5625, 5754, 5755, 5861 [Chevron]
Petrothene 320, 334, 336, 350, 353, 940, 941, 952, 953, 954, 954-1, 955, 955-0, 957,

Polyethylene *(cont'd.)*

957-0, 957-1, 962, 962L, 963, 964, 965, 980, 983, 983-6, 3401, 3401E, 3404, 3404B, 3404D, 3404H, 3407, 3407B, 3407G, 3407H, 3408, 3408A, NA 140, NA 141, NA 142, NA 143, NA 145, NA 147, NA 148, NA 152, NA 153, NA 154, NA 271, NA 272, NA 273, NA 279, NA 284, NA 301, NA 344 [Quantum/USI]

LDPE/vinyl acetate copolymer:

Chemplex 1044 (3.5% VA), 1045 (2.0% VA), 1050 (2.0% VA), 1054 (3.5% VA), 1057 (7.5% VA), 1060 (7.5% VA), 3040 (4.5% VA), 3043 (4.5% VA), 3044 (3.0% VA), 3104 (2.0% VA), 3105 (1.5% VA), 3311 (18% VA) [Chemplex]

Norchem NPE 420, NPE 480, NPE 481 [Norchem]

PE 5220 (6% VA), 5222 (6% VA), 5240 (2% VA), 5254 (4% VA), 5272 (4% VA), 5280 (8% VA), 5290 (12% VA) [Chevron]

Petrothene 420 (2.5% VA), 425 (2.5% VA), 440 (4.0% VA), 440-1 (4.0% VA), 441 (4.0% VA), 442 (4.0% VA), 445 (4.0% VA), 480 (4.5% VA), 481 (4.5% VA), 1060 (7.5% VA), 1060K (7.5% VA), 3004A (3.0% VA), 3004C (3.0% VA), 3004D (3.0% VA), 3040B (4.5% VA), 3040G (4.5% VA), 3040L (4.5% VA), 3043 (4.5% VA), 3043B (4.5% VA), 3043G (4.5% VA), 3043H (4.5% VA), 3104A (2.0% VA), 3104D (2.0% VA), 3350 (7.5% VA), NA 233 (4.5% VA), NA 234 (4.5% VA), NA 235 (4.7% VA), NA 238 (1.5% VA), NA 239 (3.5% VA), NA 289 (4.0% VA), NA 290 (7.0% VA), NA 295 (6.2% VA), NA 386 (2.0% VA), NA 387 (2.0% VA), NA 388 (2.0% VA) [Quantum/USI]

Linear polyethylene:

HDPE (high density polyethylene):

Alathon 7440, 7970, PE-5510 [DuPont]

Bapolene 2001, 2062, 2072, 2082, 2101 [Bamberger Polymers]

HiD 9300, 9301, 9327, 9346, 9347 [Chevron]

Hostalen GM 7255, GM 7746 [Hoechst Celanese]

Marlex BMN TR955, EHM TR160, HHM 4903, HHM TR401, HHM TR418, HHM TR460, HHM TR480, HXM 50100 [Phillips]

Modic H-100E, H-100F, H-400C, H-400F [Mitsubishi]

Norchem NHD 6908 [Norchem]

Paxon SS55-100, SS55-180, SS55-250, SS55-400 [Allied-Signal]

Petrothene HD 5002, HD 5003, HD 5601, HD 5602, HD 5604, HD 5703, HD 5704, HD 5705, HD 5711, HD 5712D, HD 5713, HD 6004, HD 6007, HD 6009, HD 6085, LA 203F3, LA 303F3, LA 404F3, LA 408F3, LB 748, LB 830, LB 832, LB 833, LC 732, LR 723, LR 732, LR 734, LR 920, LS 901, NA 225 [Quantum/USI]

Stat-Kon AS-F, AS-FE, BLM, FE [LNP]

Vestolen A3512, A3512 R, A3513, A3515, A4516, A5016 F, A5017, A5018, A5041 R, A5543, A5561, A5561 P, A6012, A6013, A6014, A6016, A6017, A6042 [Huls AG]

HDPE homopolymer:

Alathon 7030, 7040, 7050, 7835, 7840 [DuPont]

Bapolene 2011 [Bamberger Polymers]

Chemplex 6001, 6004, 6006, 6008, 6009, 6109 [Chemplex]

Polyethylene *(cont'd.)*

HiD 9602, 9634, 9660 [Chevron]

Hostalen GA 7960, GC 7560 [Hoechst Celanese]

Marlex EHM 6001, EHM 6003, EMN 6030, EMN TR885 [Phillips]

Paxon AA60-003, AA60-007, AD60-007 [Allied-Signal]

Petrothene LB 924 [Quantum/USI]

HDPE copolymer:

Alathon 7140, 7220, 7230, 7240, 7245, 7320, 7340, 7810, 7815, 7820, 7860, 7910, 7915, 7960, 7965, 7970, PE-5510 [DuPont]

Bapolene 2101 [Bamberger Polymers]

Chemplex 5003, 5402, 5602, 5604 [Chemplex]

HiD 9606, 9632, 9640, 9642, 9650 [Chevron]

Hostalen GB 6950, GF 7740 F2, GM 5010 T2, GM 5010 T2N [Hoechst Celanese]

Marlex BMN 53120, BMN 55200, BMN TR-880, EHM 6006, HHM 4520, HHM 5202, HHM 5502, HHM TR140 [Phillips]

Paxon AA55-003, AB50-003, AF40-003, BA50-100 [Allied-Signal]

Petrothene LR 923, LY 520, LY 600, LY 660, LY 955 [Quantum/USI]

Glass-reinforced HDPE:

RTP 701 (10% glass fibers), 703 (20% glass fibers), 705 (30% glass fibers), 707 (40% glass fibers) [Fiberite]

Thermocomp FF-1004 (20% fiber glass), FF-1006 (30% fiber glass), FF-1008 (40% fiber glass) [LNP]

HMW-HDPE (high m.w. high density polyethylene) copolymer:

Chemplex 5701, 5704, 5705 [Chemplex]

HiD 9690 [Chevron]

Hostalen GM 9255 F [Hoechst Celanese]

Paxon FD60-018 [Allied-Signal]

HMW-LDPE (high m.w. low density polyethylene):

Petrothene 3503, 3503A, 3503C, 3503E, 3503G, 3505, 3505A, 3507, 3507A [Quantum/USI]

LLDPE (linear low density polyethylene):

Bapolene 3092, 3092L [Bamberger Polymers]

Petrothene GA 601-030, 601-031, 601-032, 601-033, 601-130, 603-035, 604-040, 605-030, 605-031, 605-033, 605-133, 605-150 [Quantum/USI]

Petrothene PA 161, 162 [Quantum/USI]

LMWPE (low m.w. polyethylene):

A-C Polyethylene 6, 6A, 8, 8A, 9, 9A, 617, 617A, 629A [Allied-Signal]

Epolene C-10, C-15, N-10, N-10-P, N-11, N-11-P, N-12, N-12-P, N-14, N-14-P, N-34, N-34-P, N-45, N-45-P [Eastman]

MDPE (medium density polyethylene):

Chemplex 1101, 3015, 3024, 3052 [Chemplex]

Interflo Porous Plastic [Chromex]

Marlex HHM TR130, HHM TR400 [Phillips]

PE-25 [Washington Penn]

Polyethylene *(cont'd.)*

PE 5961 [Chevron]

UHMW-PE (ultra high m.w. polyethylene):
Hostalen GUR [Hoechst AG]
Solidur 10 100, 10 100 IV 20, 10 802 AST, 10 Color Series, 10 DS Series, 25, Ceram P, Marble [Solidur Plastics]
1900 UHMW Polymers [Hercules]
Ultra-Wear UHMWPE [Polymer Corp.]

Chlorinated polyethylene:
Bayer CM 2552 (25% Cl), CM 3610 (36% Cl), CM 3630 (36% Cl), CM 3631 (36% Cl), CM 3632 (36% Cl), CM 4230 (42% Cl) [Bayer AG]

Chlorosulfonated polyethylene:
Hypalon 20 (29% Cl, 1.4% S), 30 (43% Cl, 1.1% S), 40 (35% Cl, 1% S), 40S (35% Cl, 1% S), 4085 (36% Cl, 1% S), LD-999 (35% Cl, 1% S), 45 (24% Cl, 1% S), 48 (43% Cl, 1% S) [DuPont]

MODIFICATIONS/SPECIALTY GRADES:

Antiblock:
LBA-133 Polyethylene; LGA-563 Polyethylene; LKA-753 Polyethylene; LLA-533 Polyethylene; Norchem NPE 320 (med.), NPE 330 (high), NPE 336 (high), NPE 350 (med.), NPE 420 (med.), NPE 480 (high), NPE 481 (high), NPE 930 (high), NPE 931 (high), NPE 940 (med.), NPE 941 (high), NPE 950 (high), NPE 952 (high), NPE 954 (high); PE 2130 (high), 2151 (high), 5220 (med.), 5222 (high), 5240 (med.), 5254 (med.), 5272 (med.), 5280 (high), 5290 (high), 5325 (med.), 5554-H (med.), 5555 (low), 5561 (med.), 5565 (med.), 5613 (high), 5619 (med.), 5622 (med.), 5625 (high), 5754 (med.), 5755 (med.), 5861 (med.), 5961 (low); Petrothene 320, 336, 350, 420, 425, 440, 440-1, 441, 445, 481, 941, 952, 954, 954-1, 955, 957-1, 962, 963, 964, 980, 983, 983-6, 1060K, 3004A, 3004C, 3004D, 3040B, 3040G, 3040L, 3104A, 3104D, 3401E, 3404B, 3404D, 3404H, 3407B, 3407G, 3407H, 3408A, 3503A, 3503C, 3503E, 3503G, 3505, 3505A, 3507A, GA 601-031 (high), GA 601-032 (high), GA 601-033 (high), GA 603-035 (high), GA 605-031 (high), GA 605-033 (high), GA 605-133 (med.)

Antioxidant:
Petrothene 954-1, 962L, GA 601-030, GA 601-031, GA 601-032, GA 601-033, GA 601-130; Vestolen A3512 R, A5041 R

Antistat:
Alathon PE-5510; Solidur 10 802 AST; Stat-Kon AS-F, AS-FE

Cling agent:
Petrothene 442 (high), 3043B (low), 3043G (med.), 3043H (high)

Oxidized:
A-C Polyethylene 629A

Processing aid:
Petrothene GA 601-130, GA 605-150

Slip agent:
LBA-133 Polyethylene; LGA-563 Polyethylene; LKA-753 Polyethylene; LLA-533

Polyethylene *(cont'd.)*

Polyethylene; Norchem NPE 320 (med.), NPE 330 (high), NPE 336 (high), NPE 350 (med.), NPE 420 (med.), NPE 481 (med.), NPE 860, NPE 931 (med.), NPE 941 (med.), NPE 950 (low), NPE 952 (med.), NPE 954 (high); PE 1028 (high), 2130 (high), 2151 (med.), 5220 (low), 5222 (high), 5240 (high), 5254 (low), 5272 (high), 5280 (high), 5290 (med.), 5325 (high), 5554-H (med.), 5555 (high), 5565 (high), 5619 (med.), 5754 (med.), 5861 (high), 5961 (low); Petrothene 320 (med.), 336 (high), 350 (med.), 420 (med.), 425 (high), 440 (me4d.), 440-1 (low), 441 (med.), 445 (high), 481 (high), 941 (med.), 952 (med.), 954 (high), 954-1 (med.), 955 (med.), 957-1 (low), 962 (med.), 964 (high), 980 (med.), 1060K (med.), 3004A (high), 3004C (med.), 3004D (med.), 3040B (high), 3040G (med.), 3104A (high), 3104D (med.), 3401E (low), 3404B (low), 3404D (med.), 3407B (med.), 3407G (low), 3408A (med.), 3503A (med.), 3503E (med.-high), 3503G (high), 3505 (med.), 3505A (med.), 3507A (high), GA 601-032 (high), GA 601-033 (med.), GA 603-035 (med.), GA 605-033 (med.), GA 605-133 (med.)

Carbon black additive:
Norchem NPE 130 (2.6% carbon black), NPE 190 (2.6% carbon black); Vestolen A3512 R, A5041 R

Carbon powder grade:
Stat-Kon BLM, FE

Statically dissipative:
Stat-Kon BLM, FE

High-flow:
Bapolene 3092, 3092L; Norchem NPE 840, NPE 860, NPE 861; Paxon SS55-400; RTP 700FR

Flame-retardant:
RTP 700FR

CATEGORY:
Thermoplastic resin

PROCESSING:
Blow molding:
Bapolene 2001, 2011, 2101; Chemplex 5003, 5402, 5602, 5604, 6001, 6004, 6006, 6008, 6009; Marlex EHM 6001, EHM 6003, EHM 6006, HHM 4903, HHM 5202; Norchem NPE 810; Paxon AA55-003, AA60-003, AA60-007, AB50-003, AD60-007, AF40-003, BA50-100; Petrothene HD 5002, HD 5003, HD 5602, HD 5604, HD 5703, HD 5704, HD 5705, HD 5711, HD 5712D, HD 5713, HD 6004, HD 6007, HD 6009, HD 6085, LA 203F3, LA 303F3, LA 404F3, LA 408F3, LB 748, LB 830, LB 832, LB 833, LC 732, LR 723, LR 732, LR 734, LR 920, LS 901, NA 225, NA 279, NA 301; Stat-Kon BLM; Tenite Polyethylene

Blown film extrusion:
Alathon 7810, 7815, 7820; Chemplex 3015; Hostalen GM 9255 F; LBA-133 Polyethylene; LKA-753 Polyethylene; LLA-533 Polyethylene; LMA-000 Polyethylene; Petrothene GA 604-040, LB 924, LR 923, LY 520, LY 600, LY 660, LY 955, NA 140, NA 141, NA 142, NA 143, NA 145, NA 147, NA 148, NA 152, NA 153, NA

154, NA 233, NA 234, NA 235, NA 238, NA 239, NA 271, NA 272, NA 273, NA 284, NA 289, NA 290, NA 295, NA 344, NA 386, NA 387, NA 388, PA 161, PA 162; Vestolen A6042

Calendering:
Hypalon 40, 40S, 4085, LD-999, 45

Cast film:
Alathon 7835, 7840; Chemplex 3402; Petrothene GA 604-040, LB 924, LR 923, LY 520, LY 600, LY 660, LY 955, NA 140, NA 141, NA 142, NA 143, NA 145, NA 147, NA 148, NA 152, NA 153, NA 154, NA 233, NA 234, NA 235, NA 238, NA 239, NA 271, NA 272, NA 273, NA 284, NA 289, NA 290, NA 295, NA 344, NA 386, NA 387, NA 388, PA 161, PA 162

Chill roll casting/extrusion:
Alathon 7835, 7840, 7860

Coating:
Chemplex 1013, 1014, 1015, 1016, 1017, 1101; Hypalon 20, 30; Jonwax 26; PE 1017, 1018, 1019, 1028, 4517, 4560; Tenite Polyethylene; Vestolen A5016 F

Coextrusion:
Alathon 7810, 7815, 7820, 7835; Chemplex 3311, 6109; HiD 9632, 9634, 9640, 9660; Marlex EHM TR160; Modic H-100E, H-100F, H-400C, H-400F, L-100F, L-400F, L-400H; PE 5280; Petrothene NA 148

Compounding:
Bapolene 3092, 3092L; Chemplex 3311

Compression forming:
Paxon FD60-018

Extrusion:
Alathon 7910, 7915, 7960, 7965, 7970; Bapolene 1030; Chemplex 5003, 5402, 5602, 5604, 6001, 6004, 6006, 6008, 6009; HiD 9300, 9301, 9327, 9346, 9347; Hypalon 40, 40S, 45, 48, 4085, LD-999; Marlex EHM 6006, HHM 5202, HHM 5502, HHM TR140, HHM TR400, HHM TR401, HHM TR418, HHM TR460, HHM TR480; Norchem NPE 510, NPE 810; Paxon AA60-003, AB50-003, BA50-100; PE-25; PE 1017, 1018, 1019, 1028; Petrothene GA 601-030, GA 601-031, GA 601-032, GA 601-033, GA 601-130, GA 603-035, GA 604-040, GA 605-030, GA 605-031, GA 605-033, GA 605-133, GA 605-150; Solidur 10 100; Stat-Kon AS-FE; Tenite Polyethylene

Extrusion blow molding:
Vestolen A5543, A5561, A5561 P

Film grade:
Alathon 7810, 7815, 7820; Chemplex 1005, 1007, 1008, 1040, 1044, 1045, 1050, 1054, 1057, 1060, 3015, 3024, 3040, 3043, 3044, 3052, 3104, 3105, 3401, 3404, 3405, 6109; HiD 9632, 9634, 9640, 9650, 9660, 9690; LGA-563 Polyethylene; Marlex EHM TR160, HHM TR130, HHM TR140; Norchem NPE 320, NPE 330, NPE 333, NPE 334, NPE 336, NPE 350, NPE 353, NPE 420, NPE 480, NPE 481, NPE 930, NPE 931, NPE 940, NPE 941, NPE 950, NPE 952, NPE 953, NPE 954;

Polyethylene *(cont'd.)*

PE 1117, 2130, 2151, 5220, 5222, 5240, 5254, 5272, 5280, 5290, 5325, 5554-H, 5555, 5561, 5565, 5613, 5619, 5622, 5625, 5754, 5755, 5861, 5961; Petrothene 320, 334, 336, 350, 353 420, 425, 440, 440-1, 441, 442, 445, 480, 481, 940, 941, 952, 953, 954, 954-1, 955, 955-0, 957, 957-0, 957-1, 962, 962L, 963, 964, 965, 980, 983, 983-6, 1060, 1060K, 3004A, 3004C, 3004D, 3040B, 3040G, 3040L, 3043, 3043B, 3043G, 3043H, 3104A, 3104D, 3350, 3401, 3401E, 3404, 3404B, 3404D, 3404H, 3407, 3407B, 3407G, 3407H, 3408, 3408A, 3503, 3503A, 3503C, 3503E, 3503G, 3505, 3505A, 3507, 3507A, GA 601-030, GA 601-031, GA 601-032, GA 601-033, GA 601-130, GA 603-035, GA 605-030, GA 605-031, GA 605-033, GA 605-133, GA 605-150; Tenite Polyethylene; Vestolen A3512, A5016 F, A6042

Foam molding:

Arcel Moldable Polyethylene Copolymers; Bapolene 2062

Injection blow molding:

Marlex HHM 5502; Petrothene HD 5601

Injection foam processing:

Norchem NHD 6908, NPE 510

Injection molding:

Alathon 7030, 7040, 7050, 7140, 7220, 7230, 7240, 7245, 7320, 7340, 7440, PE-5510; Bapolene 1052, 1072, 1082, 2062, 2072, 2082, 3092, 3092L; Hostalen GA 7960, GB 6950; Hypalon 40, 40S, 45, 4085, LD-999; Marlex BMN 53120, BMN 55200, BMN TR880, EMN 6030, EMN TR885, HHM 4520; Norchem NPE 820, NPE 831, NPE 840, NPE 853, NPE 860, NPE 861; Paxon SS55-100, SS55-180, SS55-250, SS55-400; RTP 700FR, 701, 703, 705, 707; Solidur 10 100; Tenite Polyethylene; Vestolen A3512, A3513, A3515, A4516, A5017, A5018, A5543, A6013, A6014, A6016, A6017

Lamination:

Alathon 7835; Chemplex 1013, 1014, 3043, 6109; Hostalen GM 9255 F; PE 1017, 1018, 1019, 1028, 4517, 4560, 5254

Press molding:

Hostalen GF 7740 F2, GM 5010 T2, GM 5010 T2N, GUR

Rotational molding:

Marlex BMN TR955, CL-50-35, CL-100, CL-100B; Vestolen A3515

Sheet grade:

Chemplex 5705; HiD 9602, 9642, 9690; Marlex EHM 6001, EHM 6003, HHM 4903, HHM 5202, HHM 5502, HXM 50100

Thermoforming:

Chemplex 5705, 6001; HiD 9602, 9606, 9690; Marlex EHM 6001, EHM 6003, HHM 4903, HHM 5202, HHM 5502, HXM 50100

Water bath extrusion:

Alathon 7860

APPLICATIONS:

Agriculture industry: (Norchem NPE 930, NPE 931, NPE 940, NPE 941; Solidur Marble); irrigation spraying systems (Vestolen A3512 R); tanks (Marlex CL-100,

172

Polyethylene (cont'd.)

CL-100B)

Architectural applications: construction industry (Norchem NPE 930, NPE 931, NPE 940, NPE 941; Solidur Marble; Vestolen A3513); sheet roofing (Hypalon 20, 30, 40, 40S, 45, 48, 4085, LD-999)

Automotive applications: (Norchem NPE 810); battery cases (RTP 701, 703, 705, 707); bumpers (Arcel Moldable Polyethylene Copolymers); components (Hypalon 20, 30, 40, 40S, 45, 48, 4085, LD-999; Marlex BMN TR880, HHM 4903); fender liners (Thermocomp FF-1004, FF-1006, FF-1008); tire sidewalls (Hypalon 20, 30, 40, 40S, 45, 48, 4085, LD-999)

Aviation industry: aircraft ski facings (1900 UHMW Polymers)

Consumer products: bowling pins (1900 UHMW Polymers); camper tops (Marlex CL-50-35); camping equipment (Vestolen A5017, A5018, A6017); diapers (Petrothene 952, 953, 954, 954-1, LY 660); disposables (Vestolen A5018); furniture/seats (Chemplex 5704; Marlex CL-50-35, HHM 4520); household chemical containers (Paxon AA55-003; Petrothene HD 5602, HD 5604, LA 203F3, LC 732, LR 732, LR 734, LR 920); household wrap (Norchem NPE 350); housewares (Bapolene 1052, 1072, 1082, 2072, 3092, 3092L; HiD 9602; Hostalen GB 6950; Marlex BMN 55200; Norchem NPE 831, NPE 860, NPE 861; Paxon SS55-180; Vestolen A5017, A5018, A6017); ice chests and coolers (Marlex HHM 5502); ice cube trays (Bapolene 1072); novelties (Bapolene 1052; Norchem NPE 860, NPE 861); plastic nursers (Petrothene 962, 962L, 963, 964, 965); safety helmets (Marlex EMN 6030); shoes (Hypalon 20, 30, 40, 40S, 45, 48, 4085, LD-999); skids for gliders/snowmobiles (Hostalen GUR); skis (Hostalen GUR); sporting goods (Bapolene 1072); toys (Bapolene 1052, 1072, 1082, 2072; HiD 9602; Marlex BMN 55200; Norchem NPE 860, NPE 861; Paxon SS55-180); utensils (Vestolen A6017)

Electrical/electronic industry: (Hostalen GUR); cable and wire jacketing/sheaths/insulation (Bayer CM 2552, CM 3610, CM 3630, CM 3631, CM 3632, CM 4230; Hypalon 20, 30, 40, 40S, 45, 48, 4085, LD-999; Norchem NHD 6908, NPE 130, NPE 190; PE-25); electrical components (Stat-Kon AS-F, AS-FE, BLM, FE); electrical enclosures/packaging (Stat-Kon BLM, FE); insulation (Hostalen GUR; Norchem NHD 6908, NPE 130)

FDA-approved applications: (Alathon 7030, 7040, 7050, 7140, 7220, 7230, 7240, 7245, 7320, 7340, 7440, 7820, 7835, 7840, 7860, PE-5510; Bapolene 1030, 1052, 1072, 1082, 2001, 2011, 2062, 2072, 2082, 2101, 3092; Clysar ECL, EH; Marlex BMN 53120, BMN 55200, BMN TR880, BMN TR955, EHM 6001, EHM 6003, EHM 6006, EHM TR160, EMN 6030, EMN TR885, HHM 4520, HHM 4903, HHM 5202, HHM 5502, HHM TR130, HHM TR140, HXM 500100; Norchem NPE 320, NPE 330, NPE 334, NPE 336, NPE 350, NPE 420, NPE 480, NPE 481, NPE 510, NPE 810, NPE 820, NPE 930, NPE 931, NPE 940, NPE 941, NPE 950, NPE 952, NPE 953, NPE 954; Petrothene GA 601-030, GA 601-031, GA 601-032, GA 601-033, GA 601130, GA 603-035, GA 604-040, GA 605-030, GA 605-031, GA 605-033, GA 605-133, GA 605-150, HD 5002, HD 5003, HD 5601, HD 5602, HD 5604, HD 5703, HD 5704, HD 5705, HD 5711, HD 5712D, HD 5713, HD 6004,

Polyethylene *(cont'd.)*

HD 6007, HD 7009, HD 6085, LA 203F3, LA 303F3, LA 404F3, LA 408F3, LB 748, LB 830, LB 832, LB 833, LB 924, LC 732, LR 723, LR 732, LR 734, LR 920, LR 923, LS 901, LY 520, LY 600, LY 660, LY 955, NA 140, NA 141, NA 142, NA 143, NA 145, NA 147, NA 148, NA 152, NA 153, NA 154, NA 225, NA 233, NA 234, NA 235, NA 238, NA 239, NA 271, NA 272, NA 273, NA 279, NA 284, NA 289, NA 290, NA 295, NA 301, NA 344, NA 386, NA 387, NA 388; Solidur 10 100); USDA approved (Clysar ECL, EH; Norchem NPE 320, NPE 330, NPE 334, NPE 336, NPE 350, NPE 810, NPE 820, NPE 930, NPE 931, NPE 940, NPE 941, NPE 950, NPE 952, NPE 953, NPE 954; Solidur 10 100)

Food industry: (1900 UHMW Polymers; Ultra-Wear UHMWPE); beverage bottles (Chemplex 6004, 6006, 6008, 6009; Marlex EHM 6006; Paxon AD60-007; Petrothene HD 6004, HD 6007, HD 6009, LA 404F3, LA 408F3, LB 748, LB 833); food packaging (Alathon 7030), 7040, 7050, 7140, 7220, 7230, 7240, 7245, 7320, 7340, 7810, 7810, 7815, 7820, 7835, 7840, 7860, PE-5510; Bapolene 2011, 2072, 2082, 3024, 3040, 3043, 3044, 3104, 3402, 3404, 3405; Clysar ECL, EH; HiD 9634, 9660; Hostalen GUR; LKA-753 Polyethylene; Marlex EHM TR160, EMN 6030, HHM 4903, HHM TR 130, HHM TR 1140; Norchem NPE 320, NPE 350, NPE 480, NPE 481, NPE 950, NPE 952, NPE 953, NPE 954; PE 5222, 5325; Petrothene 420, 425, 441, 3004A, 3004C, 3004D, 3040B, 3040G, 3040L, 3104A, 3104D, 3404, 3404B, 3404D, 3404H, 3407, 3407B, 3407G, 3407H, 3503, 3503A, 3503C, 3503E, 3503G, 3505, 3507A, GA 601-030, GA 601-031, GA 601-032, GA 601-033, GA 601-130, GA 603-035, GA 604-040, GA 605-030, GA 605-031, GA 605-033, GA 605-133, GA 605-150, LR 923, LY 520, LY 600, NA 386, NA 387, NA 388); food processing equipment (Solidur 10 100); ice bags (PE 5222, 5280; Petrothene 480, 481, 1060, 1060K, 3350, 3507, 3507A, NA 239); milk carton coatings (Chemplex 1015, 1016, 1017, 1101)

Functional additives: antiblocking agent (Epolene C-10, C-15, N-10, N-10-P, N-11, N-11-P, N-12, N-12-P, N-14, N-14-P); binder for magnetic materials (Hypalon 45); blending resin (Chemplex 3311; Petrothene LY 660); detackifier (Epolene C-10, C-15, N-10, N-10-P, N-11, N-11-P, N-12, N-12-P, N-14, N-14-P, N-34, N-34-P, N-45, N-45-P); lubricant (A-C Polyethylene 6, 6A, 8, 8A, 9, 9A, 617, 617A, 629A; Epolene C-10, C-15, N-10, N-10-P, N-12, N-12-P, N-34, N-34-P; Peem 397, 410); melt index modifier (A-C Polyethylene 6, 6A, 8, 8A, 9, 9A, 617, 617A); mold release agent (A-C Polyethylene 6, 6A, 8, 8A, 9, 9A, 617, 617A, 629A; Epolene C-10, C-15, N-10, N-10-P, N-11, N-11-P, N-12, N-12-P, N-14, N-14-P, N-34, N-34-P, N-45, N-45-P); pigment dispersant (A-C Polyethylene 6, 6A, 8, 8A, 9, 9A, 617, 617A); processing aid (Epolene C-10, C-15, N-10, N-10-P, N-11, N-11-P, N-12, N-12-P, N-14, N-14-P, N-34, N-34-P, N-45, N-45-P); softener (Peem 122 Conc.)

Industrial applications: adhesion film (Modic H-400F, L-100F, L-400H); adhesive layer in coextrusion (Modic H-100E, H-100F, H-400C, H-400F, L-100F, L-400F, L-400H); bags (Alathon 7810, 7815, 7820; Bapolene 1030; Chemplex 1060, 1101, 3104, 3402, 3404; HiD 9632, 9640, 9650, 9690; Hostalen GM 9255 F; Marlex HHM TR130, HHM TR140; Norchem NPE 480, NPE 481, NPE 930, NPE 931, NPE 940,

Polyethylene *(cont'd.)*

NPE 941; PE 1117, 2130, 2151; Petrothene 940, 941, 952, 953, 954, 954-1, 980, 983, 983-6, 3104A, 3104D, 3404, 3404B, 3404D, 3404H, 3408, 3408A, GA 601-030, GA 601-031, GA 601-032, GA 601-033, GA 601-130, GA 605-030, GA 605-031, GA 605-033, GA 605-133, GA 605-150, LR 923, LY 520, LY 600); barrier packaging (HiD 9650, 9660; Marlex EHM TR160); bearings (Hostalen GUR; 1900 UHMW Polymers); bottles (Bapolene 2001, 2011; Norchem NPE 810; Petrothene HD 6085, LA 303F3, LB 830, LB 832, LB 833, LC 732, LS 901, NA 225, NA 279); bundling (Norchem NPE 480, NPE 481; Petrothene 962, 962L, 963, 964, 965, 3401, 3401E); cellular rubber (Bayer CM 2552, CM 3610, CM 3630, CM 3631, CM 3632, CM 4230); chemical process industry (Hostalen GUR; Hypalon 20, 30, 40, 40S, 45, 48, 4085, LD-999; Marlex HHM 5202, HHM 5502; Thermocomp FF-1004, FF-1006, FF-1008; 1900 UHMW Polymers; Ultra-Wear UHMWPE); chromatography (Interflo Porous Plastic); closures/caps (Bapolene 1052, 2082, 3092, 3092L; Marlex BMN 55200, HHM 4520; Norchem NPE 810, NPE 820, NPE 853; Vestolen A5543); coatings (Chemplex 1013, 1014; Hypalon 20, 30, 40, 40S, 45, 48, 4085, LD-999; Jonwax 26; PE 1017, 1018, 1019, 1028, 4517, 4560; Tenite Polyethylene); color concentrates (A-C Polyethylene 6, 6A, 8, 8A, 9, 9A, 617, 617A; Bapolene 1052, 1072, 3092, 3092L); containers (Bapolene 1082, 2062, 2072, 2082, 2101, 3092, 3092L; Chemplex 5402, 5704; Hostalen GB 6950; Marlex BMN 53120, BMN 55200, BMN TR955, CL-50-35, CL-100, EHM 6003, EHM 6006, EMN TR885, HHM 4520, HHM 4903; Modic H-100E, H-100F, H-400C, H-400F, L-100F, L-400F; Norchem NPE 820, NPE 831, NPE 860, NPE 861; Paxon AA55-003, AA60-003, AA60-007, AB50-003, AF40-003, SS55-100, SS55-250, SS55-400; Petrothene LR 723, NA 301; Vestolen A5017, A5018, A5561, A5561 P, A6012, A6014, A6017); conveyor belting (Bayer CM 2552, CM 3610, CM 3630, CM 3631, CM 3632, CM 4230); cordage (Vestolen A6012); crates (Hostalen GC 7560; Marlex BMN 53120, BMN TR955; Vestolen A4516, A6016); cryogenics (Hostalen GUR); detergent/bleach/solvent containers (Bapolene 2001, 2101; Chemplex 5003, 5602, 5604; Marlex HHM 5202, HHM 5502; Paxon AB50-003; Petrothene HD 5002, HD 5003, LA 203F3, LB 830, LB 832, LB 833, LR 732, LR 734, LS 901); displays (Bapolene 2072); electroplating (Hostalen GUR); fabric coating (Hypalon 20, 30, 40, 40S, 45, 48, 4085, LD-999; Vestolen A6042); fan casings (Thermocomp FF-1004, FF-1006, FF-1008); fiberglass laminate (Arcel Moldable Polyethylene Copolymers); films (Alathon 7810, 7815, 7820; Chemplex 1007, 1040, 1044, 1045, 1050, 1054, 1057, 3015, 3024, 3040, 3043, 3044, 3052, 3104, 3105, 3401, 3402, 3404, 3405, 6109; Clysar ECL, EH; HiD 9632, 9634, 9640, 9650, 9660, 9690; Hostalen GM 9255 F; LBA-133 Polyethylene; LGA-563 Polyethylene; LKA-753 Polyethylene; LLA-533 Polyethylene; LMA-000 Polyethylene; Marlex EHM TR160, HHM TR130, HHM TR140; Norchem NPE 320, NPE 330, NPE 333, NPE 334, NPE 336, NPE 350, NPE 353, NPE 420, NPE 480, NPE 481, NPE 930, NPE 931, NPE 940, NPE 941, NPE 950, NPE 952, NPE 953, NPE 954; PE 1117, 2130, 2151, 5220, 5222, 5240, 5254, 5272, 5280, 5290, 5325, 5554-H, 5555, 5561, 5565, 5613, 5619, 5622, 5625, 5754, 5755, 5861, 5961; Petrothene

175

Polyethylene *(cont'd.)*

320, 334, 336, 350, 353, 420, 425, 440, 440-1, 441, 442, 445, 480, 481, 940, 941, 952, 953, 954, 954-1, 955, 955-0, 957, 957-0, 957-1, 962, 962L, 963, 964, 965, 980, 983, 983-6, 1060, 1060K, 3004A, 3004C, 3004D, 3040B, 3040G, 3040L, 3043, 3043B, 3043G, 3043H, 3104A, 3104D, 3350, 3401, 3401E, 3404, 3404B, 3404D, 3404H, 3407, 3407B, 3407G, 3407H, 3408, 3408A, 3503, 3503A, 3503C, 3503E, 3503G, 3505, 3505A, 3507, 3507A, GA 601-030, GA 601-031, GA 601-032, GA 601-033, GA 603-035, GA 604-040, GA 605-030, GA 605-031, GA 605-033, GA 605-133, GA 605-150; Tenite Polyethylene; Vestolen A5016 F, A6042); fluidizing (Interflo Porous Plastic); foam packaging (Arcel Moldable Polyethylene Copolymers); foam structures (Norchem NPE 510); fragrance control (Interflo Porous Plastic); garment bags (LGA-563 Polyethylene; Norchem NPE 330, NPE 334, NPE 336; PE 5861; Petrothene 336); gas and liquid filtration (Interflo Porous Plastic); gaskets/seals (Hostalen GUR; Hypalon 20, 30, 40, 40S, 45, 48, 4085, LD-999; 1900 UHMW Polymers); heat-seal applications (Hostalen GM 9255 F; Norchem NPE 320, NPE 350; PE 1017, 5220, 5222, 5554-H, 5561; Petrothene 3507, 3507A); hollow articles (Norchem NPE 810; Vestolen A3512, A5543, A5561, A5561 P); hot-melt adhesive (Norcast 6368); housings (Marlex HHM 4903, HHM 5202); impact plates (Ultra-Wear UHMWPE); ink transfer (Interflo Porous Plastic); intricate/hard-to-fill moldings (RTP 700FR; Vestolen A5017, A5018); laminated paper products (Petrothene 334); laminates (Chemplex 1013, 1014, 3043, 6109; Hostalen GM 9255 F; PE 1017, 1018, 1019, 1028, 4517, 4560, 5254); large containers (Vestolen A3513); large parts (Chemplex 5705, 6001; Marlex HXM 50100; Petrothene HD 5705; RTP 701, 703, 705, 707); lids (Bapolene 1082, 3092, 3092L; Norchem NPE 840; Vestolen A6013); liners (Bapolene 1030; Chemplex 1005, 1007, 1008, 1050, 1054; HiD 9632, 9634, 9640, 9642, 9660, 9690; Hostalen GUR; Hypalon 20, 30, 40, 40S, 45, 48, 4085, LD-999; LLA-533 Polyethylene; Marlex HHM 5502, HHM TR130, HHM TR140; Norchem NPE 930, NPE 931, NPE 950, NPE 952, NPE 953, NPE 954; PE 1117, 2130, 2151; Petrothene 952, 953, 954, 954-1, 957, 957-0, 957-1, GA 601-030, GA 601-031, GA 601-032, GA 601-033, GA 601-130, LY 660, NA 143, NA 145, NA 154, NA 301, NA 344; Solidur 10 802 AST, 10 Ds Series; 1900 UHMW Polymers; Ultra-Wear UHMWPE; Vestolen A3512 R); machinery parts (Hostalen GUR); material handling (Solidur 10 802 AST; 1900 UHMW Polymers); metals processing (1900 UHMW Polymers); mining industry (Hostalen GUR; Solidur 10 DS Series; 1900 UHMW Polymers; Ultra-Wear UHMWPE); moisture release (Interflo Porous Plastic); NSF approved (Alathon 7910, 7915; HiD 9327; Hostalen GM 5010 T2; Marlex HHM TR418, HHM TR480); outdoor applications (Marlex CL-100B); overwrap (PE 5961); packaging (Chemplex 1014, 1015, 1016, 1017, 1044, 1060, 3044, 3404; Clysar ECL; HiD 9650; Hostalen GM 9255 F, GUR; Marlex HHM 5202; Norchem NPE 320, NPE 420, NPE 480, NPE 481, NPE 940, NPE 941, NPE 950, NPE 952, NPE 953, NPE 954; PE 5240, 5272, 5325, 5555, 5613, 5619, 5622, 5625; Petrothene 320, 440, 440-1, 445, 1060, 1060K, 3350, 3407, 3407B, 3407G, 3407H, 3408, 3408A, GA 601-030, GA 601-031, GA 601-032, GA 601-033, GA 601-130, GA

Polyethylene (cont'd.)

603-035, GA 604-040, GA 605-030, GA 605-031, GA 605-033, GA 605-133, GA 605-150, NA 140, NA 141, NA 142, NA 147, NA 148, NA 152, NA 153, NA 154, NA 233, NA 234, NA 235, NA 238, NA 239, NA 271, NA 272, NA 284, NA 289, NA 290, NA 295, NA 344, NA 386, NA 387, NA 388, PA 161, PA 162; Vestolen A5017, A5018, A5543; Vestolen A6017); pallets (HiD 9690; Marlex BMN TR955, HXM 50100; Vestolen A6013); paper coatings (Chemplex 1015, 1016, 1017; PE 4517, 4560; Vestolen A5016 F); paper industry (Solidur 10 Ds Series, Ceram P; 1900 UHMW Polymers; Ultra-Wear UHMWPE); paper products overwrap/laminates (Chemplex 3024; Norchem NPE 333); petroleum industry (Marlex HHM TR400, HHM TR401); pipe fittings (Marlex HHM TR400, HHM TR401, HHM TR418, HHM TR460, HHM TR480; Thermocomp FF-1004, FF-1006, FF-1008; Vestolen A3512 R); pipes (Alathon 7910, 7915, 7960, 7965, 7970; Chemplex 5402; HiD 9300, 9301, 9327, 9346, 9347; Hostalen GM 5010 T2, GM 5010 T2N; Marlex HHM TR400, HHM TR401, HHM TR418, HHM TR460, HHM TR480; PE-25; Tenite Polyethylene; Vestolen A3512 R, A5041 R); plastics (A-C Polyethylene 6, 6A, 8, 8A, 9, 9A, 617, 617A, 629A; Epolene N-12); potable water applications (HiD 9301, 9327, 9347; Hostalen GM 5010 T2); profiles (Bapolene 1030); pumps (Hostalen GUR; Thermocomp FF-1004, FF-1006, FF-1008); release paper (Hostalen GM 9255 F); roll covers (Bayer CM 2552, CM 3610, CM 3630, CM 3631, CM 3632, CM 4230); rolls (Hypalon 20, 30, 40, 40S, 45, 48, 4085, LD-999); rubber (Bayer CM 2552, CM 3610, CM 3630, CM 3631, CM 3632, CM 4230; Epolene C-10, C-15, N-10, N-10-P, N-11, N-11-P, N-12, N-12-P, N-14, N-14-P, N-34, N-34-P, N-45, N-45-P); rubberized fabrics (Bayer CM 2552, CM 3610, CM 3630, CM 3631, CM 3632, CM 4230); sheet (Alathon 7810, 7815, 7820; Chemplex 5705, 6001; HiD 9602, 9606, 9690; Marlex EHM 6001, EHM 6003, EHM TR160, HHM 4903, HHM 5202, HHM 5502, HXM 50100; Modic H-100F, H-400F, L-100F, L-400F; Paxon AB50-003, BA50-100, FD60-018; Vestolen A3512 R, A5041 R); shrink and stretch applications (Chemplex 1040, 1045, 1054, 3043, 3401; Clysar ECL, EH; PE 5613, 5619, 5622, 5625; Petrothene 442, 980, 983, 983-6, 3043, 3043B, 3043G, 3043H, 3503, 3503A, 3503C, 3503E, 3503G, 3505, 3505A, 3507, 3507A, GA 604-040, NA 273); slide plates (Hostalen GUR); slit film (Hostalen GF 7740 F2); structural foam (Bapolene 2062); tanks, drums (Chemplex 5701, 5704; Marlex CL-50-35, CL-100, CL-100B, EHM 6003, HHM 4520, HHM 4903; Paxon BA50-100; Petrothene HD 5704, HD 5711, HD 5712D, HD 5713, LR 723; Vestolen A6013, A6014); technical/industrial parts (Bayer CM 2552, CM 3610, CM 3630, CM 3631, CM 3632, CM 4230; Marlex BMN 53120, BMN TR880, EMN 6030, HHM 5502; Norchem NPE 820, NPE 831; Paxon SS55-100; Vestolen A6014); textile applications (Peem 122 Conc., 397, 410; 1900 UHMW Polymers; Vestolen A6012, A6042); textile industry parts (Hostalen GUR); thin sections (Bapolene 1082, 2082; Hostalen GA 7960; Marlex BMN 55200, EMN TR885; Paxon SS55-250, SS55-400; Petrothene LB 748; Vestolen A4516, A5017, A5018); tote bins, boxes (Bapolene 2062; HiD 9690; Marlex BMN TR880, BMN TR955, EHM 6003, HHM 4903, HHM 5202, HHM 5502; Thermocomp FF-1004, FF-1006, FF-1008;

Polyethylene (cont'd.)

Vestolen A6013); trays (HiD 9690; Marlex HHM 4903, HHM 5202, HHM 5502; RTP 701, 703, 705, 707); tubing/hoses (Bapolene 1030; Bayer CM 2552, CM 3610, CM 3630, CM 3631, CM 3632, CM 4230; Hypalon 20, 30, 40, 40S, 45, 48, 4085, LD-999; Modic H-100E, H-100F, H-400C, H-400F, L-100F, L-400F; Norchem NPE 810); tubular film (Modic H-100F, L-100F, L-400F, L-400H; Vestolen A3512, A6042); typewriter ribbon and ink/carbon coating substrates (Petrothene LY 955); valves (Hostalen GUR); wear parts (1900 UHMW Polymers; Ultra-Wear UHMWPE)

Marine equipment: boats (Marlex CL-50-35, HXM 50100); buoys (Arcel Moldable Polyethylene Copolymers); flotation toys (Arcel Moldable Polyethylene Copolymers); surfboards (Arcel Moldable Polyethylene Copolymers)

Medical applications: hospitalwares (Bapolene 2072); orthopedics (Hostalen GUR); pharmaceutical processing equipment (Solidur 10 100); sterilizable packaging (Hostalen GM 9255 F)

Military packaging: (Marlex CL-100, CL-100B)

PROPERTIES:

Form:

Liquid (Peem 122 Conc.)

Emulsion (Jonwax 26; Peem 397, 410)

Viscous (Vestolen A3512 R, A5041 R)

Solid (Interflo Porous Plastic)

Pellets (Bapolene 3092, 3092L; Epolene C-10, C-15, N-10, N-11, N-12, N-14, N-34, N-45; Hostalen GM 9255 F; Marlex CL-100, CL-100B; Paxon AA55-003, AA60-003, AA60-007, AB50-003, AD60-007, AF40-003, BA50-100, FD60-018, SS55-100, SS55-180, SS55-250, SS55-400; Petrothene GA 601-030, GA 603-035, GA 604-040, GA 605-030)

Powder (A-C Polyethylene 6A, 8A, 9A, 617A, 629A; Epolene N-10-P, N-11-P, N-12-P, N-14-P, N-34-P, N-45-P; Hostalen GUR; Marlex BMN TR955, CL-50-35)

Fine powder (Bayer CM 2552, CM 3610, CM 3630, CM 3631, CM 3632, CM 4230)

Free-flowing powder (Vestolen A5561 P)

Prilled or diced (A-C Polyethylene 6, 8, 9, 617)

Chips (Hypalon 20, 30, 40, 40S, 45, 48, 4085, LD-999)

Avail. in rod, plate, bar, strip, etc. (Ultra-Wear UHMWPE)

Biaxially oriented film; avail. in 0.60-, 0.75-, 1.00-, and 1.50-mil gauges as flat or folded film (Clysar ECL)

Biaxially oriented film; avail. in 0.60-, 0.75-, 1.00-, 1.50-, and 2.00-mil gauges as flat or folded film (Clysar EH)

Color:

Translucent (Paxon AA55-003, AA60-003, AA60-007, AB50-003, AD60-007, AF40-003, BA50-100, FD60-018, SS55-100, SS55-180, SS55-250, SS55-400)

Avail. unpigmented and in black, gray, red, yellow, green, and blue (Hostalen GUR)

Avail. in white, red, brown, gray, yellow, blue, black, orange (Solidur 10 Color Series)

Avail. in white, red, green, brown, gray, blue (Solidur 10 DS Series)

Avail. in green, gray, black (Solidur 25)

White (Epolene C-10, C-15, N-10, N-10-P, N-11, N-11-P, N-12, N-12-P, N-14, N-14-P, N-34, N-34-P, N-45, N-45-P; Hypalon 20, 30, 40, 40S, 45, 48, 4085, LD-999)

Light (Bayer CM 2552, CM 3610, CM 3630, CM 3631, CM 3632, CM 4230)

Natural (Bapolene 3092; HiD 9301, 9346; Marlex CL-50-35, HHM TR400, HHM TR401; RTP 700FR)

Natural white, colorable (Interflo Porous Plastic)

Natural and standard colors (Marlex BMN TR955, CL-100)

Natural or custom colors (PE005)

Natural, black (RTP 701, 703, 705, 707)

Light amber (Norcast 6368)

Light tan (Jonwax 26)

Orange (HiD 9300; Marlex HHM TR460)

Lime green (Solidur Ceram P)

Black (HiD 9327, 9347; Hostalen GM 5010 T2; Marlex CL-100B, HHM TR418, HHM TR480; Norchem NPE 130, NPE 190; PE012, PE012/C; Solidur 10 802 AST)

Odor:

None (Hypalon 20, 30, 40, 40S, 45, 48, 4085, LD-999; Norcast 6368)

Virtually none (Vestolen A3512 R), A5041 R

Composition:

25% nonvolatile in water (Jonwax 26)

100% natural resin (Paxon AA55-003, AA60-003, AA60-007, AB50-003, AD60-007, AF40-003, BA50-100, FD60-018, SS55-100, SS55-180, SS55-250, SS55-400)

GENERAL PROPERTIES:

Solubility:

Readily sol. in common solvents (Hypalon 20, 30)

Excellent water dispersibility (Jonwax 26)

M.W.:

1800 (Epolene N-14, N-14-P)

2100 (Epolene N-45, N-45-P)

2200 (Epolene N-11, N-11-P)

2300 (Epolene N-12, N-12-P)

2900 (Epolene N-34, N-34-P)

3000 (Epolene N-10, N-10-P)

4000 (Epolene C-15)

8000 (Epolene C-10)

37,000 (Vestolen A5018)

46,000 (Vestolen A6017)

47,000 (Vestolen A5017)

52,000 (Vestolen A4516, A5016 F, A6016)

55,000 (Vestolen A3515)

60,000 (Vestolen A6014)

68,000 (Vestolen A6013)

Polyethylene *(cont'd.)*

78,000 (Vestolen A3513, A5543)
92,000 (Vestolen A3512, A3512 R)
96,000 (Vestolen A6012)
110,000 (Vestolen A6042)
147,000 (Vestolen A5041 R)
178,000 (Vestolen A5561, A5561 P)

Melt Flow:

0.25 dg/min (Alathon 7810 film)
0.45 dg/min (Alathon 7815, 7820)
1.2 dg/min (Alathon 7320)
2.8 dg/min (Alathon 7030)
3.0 dg/min (Alathon 7835)
4.0 dg/min (Alathon 7220, 7340, 7440)
4.5 dg/min (Alathon 7140)
6.0 dg/min (Alathon 7040, 7230, 7840 film, 7860 film)
12.0 dg/min (Alathon 7240)
17.5 dg/min (Alathon 7050)
20.0 dg/min (Alathon 7245)
0.04 g/10 min (HiD 9690)
0.05 g/10 min (Petrothene LY 600)
0.10 g/10 min (Paxon AF40-003; Vestolen A5561, A5561 P)
0.14 g/10 min (Petrothene LB 830, LY 520)
0.15 g/10 min (Marlex EHM 6001; Vestolen A5041 R)
0.17 g/10 min (Chemplex 5402; HiD 9632)
0.18 g/10 min (HiD 9327; Petrothene NA 140, NA 238)
0.20 g/10 min (HiD 9300, 9301, 9346, 9347, 9640, 9642; Marlex HHM TR418; Modic H-100E)
0.21 g/10 min (Petrothene HD 5002)
0.22 g/10 min (Bapolene 2001; Chemplex 5602)
0.24 g/10 min (Petrothene HD 5602)
0.25 g/10 min (Alathon 7910, 7960; Chemplex 6001; Marlex HHM TR400, HHM TR401, HHM TR460, HHM TR480; Norchem NPE 480, NPE 481, NPE 940, NPE 941; Petrothene 480, 481, 940, 941, 980, 983, 983-6, 3507, 3507A)
0.26 g/10 min (Petrothene LB 832, LB 833)
0.28 g/10 min (Petrothene LC 732)
0.30 g/10 min (Bapolene 2101; Chemplex 3311, 5003; HiD 9650; Marlex EHM 6003, HHM 4903, HHM TR130, HHM TR140; Modic H-400C; Paxon AA55-003, AA60-003, AB50-003; Petrothene 3503, 3503A, 3503C, 3503E, 3503G, HD 5003, LR 732, NA 141)
0.32 g/10 min (Chemplex 6004; Petrothene HD 6004, NA 273)
0.35 g/10 min (Chemplex 5604; HiD 9602, 9606; Marlex HHM 5202, HHM 5502; Petrothene HD 5604, LA 203F3, LA 303F3, LA 404F3, NA 235, NA 239)
0.38 g/10 min (Petrothene LR 734)

0.40 g/10 min (PE 5619; Petrothene 3350; Vestolen A6042)

0.42 g/10 min (Petrothene NA 289)

0.45 g/10 min (Alathon 7915, 7965, 7970; Petrothene LR 920, LR 923)

0.5 g/10 min (Chemplex 1040, 1044, 1045, 1060; Norchem NPE 930, NPE 931; PE005; PE 5280, 5613, 5622; Petrothene 3505, 3505A; Vestolen A3512, A3512R)

0.6 g/10 min (Chemplex 6006; PE 5290; Petrothene 1060, 1060K)

0.65 g/10 min (PE 5625; Petrothene LB 748)

0.7 g/10 min (HiD 9660; Marlex EHM TR160; Paxon AA60-007, AD60-007; Petrothene HD 6007, LB 924)

0.7–40 g/10 min (Tenite Polyethylene)

0.75 g/10 min (Marlex EHM 6006)

0.8 g/10 min (Bapolene 2011; Chemplex 6008, 6009; HiD 9634; PE 5754, 5755; Petrothene GA 605-030, GA 605-031, GA 605-033, GA 605-133, GA 605-150, HD 6009, LA 408F3; Vestolen A6012)

0.85 g/10 min (Chemplex 6109; Petrothene 962L, 963, HD 6085)

0.9 g/10 min (Petrothene NA 142)

1.0 g/10 min (Chemplex 3040, 3401; Modic H-100F, L-400F; PE-25; PE 5220, 5222; Petrothene 441, 962, 964, 965, 3040B, 3040G, 3040L, GA 601-030, GA 601-031, GA 601-032, GA 601-033, GA 601-130, HD 5601, PA 161)

1.1 g/10 min (Petrothene NA 147)

1.2 g/10 min (Alathon PE-5510; Chemplex 1057; Modic H-400F; PE 2151; Petrothene 3401, 3401E, NA 234, NA 271, NA 290, NA 301)

1.3 g/10 min (Norchem NPE 810; PE 5254, 5272, 5554-H, 5561; Vestolen A3513)

1.4 g/10 min (Modic L-100F; Petrothene LS 901, NA 148, NA 233, NA 272)

1.5 g/10 min (Chemplex 3044; Petrothene NA 386)

1.6 g/10 min (PE 5240)

1.7 g/10 min (Chemplex 3015; Petrothene PA 162)

1.8 g/10 min (Chemplex 3404; Paxon FD60-018; Petrothene 3404, 3404B, 3404DE, 3404H, NA 225; Vestolen A5543)

2.0 g/10 min (Bapolene 1030; Chemplex 1054, 3043, 3104, 3105, 3402; Marlex HHM 4520; Norchem NPE 350, NPE 353, NPE 853, NPE 950, NPE 952, NPE 953, NPE 954; PE 1117, 2130, 5325, 5555, 5565; Petrothene 350, 353, 952, 953, 954, 954-1, 3004A, 3004C, 3004D, 3043, 3043B, 3043G, 3043H, 3104A, 3104D, 3407, 3407B, 3407G, 3407H, GA 603-035, NA 143, NA 153, NA 388; Vestolen A6013)

2.1 g/10 min (Norchem NPE 510; Petrothene NA 279)

2.2 g/10 min (Modic L-400H; Norchem NPE 820; Petrothene NA 154, NA 344)

2.3 g/10 min (Petrothene 3408, 3408A)

2.4 g/10 min (Petrothene NA 152)

2.5 g/10 min (Chemplex 1007; Norchem NPE 320, NPE 420; Petrothene 320, 420, 425, 440, 440-1, 442, 445, 955, 955-0, GA 604-040)

2.6 g/10 min (Petrothene 957, 957-0, 957-1, NA 295, NA 387)

2.7 g/10 min (PE 5961)

3.0 g/10 min (Chemplex 3024; Marlex EMN 6030)

Polyethylene (cont'd.)

3.5 g/10 min (Chemplex 1005, 1050, 3405; Petrothene NA 145; Vestolen A6014)

3.7 g/10 min (Chemplex 1016)

4.5 g/10 min (Chemplex 1015; Norchem NPE 333)

5.0 g/10 min (Chemplex 1008; PE 4517; Petrothene NA 284; Vestolen A3515)

5.2 g/10 min (Chemplex 3052)

5.5 g/10 min (Chemplex 1017)

6.0 g/10 min (Chemplex 1101; Norchem NPE 330, NPE 334, NPE 336; PE 5861; Petrothene 334, 336)

6.2 g/10 min (Petrothene LY 660)

6.5 g/10 min (Marlex BMN TR955)

7.0 g/10 min (Chemplex 1014; PE 1017; Vestolen A4516, A5016 F, A6016)

8.0 g/10 min (Bapolene 1052)

8.5 g/10 min (Norchem NHD 6908)

8.8 g/10 min (Petrothene LY 955)

9.0 g/10 min (Norchem NPE 831; PE 1028; Vestolen A5017)

10 g/10 min (Marlex HXM 50100; Paxon BA50-100, SS55-100; PE 4560)

11 g/10 min (Chemplex 5701; Vestolen A6017)

12 g/10 min (Bapolene 2062; Marlex BMN 53120; PE 1018)

13 g/10 min (Chemplex 1013)

14.5 g/10 min (PE 1019)

18 g/10 min (Marlex BMN TR880; Paxon SS55-180)

20 g/10 min (Chemplex 5704, 5705; Marlex BMN 55200; Vestolen A5018)

22 g/10 min (Bapolene 1072)

23 g/10 min (Bapolene 2072)

24 g/10 min (Norchem NPE 860, NPE 861

25 g/10 min (Paxon SS55-250; PE012/C)

30 g/10 min (Bapolene 1082, 2082; Marlex EMN TR885)

40 g/10 min (Paxon SS55-400; PE012)

45 g/10 min (Norchem NPE 840)

50 g/10 min (Bapolene 3092)

Sp. Vol.:

21.6 in.³/lb (Thermocomp FF-1008)

23.7 in.³/lb (Thermocomp FF-1006)

25.2 in.³/lb (Thermocomp FF-1004)

Density:

0.906 (Epolene C-10, C-15)

0.910 (A-C Polyethylene 617, 617A; Epolene N-34, N-34-P)

0.915–0.924 (Tenite Polyethylene)

0.916 (Petrothene GA 604-040)

0.917 (PE 1017, 1018, 1019, 1028)

0.918 (Chemplex 1013, 1014; LMA-000; Norchem NPE 330, NPE 334, NPE 336, NPE 510, NPE 810, NPE 820, NPE 831, NPE 950, NPE 952, NPE 953, NPE 954; Petrothene 334, 336, 952, 953, 954, 954-1, 957, 957-0, 957-1, GA 601-030, GA

601-031, GA 601-032, GA 601-033, GA 601-130, NA 140, NA 225, NA 301)

0.919 (Chemplex 1005, 1007, 1008; Norchem NPE 333, NPE 930, NPE 931, NPE 940, NPE 941; Petrothene PA 161)

0.920 (A-C Polyethylene 6A; Bapolene 1030, 1082; Epolene N-14, N-14-P; LLA-533; Modic L-100F; PE 1117, 2151; Petrothene 940, 941, 962, 962L, 963, 964, 965, NA 142, NA 145, NA 154, NA 238, NA 344, PA 162)

0.921 (Epolene N-11, N-11-P; LGA-563; Petrothene NA 143)

0.922 (Chemplex 3404; Norchem NPE 350, NPE 353; PE 5625; Petrothene 980, 983, 983-6, 3401, 3401E, 3404, 3404B, 3404D, 3404H, 3408, 3408A, GA 603-035, NA 147, NA 148, NA 273)

0.923 (Chemplex 1015, 1016, 1017, 1040, 3401, 3405; Norchem NPE 853, NPE 860, NPE 861; PE 2130, 4517, 4560, 5613, 5619, 5622; Petrothene 350, 353, NA 141, NA 386, NA 387, NA 388)

0.924 (LBA-133; LKA-753; Norchem NPE 320, NPE 840; PE 5754, 5755, 5861; Petrothene 320, 955, 955-0, 3407, 3407B, 3407G, 3407H, NA 152, NA 153, NA 239, NA 279)

0.925 (Bapolene 1052, 1072, 3105, 3402; Epolene N-10, N-10-P; PE 5325, 5554-H, 5561; Petrothene 3503, 3503A, 3503C, 3503E, 3503G, 3505, 3505A, NA 233, NA 234, NA 235)

0.926 (Bapolene 3092; Chemplex 1045, 1050, 3052, 3104; Norchem NPE 420, NPE 480, NPE 481; PE 5555; Petrothene 441, 442, 480, 481, 3040B, 3040G, 3040L, 3104A, 3104D, GA 605-030, GA 605-031, GA 605-033, GA 605-133, GA 605-150, NA 289, NA 290, NA 295)

0.927 (Chemplex 1054, 3044; PE 5240, 5565; Petrothene 420, 425, 3004A, 3004C, 3004D, 3043, 3043B, 3403G, 3043H)

0.928 (Chemplex 1057, 1060, 3015, 3040, 3043; PE 5272; Petrothene 1060, 1060K, 3350, 3507, 3507A)

0.929 (Chemplex 1044; PE 5254; Petrothene 440, 440-1, 445)

0.930 (Modic H-400F, L-400F, L-400H; Norchem NPE 130; PE 5220, 5222; Petrothene NA 271, NA 284; Solidur 10 100 IV 20)

0.930–0.933 (Marlex CL-100, CL-100B)

0.931 (Chemplex 1101; PE 5280; Petrothene NA 272)

0.932 (Chemplex 3024; PE 5961)

0.934 (Norchem NPE 190)

0.935 (PE 5290; Vestolen A3512, A3513, A3515)

0.937 (Marlex HHM TR130)

0.937–0.940 (Marlex CL-50-35)

0.938 (Epolene N-12, N-12-P)

0.939 (Chemplex 3311; HiD 9301; Marlex HHM TR400)

0.940 (Alathon 7440; HiD 9300, 9640, 9642; Hostalen GUR; Modic H-400C; PE-25; Solidur 25, Marble; Stat-Kon AS-F, AS-FE; Ultra-Wear UHMWPE)

0.942 (Chemplex 5402; Hostalen GM 7746; Paxon AF40-003)

0.943 (Hostalen GF 7740 F2; Marlex HHM TR401)

Polyethylene *(cont'd.)*

0.944 (Alathon 7810, 7910, 7960; Hostalen GM 5010 T2N; Vestolen A3512 R)

0.945 (HiD 9346, 9632; Vestolen A4516)

0.946 (Marlex HHM TR140, HHM TR460; Petrothene LR 920, LR 923)

0.947 (Epolene N-45, N-45-P; Marlex HHM 4520; Petrothene HD 5705)

0.948 (Alathon PE-5510; Chemplex 5705)

0.949 (Marlex HHM 4903; Petrothene HD 5002, HD 5712D, LY& 520)

0.950 (Alathon 7140, 7320, 7340, 7815, 7915, 7965; Bapolene 2062; HiD 9606, 9650; Marlex HHM TR418, HXM 50100; Modic H-100E, H-100F; Norcast 6368; Paxon AB50-003, BA50-100; Petrothene LR 723, LY 660; Vestolen A5016 F, A5017, A5018)

0.950–0.954 (Hostalen GM 9255 F)

0.951 (Marlex HHM 5202; Petrothene LA 203F3, LY 600)

0.952 (Alathon 7220; Bapolene 2101; Chemplex 5003, 5704; HiD 9690; Hostalen GB 6950; Paxon SS55-100, SS55-180; Petrothene HD 5003, HD 5704, LS 901)

0.953 (Chemplex 5701; Hostalen GM 7255; Marlex BMN 53120; Paxon SS55-400; Petrothene HD 5713, LC 732, LR 732, LY 955)

0.954 (Chemplex 5604; Marlex BMN 55200; Paxon AA55-003, FD60-018; PE012/C; Petrothene HD 5601, HD 5604, HD 5703, LR 734)

0.955 (Alathon 7230, 7240, 7245, 7860; Bapolene 2001, 2072; Chemplex 5602; Hostalen GM 5010 T2; Marlex MBN TR880, BMN TR955, HHM 5502, HHM TR480; Paxon SS55-250; PE005; Petrothene HD 5602, LA 303F3; Vestolen A5041 R, A5543, A5561, A5561 P, A6014)

0.956 (HiD 9327, 9347)

0.957 (Alathon 7820, 7970; Petrothene HD 5711, LB 830, LB 832; Vestolen A6012)

0.958 (Vestolen A6013)

0.959 (Hostalen GC 7560; Petrothene LB 748, LB 833; Vestolen A6042)

> 0.959 (Petrothene LB 924)

0.960 (Alathon 7030, 7040, 7050, 7835, 7840; Bapolene 2011, 2082; Chemplex 6001, 6004, 6006, 6008, 6009; HiD 9634; Hostalen GA 7960; Marlex EHM 6001; Paxon AA60-003, AA60-007, AD60-007; PE012; Petrothene HD 6004, HD 6007, HD 6009, HD 6085; Solidur 10 DS Series)

0.960+ (Chemplex 6109; Norchem NHD 6908)

0.961 (Petrothene LA 404F3)

0.962 (HiD 9660; Petrothene LA 408F3; Vestolen A6016)

0.963 (HiD 9602; Marlex EHM 6003;   Vestolen A6017)

0.964 (Marlex EHM 6006, EHM TR160, EMN 6030)

0.967 (Marlex EMN TR885)

0.97 (Solidur 10 802 AST)

0.98 (Solidur Ceram P)

1.02 (RTP 701)

1.06 (Stat-Kon BLM, FE)

1.10 (Bayer CM 2552; RTP 703; Thermocomp FF-1004)

1.11 (Hypalon 45)

1.14 (Hypalon 20)
1.17 (Thermocomp FF-1006)
1.18 (Hypalon 40, 40S, 4085, LD-999; RTP 705)
1.19 (Bayer CM 3610, CM 3630, CM 3631, CM 3632)
1.25 (Bayer CM 4230)
1.26 (Hypalon 48)
1.27 (Hypalon 30)
1.28 (RTP 707; Thermocomp FF-1008)
1.43 (RTP 700FR)

Bulk Density:
380–420 g/l (Hostalen GUR)
8.2 lb/gal (Jonwax 26)

Visc.:
12 cps (5% in xylene) (Hypalon 48)
18 cps (5% in xylene) (Hypalon LD-999)
20 cps (5% in xylene) (Hypalon 40S)
25 cps (5% in xylene) (Hypalon 40)
50 cps (5% in xylene) (Hypalon 4085)
60 cps (5% in xylene) (Hypalon 45)
180 cps (140 C) (A-C Polyethylene 617, 617A)
230 cps (125 C) (Epolene N-14, N-14-P)
350 cps (140 C) (A-C Polyethylene 6A)
400 cps (25% in toluene) (Hypalon 30)
430 cps (125 C) (Epolene N-12-P)
450 cps (125 C) (Epolene N-12, N-45, N-45-P)
480 cps (125 C) (Epolene N-11, N-11-P)
660 cps (125 C) (Epolene N-34, N-34-P)
1300 cps (25% in toluene) (Hypalon 20)
1800 cps (125 C) (Epolene N-10, N-10-P)
4300 cps (150 C) (Epolene C-15)
7000 cps (465 F) (Norchem 6368)
9400 cps (150 C) (Epolene C-10)
56,000 cps (325 F) (Norcast 6368)
≈ 35 (Mooney, ML1+4, 120 C) (Bayer CM 3610)
≈ 85 (Mooney, ML1+4, 120 C) (Bayer CM 3630, CM 3631, CM 4230)
≈ 110 (Mooney, ML1+4, 120 C) (Bayer CM 3632)
≈ 130 (Mooney, ML5+4, 120 C) (Bayer CM 2552)

M.P.:
114 C (Modic L-400H)
116 C (Modic L-100F, L-400F)
121–124 C (Vestolen A3512, A3512 R, A3513, A3515)
125 C (Hostalen GF 7740 F2)
125–133 C (Hostalen GM 9255 F)

Polyethylene *(cont'd.)*

126–130 C (Hostalen GM 7746; Vestolen A4516, A5041 R)
127–131 C (Hostalen GB 6950; Vestolen A5017, A5018)
128 C (Modic H-400F)
128–132 C (Hostalen GM 7255; Vestolen A5016 F)
129–132 C (Vestolen A5543, A6014)
130–132 C (Hostalen GC 7560)
130–133 C (Hostalen GA 7960)
130–135 C (Vestolen A5561, A5561 P)
132–135 C (Vestolen A6012, A6013, A6042)
133 C (Modic H-100E, H-100F, H-400C)
133–136 C (Vestolen A6016, A6017)
135–138 C (Hostalen GUR)
127 F (Hostalen GM 5010 T2, GM 5010 T2N)
266 F (Ultra-Wear UHMWPE)

Softening Pt.:

102 C (A-C Polyethylene 617, 617A; Epolene C-15)
103 C (Epolene N-34, N-34-P)
104 C (Epolene C-10)
106 C (A-C Polyethylene 6A; Epolene N-14, N-14-P)
108 C (Epolene N-11, N-11-P)
111 C (Epolene N-10, N-10-P)
113 C (Epolene N-12, N-12-P)
118 C (Epolene N-45, N-45-P)
210–220 F (Norcast 6368)

Cloud Pt.:

69 C (Epolene N-34, N-34-P)
75 C (Epolene C-15)
77 C (Epolene C-10, N-14, N-14-P)
79 C (Epolene N-11, N-11-P)
85 C (Epolene N-10, N-10-P)
87 C (Epolene N-12, N-12-P)
97 C (Epolene N-45, N-45-P)

Stability:

Good resistance to aging in oils and hot air, to ozone, weathering, color change, to oils and liquid fuels; very good resistance to chemicals (Bayer CM 2552, CM 3610, CM 3630, CM 3631, CM 3632, CM 4230)

Remains durable at freezer temps.; does not embrittle with age (Clysar EH)

Remains durable at freezer temps.; will not become brittle with age; good burn-through resistance Clysar ECL)

Very good resistance to corrosive media except for strong oxidizing acids; good resistance to stress cracking (Hostalen GUR)

Excellent storage and color stability; excellent resistance to ozone and weathering; very good resistance to abrasion, heat aging (Hypalon 20)

Polyethylene *(cont'd.)*

Excellent storage and color stability; excellent resistance to chemicals, ozone, petroleum oils, weathering; very good resistance to abrasion, heat aging (Hypalon 30)

Excellent storage and color stability; excellent resistance to abrasion, chemicals, ozone, weathering; good resistance to compression set, flame, heat aging, petroleum oils (Hypalon 40, 40S, 4085, LD-999)

Excellent storage and color stability; excellent resistance to abrasion, ozone, weathering; good resistance to chemicals, compression set, heat aging (Hypalon 45)

Excellent storage and color stability; excellent resistance to abrasion, chemicals, ozone, petroleum oils, weathering; good resistance to flame, heat aging (Hypalon 48)

Substantially inert; slight swelling in chlorinated hydrocarbons; attacked by strong acids (Interflo Porous Plastic)

Excellent freeze/thaw stability (Jonwax 26)

Good weatherability and stress crack resistance (Norchem NPE 130)

Excellent weather and environmental stress crack resistance (Norchem NPE 190)

Superior resistance to tear and puncture (Norchem NPE 330, NPE 334, NPE 336)

Good chemical and environmental stress crack resistance (Norchem NPE 860)

Excellent chemical and corrosion resistance, abrasion resistance, and high impact resistance (Ultra-Wear UHMWPE)

pH:

9.6 (Jonwax 26)

MECHANICAL PROPERTIES:

Tens. Str.:

2000–3000 g/in. (break) (Clysar EH)

2700–6800 g/in. (break) (Clysar ECL)

8.3–12.4 MPa (yield) (Tenite Polyethylene)

8.6 MPa (yield) (Norchem NPE 510, NPE 820)

9.6 MPa (yield) (Norchem NPE 190, NPE 810, NPE 831)

9.65 MPa (yield) (Norchem NPE 930, NPE 931)

9.7 MPa (yield) 89Norchem NPE 940, NPE 941, NPE 950, NPE 952, NPE 953, NPE 954)

10.3 MPa (yield) (Norchem NPE 130)

12.4 MPa (yield) (Norchem NPE 350)

13.8 MPa (Norchem NPE 330 film, NPE 334 film, NPE 336 film); (yield) (Norchem NPE 860, NPE 861)

16.6 MPa (Norchem NPE 420 film)

17.2 MPa (Norchem NPE 320 film)

18 MPa (yield) (Vestolen A3512, A3512 R, A3515)

18.6 MPa (yield) (Alathon 7860 film)

19 MPa (yield) (Vestolen A3513)

19.3 MPa (yield) (Marlex HHM TR400, HHM TR418)

20.0 MPa (yield) (Alathon 7835 film)

≤ 20.6 MPa (Hypalon 20 vulcanizate, carbon black stocks)

Polyethylene *(cont'd.)*

20.7 MPa (Norchem NPE 480 film, NPE 481 film); (yield) (Alathon 7840 film)

21.4 MPa (yield) (Marlex HHM TR130 film, HHM TR401, HHM TR460, HHM TR480)

22 MPa (yield) (Hostalen GB 6950, GUR)

22.2 MPa (yield) (Marlex EHM TR160 film)

22.9 MPa (yield) (Alathon 7910, 7960)

23 MPa (yield) (Hostalen GF 7740 F2)

23.5 MPa (yield) (Alathon 7810 film)

24 MPa (yield) (Vestolen A4516, A5041 R)

24.1 MPa (yield) (Marlex HHM 4520)

≤ 24.2 MPa (Hypalon 30 vulcanizate, carbon black stocks)

24.8 MPa (yield) (Marlex HHM 4903)

25 MPa (yield) (Hostalen GM 5010 T2, GM 7746; Vestolen A5016 F)

25.2 MPa (yield) (Alathon 7915, 7965)

25.5 MPa (yield) (Alathon 7815 film, PE-5510)

26 MPa (yield) (Hostalen GA 7960, GC 7560; Vestolen A5017)

26.1 MPa (Marlex HHM TR140 film); (yield) (Marlex HXM 50100)

26.2 MPa (yield) (Marlex BMN 53120)

26.9 MPa (yield) (Alathon 7820 film, 7970; Marlex HHM 5202)

27.0 MPa (yield) (Marlex BMN 55200; Vestolen A5018)

≤27.6MPa (Hypalon 40, 40S, 45, 48, 4085, LD-999 vulcanizates, carbon black stocks)

27.6 MPa (yield) (Marlex HHM 5502)

28 MPa (yield) (Hostalen GM 7255; Vestolen A5543, A5561, A5561 P)

28.9 MPa (yield) (Marlex BMN TR880, EMN TR885)

29 MPa (yield) (Vestolen A6012)

29.7 MPa (yield) (Marlex EHM 6001)

30 MPa (yield) (Vestolen A6013, A6014, A6042)

30.2 MPa (yield) (Marlex EHM 6002, EHM 6006)

32 MPa (yield) (Vestolen A6016)

32.4 MPa (yield) (Marlex EMN 6030)

33 MPa (yield) (Vestolen A6017)

120 kg/cm² (yield) (Modic L-100F)

130 kg/cm² (yield) (Modic L-400F)

140 kg/cm² (Modic L-400H)

150 kg/cm² (yield) (Modic H-400F)

210 kg/cm² (yield) (Modic H-400C)

260 kg/cm² (yield) (Modic H-100E, H-100F)

35 psi (Arcel Moldable Polyethylene Copolymers 1.5 pcf molded foam)

1200 psi (yield) (Chemplex 1057 film, 1060 film)

1320 psi (yield) (LLA-533 film; LMA-000)

1350 psi (yield) (Chemplex 1044 film, 3043 film)

1400 psi (yield) (Chemplex 1005 film, 1007 film, 1008 film, 1054 film, 3040 film, 3044 film)

1500 psi (yield) (Chemplex 1050 film, 3105 film)
1600 psi (yield) (Chemplex 1045 film, 3104 film, 3402 film, 3405 film)
1640 psi (yield) (LBA-133 film; LKA-753 film)
1660 psi (yield) (Chemplex 3401 film, 3404 film)
1700 psi (yield) (Bapolene 1030 film, 1052, 1072, 1082, 1040 film)
1720 psi (Petrothene NA 143, NA 153)
1760 psi (break) (Bapolene 3092)
1780 psi (Petrothene NA 154)
1800 psi (Petrothene NA 145)
1820 psi (Petrothene NA 152, NA 279)
1900 psi (Petrothene NA 148, NA 225)
1910 psi (Petrothene NA 388)
1970 psi (Petrothene NA 344)
2000 psi (yield) (Chemplex 3052 film)
2050 psi (Petrothene NA 273)
2060 psi (Petrothene NA 284)
2100 psi (Petrothene NA 142, NA 233, NA 271, NA 301, NA 386)
2150 psi (Petrothene NA 147)
2200 psi (RTP 700FR); (yield) (Chemplex 3015 film, 3024 film; PE-25)
2300 psi (Petrothene NA 141, NA 272, NA 387, PA 162)
2340 psi (Petrothene NA 234)
2390 psi (Petrothene NA 290)
2400 psi (Petrothene NA 140, NA 295)
2440 psi (Petrothene NA 238)
2500 psi (break) (LGA-563 film)
2600 psi (Petrothene NA 235, NA 239, NA 289); (yield) (Marlex CL-50-35, CL-100, CL-100B; PE012/C)
3000 psi (Petrothene PA 161); (yield) (Paxon PE012)
3100 psi (Petrothene LY 955); (yield) (HiD 9300, 9301)
3200 psi (yield) (Chemplex 5402)
3300 psi (Petrothene LY 660; Stat-Kon AS-FE); (yield) (Alathon 7440)
3400 psi (yield) (Bapolene 2062; HiD 9346)
3450 psi (Petrothene LR 920)
3500 psi (Stat-Kon AS-F); (yield) (HiD 9327, 9347)
3600 psi (Petrothene HD 5002, HD 5705, HD 5712D, LA 203F3); (yield) (Chemplex 6109 film)
3625 psi (yield) (Hostalen GM 5010 T2N)
3700 psi (yield) (Alathon 7140, 7320, 7340); (break) (Chemplex 3311 film)
3780 psi (Petrothene HD 5003, HD 5704)
3800 psi (yield) (Alathon 7220; Bapolene 2072; Chemplex 5705; PE005)
3840 psi (Petrothene HD 5703, HD 5713)
3850 psi (Petrothene LC 732, LR 734)
3900 psi (Marlex BMN TR955; Petrothene HD 5601, HD 5604, LA 303F3, LR 732);

Polyethylene *(cont'd.)*

    (yield) (Bapolene 2101; Chemplex 5003, 5701, 5704)

3960 psi (Petrothene HD 5602)

3970 psi (Petrothene LR 723)

4000 psi (yield) (Alathon 7230, 7240, 7245)

4000–5500 psi (Ultra-Wear UHMWPE)

4080 psi (Petrothene HD 5711)

4100 psi (Petrothene LS 901); (yield) (Bapolene 2082)

4200 psi (yield) (Alathon 7030; Bapolene 2001; Chemplex 5602, 5604)

4300 psi (Stat-Kon BLM); (yield) (Alathon 7040, 7050)

4350 psi (Petrothene HD 6085)

4390 psi (Petrothene LB 830)

4400 psi (Chemplex 6009; Petrothene HD 6004, HD 6007, HD 6009); (yield) (Bapolene 2011; Chemplex 6004, 6006, 6008; Norchem NHD 6908); (break) (HiD 9660)

4450 psi (Petrothene LB 832, LB 833, LR 923)

4500 psi (Petrothene LA 404F3, LB 748; Stat-Kon FE); (yield) (Chemplex 6001); (break) (HiD 9634)

4510 psi (Petrothene LB 924)

4600 psi (Petrothene LA 408F3)

4900 psi (break) (Solidur 25, Marble)

5000 psi (Petrothene LY 600)

5600 psi (break) (Solidur 10 802 AST)

5700 psi (RTP 701)

6000 psi (break) (Petrothene GA 603-035 film; Solidur 10 100 IV 20)

6100 psi (break) (Petrothene GA 601-030 film, GA 601-031 film, GA 601-032 film, GA 601-033 film, GA 601-130 film; Solidur 10 DS Series)

6400 psi (break) (Petrothene GA 604-040; Solidur Ceram P)

6600 psi (break) (HiD 9640; Petrothene GA 605-030 film, GA 605-031 film, GA 605-033 film, GA 605-133 film, GA 605-150 film)

7000 psi (RTP 703)

7400 psi (break) (HiD 9650, 9690)

8000 psi (Thermocomp FF-1004); (break) (HiD 9632)

9000 psi (RTP 705)

10,000 psi (RTP 707; Thermocomp FF-1006)

11,500 psi (Thermocomp FF-1008)

Tens. Elong.:

Broke (yield) (Bapolene 2082)

1.5% (RTP 705, 707)

2–3% (Thermocomp FF-1004, FF-1006, FF-1008)

2.5% (RTP 703)

3.4% (RTP 701)

10% (Marlex EMN TR885)

10+% (RTP 700FR)

15–25% (Arcel Moldable Polyethylene Copolymers 1.5 pcf molded foam)

19% (Stat-Kon FE)

22% (Stat-Kon BLM)

60% (break) (Petrothene LY 955)

100% (Alathon 7245; Bapolene 2062, 2072); (yield) (Norchem NPE 930, NPE 931, NPE 940, NPE 941, NPE 950, NPE 952, NPE 953, NPE 954)

100–600% (break) (Tenite Polyethylene)

115% (Clysar ECL; LLA-533 film; LMA-000)

130% (Clysar EH)

150% (Bapolene 1082; Marlex BMN 55200); (break) (Bapolene 3092; Norchem NPE 860, NPE 861)

180% (yield) (Norchem NPE 350); (break) (Petrothene NA 272)

200% (Bapolene 1052); (break) (Hostalen GM 7255; Norchem NPE 831)

200–450% (Ultra-Wear UHMWPE)

210% (LBA-133 film)

220% (LGA-563)

250% (Chemplex 6109 film; Stat-Kon AS-F)

260% (break) (Hostalen GM 7746; Solidur Ceram P)

265% (Marlex EHM TR160 film)

275% (LKA-753)

300% (Marlex HHM 4520; Stat-Kon AS-FE)

321–400% (break) (Solidur 25, Marble)

333% (break) (Solidur 10 DS Series)

340% (Marlex HHM TR140 film)

350% (HiD 9660; Norchem NHD 6908)

370% (break) (Solidur 10 802 AST)

390% (break) (Petrothene NA 284)

400% (Alathon 7050; Bapolene 1072); (break) (Marlex CL-50-35; Norchem NPE 820)

450% (Chemplex 1005 film, 1007 film, 1008 film; HiD 9632; Norchem NPE 330 film, NPE 334 film, NPE 336 film); (break) (Hostalen GUR; Marlex CL-100, CL-100B; Solidur 10 100 IV 20)

460% (Chemplex 3015 film; HiD 9650)

> 465% (break) (Hostalen GF 7740 F2)

480% (Chemplex 1040 film); (break) (Hostalen GA 7960)

490% (Chemplex 1045 film, 1054 film; HiD 9634)

500% (Bapolene 2011; Chemplex 1044 film, 1057 film, 6008; HiD 9640; Marlex HHM TR130 film); (break) (Norchem NPE 810; Petrothene GA 601-030 film, GA 601-031 film, GA 601-032 film, GA 601-033 film, GA 601-130 film, GA 605-030 film, GA 605-031 film, GA 605-033 film, GA 605-133 film, GA 605-150 film; Vestolen A5543)

> 500% (Marlex HHM TR400, HHM TR401, HHM TR418, HHM TR460, HHM TR480); (yield) (PE-25)

520% (HiD 9690)

Polyethylene *(cont'd.)*

530% (Chemplex 1050 film, 3402 film)

540% (Chemplex 3024 film, 3040 film, 3043 film); (break) (Petrothene NA 145)

550% (Bapolene 1030 film; Chemplex 3044 film, 3104 film, 3401 film, 3404 film, 3405 film; Norchem NPE 320 film, NPE 420 film); (break) (Petrothene GA 603-035 film, NA 271, NA 388)

560% (Chemplex 3105 film)

570% (Chemplex 1060 film); (break) (Petrothene NA 147, NA 225, NA 386)

580% (Modic L-400H); (break) (Petrothene NA 141)

590% (break) (Petrothene NA 152, NA 301)

600% (Chemplex 3052 film, 3311 film; Marlex BMN TR880, EMN 6030; Modic H-100F; Norchem NPE 480 film, NPE 481 film); (break) (Norchem NPE 350; Petrothene HD 5713, LA 408F3, NA 289, NA 344; Vestolen A5016 F, A5017, A5018, A5041 R, A5561, A5561 P, A6012, A6013, A6014, A6016, A6017, A6042)

> 600% (Bapolene 2001, 2101; Chemplex 5003, 5402, 5602, 5604, 5701, 5704, 5705, 6001, 6004, 6006, 6009; Marlex EHM 6001, EHM 6003, EHM 6006, HHM 4903, HHM 5202, HHM 5502, HXM 50100); (break) (Petrothene HD 5002, HD 5003, HD 5601, HD 5602, HD 5604, HD 5703, HD 5704, HD 5705, HD 5711, HD 5712D, HD 6004, HD 6007, HD 6009, HD 6085)

610% (break) (Petrothene NA 148)

615% (break) (Petrothene NA 387)

620% (break) (Petrothene NA 279)

625% (break) (Petrothene NA 273)

630% (break) (Norchem NPE 930, NPE 931; Petrothene NA 238)

640% (break) (Petrothene NA 153)

645% (break) (Norchem NPE 130)

650% (Modic L-100F); (break) (Hostalen GB 6950; Norchem NPE 950, NPE 952, NPE 953, NPE 954; Petrothene LR 923, NA 143, NA 154)

> 650% (Marlex BMN 53120, BMN TR955; Petrothene GA 604-040)

660% (break) (Petrothene LR 920, NA 142, NA 235)

670% (break) (Petrothene NA 140, PA 162)

680% (break) (Petrothene NA 239)

700% (Modic H-400C; Norchem NPE 190); (break) (Petrothene LA 203F3, LA 303F3; Vestolen A3512, A3512 R, A3513, A3515, A4516)

710% (break) (Petrothene NA 290)

720% (break) (Petrothene NA 233)

730% (Modic L-400F)

740% (break) (Petrothene NA 295)

750% (Alathon 7835 film; HiD 9327); (break) (Hostalen GC 7560; Norchem NPE 510, NPE 940, NPE 941; Petrothene NA 234)

760% (break) (Petrothene PA 161)

800% (Alathon 7440, 7810 film, 7815 film, 7915, 7965, 7970; HiD 9300, 9301); (break) (Hostalen GM 5010 T2, GM 5010 T2N)

810% (break) (Petrothene LR 723)

850% (Alathon 7240, 7820 film, 7840 film, 7860 film; HiD 9347); (break) (Petrothene LC 732, LR 734)

870% (HiD 9346)

880% (break) (Petrothene LB 830)

890% (Alathon 7910, 7960)

900% (Alathon 7030, 7040, 7140; Modic H-400F); (break) (Petrothene LA 404F3, LR 732)

910% (break) (Petrothene LB 748, LB 924)

950% (break) (Petrothene LY 600)

960% (break) (Petrothene LB 832, LB 833)

1000% (Alathon 7220; Modic H-100E); (break) (Petrothene LS 901)

1100% (Alathon 7320, 7340, PE-5510); (break) (Petrothene LY 660)

1300% (Alathon 7230)

Tens. Mod.:

186 MPa (Norchem NPE 810)

650 MPa (Vestolen A3512, A3515)

680 MPa (Hostalen GF 7740 F2)

690 MPa (Alathon 7810 film)

700 MPa (Vestolen A3512 R, A3513)

725 MPa (Alathon 7835 film, 7860 film)

793 MPa (Alathon 7815 film)

897 MPa (Alathon 7840 film)

966 MPa (Alathon 7820 film)

1000 MPa (Vestolen A4516)

1100 MPa (Vestolen A5016 F, A5017, A5041 R)

1200 MPa (Vestolen A5018)

1300 MPa (Vestolen A6014)

1400 MPa (Vestolen A5543, A5561, A5561 P)

1500 MPa (Vestolen A6012, A6013, A6042)

1700 MPa (Vestolen A6016, A6017)

7200 psi (Chemplex 3311 film)

16,000 psi (Chemplex 1057 film, 1060 film)

21,000 psi (Chemplex 3052 film)

22,000 psi (LLA-533 film; LMA-000 film)

24,000 psi (Chemplex 3040 film, 3043 film)

25,000 psi (Chemplex 1044 film, 3044 film)

26,000 psi (LGA-563 film; Petrothene GA 604-040 film)

27,000 psi (Chemplex 1008 film, 3402 film)

28,000 psi (Chemplex 1005 film, 1007 film, 1054 film, 3104 film, 3105 film)

29,000 psi (Chemplex 3401 film, 3404 film, 3405 film; LBA-133 film; LKA-753 film)

29,500 psi (Petrothene GA 601-030 film, GA 601-031 film, GA 601-032 film, GA 601-033 film, GA 601-130 film)

30,000 psi (Bapolene 1082; Chemplex 1050 film)

Polyethylene (cont'd.)

31,000 psi (Bapolene 1072; Chemplex 1045 film)
32,000 psi (Bapolene 1052)
35,000 psi (Petrothene GA 603-035 film)
36,000 psi (Chemplex 1040 film)
41,000 psi (Chemplex 3015 film)
42,000 psi (Petrothene GA 605-030 film, GA 605-031 film, GA 605-033 film, GA 605-133 film, Ga 605-150 film)
43,000 psi (Chemplex 3024 film)
49,000 psi (Bapolene 3092)
80,000–100,000 psi (Ultra-Wear UHMWPE)
85,000 psi (PE012/C)
110,000 psi (PE012)
113,000 psi (Hostalen GM 5010 T2, GM 5010 T2N)
120,000 psi (PE005)
175,000 psi (Chemplex 6109 film)
230,000 psi (RTP 700FR)
400,000 psi (RTP 701)
600,000 psi (RTP 703)
900,000 psi (RTP 705)
1.25×10^6 psi (RTP 707)

Flex. Str.:

27 MPa (Hostalen GUR)
18 psi (5% strain) (Arcel Moldable Polyethylene Copolymers 1.5 pcf molded foam)
1500 psi (PE012/C)
2800 psi (Stat-Kon AS-FE)
3000 psi (PE005, PE012; Stat-Kon AS-F)
3600 psi (RTP 700FR)
3700 psi (Stat-Kon BLM)
3900 psi (Stat-Kon FE)
6500 psi (RTP 701)
9000 psi (RTP 703)
10,000 psi (Thermocomp FF-1004)
11,000 psi (RTP 705)
11,500 psi (Thermocomp FF-1006)
12,000 psi (RTP 707)
14,000 psi (Thermocomp FF-1008)

Flex. Mod.:

689 MPa (Marlex HHM TR400, HHM TR418)
790 MPa (Hostalen GF 7740 F2)
861 MPa (Marlex HHM TR401, HHM TR460, HHM TR480)
882 MPa (Marlex HHM TR130)
931 MPa (Alathon 7910, 7960)
1030 MPa (Marlex HHM 4520)

1069 MPa (Alathon 7915, 7965)
1157 MPa (Marlex HHM TR140)
1165 MPa (Marlex HHM 4903)
1207 MPa (Alathon PE-5510)
1276 MPa (Alathon 7970)
1309 MPa (Marlex BMN 53120, HHM 5202)
1340 MPa (Marlex BMN 55200)
1380 MPa (Marlex BMN TR880, HHM 5502)
1580 MPa (Hostalen GA 7960, GB 6950)
1653 MPa (Marlex EMN 6030)
1654 MPa (Marlex EHM 6001, EHM 6003)
1790 MPa (Hostalen GC 7560)
1791 MPa (Marlex EHM 6006)
1795 MPa (Marlex EMN TR885)
1846 MPa (Marlex EHM TR160)
2080 MPa (Marlex HXM 50100)
75,000 psi (PE012/C; Ultra-Wear UHMWPE)
80,000 psi (Stat-Kon AS-FE)
90,000 psi (PE012; Stat-Kon AS-F)
100,000 psi (Marlex CL-100, CL-100B; PE005)
110,000 psi (Marlex CL-50-35)
120,000 psi (Chemplex 5402)
125,000 psi (Bapolene 2062)
130,000 psi (Bapolene 2072)
136,000 psi (Hostalen GM 5010 T2, GM 5010 T2N)
140,000 psi (Alathon 7440)
150,000 psi (Bapolene 2082; Chemplex 5705; Stat-Kon BLM)
160,000 psi (Bapolene 2101; Chemplex 5003, 5701, 5704)
165,000 psi (Stat-Kon FE)
170,000 psi (Chemplex 5604)
175,000 psi (Alathon 7140, 7320, 7340)
180,000 psi (Bapolene 2001; Chemplex 5602; RTP 700FR)
185,000 psi (Alathon 7220)
195,000 psi (Marlex BMN TR955)
205,000 psi (Alathon 7230, 7240, 7245)
220,000 psi (Alathon 7030, 7040, 7050; Chemplex 6001, 6004, 6006, 6008, 6009)
225,000 psi (Bapolene 2011)
230,000 psi (Norchem NHD 6908)
260,000 psi (RTP 701)
550,000 psi (RTP 703)
600,000 psi (Thermocomp FF-1004)
800,000 psi (RTP 705)
900,000 psi (Thermocomp FF-1006)

Polyethylene *(cont'd.)*

10^6 psi (RTP 707)
1.1×10^6 psi (Thermocomp FF-1008)

Compr. Str.:

10 psi (5% deform.) (Arcel Moldable Polyethylene Copolymers 1.5 pcf molded foam)
2400 psi (RTP 700FR)
4000 psi (RTP 701)
5000 psi (RTP 703)
7000 psi (RTP 705)
7500 psi (RTP 707)

ESCR:

4 h (Bapolene 3092)
15 h (Bapolene 2011)
50 h (Bapolene 2001)
60 h (Bapolene 2101)
65 h (Alathon 7970)
100 h (Alathon PE-5510)
> 200 h (Alathon 7915, 7965)
> 1000 h (Alathon 7910, 7960)

Tear Str. (Elmendorf):

58–116 mN/μm (Alathon 7815 film)
96–170 mN/μm (Alathon 7820 film)
115 mN/μm (Alathon 7860 film)
116–193 mN/μm (Alathon 7810 film)
135 mN/μm (Alathon 7840 film)
190 mN/μm (Alathon 7835 film)
4–15 g (Clysar EH)
10–30 g (Clysar ECL)
100 g (Marlex EHM TR160 film)
250 g (Bapolene 1030 film)
560 g (Marlex HHM TR140 film)
620 g (Marlex HHM TR130 film)
13 g/mil (MD) (HiD 9660)
15 g/mil (MD) (HiD 9632)
16 g/mil (MD) (HiD 9650)
18 g/mil (MD) (HiD 9690)
20 g/mil (MD) (HiD 9634, 9640)
170 g/mil (MD) (LKA-753 film)
220 g (MD) (Petrothene GA 603-035)
225 g/mil (MD) (LLA-533 film; LMA-000 film)
230 g (MD) (Petrothene GA 605-030, GA 605-031, GA 605-033, GA 605-133, GA 605-150)
255 g/mil (TD) (HiD 9660)
280 g/mil (MD) (LGA-563 film)

345 g/mil (TD) (HiD 9634)
350 g (MD) (Petrothene GA 604-040)
400 g (MD) (Petrothene GA 601-030, GA 601-031, GA 601-032, GA 601-033, GA 601-130)
400 g/mil (TD) (HiD 9650, 9690)
580 g/mil (TD) (HiD 9640)
590 g/mil (TD) (HiD 9632)
Shear Str.:
3500 psi (Ultra-Wear UHMWPE)
4000 psi (Thermocomp FF-1004)
4400 psi (Thermocomp FF-1006)
4900 psi (Thermocomp FF-1008)
Impact Str. (Spencer):
53.4 kJ/μm (Alathon 7840 film, 7860 film)
53.4–85 kJ/μm (Alathon 7820 film)
58–85 kJ/μm (Alathon 7815 film)
89 kJ/μm (Alathon 7835 film)
93.4–133 kJ/μm (Alathon 7810 film)
217 MPa/mm (Marlex EHM TR160 film)
407 MPa/mm (Marlex HHM TR140 film)
461 MPa/mm (Marlex HHM TR130 film)
Impact Str. (Dart):
30 g/mil (HiD 9660)
35 g/mil (HiD 9634)
50 g (Petrothene NA 152)
60 g (Petrothene NA 272, NA 284)
65 g (Petrothene NA 271)
70 g (Chemplex 3015 film; Petrothene 3104A film, 3104D film)
75 g (Norchem NPE 320 film; Petrothene 320 film, 3407 film, 3407B film, 3407G film, 3407H film, NA 145, NA 153)
80 g (Chemplex 3024 film; Petrothene 3408 film, 3408A film)
85 g (Petrothene NA 143)
85 g/mil (HiD 9632)
90 g (Norchem NPE 330 film, NPE 334 film, NPE 336 film; Petrothene 334 film, 336 film, 955 film, 955-0 film, 3004A film, 3004C film, 3004D, 3404 film, 3404B film, 3404D film, 3404H film)
90 g/mil (HiD 9640, 9650)
95 g (Petrothene 420 film, 425 film, 3401 film, 3401E film)
100 g (Norchem NPE 350 film; Petrothene 350 film, 353 film, LY 520, NA 154, NA 387, PA 162)
105 g (Chemplex 3405 film; Petrothene 442 film, 3043 film, 3043B film, 3043G film, 3043H film)
110 g (Norchem NPE 420 film; Petrothene 3040B, 3040G film, 3040L film)

Polyethylene (cont'd.)

115 g (Chemplex 3104; Petrothene NA 388)

120 g (Chemplex 1008 film, 3105; Petrothene 440 film, 440-1 film, 445 film, 957 film, 957-0 film, 957-1 film)

125 g (Chemplex 1005 film)

130 g (Chemplex 1050 film, 3044 film; Norchem NPE 950 film, NPE 952 film, NPE 953 film, NPE 954 film; Petrothene 952 film, 953 film, 954 film, 954-1 film, NA 386, PA 161)

135 g (Chemplex 1007 film, 3404 film); (50 F) (Bapolene 1030 film)

140 g (Chemplex 3043 film; Petrothene 441 film)

150 g (Chemplex 1054 film; Petrothene GA 603-035 film, GA 605-030 film, GA 605-031 film, GA 605-033 film, GA 605-133 film, GA 605-150 film, NA 147, NA 148, NA 273, NA 344)

155 g (Petrothene 962 film, 962L film, 963 film, 964 film, 965 film)

160 g (Chemplex 3040 film; Norchem NPE 930 film, NPE 931 film)

170 g (Chemplex 3401 film; Petrothene NA 142)

175 g (Petrothene GA 604-040)

180 g (Petrothene 980 film, 983 film, 983-6 film)

200 g (Petrothene 3505 film, 3505A film, GA 601-030 film, GA 601-031 film, GA 601-032 film, GA 601-033 film, GA 601-130 film, NA 233)

210 g (Petrothene NA 141)

210 g/mil (HiD 9690)

220 g (Chemplex 1040 film; Norchem NPE 940 film, NPE 941 film; Petrothene 940 film, 941 film)

225 g (Petrothene NA 140, NA 239)

235 g (Chemplex 1044 film; Petrothene LY 600)

250 g (Norchem NPE 480 film, NPE 481 film; Petrothene 3503 film, 3503A film, 3503C film, 3503E film, 3503G film)

260 g (Chemplex 1045 film)

275 g (Petrothene NA 289)

290 g (Petrothene NA 234)

300 g (Petrothene NA 238)

320 g (Petrothene NA 235)

340 g (Chemplex 1057 film; Petrothene NA 290, NA 295)

360 g (Petrothene 3350 film)

370 g (Petrothene 1060 film, 1060K film)

380 g (Petrothene 480 film, 481 film)

400 g (Chemplex 1060 film)

430 g (Petrothene 3507 film, 3507A film)

Impact Str. (Charpy):

> 30 kg cm/cm² (Modic H-100E, H-100F, H-400C, H-400F, L-100F, L-400F, L-400H)

Impact Str. (Izod):

30 J/m notched (Vestolen A6017)

40 J/m notched (Vestolen A5018)

50 J/m notched (Vestolen A5017)

60 J/m notched (Vestolen A5016 F, A6016)

80 J/m notched (Vestolen A4516)

100 J/m notched (Vestolen A5543, A6014)

120 J/m notched (Vestolen A3515)

300 J/m notched (Vestolen A6013)

400 J/m notched (Vestolen A3513)

500 J/m notched (Vestolen A3512, A3512 R, A5041 R)

700 J/m notched (Vestolen A5561, A5561 P, A6012, A6042)

0.4 ft lb/in. (PE012)

0.5 ft lb/in. (Bapolene 2082); notched (Alathon 7240, 7245; Hostalen GA 7960; RTP 700FR)

0.6 ft lb/in. (Bapolene 2072)

0.8 ft lb/in. notched (Alathon 7050; Bapolene 2062; Hostalen GB 6950, GC 7560)

1.0 ft lb/in. notched (Alathon 7230; RTP 701; Thermocomp FF-1004)

1.1 ft lb/in. notched (Thermocomp FF-1006)

1.2 ft lb/in. notched (Alathon 7340; RTP 703; Stat-Kon FE)

1.3 ft lb/in. notched (Alathon 7220; RTP 705; Stat-Kon BLM; Thermocomp FF-1008)

1.4 ft lb/in. notched (Alathon 7140; RTP 707)

1.5 ft lb/in. notched (Alathon 7040, 7440)

1.8 ft lb/in. notched (Alathon 7030)

2.0 ft lb/in. (PE012/C); notched (Alathon 7320)

2.5 ft lb/in. notched (Bapolene 2101; Chemplex 5003; Petrothene HD 5003, LA 203F3)

3.0 ft lb/in. notched (Chemplex 6006, 6009; Petrothene HD 6009, HD 6085, LA 303F3, LA 408F3)

3.2 ft lb/in. notched (Chemplex 5704, 5705; Petrothene HD 5704, HD 5705)

3.5 ft lb/in. notched (Petrothene HD 5002, HD 5601)

4.0 ft lb/in. notched (Bapolene 2011; Chemplex 6008)

4.1 ft lb/in. notched (Petrothene HD 5712D)

4.5 ft lb/in. notched (Chemplex 5701; Petrothene HD 5604, HD 5713; Stat-Kon AS-F)

5.0 ft lb/in. (PE005); notched (Chemplex 5402; Petrothene HD 5602, HD 6007, LA 404F3; Stat-Kon AS-FE)

5.5 ft lb/in. notched (Petrothene HD 5711)

6.0 ft lb/in. notched (Bapolene 2001; Chemplex 5602, 5604, 6004; Petrothene HD 6004)

8.0 ft lb/in. notched (Chemplex 6001)

12 ft lb/in. notched (Hostalen GM 7255, GM 7746)

14.5 ft lb/in. notched (PE-25)

18 ft lb/in. notched (Solidur 25, Marble)

25–28 ft lb/in. notched (Solidur Ceram P)

28–30 ft lb/in. notched (Solidur 10 DS Series)

30 ft lb/in. notched (Solidur 10 802 AST)

Polyethylene *(cont'd.)*

32 ft lb/in. notched (Solidur 10 100 IV 20)

No break notched (Bapolene 1052, 1072, 1082; Petrothene HD 5703)

Tens. Impact:

64–67 mJ/mm^2 notched (Hostalen GUR)

158 kJ/m^2 (Alathon 7970)

231 kJ/m^2 (Alathon 7915, 7965)

284 kJ/m^2 (Alathon PE-5510)

514 kJ/m^2 (Alathon 7910, 7960)

25 ft lb/in.^{2} (Bapolene 2082; Thermocomp FF-1004)

28 ft lb/in.^{2} (Thermocomp FF-1006)

31 ft lb/in.^{2} (Thermocomp FF-1008)

35 ft lb/in.^{2} (Bapolene 2072)

42 ft lb/in.^{2} (Bapolene 2062)

80 ft lb/in.^{2} (Bapolene 2011)

120 ft lb/in.^{2} (Bapolene 2001, 2101)

1000 ft lb/in.^{2} (Ultra-Wear UHMWPE)

Hardness:

Ball Indentation 30 MPa (Vestolen A3515)

Ball Indentation 31 MPa (Vestolen A3512, A3513)

Ball Indentation 32 MPa (Vestolen A3512 R)

Ball Indentation 38 MPa (Hostalen GUR)

Ball Indentation 41 MPa (Vestolen A4516)

Ball Indentation 43 MPa (Vestolen A5041 R)

Ball Indentation 44 MPa (Vestolen A5016 F)

Ball Indentation 47 MPas (Vestolen A5017, A5018)

Ball Indentation 50 MPa (Vestolen A6014)

Ball Indentation 51 MPa (Vestolen A5561, A5561 P)

Ball Indentation 52 MPa (Vestolen A5543, A6012)

Ball Indentation 53 MPa (Vestolen A6013)

Ball Indentation 54 MPa (Vestolen A6042)

Ball Indentation 61 MPa (Vestolen A6016)

Ball Indentation 63 MPa (Vestolen A6017)

Rockwell R60 (RTP 700FR, 701)

Rockwell R64 (Ultra-Wear UHMWPE)

Rockwell R65 (RTP 703)

Rockwell R75 (RTP 705)

Rockwell R80 (Thermocomp FF-1004)

Rockwell R85 (RTP 707; Thermocomp FF-1006)

Rockwell R90 (Thermocomp FF-1008)

Shore A40–95 (Hypalon 40, 40S, 4085, LD-999)

Shore A45–95 (Hypalon 20)

Shore A60–95 (Hypalon 48)

Shore A65–98 (Hypalon 45)

Shore A69–95 (Hypalon 30)
Shore D42 (Norchem NPE 930, NPE 931, NPE 940, NPE 941, NPE 950, NPE 952, NPE 953, NPE 954)
Shore D44 (Norchem NPE 810)
Shore D45 (Norchem NPE 820, NPE 831)
Shore D50 (Bapolene 1052, 1082; Norchem NPE 350, NPE 860, NPE 861)
Shore D51 (Bapolene 1072)
Shore D59 (Alathon 7440; PE012/C)
Shore D60 (HiD 9642)
Shore D61 (Hostalen GM 5010 T2, GM 5010 T2N; PE012)
Shore D62 (Alathon 7140, 7220, 7320, 7340, PE-5510; Chemplex 5402; Hostalen GB 6950, GF 7740 F2; PE005)
Shore D63 (Alathon 7230, 7240, 7245; Chemplex 5705)
Shore D64 (Chemplex 5701, 5704; Petrothene HD 5002, HD 5705, HD 5712D)
Shore D65 (Alathon 7030, 7040, 7050; Bapolene 2001, 2101; Chemplex 5003, 5602, 5604; HiD 9606; Hostalen GA 7960, GC 7560, GM 7746; Petrothene HD 5003, HD 5703, HD 5704, HD 5713; Solidur 10 100 IV 20, 10 802 AST)
Shore D66 (HiD 9690; Hostalen GM 7255; Petrothene HD 5601, HD 5602, HD 5604, HD 5711)
Shore D67 (Petrothene HD 6085, LA 203F3)
Shore D68 (Bapolene 2011; Chemplex 6001, 6004, 6006, 6008, 6009; Petrothene HD 6004, HD 6007, HD 6009, LA 303F3; Solidur 10 DS Series, 25, Marble)
Shore D69 (HiD 9602; Norchem NHD 6908; Petrothene LA 404F3, LA 408F3)
Shore D70 (Solidur Ceram P)

Mold Shrinkage:
0.020 in./in. (PE005)
0.022 in./in. (Stat-Kon BLM, FE)
0.026 in./in. (Stat-Kon AS-F)
0.028 in./in. (PE012/C; Stat-Kon AS-FE)
0.030 in./in. (PE012)

Water Absorp.:
0.01% (Stat-Kon BLM, FE)
0.5–1.0% vol. (Arcel Moldable Polyethylene Copolymers 1.5 pcf molded foam)

THERMAL PROPERTIES:

Soften. Pt. (Vicat):
74 C (Hostalen GUR)
80 C (Norchem NPE 510, NPE 820, NPE 950, NPE 952, NPE 953, NPE 954)
82 C (Norchem NPE 930, NPE 931)
86 C (Petrothene NA 290)
88 C (Petrothene NA 295)
89 C (Petrothene NA 233)
90 C (Modic L-100F; Norchem NPE 350; Petrothene NA 154, NA 234, NA 289, NA 344)

Polyethylene *(cont'd.)*

93 C (Petrothene NA 225, NA 235)

94 C (Petrothene NA 142, NA 152, NA 301, NA 387)

96 C (Modic L-400F, L-400H; Norchem NPE 831; Petrothene NA 143, NA 145, NA 239, NA 386, NA 388)

97 C (Petrothene NA 238)

98 C (Petrothene NA 140, NA 153, NA 273)

99 C (Norchem NPE 860, NPE 861; Petrothene NA 148)

100 C (Petrothene NA 279)

101 C (Petrothene NA 147)

102 C (Petrothene NA 141)

103 C (Petrothene PA 161, PA 162)

105 C (Modic H-400F)

107 C (Petrothene NA 271)

110 C (Petrothene NA 272)

113 C (Petrothene NA 284)

119 C (Modic H-400C)

120 C (Chemplex 5402)

121 C (Petrothene HD 5002, HD 5705, HD 5712D)

122 C (Petrothene HD 5003, HD 5603, HD 5704, HD 5713, LR 920, LY 660)

123 C (Alathon PE-5510; Modic H-100E; Petrothene HD 5601, HD 5602, HD 5604, LY 955)

124 C (Chemplex 5604, 5701, 5704, 5705; Modic H-100F; Petrothene HD 5711, LY 520)

125 C (Chemplex 5003, 5602; Hostalen GF 7740 F2; Norchem NHD 6908; Petrothene HD 6004, HD 6007, HD 6009, HD 6085)

126 C (Petrothene LA 203F3, LC 732, LR 923, LY 600)

127 C (Chemplex 6001, 6004, 6006, 6008, 6009; Hostalen GM 7746; Petrothene LA 303F3, LR 734, LS 901)

128 C (Petrothene LR 723, LR 732)

129 C (Hostalen GM 7255; Petrothene LA 408F3)

130 C (Petrothene LA 404F3, LB 748, LB 924)

131 C (Petrothene LB 832, LB 833)

133 C (Petrothene LB 830)

210 F (Bapolene 1052)

212 F (Bapolene 1082)

215 F (Bapolene 1072)

239 F (Alathon 7440)

240 F (Marlex CL-50-35, CL-100, CL-100B)

250 F (Bapolene 2101; HiD 9642)

255 F (Alathon 7140, 7320, 7340; Hostalen GM 5010 T2, GM 5010 T2N)

256 F (Alathon 7220)

257 F (Bapolene 2001; HiD 9690)

258 F (Alathon 7230, 7240, 7245; HiD 9606)

260 F (Alathon 7030, 7040, 7050; Bapolene 2011)

262 F (HiD 9602)

Conduct.:

0.41 W/mK (Vestolen A3512, A3512 R, A3513, A3515)

0.42 W/mK (Hostalen GUR; Vestolen A5041 R)

0.43 W/mK (Vestolen A4516, A5016 F)

0.44 W/mK (Vestolen A5017, A5018)

0.49 W/mK (Vestolen A5561, A5561 P, A6012, A6013, A6042)

0.51 W/mK (Vestolen A5543, A6014, A6016, A6017)

0.3 Btu/h/ft²/F/in. (Arcel Moldable Polyethylene Copolymers 1.5 pcf molded foam)

1.9 Btu/h/ft²/F/in. (Stat-Kon AS-F, AS-FE, BLM, FE)

2.05 Btu/h/ft²/F/in. (RTP 701)

2.2 Btu/h/ft²/F/in. (RTP 700FR)

2.3 Btu/h/ft²/F/in. (RTP 703)

2.4 Btu/h/ft²/F/in. (Thermocomp FF-1004)

2.5 Btu/h/ft²/F/in. (RTP 705)

2.6 Btu/h/ft²/F/in. (Thermocomp FF-1006)

2.7 Btu/h/ft²/F/in. (RTP 707)

2.8 Btu/h/ft²/F/in. (Thermocomp FF-1008)

Distort. Temp.:

39 C (Vestolen A3512, A3512 R, A3513, A3515)

40 C (264 psi) (PE012/C)

42 C (Vestolen A4516, A5041 R); (264 psi) (PE012)

43 C (Hostalen GB 6950)

44 C (Vestolen A5016 F, A5017, A5018, A5543, A6014)

45 C (Hostalen GC 7560; Vestolen A6012, A6042)

46 C (Hostalen GA 7960; Vestolen A5561, A5561 P, A6013)

47 C (Vestolen A6016, A6017)

49 C (0.455 MPa) (Norchem NPE 810, NPE 831)

51 C (0.455 MPa) (Norchem NPE 860, NPE 861)

54 C (264 psi) (PE005)

65 C (66 psi) (Chemplex 5402)

66 C (0.455 MPa) (Alathon PE-5510)

68 C (66 psi) (Chemplex 5705)

69 C (66 psi) (Chemplex 5704)

70 C (Hostalen GF 7740 F2); (66 psi) (Chemplex 5701; Petrothene HD 5002, HD 5705, HD 5712D)

72 C (66 psi) (Petrothene HD 5003, HD 5704)

73 C (66 psi) (Petrothene HD 5703, HD 5713)

74 C (66 psi) (Petrothene HD 5601, HD 5604, HD 5711)

75 C (66 psi) (Chemplex 5003; Petrothene HD 5602)

76 C (66 psi) (Chemplex 5602, 5604)

78 C (66 psi) (Petrothene HD 6004, HD 6007, HD 6009, HD 6085)

Polyethylene *(cont'd.)*

80 C (66 psi) (Chemplex 6001, 6004, 6006, 6008, 6009; Norchem NHD 6908)
95 C (Hostalen GUR)
100 F (264 psi) (PE-25)
130 F (264 psi) (Stat-Kon AS-F, AS-FE, BLM, FE)
150 F (66 psi) (Alathon 7140, 7320, 7340)
152 F (66 psi) (Alathon 7220)
155 F (66 psi) (Alathon 7230, 7240)
158–174 F (66 psi) (Ultra-Wear UHMWPE)
160 F (66 psi) (Alathon 7030, 7040, 7050)
165 F (66 psi) (Bapolene 2101)
169 F (66 psi) (Bapolene 2001)
172 F (Hostalen GM 5010 T2, GM 5010 T2N)
176 F (66 psi) (Bapolene 2011)
200 F (264 psi) (RTP 700FR)
210 F (264 psi) (RTP 701)
240 F (264 psi) (RTP 703)
250 F (264 psi) (RTP 705, 707; Thermocomp FF-1004)
260 F (264 psi) (Thermocomp FF-1006, FF-1008)

Brittle Temp.:

< –76 C (Alathon 7915, 7965, 7970, PE-5510)
< –75 to –15 C (Tenite Polyethylene)
> –70 C (Modic H-100E, H-100F, H-400C, H-400F, L-100F, L-400F, L-400H)
< 76 C (Alathon 7910, 7960)
76 C (Norchem NHD 6908)
< –105 F (Bapolene 1072, 2001, 2011, 2101)
< –95 F (Bapolene 1052)
< –30 F (Bapolene 1082)

Coeff. of Linear Exp.:

1.5×10^{-4} K^{-1} (Vestolen A5543, A5561, A5561 P, A6012, A6013, A6014, A6016, A6017, A6042)
1.7×10^{-4} K^{-1} (Vestolen A5017, A5018)
2×10^{-4} K^{-1} (Vestolen A3512, A3512 R, A3513, A3515, A4516, A5016 F, A5041 R); (20–100 C) (Hostalen GUR)
2.4×10^{-5} in./in./F (Thermocomp FF-1008)
2.7×10^{-5} in./in./F (Thermocomp FF-1006)
3.0×10^{-5} in./in./F (Thermocomp FF-1004)
5.7×10^{-5} in./in./F (Stat-Kon AS-F)
6.0×10^{-5} in./in./F (Stat-Kon AS-FE, FE)
6.1×10^{-5} in./in./F (Stat-Kon BLM)
7.2×10^{-5} in./in./F (Ultra-Wear UHMWPE)
8×10^{-5} in./in./F (Hostalen GF 7740 F2, GM 5010 T2, GM 5010 T2N)

Sp. Heat:

1.7 kJ/kgK (Vestolen A3512, A3512 R, A3513, A3515, A5041 R, A5561, A5561 P,

A6012, A6013, A6014, A6016, A6017, A6042)

1.9 kJ/kgK (Vestolen A4516, A5016 F, A5017, A5018, A5543)

1.84 kJ/kg K (Hostalen GUR)

0.5 Cal/g/C (Norchem NPE 350, NPE 810, NPE 930, NPE 931, NPE 940, NPE 941, NPE 950, NPE 952, NPE 953, NPE 954)

Flamm.:

V-0 (RTP 700FR)

HB (RTP 701, 703, 705, 707; Stat-Kon AS-F, AS-FE, BLM, FE; Thermocomp FF-1004, FF-1006, FF-1008)

Combustible (Hostalen GUR)

ELECTRICAL PROPERTIES:

Dissip. Factor:

0.000062 (100 kHz) (Norchem NHD 6908)

0.00007 (1 MHz) (Norchem NPE 510)

0.00019 (50 Hz) (Hostalen GUR)

0.0004 (1 MHz) (Norchem NPE 130)

< 0.0005 (50 Hz) (Vestolen A3512, A3512 R, A3513, A3515, A4516, A5016 F, A5017, A5018, A5041 R, A5543, A5561, A5561 P, A6012, A6013, A6014, A6016, A6017, A6042); (1 MHz) (Norchem NPE 950, NPE 952, NPE 953, NPE 954)

0.002 (1 MHz) (RTP 701)

0.0055 (1 MHz) (Norchem NPE 190)

0.007 (1 MHz) (RTP 703)

0.008 (1 MHz) (RTP 705, 707)

0.01 (1 Mhz) (RTP 700FR)

Dielec. Str.:

50 kV/cm (Solidur 10 802 AST)

900 kV/cm (Hostalen GUR; Solidur 10 100 IV 20, 10 DS Series, 25, Ceram P, Marble)

70 kV/mm (Vestolen A3512 R, A5041 R)

80 kV/mm (Vestolen A3512, A3513, A3515, A4516, A5016 F, A5017, A5018, A5543, A5561, A5561 P, A6012, A6013, A6014, A6016, A6017, A6042)

450–500 V/mil (Ultra-Wear UHMWPE)

500 V/mil (RTP 700FR, 701, 703, 705, 707)

780 V/mil (Norchem NPE 510)

Dielec. Const.:

2.27 (1 MHz) (Norchem NPE 510)

< 2.30 (1 MHz) (Norchem NPE 950, NPE 952, NPE 953, NPE 954)

2.30 (2×10^6 Hz) (Hostalen GUR)

2.36 (100 kHz) (Norchem NHD 6908)

2.52 (1 MHz) (Norchem NPE 130)

2.58 (1 MHz) (Norchem NPE 190)

2.7 (1 MHz) (RTP 701, 703, 705)

2.8 (1 MHz) (RTP 700FR, 707)

Polyethylene *(cont'd.)*

Vol. Resist.:

10,000 ohm-cm (Stat-Kon BLM, FE)

10^{15} ohm-cm (RTP 700FR)

> 10^{15} ohm-cm (Solidur 10 100 IV 20, 10 DS Series, 25, Ceram P, Marble)

10^{16} ohm-cm (RTP 701, 703, 705, 707)

> 5×10^{16} ohm-cm (Hostalen GUR)

10^{18} ohm-cm (Ultra-Wear UHMWPE)

> 10^{18} ohm-cm (Vestolen A3512, A3512 R, A3513, A3515, A4516, A5016 F, A5017, A5018, A5041 R, A5543, A5561, A 5561 P, A6012, A6013, A6014, A6016, A6017, A6042)

Surf. Resist.:

10,000 ohm/sq. (Stat-Kon BLM, FE)

10^{10}–10^{13} ohm/sq. (Stat-Kon AS-F, AS-FE)

> 10^{13} ohm (Hostalen GUR; Solidur 10 100 IV 20, 10 DS Series, 25, Ceram P, Marble)

> 10^{14} ohm (Vestolen A3512, A3512 R, A3513, A3515, A4516, A5016 F, A5017, A5018, A5041 R, A5543, A5561, A 5561 P, A6012, A6013, A6014, A6016, A6017, A6042)

Arc Resist.:

100 s (RTP 700FR)

140 s (RTP 701, 703, 705, 707)

TOXICITY/HANDLING:

Low concs. of fumes can evolve at temps. above 260 C—use with adequate ventilation (Alathon 7810, 7815, 7820, 7835, 7840, 7860, 7970, PE-5510)

Do not touch molten glue as serious burns can result (Norcast 6368)

STD. PKGS.:

Bags (Epolene C-10, C-15, N-10, N-10-P, N-11, N-11-P, N-12, N-12-P, N-14, N-14-P, N-34, N-34-P, N-45, N-45-P)

Bags and bulk hopper cars (Marlex CL-100, CL-100B)

Boxes and bulk (Marlex BMN TR955, CFL-50-35)

20-kg bags in crate pallets of 450 kg and loose material pallets of 900 kg (Vestolen A3512)

20-kg paper bags, polyethylene-lined, on 600-kg pallets (Bayer CM 2552, CM 3610)

25-kg bags (A-C Polyethylene 6A, 617, 617A)

25-kg bags in crate pallets of 450 kg and loose material pallets of 900 kg (Vestolen A3513, A3515, A4516, A5016 F, A5017, A5018, A5041 R, A5543, A5561, A5561 P, A6012, A6013, A6014, A6016, A6017, A6042)

25-kg paper bags, polyethylene-lined, on 750-kg pallets (Bayer CM 3630, CM 3631, CM 3632, CM 4230)

50-lb bags, 1000-lb boxes, bulk hopper trucks or hopper cars (Paxon AA55-003, AA60-003, AA60-007, AB50-003, AD60-007, AF40-003, BA50-100, FD60-018, SS55-100, SS55-180, SS55-250, SS55-400)

SYNONYMS:
Polybutene (CTFA)
Polybutylene
Polyisobutylene

TRADENAME EQUIVALENTS:
Glissoviscal B [BASF AG]
Olict C [Alox]
Shell Polybutylene 0200, 0300, 0400, 0700, 1600A, 4101, 4103, 4110, 4121, 4127, 4128, 8240, 8640 [Shell]

MODIFICATIONS/SPECIALTY GRADES:
Film grade:
Shell Polybutylene 1600A
Pipe grade:
Shell Polybutylene 4101, 4103, 4110, 4121, 4127, 4128

CATEGORY:
Thermoplastic resin

PROCESSING:
Blown film:
Shell Polybutylene 1600A
Compression molding:
Shell Polybutylene 0200, 0300, 0400, 0700, 4101, 4103, 4110, 4121, 4127, 4128, 8240, 8640
Extrusion:
Shell Polybutylene 1600A

APPLICATIONS:
Food-contact applications: food packaging (Shell Polybutylene 1600A)
Functional additives: blending resin (Shell Polybutylene 0200, 0300, 0400, 0700, 8240, 8640); lubricant (Olicat C); tackiness agent (Olicat C); thickener (Glissoviscal B)
Industrial applications: films (Shell Polybutylene 1600A); lubricating oils (Glissoviscal B); pipes (Shell Polybutylene 4101, 4103, 4110, 4121, 4127, 4128)

PROPERTIES:
Form:
Viscous liquid (Olicat C)
Solid (block) (Glissoviscal B)
Pellets (Shell Polybutylene 0200, 0300, 0400, 0700, 1600A, 4101, 4103, 4110, 4121, 4127, 4128, 8240, 8640)
Color:
Clear (Shell Polybutylene 4110)
Natural (Shell Polybutylene 0200, 0300, 0400, 0700, 1600A)
Blue (Shell Polybutylene 4103)
Black (Shell Polybutylene 4101, 4121, 4127, 4128)

Polyisobutene *(cont'd.)*

Composition:
100% conc. (Glissoviscal B)

GENERAL PROPERTIES:

Melt Flow:
0.4 g/10 min (Shell Polybutylene 4101, 4103, 4110, 4121, 4127, 4128)
1.0 g/10 min (Shell Polybutylene 8640); (film) (Shell Polybutylene 1600A)
1.8 g/10 min (Shell Polybutylene 0200)
2.0 g/10 min (Shell Polybutylene 8240)
4.0 g/10 min (Shell Polybutylene 0300)
10 g/10 min (Shell Polybutylene 0700)
20 g/10 min (Shell Polybutylene 0400)

Sp. Gr.:
0.908 (Shell Polybutylene 8240, 8640)
0.910 (film) (Shell Polybutylene 1600A)
0.915 (Shell Polybutylene 0200, 0300, 0400, 0700, 4110)
0.925 (Shell Polybutylene 4101, 4121)
0.930 (Shell Polybutylene 4103, 4127, 4128)

M.P.:
255–259 F (Shell Polybutylene 0200, 0300, 0400, 0700, 4101, 4103, 4110, 4121, 4127, 4128, 8240, 8640)

MECHANICAL PROPERTIES:

Tens. Str.:
4200 psi (break) (Shell Polybutylene 0400, 0700)
4500 psi (break) (Shell Polybutylene 0200, 0300, 8240, 8640)
4800 psi (break) (Shell Polybutylene 4101)
5500 psi (break) (film) (Shell Polybutylene 1600A)

Tens. Elong.:
200% (film) (Shell Polybutylene 1600A)
280% (Shell Polybutylene 4101)
350% (Shell Polybutylene 0200, 0300, 0400, 0700, 8240, 8640)

Impact Str. (Dart):
200 g (film) (Shell Polybutylene 1600A)

Impact Str. (Izod):
No break (Shell Polybutylene 0200, 0300, 0400, 0700, 8240, 8640)

Hardness:
Shore D50 (Shell Polybutylene 8240, 8640)
Shore D55 (Shell Polybutylene 0200, 0300, 0400, 0700)
Shore D60 (Shell Polybutylene 4101)

THERMAL PROPERTIES:

Soften. Pt. (Vicat):
235 F (Shell Polybutylene 4101, 4103, 4110, , 4121, 4127, 4128)

Polypropylene

SYNONYMS:
PP

Propylene polymer

STRUCTURE:

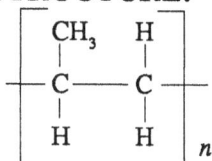

CAS No.:
9003-07-0

TRADENAME EQUIVALENTS:
EmPee PP 401, PP 401CS, PP 402 [Monmouth Plastics]

Fortilene 14X01, 16X01, 18X02, 18X03, 19X02, 19X03, 1401, 1602, 1602A [Soltex Polymer]

Hostalen PP920, PP933, PP934, PP936, PP941, PP942, PP975 [Hoechst Celanese]

Marlex CHM-040-01, CHM-040-02 [Phillips]

Modic P-110F, P-300F, P-300M [Mitsubishi]

MWB Film [Hercules]

Nortuff NFA 1700 MO, NFA 1800 TO, NFA 2400 CO, NFA 2600 CO, NFA 4400 TO, NFC 1700 MO, NFC 1800 TO, NFC 2200 CO, NFC 2400 CO, NFC 2440 CF, NFC 2600 CO, NFC 4000 FR, NFC 4400 CO, NFC 4600 CO [Norchem]

P012, P032, P04, P0113, P0119 [M.A. Industries]

PP-C2IM, PP-C4IM, PP-HIFR [Washington Penn]

RTP 150, 199x22898, 199x23835, 199x28016 [Fiberite]

SB522/1S Film [Hercules]

Stat-Kon M [LNP]

Tenite Polypropylene [Eastman]

Amorphous polypropylene:

A-Fax 500, 600, 800, 940 [Hercules]

Polypropylene homopolymer:

Amoco 10-4017, 1012, 1046, 1088, 4018, 4036, 4039, 4222, 4228, 5219, 6200P, 6400P, 6420P, 6431, 6800P, 7000P, 7200P, 7220P, 7232, 7233, 7234, CR22N, CR22NA, CR35, CR35A, CR35N, CR35NA [Amoco]

Bapolene 4042, 4062, 4072, 4082 [Bamberger Polymers]

Fortilene 10X01, 21X04, 38X01, 40X05, 41X03, 2251, 3151, 3250, 3251, 3606, 3907, 9205, 9605 [Soltex Polymer]

Hostalen PP927, PP989 [Hoechst Celanese]

Marlex HGH-050, HGN-120-01, HGN-120-A, HGN-350, HGV-040-01, HGX-030, HGZ-040, HGZ-050-02, HGZ-080-02, HGZ-120-02, HGZ-120-04, HGZ-350, HLN-020-01, HNS-080 [Phillips]

Norchem NPP 1001-LF, NPP 1006-GF, NPP 1008-AK, NPP 1010-LC, NPP 2000-GJ,

Polypropylene *(cont'd.)*

NPP 2003-GJ, NPP 2004-MR, NPP 2013-UJ, NPP 3007-GO, NPP 3010-SO, NPP 7000-GF, NPP 7000-ZF, NPP 8000-GK, NPP 8001-LK, NPP 8002-HK, NPP 8004-MR, NPP 8005-AR, NPP 8006-GF, NPP 8007-GO, NPP 8004-ZR [Norchem]

Petrothene PP 2004-MR, PP 8085-GU, PP 8000-GK, PP 8001-LK, PP 8004-MR, PP 8004-ZR, PP 8005-AR, PP 8020-AU, PP 8020-GU, PP 8020-ZU, PP 8080-AW, PP 8080-GW, PP 8080-ZW [Quantum/USI]

Pro-fax 6131, 6323, 6323F, 6329, 6331NW, 6523F, 6524, 6532F, 7531, PC-072, PC942, PC968, PD064, PD195, PD401, PD626, PD701, PF101, PF151 [Hercules]

Rexene 6310, PP41E2, PP41E4, XO-325 [Rexene Products]

Shell 5225, 5384, 5419, 5431, 5520, 5524, 5550, 5610, 5820, 5824S, 5840, 5864, 5944S [Shell]

Vestolen P1200, P1200F, P2200, P2200F, P3200, P3200F, P3230 F, P4200, P4200F, P5200, P5200F, P5202, P5204, P5206S, P5212 LF, P6200, P6202, P6206S [Huls AG]

Polypropylene copolymer:

Bapolene 5042, 5052, 5072 [Bamberger Polymers]

Fortilene 41X04, 42X07, 54X02, 4141, 4141F [Soltex Polymer]

Hostalen PP996, PP998 [Hoechst Celanese]

Petrothene PP 1510-HC, PP 1510-LC, PP 8402-HO, PP 8402-TO, PP 8403-HO, PP 8403-TO, PP 8404-HJ, PP 8404-ZJ, PP 8410-ZR, PP 8411-ZR, PP 8412-HK, PP 8412-TK, PP 8420-HK, PP 8462-HR, PP 8470-HU, PP 8470-ZU, PP 8502-HK, PP 8602-HJ, PP 8752-HF, PP 8755-HK, PP 8762-HR, PP 8770-HU, PP 8802-HO, PP 8815-ZR, PP 8820-HU [Quantum/USI]

Pro-fax 7131, 7523, 7823, 8523, 8623, SA-595, SA-752, SA-861, SA-862, SA-868M, SA-878, SB-661, SB-751, SB-782, SB-786, SB-787, SD-062, SD-101, SE-191 [Hercules]

Rexene 9234, 9400, 9401, 9402, 9403, 4903E, 9500 [Rexene Products]

Shell 7129, 7221, 7328, 7521, 7522, 7525, 7623, 7627, 7635, 7912 [Shell]

Vestolen P4700, P4702L, P4800, P4802L, P5800, P5802 L, P6500, P6502, P6503, P6522 [Huls AG]

Polypropylene random copolymer:

Norchem NPP 7300-GF, NPP 7300-AF [Norchem]

Petrothene PP 7200-AF, PP 7200-GF, PP 7200-MF, PP 7300-KF, PP 7300-MF, PP 8310-GO, PP 8310-KO [Quantum/USI]

Shell DX6016, DX6020 [Shell]

Vestolen P2300, P2300F, P2330 F, P5400, P5400F, P6421 [Huls AG]

Chlorinated polypropylene:

Parlon P [Hercules]

Aluminum-filled polypropylene:

EMI-X MA-40 (40% aluminum flake) [LNP]

Calcium carbonate-filled polypropylene:

PP-C3CC-4 (40% calcium carbonate), C5CC-2 (20% copolymers), -C6CC-4 (40% copolymers), H1CC-1 (10% homopolymers), -H2CC-2 (20% homopolymers),

H6CC-4 (40% homopolymers), H7CC-4 (40% homopolymers) [Washington Penn]
Pro-fax 65F5-4 (40% calcium carbonate) [Hercules]
Rexene 6310C25, 9401C2 [Rexene Products]
RTP 140 (40% calcium carbonate), 141 (20% calcium carbonate), 142 (10% calcium carbonate), 143 (30% calcium carbonate) [Fiberite]

Carbon-reinforced polypropylene:
Stat-Kon M-1 HI (carbon powder grade), M-2 (carbon powder grade), ME (carbon powder grade) [LNP]

Carbon/glass-reinforced polypropylene:
Stat-Kon MF-15 (carbon powder grade, 15% glass fiber) [LNP]

Carbon/mica-reinforced polypropylene:
Stat-Kon MM-3340 (carbon powder grade, 20% mica) [LNP]

Glass-reinforced polypropylene:
PP-HFR-2 (20% fiberglass), -HFR-3 (30% fiberglass) [Washington Penn]
RTP 100GB10 (10% glass beads), 100GB20 (20% glass beads), 100GB30 (30% glass beads), 100GB40 (40% glass beads), 101 (10% glass fiber), 101CC (10% glass fiber, chemically combined), 101FR (10% glass fiber), 101 SP Foamed (10% glass fiber), 103 (20% glass fiber), 103CC (20% glass fiber, chemically combined), 103 SP Foamed (20% glass fiber), 105 (30% glass fiber), 105CC (30% glass fiber, chemically combined), 105CC FR (30% glass fiber, chemically combined), 105 SP Foamed (30% glass fiber), 107 (40% glass fiber), 107CC (40% glass fiber) [Fiberite]
Thermocomp MF-1002 (10% fiber glass), MF-1002HI (10% fiber glass), MF-1004 (20% fiber glass), MF-1004FR (20% glass fiber), MF-1004HI (20% fiber glass), MF-1006 (30% fiber glass), MF-1006HI (30% fiber glass), MF-1008 (40% fiber glass), MF-1008HI (40% fiber glass), MFX-1004HS (20% fiber glass, chemically coupled), MFX-1006HS (30% fiber glass, chemically coupled), MFX-1008HS (40% fiber glass, chemically coupled) [LNP]
Vestolen P5232G (20% glass fiber) [Huls AG]

Glass/mineral-reinforced polypropylene:
RTP 175 (glass, blass beads, mineral), 175X (glass, glass beads, mineral), 177 (glass, glass beads, mineral), 178 (glass, glass beads, mineral), 178X (glass, glass bead, mineral) [Fiberite]

Mica-filled polypropylene:
Micalite PP-G2MF-4, PP-G2MF-5, PP-G2MF-6, PP-H2MFS-4, PP-H2MFS-5, PP-H2MFS-6 [Washington Penn]
PP-G2MF-4 (40% mica), -G2MF-5 (50% mica), -H2MF-4 (40% mica), -H2MF-5 (50% mica), -H3MF-2 (20% coupled), -H2MFQ-3 (30% coupled), -H2MFQ-4 (40% coupled), -H3MFQ-1 (12% coupled), -H3MFQ-5 (50% coupled), -H2MFS-3 (30% uncoupled), -H2MFS-4 (40% mica), -H2MFS-5 (50% mica) [Washington Penn]

Mineral-filled polypropylene:
P2120 (20% mineral), P2130 (30% mineral), P2140 (40% mineral), P2230 (30%

Polypropylene (cont'd.)

mineral), P9125 (25% mineral) [M.A. Industries]

Talc-filled polypropylene:

Bapolene 4112 (20% talc), 4114 (40% talc) [Bamberger Polymers]

PP-C2TF-1 (10% talc, copolymer), -C2TF-4 (40% talc, copolymer), -C3TF-2 (20% copolymers), -C3TFA-2 (20% talc, copolymer), -H1TF-4 (40% talc), -H2LTF-4 (40% homopolymers), -H2TF-2 (20% talc), -H2TF-4 (40% talc), -H4TF-3 (30% homopolymers), -H6TF-4 (40% talc), -VB-511 (10% talc) [Washington Penn]

Pro-fax 65F4-4 (40% talc) [Hercules]

Rexene 6310T2, 6310T4, 9400T2 [Rexene Products]

RTP 100T10 or 131 (10% talc), 100T20 or 128 (20% talc), 100T30 or 132 (30% talc), 100T40 or 127 (40% talc) [Fiberite]

Vestolen P5232T (20% talc), P5272T (40% talc) [Fiberite]

MODIFICATIONS/SPECIALTY GRADES:

Conductive grade/statically dissipative:

EMI-X MA-40; RTP 199x22898, 199x23835, 199x28016; Stat-Kon M, M-1 HI, M-2, ME, MF-15, MM-3340; Vestolen P5204

Impact-modified/medium impact:

Bapolene 5072; Hostalen PP920, PP996; Marlex CHM-040-01, CHM-040-02; Norchem NPP 7300-GF, NPP 7300-AF; Nortuff NFA 1700 MO, NFA 4400 TO, NFC 1800 TO, NFC 2200 CO, NFC 2400 CO, NFC 2440 CF, NFC 2600 CO, NFC 4000 FR, NFC 4400 CO, NFC 4600 CO; P012; Petrothene PP 8402-HO, PP 8402-TO, PP 8403-HO, PP 8403-TO, PP 8404-HJ, PP 8404-ZJ, PP 8410-ZR, PP 8411-ZR, PP 8412-HK, PP 8412-TK, PP 8420-HK, PP 8502-HK, PP 8802-HO; PP-C2IM, -C2TF-1, -C4IM, -VB-511; Pro-fax 7131, 7523, PD195, SA595, SA862, SB661, SB751, SB782, SB786, SB787; RTP 101CC, 103CC, 105CC, 107CC; Shell 7521, 7522, 7525, 7623, 7627, 7635, 7912, DX6020; Stat-Kon MF-15; Thermocomp MF-1004, MF-1006, MF-1008, MF-1004HS, MFX-1006HS, MFX-1008HS; Vestolen P4800, P5800, P6500

High-impact:

Bapolene 5042, 5052; Nortuff NFA 1800 TO, NFA 2400 CO, NFA 2600 CO; P032, P04; Petrothene PP 1510-HC, PP 1510-LC, PP 8462-HR, PP 8470-HU, PP 8470-ZU, PP 8602-HJ, PP 8752-HF, PP 8755-HK, PP 8762-HR, PP 8770-HU; Pro-fax 7823, 8523, 8623, PC-072, SA868M, SA878, SE191; Rexene 9401C, 9402, 9403, 9403E, 9500; Shell 7129, 7221, 7328; Stat-Kon M, M-1 HI, M-2, ME, MM-3340; Thermocomp MF-1002HI, MF-1004HI, MF-1006HI, MF-1008HI; Vestolen P5400, P5400 F, P6421, P6522

High-flow:

Amoco 5219, 7220P, 7232, 7233, 7234, CR22N, CR22NA, CR35, CR35A, CR35N, CR35NA; Bapolene 4062, 4072, 4082, 5072; Fortilene 16X01, 18X02, 18X03, 19X02, 19X03, 38X01, 1602, 1602A, 3606, 3907, 9605; Hostalen PP941, PP942, PP975, PP989; Marlex HGN-120-01, HGN-120-A, HGN-350, HGZ-120-02, HGZ-120-04, HGZ-350; Modic P-300M; Norchem NPP 8004-MR, NPP 8005-AR, NPP 8004-ZR; Nortuff NFC 4000 FR; P0119, P2120, P2130, P2140, P2230,

P9125; Petrothene PP 2004-MR, PP 2085-GU, PP 8004-MR, 8004-ZR, PP 8005-AR, PP 8020-AU, PP 8020-GU, PP 8020-ZU, PP 8080-AW, PP 8080-GW, PP 8080-ZW, PP 8462-HR, PP 8470-HU, PP 8470-ZU, PP 8762-HR, PP 8770-HU, PP 8820-HU; PP-C5CC-2, -C6CC-4, -H6CC-4, H7CC-4, -H6TF-4; Pro-fax 6131, 6323, 6323F, 6329, 6331NW, 6523F, 7131, PC968, PD626, PD701, PF151, SB751, SB787, SD062; Shell 5820, 5824S, 5840, 5864, 5944S, 7912; Vestolen P1200, P1200 F, P2200, P2200 F, P2300, P2300 F

Controlled rheology grade:
Amoco 4222, 4228, 5219, CR22N, CR22NA, CR35, CR35A, CR35N, CR35NA; Fortilene 18X02, 18X03, 19X02, 19X03, 38X01, 3907; Marlex HGN-350, HGZ-350

Nucleated:
Amoco 7232, 7234, CR22N, CR22NA, CR35N, CR35NA; Fortilene 16X01; Marlex HGN-120-01, HGN-120-A, HGN-350, HLN-020-01; Norchem NPP 7000-ZF, NPP 8004-ZR; Petrothene PP 8004-ZR, PP 8020-ZU, PP 8080ZW, PP 8404-ZJ, PP 8410-ZR, PP 8411-ZR, PP 8470-ZU, PP 8815-ZR; Pro-fax 6331NW; Shell 5384, 5524, 5824S, 5864, 5944S

Antistat:
Amoco 4036, 4039, 7232, 7233, 7234, CR22NA, CR35A, CR35NA; Fortilene 16X01, 18X02, 41X04, 1602A; Marlex GHZ-350, HLN-020-01; Norchem NPP 1008-AK, NPP 7000-ZF, NPP 7300-AF, NPP 8005-AR, NPP 8004-ZR; Petrothene PP 7200-AF, PP 8004-ZR, PP 8005-AR, PP 8020-AU, PP 8020-ZU, PP 8080-AW, PP 8080-ZW; PP-C3TFA-2; Pro-fax 6331NW; Shell 5824S, 5944S

Slip agent:
MWB Film; Norchem NPP 3010-SO

Antiblock additive:
Norchem NPP 3010-SO

Copper-stabilized:
Empee PP 401CS

Flame-retardant:
Nortuff NFA 4400 TO, NFC 4000 FR, NFC 4400 CO, NFC 4600 CO; PP-HIFR; Pro-fax SA595; RTP 101FR, 105CC FR, 150; Thermocomp MF-1004FR; Vestolen P5206 S, P6206 S

Heat-resistant:
Amoco 1046; Shell 5550; Vestolen P6421

Heat-stabilized:
MWB Film; Norchem NPP 8002-HK, NPP 8004-MR; Petrothene PP 1510-HC, PP 1510-LC, PP 8402-HO, PP 8402-TO, PP 8404-HJ, PP 8412-HK, PP 8412-TK, PP 8420-HK, PP 8462-HR, PP 8470-HU, PP 8502-HK, PP 8602-HJ, PP 8802-HO, PP 8820-HU; Pro-fax 6323F, 6523F, 7523, 7823, 8523, 8623; RTP 100 GB10, 100 GB20, 100 GB30, 100 GB40, 100T10 or 131, 100T20 or 128, 100T30 or 132, 100T40 or 127, 101, 101CC, 103, 103CC, 105, 105CC, 107, 107CC, 140, 141, 142, 143, 175, 175X, 177, 178, 178X; Shell 5384, 5419, 5610, 7912, DX6016; Vestolen

Polypropylene (cont'd.)

P1200, P2200, P2300, P4200, P4700, P4702L, P4800, P4802L, P5200, P5202, P5232 G, P5232 T, P5272 T, P5400, P5400 F, P5800, P5802 L, P6200, P6202, P6500, P6502, P6503, P6522

Heat-stabilized, long-term:
Fortilene 14X01; Hostalen PP920, PP942, PP996, PP998; Norchem NPP 8001-LK; Pro-fax 6524; Shell 5225, 5520, 5524, 5820, 5824S, 7129, 7221, 7328, 7521, 7522, 7525, 7623, 7627, DX6020

UV-stabilized:
Marlex HGV-040-01; Norchem NPP 2013-UJ; Pro-fax PC968; Shell 5431, 7635; Vestolen P4702L, P4802L, P5212 LF, P5802 L, P6503

Radiation-resistant:
Pro-fax PD626; Rexene 9234

PROCESSING:

Blow molding:
Fortilene 41X03, 4141; Hostalen PP920; Marlex HLN-020-01; Norchem NPP 7000-GF, NPP 7000-ZFR, NPP 7300-GF, NPP 7300-AF; Petrothene PP 7200-AF, PP 7200-MF, PP 7300-KF, PP 7300-MF; Pro-fax SA878; Shell 5225, 5384, 7525, 7627, 7635, DX6020

Blown film:
Amoco 1088; Marlex HNS-080; Vestolen P5200 F

Cast film:
Marlex HNS-080; Modic P-300M; Norchem NPP 3007-GO, NPP 3010-SO; Pro-fax PC942, SA752, SA861; Vestolen P2200 F, P2300 F, P3200 F

Coextrusion:
Modic P-110F, P-300F, P-300M

Compounding:
Fortilene 19X02, 9205, 9605

Compression molding:
Pro-fax PD195

Extrusion:
Amoco 10-4017, 1012, 1046, 4018, 4036, 40396431, 7232, 7233, 7234; Bapolene 4042; Fortilene 10X01, 40X05, 9205, 9605; Hostalen PP920, PP927, PP933, PP934, PP936; Marlex HGZ-050-02, HGZ-080-02, HNS-080; Norchem NPP 1001-LF, NPP 1006-GF, NPP 1010-LC, NPP 2000-GJ, NPP 2003-GJ, NPP 2004-MR, NPP 2013-UJ; Nortuff NFA 1700 MO, NFA 1800 TO, NFA 2400 CO, NFA 2600 CO, NFA 4400 TO; Petrothene PP 1510-HC, PP 1510-LC, PP 2004-MR, PP 2085-GU, PP 7200-AF, PP 7200-GF, PP 7200-MF, PP 7300-KF, PP 7300-MF; Pro-fax 6323, 7523, 8623, PD195; Rexene 6310, 6310C25, 6310T2, 6310T4, 9234, 9400, 9400T2, 9401, 9401C, 9402, 9403, 9403E, 9500, PP41E2, PP41E4, XO-325; Shell 5225, 5419, 5520, 5550, 5610, 5820, 5840, 7129, 7221, 7328, 7525, 7627, 7635, DX6016; Stat-Kon ME; Vestolen P6206 S, P6421

Extrusion blow molding:
Vestolen P5400

Fiber and filament grade:

Fortilene 38X01, 3151, 3230, 3251, 3606, 3907; Hostalen PP933; Marlex HGV-040-01, HGX-030, HGZ-040, HGZ-120-04; Norchem NPP 2000-GJ, NPP 2003-GJ, NPP 2004-MR, NPP 2013-UJ; Petrothene PP 2004-MR, PP 2085-GU; Pro-fax 6323F, 6523F, 7523, PC968, PD401, PF151; Vestolen P1200, P1200 F, P2200 F, P3200, P3200 F, P4200 F, P5200 F

Film grade:

Fortilene 21X04, 42X07, 2251, 4141F; Hostalen PP934; Norchem NPP 3007-GO, NPP 3010-SO; Petrothene PP 1510-HC, PP 1510-LC, PP 2004-MR, PP 2085-GU; Pro-fax 6131, 6331NW, 7531, PC942, PD064, SA752, SA861; Rexene PP41E2, PP41E4, XO-325; Shell DX6016; Vestolen P5400 F

Injection blow molding:

Fortilene 41X04; Petrothene PP 7200-GF

Injection molding:

Amoco 10-4017, 1046, 4018, 4036, 4039, 4222, 4228, 5219, 6431, 7232, 7233, 7234, CR22N, CR22NA, CR35, CR35A, CR35N, CR35NA; Bapolene 4042, 4062, 4072, 4082, 4112, 4114, 5042, 5052, 5072; EMI-X MA-40; Empee PP 401, PP 401CS, PP 402; Fortilene 14X01, 16X01, 18X02, 18X03, 19X02, 19X03, 54X02, 1401, 1602, 1602A; Hostalen PP941, PP942, PP975, PP989, PP996, PP998; Marlex CHM-040-01, CHM-040-02, HGH-050, HGN-120-01, HGN-120-A, HGN-350, HGZ-050-02, HGZ-080-02, HGZ-120-02; Micalite PP-G2MF-4, PP-G2MF-5, PP-G2MF-6, PP-H2MFS-4, PP-H2MFS-5, PP-H2MFS-6; Norchem NPP 8000-GK, NPP 8001-LK, NPP 8002-HK, NPP 8004-MR, NPP 8005-AR, NPP 8006-GF, NPP 8007-GO, NPP 8004-ZR; Nortuff NFA 2400 CO, NFA 2600 CO, NFA 4400 TO, NFC 1700 MO, NFC 1800 TO, NFC 2200 CO, NFC 2400 CO, NFC 2440 CF, NFC 2600 CO, NFC 4000 FR, NFC 4400 CO, NFC 4600 CO; Petrothene PP 8000-GK, PP 8001-LK, PP 8004-MR, PP 8004-ZR, PP 8005-AR, PP 8020-AU, PP 8020-GU, PP 8020-ZU, PP 8080-AW, PP 8080-GW, PP 8080-ZW, PP 8310-GO, PP 8310-KO, PP 8402-HO, PP 8402-TO, PP 8403-HO, PP 8403-TO, PP 8404-HJ, PP 8404-ZJ, PP 8410-ZR, PP 8411-ZR, PP 8412-HK, PP 8412-TK, PP 8420-HK, PP 8462-HR, PP 8470-ZU, PP 8502-HK, PP 8602-HJ, PP 8752-HF, PP 8755-HK, PP 8762-HR, PP 8770-HU, PP 8802-HO, PP 8815-ZR, PP 8820-HU; Pro-fax 6131, 6323, 6331NW, 7131, 7523, 7531, 8523, 8623, PC-072, PD626, PD701, SA595, SA862, SA868M, SB661, SB751, SB782, SB786, SB787; RTP 100 GB10, 100 GB20, 100 GB30, 100 GB40, 100 T10 or 131, 100 T20 or 128,, 100 T30 or 132, 100 T40 or 127, 101, 101CC, 101FR, 101 SP Foamed, 103, 103CC, 103 SP Foamed, 105, 105CC, 105CC FR, 105 SP Foamed, 107, 107CC, 140, 141, 142, 143, 150, 175, 175X, 177, 178, 178X, 199x22898, 199x23835, 199x28016; Shell 5419, 5431, 5520, 5524, 5550, 5610, 5820, 5824S, 5840, 5864, 5944S, 7129, 7328, 7521, 7522, 7623, 7912; Tenite Polypropylene; Vestolen P2200, P2300, P3200, P4200, P4700, P4800, P5200, P5206S, P5232 G, P5232 T, P5272 T, P5800, P6200, P6500

Laminating:

Pro-fax SD062; SB522/1S Film

Polypropylene (cont'd.)

Melt spinning/drawing:
Fortilene 3606, 3907; Marlex HGZ-120-04; Vestolen P1200, P1200 F
Solid phase forming:
Shell 5384, 7221
Structural foam molding:
RTP 101 SP Foamed, 103 SP Foamed, 105 SP Foamed
Thermoforming:
Nortuff NFA 1700 MO, NFA 1800 TO, NFA 2400 CO, NFA 2600 CO; Pro-fax 7531, 7823, PF101, SA878, SD101; Rexene 6310, 6310C25, 6310T2, 6310T4, 9234, 9400, 9400T2, 9401, 9401C, 9402, 9403, 9403E, 9500

CATEGORY:
Thermoplastic resin

APPLICATIONS:
Automotive applications: (Hostalen PP996, PP998; Marlex CHM-040-01; Micalite PP-G2MF-4, PP-G2MF-5, PP-G2MF-6, PP-H2MFS-4, PP-H2MFS-5, PP-H2MFS-6; Norchem NPP 8000-GK, NPP 8006-GF, NPP 8007-GO; PP-C2IM, –C2TF-1, -C3CC-4, -C4IM, -H1TF-4, -H2MF-5, -H2MFQ-4, -H2TF-2, -H2TF-4, -H3MF-2, -H3MFQ-1, -H3MFQ-5, -HFR-2, -HFR-3; Pro-fax 65F4-4, 6323, 6524, 7523, 8523, 8623, PC072, SA595, SB661; RTP 100T10, 100T20, 100T30, 100T40, 101, 101CC, 101FR, 103, 103CC, 105, 105CC, 105CC FR, 107, 107CC; Thermo-comp MF-1004FR, MFX-1004HS, MFX-1006HS, MFX-1008HS; Vestolen P5232 G, P5232 T, P5272 T); battery cases (Bapolene 5052; Pro-fax SB661; Vestolen P4700); consoles (PP-VB-511); crash panels (PP-H2MF-4, -H2MFS-4, –H2MFS-5); door panels (PP-VB-511); fender liners and load floors (Rexene 6310, 6310C25, 6310T2, 6310T4, 9234, 9400, 9400T2, 9401, 9401C, 9402, 9403, 9403E, 9500); instrument panels (Bapolene 4112, 4114; Micalite PP-G2MF-4, PP-G2MF-5, PP-G2MF-6, PP-H2MFS-4, PP-H2MFS-5, PP-H2MFS-6; PP-H2MF-4, –H2MFS-4, -H2MFS-5, -H2TF-2); interior parts (PP-VB-511); light housings (Thermocomp MF-1002, MF-1004, MF-1006, MF-1008); trim (Pro-fax SB782, SB786, SB787); under-the-hood parts (PP-H2TF-4)

Aviation industry: avionics housings (EMI-X MA-40)

Consumer products: (Bapolene 4062); appliance housings (Nortuff NFA 4400 TO); appliance parts (Marlex HGH-050; Norchem NPP 8006-GF; Nortuff NFA 4400 TO; RTP 100T10, 100T20, 100T30, 100T40, 101, 101CC, 101FR, 103, 103CC, 105, 105CC, 105CC FR, 107, 107CC); appliances (Bapolene 4112, 4114, 5042, 5052, 5072; Hostalen PP942, PP996, PP998; Norchem NPP 8001-LK, NPP 8002-HK, NPP 8007-GO; Nortuff NFC 1700 MO; PP-C2TF-4, -C3TFA-2, -H2TF-2, –HFR-2, -HFR-3; Pro-fax 65F4-4, 65F5-4; Pro-fax 6524, 8523, 8623, PC072, SA595, SB782, SB786, SB787; Thermocomp MF-1004FR, MFX-1004HS, MFX-1006HS, MFX-1008HS); cabinets (PP-C3TF-2); carrying cases (Marlex CHM-040-01; PP-C2IM); clothing (Vestolen P1200, P1200 F); cosmetic packaging (Pro-fax SA878); cutlery/utensils (Bapolene 4072; Norchem NPP 8001-LK, NPP 8002-HK); decorative containers (Norchem NPP 8005-AR); drinkware (Bapolene 4082;

Polypropylene *(cont'd.)*

Pro-fax 7531); furniture (Bapolene 4062, 5042, 5072; Hostalen PP942, PP996, PP998; Marlex CHM-040-01, CHM-040-02; PP-C2TF-4, -H2TF-2; Pro-fax 6329, 7523, SB782, SB786, SB787; Vestolen P4800); grilles (Nortuff NFA 4400 TO); home furnishings (Vestolen P1200, P1200 F); housewares (Bapolene 4042, 4062, 5042, 5052, 5072;Fortilene 14X01, 1401; Hostalen PP941, PP975, PP989; Marlex CHM-040-02, HGN-120-01, HGN-350, HGZ-050-02, HGZ-080-02; Norchem NPP 8000-GK; Nortuff NFC 1700 MO; PP-C2IM, -C3CC-4, -C3TFA-2, -C6CC-4, -H7CC-4; Pro-fax 65F5-4, 6323, 6329, 6331NW, PD701, SA862, SA868M; Vestolen P5200, P5206 S); kitchenware (Norchem NPP 8005-AR); lawn and garden equipment (Nortuff NFC 1800 TO; PP-H2CC-2); laundry tubs (PP-C3TFA-2); luggage (Hostalen PP998; Pro-fax 8523, 8623); microwave trays (Micalite PP-G2MF-4, PP-G2MF-5, PP-G2MF-6, PP-H2MFS-4, PP-H2MFS-5, PP-H2MFS-6; Rexene 9234, 9400, 9400T2, 9401, 9401C, 9402, 9403, 9403E, 9500); novelties (Bapolene 4072); outdoor applications (PP-C5CC-2); personal care products (Norchem NPP 8001-LK, NPP 8002-HK, NPP 8007-GO); pool covers (Pro-fax PC968); pool equipment (PP-H2MFQ-3, -H2MFS-3); smoke detectors (Nortuff NFA 4400 TO); thermos bottles (Pro-fax 6131); toys (Bapolene 4062, 4072, 4082, 5042, 5052, 5072; Hostalen PP941, PP996, PP998; Marlex CHM-040-02, HGZ-050-02, HGZ-080-02; PP-C2IM, -C2TF-4, -C3CC-4, -C3TFA-2, -H2TF-2; Pro-fax 6331NW)

Electrical/electronic industry: (A-Fax 600; Empee PP 401, PP 401CS, PP 402; Nortuff NFC 4400 CO, NFC 4600 CO; Thermocomp MF-1004FR; Vestolen P5232 G, P5232 T, P5272 T); air conditioners (PP-H2LTF-4, -H2TF-4); business machines/office equipment (Bapolene 4112, 4114; EMI-X MA-40; Micalite PP-G2MF-4, PP-G2MF-5, PP-G2MF-6, PP-H2MFS-4, PP-H2MFS-5, PP-H2MFS-6; Thermocomp MF-1004FR); cable and wire insulation (Pro-fax SE191); cable and wire saturants (A-Fax 940); cassette housings (Bapolene 5042, 5052, 5072); connectors (PP-H2TF-4); electrical appliances (PP-HIFR); electrical components (Stat-Kon M-1 HI, M-2, ME, MF-15, MM-3340); electrical enclosures/packaging (Nortuff NFC 2200 CO, NFC 2400 CO, NFC 2440 CF, NFC 2600 CO; Stat-Kon M-1 HI, M-2, ME, MF-15, MM-3340); electronic devices (EMI-X MA-40); heating ducts/fan shrouds (Bapolene 4112, 4114; PP-C4IM, -H2LTF-4, -H2TF-4); housings (EMI-X MA-40; Micalite PP-G2MF-4, PP-G2MF-5, PP-G2MF-6, PP-H2MFS-4, PP-H2MFS-5, PP-H2MFS-6; Nortuff NFC 4400 CO, NFC 4600 CO; PP-H2LTF-4, H2TF-4); phonograph covers (Norchem NPP 8005-AR); power tools (Micalite PP-G2MF-4, PP-G2MF-5, PP-G2MF-6, PP-H2MFS-4, PP-H2MFS-5, PP-H2MFS-6; Nortuff NFC 1800 TO, NFC 2200 CO, NFC 2400 CO, NFC 2440 CF, NFC 2600 CO, NFC 4400 CO, NFC 4600 CO); underground and marine cable (A-Fax 600)

FDA-approved applications: (Amoco 10-4017, 1012, 1046, 1088, 4018, 4036, 4039, 4222, 4228, 5219, 6431, 7232, 7233, 7234, CR22N, CR22NA, CR35, CR35A, CR35N, CR35NA; Bapolene 4042, 4062, 4072, 4082, 5042, 5052, 5072; Fortilene 10X01, 14X01, 16X01, 18X02, 18X03, 19X02, 19X03, 21X04, 40X05, 41X03, 41X04, 42X07, 54X02, 1401, 1602, 1602A, 2251, 3151, 3230, 3251, 3606, 4141,

Polypropylene *(cont'd.)*

4141F, 9205, 9605; Hostalen PP975, PP989; Marlex CHM-040-01, CHM-040-02, HGH-050, HGN-120-01, HGN-120-A, HGN-350, HGZ-040, HGZ-050-02, HGZ-080-02, HGZ-120-02, HGZ-120-04, HNS-080; MWB Film; Norchem NPP 1001-LF, NPP 1006-GF, NPP 1008-AK, NPP 1010-LC, NPP 2000-GJ, NPP 2003-GJ, NPP 2004-MR, NPP 2013-UJ, NPP 3007-GO, NPP 3010-SO, NPP 7000-GF, NPP 7300-GF, NPP 7300-AF, NPP 8000-GK, NPP 8001-LK, NPP 8002-HK, NPP 8004-MR, NPP 8005-AR, NPP 8006-GF, NPP 8007-GO, NPP 8004-ZR; Nortuff NFA 1700 MO, NFA 1800 TO, NFC 1700 MO, NFC 1800 TO; Petrothene PP 8000-GK, PP 8001-LK, PP 8004-MRPP-H6TF-4; Pro-fax 6323, 6323F, 6523F, 6331NW, 6524, 6532F, 7131, 7523, 7531, 7823, PC942, PD064, PD401, PF101, SA861, SA862, SA878, SD062, SD101; Rexene 6310, 6310C25, 6310T2, 6310T4, 9234, 9400, 9400T2, 9401, 9401C, 9402, 4903, 4903E, 9500; Shell 5225, 5384, 5431, 5520, 5524, 5550, 5610, 5820, 5824S, 5840, 5864, 5944S, 7129, 7328, 7522, 7912, DX6016, DX6020)

Food-contact applications: (Fortilene 40X05, 41X03, 41X04, 42X07, 54X02, 4141, 4141F; Marlex CHM-040-01, CHM-040-02, HGH-050, HGZ-040, HGZ-120-04; Norchem NPP 8006-GF, NPP 8007-GO; Shell 5225, 5384, 5431, 5520, 5524, 5550, 5610, 5820, 5824S, 5840, 5864, 5944S, 7522, 7912, DX6016, DX6020); food containers/packaging (Bapolene 4062, 4072; Hostalen PP941, PP975, PP989; Marlex HGN-120-A, HGZ-050-02, HGZ-080-02, HNS-080; MWB Film; Norchem NPP 8000-GK, NPP 8002-HK; PP-H6CC-4, -H6TF-4; Pro-fax 6131, PC942, PF101, SA861, SA862, SA868M, SA878, SD101; Rexene 6310, 6310C25, 6310T2, 6310T4, 9234, 9400, 9400T2, 9401, 9401C, 9402, 9403, 9403E, 9500; SB522/1S Film; Vestolen P2200, P2330 F, P3200F, P3230 F)

Functional additives: adhesive layer in coextrusions (Modic P-110F, P-300F, P-300M); modifier for polyolefin, wax, asphalt (A-Fax 940)

Industrial applications: (Thermocomp MFX-1004HS, MFX-1006HS, MFX-1008HS); adhesives (A-Fax 500, 600, 940; Vestolen P5400 F); bags (Fortilene 3230; MWB Film; Pro-fax PC942, SA861; Vestolen P5400 F); batteries (Pro-fax SB782, SB786, SB787); bearing applications (Stat-Kon M); biaxially oriented bottles (Marlex HLN-020-01); biaxially oriented films (Fortilene 21X04; Norchem NPP 3007-GO, NPP 3010-SO; Pro-fax PD064; Rexene PP41E2, PP41E4, XO-325; Vestolen P5200 F) ; bottles (Norchem NPP 7000-ZF; Vestolen P5400); bristles (Hostalen PP927; Pro-fax 6323F, 6523F); cams (Nortuff NFC 2200 CO, NFC 2400 CO, NFC 2440 CF, NFC 2600 CO, NFC 4400 CO, NFC 4600 CO); carpet backing (A-Fax 940; Fortilene 3151; Norchem NPP 2004-MR; Vestolen P4200 F, P5200 F); carpets (Vestolen P1200, P1200 F); cast film (Norchem NPP 3007-GO, NPP 3010-SO; Pro-fax PC942, SA752, SA861; Vestolen P2200 F, P2330 F, P3200 F, P3230 F); caulks/sealants (A-Fax 800, 940); chemical equipment/containers (Bapolene 4042; Marlex HGH-050; Pro-fax 6524, PC072, SB751); closures (Bapolene 4062, 4072, 5052, 5072; Fortilene 14X01, 16X01, 18X02, 18X03, 19X02, 19X03, 54X02, 1401, 1602, 1602A; Hostalen PP975, PP989; Marlex HGZ-120-2, HGZ-350; Norchem NPP 8000-GK, NPP 8007-GO, NPP 8004-ZR; PP-H6TF-4; Pro-fax

Polypropylene (cont'd.)

6323, 6329, 6331NW, 7823, PD701); coatings (Petrothene PP 1510-HC, PP 1510-LC); color concentrates (Bapolene 4072; Fortilene 9605); compounding (Fortilene 19X02, 9205, 9605); compounding with glass (Pro-fax PC072); containers (Bapolene 5042, 5052, 5072; Fortilene 18X02, 18X03, 19X02, 19X03, 41X03, 41X94, 4141; Hostalen PP989, PP996, PP998; Marlex HGN-120-01, HGN-350, HGZ-120-02, HGZ-350; Norchem NPP 7000-GF, NPP 7000-ZF, NPP 7300-GF, NPP 7300-AF; Pro-fax 6331NW, 7131, 7523, SB751, SB782, SB786); ducts (Vestolen P6206 S); fabric coatings (Pro-fax SD062); fiber and filaments (Petrothene PP 2004-MR, PP 2085-GU; Shell 5840); fibers for outdoor exposure (Norchem NPP 2013-UJ); films (Fortilene 40X05, 42X07, 2251, 4141F; Hostalen PP934; Marlex HNS-080; Modic P-300M; Petrothene PP 1510-HC, PP 1510-LC, PP 2004-MR, PP 2085-GU; Pro-fax 6131); flat parts (RTP 100 GB10, 100 GB20, 100 GB30, 100 GB40); foamed parts (RTP 100T10, 100T20, 100T30, 100T40, 101, 101CC, 101FR, 103, 103CC, 105, 105CC, 105CC FR, 107, 107CC); foamed ribbon (Pro-fax 6323F, 6523F); frictional/sliding applications (Stat-Kon M); gears (Nortuff NFC 2200 CO, NFC 2400 CO, NFC 2440 CF, NFC 2600 CO, NFC 4400 CO, NFC 4600 CO; Stat-Kon M); general purpose moldings (Fortilene 14X01, 16X01, 1602, 1602A); heater components (Thermocomp MF-1002, MF-1004, MF-1006, MF-1008); hollow articles (Vestolen P5200, P5400, P6200, P6500); housings (PP-H2TF-2; RTP 100T10, 100T20, 100T30, 100T40, 101, 101CC, 101FR, 103, 103CC, 105, 105CC, 105CC FR, 107, 107CC; Thermocomp MF-1002, MF-1004, MF-1006, MF-1008); industrial parts (Hostalen PP942; Marlex HGH-050; Pro-fax 65F4-4); intricate/hard-to-fill moldings (Norchem NPP 8004-MR, NPP 8004-ZR); laminates (SB522/1S Film); large parts (Bapolene 4072, 4082); monofilament (Hostalen PP933; Marlex HGV-040-01, HGX-030, HGZ-040; Norchem NPP 2000-GJ, NPP 2003-GJ; Pro-fax 6323F, 6523F; Vestolen P5200 F); multifilament (Marlex HGZ-120-04; Norchem NPP 2004-MR; Pro-fax 6323F, 6523F, PC968); netting (Pro-fax PD401); NSF-approved applications: (Pro-fax 6524, 7523, 7823); office supplies (Vestolen P2330 F); oriented film (Shell DX6016); packaging/displays (Bapolene 4072, 4082; Norchem NPP 8006-GF); packaging film (MWB Film; Vestolen P2000 F, P2330 F, P3200 F, P3230 F); paper coatings (Pro-fax SD062); pipe fittings (Thermocomp MF-1002, MF-1004, MF-1006, MF-1008); pipes (Norchem NPP 1006-GF; Vestolen P6206 S, P6421, P6522); plates (Pro-fax PD195); profiles (Fortilene 10X01, 40X05; Hostalen PP933; Norchem NPP 1001-LF, NPP 1006-GF, NPP 1010-LC; Nortuff NFA 1700 MO, NFA 1800 TO, NFA 2400 CO, NFA 2600 CO; Petrothene PP 1510-HC, PP 1510-LC); pumps (PP-H2TF-2; RTP 100T10, 100T20, 100T30, 100T40, 101, 101CC, 101FR, 103, 103CC, 105, 105CC, 105CC FR, 107, 107CC; Thermocomp MF-1002, MF-1004, MF-1006, MF-1008); rods (Nortuff NFA 2400 CO, NFA 2600 CO); rope/twine/cordage (Bapolene 4042; Fortilene 3251; Hostalen PP933, PP934, PP936; Pro-fax 6532F; Vestolen P4200 F, P5200 F); sheet (Empee PP 401, PP 401CS; Fortilene 10X01, 40X05; Hostalen PP920; Norchem NPP 1010-LC; Nortuff NFA 1700 MO, NFA 1800 TO; PP-H1CC-1; Pro-fax 7531, 7823, PD195, SD101; Rexene 6310,

Polypropylene *(cont'd.)*

6310C25, 6310T2, 6310T4, 9234, 9400, 9400T2, 9401, 9401C, 9402, 9403, 9403E, 9500; Shell 5384; Vestolen P6421); slit film (Fortilene 3151, 3230, 3251; Marlex HGV-040-1, HGX-030, HGZ-040; Norchem NPP 2000-GJ); slit tape (Pro-fax 6532F); staple fibers and continuous filament (Fortilene 38X01, 3606; Marlex HGX-030, HGZ-120-04; Vestolen P1200, P1200 F, P200 F, P3200); strapping (Fortilene 10X01, 40X05; Hostalen PP920; Norchem NPP 1001-LF, NPP 1010-LC; Pro-fax PD401; Shell 5384); stretched tape yarn (Bapolene 4042; Shell 5419, 5431); structural foam (RTP 101 SP Foamed, 103 SP Foamed, 105 SP Foamed); tapes and ribbons (Fortilene 3251; Hostalen PP927, PP936; Vestolen P4200 F, P5200 F, P5400 F); textile applications (Fortilene 3230, 3907; Hostalen PP934; Norchem NPP 2004-MR; Pro-fax PC968, PF151; Vestolen P1200, P1200 F); thick sections (Pro-fax PD195); thin sections (Bapolene 4072, 4082; Fortilene 18X02, 18X03, 19X02, 19X03; Marlex HGN-120-01, HGN-350, HGZ-120-02, HGZ-350; Norchem NPP 8004-MR; Pro-fax 7131, SB751; RTP 100 GB10, 100 GB20, 100 GB30, 100 GB40; Vestolen P2200, P2300, P3200, P4200); totes/bins (Pro-fax 7523, 8523, 8623, SB782, SB786, SB787; Vestolen P4800, P5800); trays (Norchem NPP 8000-GK; Pro-fax 7131, 7523, SB751; Vestolen P4800, P5800); tubing/hoses (Hostalen PP933; Nortuff NFA 2400 CO, NFA 2600 CO; Petrothene PP 1510-HC, PP 1510-LC)

Medical applications: (Fortilene 54X02; Marlex HGN-120-01, HGN-350, HGZ-120-02, HGZ-350; Norchem NPP 8004-MR); drug packaging/closures (Marlex HGN-120-A; PP-H6TF-4; Pro-fax SA878; Rexene 6310, 6310C25, 6310T2, 6310T4, 9234, 9400, 9400T2, 9401, 9401C, 9402, 9403, 9403E, 9500); hospitalware (Bapolene 4042; Norchem NPP 8004-MR; Pro-fax 6323); medical devices (Bapolene 4082; PP-H4TF-3; Pro-fax PD626); syringes (Bapolene 4062, 4082; Marlex HGZ-120-02, HGZ-350); vials (Hostalen PP975; Marlex HGN-120-A)

PROPERTIES:

Form:

Powder (Amoco 6200P, 6400P, 6420P, 6800P, 7000P, 76200P, 7220P)

Waxy solid (A-Fax 500, 600, 940

Strand-cut pellets (Micalite PP-G2MF-4, PP-G2MF-5, PP-G2MF-6, PP-H2MFS-4, PP-H2MFS-5, PP-H2MFS-6)

Granules (Vestolen P6522)

Spheres (Fortilene 10X01, 14X01, 16X01, 18X02, 18X03, 19X02, 19X03, 21X04, 38X01, 40X05, 41X03, 41X04, 42X07, 54X02, 1401, 1602, 1602A, 2251, 3151, 3230, 3251, 3606, 3907, 4141, 4141F)

Free-flowing spheres (Fortilene 9205, 9605)

Film avail. in 45-, 75-, and 100-gauge thicknesses in various widths (MWB Film)

Film avail. in 70- and 90-gauge thicknesses in widths from 1–54 in. (SB522/1S Film)

Fineness:

99% retained 60–140 mesh; 1% retained 200 mesh (Amoco 6200P, 6400P, 6420P, 6800P, 7000P, 7200P, 7220P)

Color:
Off-white (A-Fax 500, 600, 940)
Trans-white (Petrothene PP 8402-TO, PP 8403-TO, PP 8412-TK)
Earthtone color (P012, P032)
Black (Vestolen P6503; RTP 199x22898, 199x23835, 199x28016)
Colors avail. (P04, P0113)
Natural (Vestolen P1200, P1200 F, P2200, P2200 F, P2300, P2300 F, P3200, P3200
 F, P4200, P4200 F, P4700, P4800, P5400, P5400 F, P5800, P6200, P6421, P6500)
Natural and standard colors (Vestolen P5200, P5200 F)
Natural or custom colors (P2120, P2130, P2140, P2230, P9125)
Natural, white, gray, black (Vestolen P5232 G)
Natural, gray, black (Vestolen P5232 T, P5272 T)
Natural and black (RTP 100 GB10, 100 GB20, 100 GB30, 100 GB40, 100T10,
 100T20, 100T30, 100T40, 101, 101CC, 101FR, 101 SP Foamed, 103, 103CC, 103
 SP Foamed, 105, 105CC, 105CC FR, 105 SP Foamed, 107, 107CC, 140, 141, 142,
 143, 150, 175, 175X, 177, 178, 178X)
Gray (Vestolen P5206 S, P6206 S, P6522)
Gardner 5.5 (Rexene PP41E2, PP41E4, XO-325)

GENERAL PROPERTIES:
M.W.:
12,000 (A-Fax 600)
23,000 (A-Fax 500)
43,000 (A-Fax 940)
200,000 (Vestolen P1200, P1200 F)
220,000 (Vestolen P2200, P2200 F, P2300, P2300 F)
240,000 (Vestolen P3200, P3200 F)
340,000 (Vestolen P4200, P4200 F)
370,000 (Vestolen P4700)
380,000 (Vestolen P4800, P5200, P5200 F, P5206 S, P5232 G, P5232 T, P5272 T)
440,000 (Vestolen P5400, P5400 F)
470,000 (Vestolen P5800, P6200, P6206 S, P6421)
500,000 (Vestolen P6500)
540,000 (Vestolen P6522)

Melt Flow:
0.9 dg/min (Rexene 6310)
1.0 dg/min (Rexene 6310T2, 9400)
1.1 dg/min (Rexene 9401, 9401C)
1.4 dg/min (Rexene 9402, 9500)
1.5 dg/min (Rexene 9234, 9400T2)
1.6 dg/min (Rexene 6310C25, 9403)
1.9 dg/min (Rexene 9403E)
2.0 dg/min (Rexene 6310T4, PP41E2)
4.0 dg/min (Rexene PP41E4, XO-325)

Polypropylene *(cont'd.)*

0.3 g/10 min (Fortilene 40X05; Vestolen P6522)

0.4 g/10 min (Pro-fax 7823)

0.5 g/10 min (Vestolen P6200, P6206 S, P6500)

0.5–2.5 g/10 min (PP-H1CC-1)

0.6 g/10 min (Norchem NPP 1010-LC; Shell 5225, 7129; Vestolen P6421)

0.7 g/10 min (Fortilene 10X01; Petrothene PP 1510-HC, PP 1510-LC; Shell 7221)

0.9 g/10 min (Pro-fax SA595)

1.0 g/10 min (Nortuff NFA 1800 TO; PP-H1TF-4; Vestolen P5400, P5400 F)

1.2 g/10 min (Amoco 1012; Norchem NPP 1001LF)

1.3 g/10 min (Modic P-110F, P-300F)

1.5 g/10 min (Norchem NPP 7300-GF, NPP 7300-AF; Nortuff NFA 1700 MO, NFA 2600 CO; Pro-fax PD195, PD401)

1.7 g/10 min (Nortuff NFA 4400 TO)

2 g/10 min (Amoco 6200P; Fortilene 41X03, 41X04, 4141, 4141F; Marlex HLN-020-01; Norchem NPP 1006-GF, NPP 7000-GF, NPP 7000-ZF, NPP 8006-GF; Petrothene PP PP 7200-AF, PP 7200-GF, PP 7200-MF, PP 7300-KF, PP 7300-MF, PP 8752-HF; Pro-fax 8623, PF101, SA878, SD101; Shell 5384, 7328, DX6020; Vestolen P5800)

2.4 g/10 min (Pro-fax SE191)

2.5 g/10 min (Fortilene 21X04; Nortuff NFA 2400 CO; Petrothene PP 8602-HJ)

2.5–90 g/10 min (Tenite Polypropylene)

2.8 g/10 min (Shell 5431)

3 g/10 min (Fortilene 3151; Marlex HGX-030; Pro-fax SA752, SA861; Shell 5419; Vestolen P5200, P5200 F, P5206 S, P5232 G, P5232 T, P5272 T)

3–7 g/10 min (PP-H2CC-2, -H2LTF-4, -H2MFQ-3, -H2MFQ-4, -H2MFS-3)

3.1 g/10 min (Norchem NPP 2003-GJ)

3.5 g/10 min (Norchem NPP 2013-UJ, 7521)

3.7 g/10 min (Norchem NPP 2000-GJ)

4 g/10 min (Amoco 4036, 6400P, 6420P, 6431; Bapolene 4042, 5042; Fortilene 42X07, 2251, 3230, 3251, 9205; Marlex CHM-040-01, CHM-040-02, HGV-040-01, HGZ-040; Nortuff NFC 1800 TO; PP-C2TF-1; Pro-fax 65F4-4, 65F5-4, 6523F, 6524, 6532F, 7523, 7531, 8523, PD064; Shell DX6016)

4–10 g/10 min (PP-H3MF-2)

4.5 g/10 min (Petrothene PP 8404-HJ, PP 8404-ZJ, 7522)

5 g/10 min (Amoco 1046; Marlex HGH-050, HGZ-050-02; Norchem NPP 1008-AK, NPP 8000-GK, NPP 8001-LK, NPP 8002-HK; Nortuff NFC 2440 CF, NFC 4400 CO, NFC 4600 CO; P04; Petrothene PP 8000-GK, PP 8001-LK, PP 8502-HK, PP 8755-HK; PP-C2IM, -C3CC-4; Pro-fax SA862; Shell 5520, 5524, 5550, 7525; Vestolen P4200, P4200 F, P4700, P4800)

5–7 g/10 min (Empee PP 401, PP 401CS, PP 402; PP-C2TF-4)

5–8 g/10 min (PP-VB-511)

6 g/10 min (Nortuff NFC 2600 CO; P032; Petrothene PP 8420-HK; PP-C3TFA-2)

6–8 g/10 min (PP-C4IM)

6–10 g/10 min (PP-C3TF-2, -H3MFQ-1, -H3MFQ-5)

7 g/10 min (Bapolene 5052; Nortuff NFC 1700 MO, NFC 2400 CO; Petrothene PP 8412-HK, PP 8412-TK; PP-H2TF-4; Pro-fax PCF942, SA868M, SB782; Shell 7627)

7–11 g/10 min (PP-H2TF-2)

8 g/10 min (Amoco 10-4017, 1088, 6800P; Bapolene 4112, 4114; Fortilene 14X01, 54X02, 1401; Marlex HGZ-080-02, HNS-080; Norchem NPP 3007-GO, NPP 3010-SO, NPP 8007-GO; P012; Petrothene PP 8402-HO, PP 8402-TO, PP 8403-HO, PP 8403-TO, PP 8802-HO; Pro-fax SB661, SB786; Shell 5610)

8–12 g/10 min (PP-H4TF-3)

8.5 g/10 min (Shell 7635)

9 g/10 min (Nortuff NFC 2200 CO; Shell 7623)

10 g/10 min (Amoco 7000P; P0113; Petrothene PP 8310-GO, PP 8310-KO, PP 8411-ZR; Vestolen P3200, P3200 F)

11 g/10 min (Petrothene PP 8410-ZR)

12 g/10 min (Amoco 4018, 7200P, 7232, 7233, 7234; Bapolene 4062; Fortilene 16X01, 1602, 1602A, 3606, 9605; Marlex HGN-120-01, HGN-120-A, HGZ-120-02, HGZ-120-04; Modic P-300M; Norchem NPP 2004-MR, NPP 8004-MR, NPP 8005-AR, NPP 8004-ZR; Nortuff NFC 4000 FR; P2120, P2130, P2140, P2230, P9125; Petrothene PP 2004-MR, PP 8004-MR, PP 8004-ZR, PP 8005-AR, PP 8462-HR, PP 8762-HR, PP 8815-ZR; Pro-fax 6323, 6323F, 6329, 6331NW, PD626; Shell 5820, 5824S, 5864)

12–18 g/10 min (PP-C5CC-2)

13 g/10 min (Shell 5840)

14 g/10 min (Amoco 5219)

15 g/10 min (Amoco 4039; PP-H6TF-4; Pro-fax SD062)

18 g/10 min (Fortilene 38X01)

18–25 g/10 min (PP-C6CC-4, -H6CC-4)

20 g/10 min (Bapolene 4072; Fortilene 18X02, 18X03; Petrothene PP 8020-AU, PP 8020-GU, PP 8020-ZU, PP 8470-HU, PP 8470-ZU, PP 8770-HU, PP 8820-HU; Pro-fax SB787; Vestolen P2200, P2200 F, P2300, P2300 F)

22 g/10 min (Amoco CR22N, CR22NA; Shell 7912)

23 g/10 min (Bapolene 5072)

24 g/10 min (Amoco 4222)

25–35 g/10 min (PP-H7CC-4)

28 g/10 min (Amoco 4228)

30 g/10 min (Fortilene 19X02, 19X03; P0119; Shell 5944S)

33 g/10 min (Fortilene 3907; Vestolen P1200, P1200 F)

35 g/10 min (Amoco CR35, CR35A, CR35N, CR35NA; Bapolene 4082; Marlex HGN-350, HGZ-350; Petrothene PP PP 2085-GU, PP 8080-AW, PP 8080-GW, PP 8080-ZW; Pro-fax 6131, 7131, PC968, PD701, PF151, SB751)

Sp. Gr.:

0.77 (RTP 101 SP Foamed)

Polypropylene *(cont'd.)*

0.84 (RTP 103 SP Foamed)

0.89 (Modic P-300F, P-300M; Shell DX6020)

0.895 (Vestolen P5400, P5400 F, P6421)

0.895–0.905 (PP-C2IM)

0.897 (Pro-fax 7131, 7823, SA862, SA878, SB751)

0.899 (Pro-fax 6329, 7523, SB661, SB782)

0.900 (Amoco 10-4017, 1012, 1046, 1088, 4018, 4036, 4039, 4222, 4228, 5219, 6200P, 6400P, 6420P, 6431, 6800P, 7000P, 7200P, 7220P, 7232, 7233, 7234, CR22N, CR22NA, CR35, CR35A, CR35N, CR35NA; Hostalen PP920, PP927, PP933, PP934, PP936, PP941, PP942, PP975, PP989, PP996, PP998; Modic P-110F; Pro-fax 8623, SA868M, SE191; Rexene 6310, 9234, 9400, 9401, 9402, 9403, 9403E, 9500; RTP 105 SP Foamed; Shell 7522, 7912, DX6016)

0.900–0.904 (Tenite Polypropylene)

0.901 (Bapolene 5042, 5052, 5072; Pro-fax 8523, PD626, SB787; Shell 7129, 7221, 7328; Vestolen P2300, P2300 F)

0.902 (Bapolene 4042; Norchem NPP 1001-LF, NPP 1006-GF, NPP 1008-AK, NPP 1010-LC, NPP 2000-GJ, NPP 2003-GJ, NPP 2004-MR, NPP 3007-GO, NPP 3010-SO, NPP 7000-ZF, NPP 7300-GF, NPP 7300-AF, NPP 8000-GK, NPP 8001-LK, NPP 8002-HK, NPP 8004-MR, NPP 8005-AR, NPP 8006-GF, NPP 8007-GO, NPP 8004-ZR; Pro-fax 6131, PD195, PD701, SB786; Shell 5225, 7525, 7623, 7635)

0.903 (Bapolene 4062; Pro-fax 6323, 6331NW, 6524; Shell 5384, 5419, 5431, 5520, 5550, 5610, 5820, 5840, 7521)

0.904 (Shell 5524, 5824S, 5864; Vestolen P6200, P6500)

0.905 (Bapolene 4072, 4082; Marlex HGH-050, HGV-040-01, HGX-030, HGZ-040, HGZ-050-02; Rexene PP41E2, PP41E4, XO-325; Shell 5944S, 7627; Vestolen P5800)

0.906 (Marlex HLN-020-01; Vestolen P4200, P4200 F, P4800, P5200, P5200 F)

0.907 (Vestolen P1200, P1200 F, P2200, P2200 F, P3200, P3200 F)

0.908 (Marlex CHM-040-01, CHM-040-02, HGZ-080-02, HGZ-120-02, HGZ-120-04, HGZ-350, HNS-080)

0.909 (Marlex HGN-120-01, HGN-350)

0.910 (PP-C4IM; Vestolen P4700, P6522)

0.913 (Marlex HGN-120-A; PP-HIFR)

0.938 (Empee PP 401, PP 401CS, PP 402)

0.94 (PP-C2TF-1; Stat-Kon M)

0.945 (Vestolen P5206 S, P6206 S)

0.96 (Thermocomp MF-1002, MF-1002HI)

0.97 (Nortuff NFC 2200 CO; RTP 100 GB10)

0.975 (PP-H1CC-1)

0.98 (Nortuff NFC 4000 FR; RTP 100T10, 101, 101CC, 142)

0.99 (PP-H3MFQ-1)

1.00 (Stat-Kon ME)

1.02 (PP-VB-511; RTP 199x28016)

1.03 (PP-C3TF-2, -H3MF-2; Thermocomp MF-1004HI)
1.04 (Nortuff NFA 2400 CO, NFC 2400 CO; PP-H2CC-2, -H2TF-2, -HFR-2; RTP 100 GB20; Stat-Kon M-2; Thermocomp MF-1004, MFX-1004HS; Vestolen P5232 G, P5232 T)
1.05 (Bapolene 4112; Rexene 6310T2, 9400T2, 9401C; RTP 100T20, 103, 103CC, 141)
1.06 (PP-C3TFA-2, -C5CC-2; Stat-Kon M-1 HI)
1.10 (PP-HFR-3; Rexene 6310C25)
1.12 (Nortuff NFA 2600 CO, NFC 2600 CO; Pro-fax SA595; RTP 100 GB30, 175, 175X; Thermocomp MF-1006HI)
1.13 (PP-H2MFQ-3, -H2MFS-3, -H4TF-3; RTP 105, 105CC, 178, 178X, 199x22898; Thermocomp MF-1006)
1.14 (RTP 100T30, 143)
1.17 (Nortuff NFA 1700 MO, NFC 1700 MO)
1.18 (Stat-Kon MF-15)
1.20 (Nortuff NFC 4400 CO; Rexene 6310T4)
1.21 (Stat-Kon MM-3340; Thermocomp MF-1008HI)
1.22 (PP-H2MF-4; Pro-fax 65F4-4, 65F5-4; RTP 100 GB40; Thermocomp MF-1008)
1.23 (EMI-X MA-40; Micalite PP-G2MF-4, PP-H2MFS-4; Nortuff NFA 1800 TO, NFC 1800 TO, NFC 2440 CF; PP-C3CC-4, -C6CC-4, -G2MF-4, -H1TF-4, –H2MFQ-4, -H2MFS-4; RTP 107, 107CC; Vestolen P5272 T)
1.24 (Bapolene 4114; PP-H2LTF-4, -H2TF-4, -H6TF-4; RTP 140, 177)
1.25 (PP-C2TF-4, -H7CC-4; RTP 100T40, 150)
1.26 (Nortuff NFA 4400 TO)
1.28 (PP-H6CC-4)
1.33 (PP-H2MF-5)
1.34 (Micalite PP-H2MFS-5; Nortuff NFC 4600 CO; PP-H2MFS-5)
1.35 (PP-H3MFQ-5)
1.37 (Micalite PP-G2MF-5; PP-G2MF-5)
1.40 (RTP 105CC FR)
1.41 (RTP 101FR)
1.52 (Micalite PP-H2MFS-6; Thermocomp MF-1004FR)
1.53 (Micalite PP-G2MF-6)

Sp. Vol.:
22.7 in.³/lb (Thermocomp MFX-1008HS)
24.5 in.³/lb (Thermocomp MFX-1006HS)
26.6 in.³/lb (Thermocomp MFX-1004HS)

Density:
0.86 kg/l (A-Fax 500, 600, 940)
0.905 g/ml (P012, P032, P04, P0113, P0119)
1.05 g/ml (P2120)
1.15 g/ml (P9125)
1.17 g/ml (P2130, P2230)

Polypropylene *(cont'd.)*

1.25 g/ml (P2140)

Visc.:

240 cps (175 C) (A-Fax 600)

500 cps (100 C) (A-Fax 800)

2000 cps (160 C) (A-Fax 500); (190 C) (A-Fax 940)

190 cc/g (Vestolen P1200, P1200 F)

200 cc/g (Vestolen P2200, P2200 F, P2300, P2300 F)

220 cc/g (Vestolen P3200, P3200 F)

300 cc/g (Vestolen P4200, P4200 F)

320 cc/g (Vestolen P4700)

330 cc/g (Vestolen P4800, P5200, P5200 F)

370 cc/g (Vestolen P5400, P5400 F)

400 cc/g (Vestolen P5800, P6200, P6421)

420 cc/g (Vestolen P6500)

450 cc/g (Vestolen P6522)

M.P.:

130–164 C (Vestolen P4700, P4800, P5800)

140–150 C (Vestolen P5400, P5400 F, P6421)

150–158 C (Vestolen P2300, P2300 F)

153 C (Modic P-300M)

158–164 C (Vestolen P1200, P1200 F, P2200, P2200 F, P3200, P3200 F, P4200, P4200 F, P5200, P5200 F, P5206 S, P6200, P6500, P6522)

168 C (Modic P-110F, P-300F; Shell 5225, 5384, 5419, 5431, 5520, 5524, 5550, 5610, 5820, 5824S, 5840, 5864, 5944S)

Soften. Pt. (R&B):

20 C (A-Fax 800)

120 C (A-Fax 600)

150 C (A-Fax 500)

155 C (A-Fax 940)

Fire Pt.:

280 C (A-Fax 500, 600, 940)

Stability:

Excellent grease and oil resistance (MWB Film)

Superior stress crack and abrasion resistance, and resistance to chemical solutions (Norchem NPP 1006-GF)

Good heat and process stability (Norchem NPP 1010-LC)

Superior resistance to scuffing and abrasion (Norchem NPP 7000-ZF)

Good stress crack and solvent resistance (Norchem NPP 7300-AF, NPP 7300-GF)

Good processing stability; excellent resistance to elevated temp. environments (Norchem NPP 8006-GF)

Resistant to chemical and detergent attack over a wide temp. range (Norchem NPP 8004-ZR)

Excellent heat age stability (PP-H1TF-4)

Excellent dimensional stability (PP-H6TF-4)

MECHANICAL PROPERTIES:

Tens. Str.:

16.5 MPa (Nortuff NFC 4000 FR)

17.2 MPa (Nortuff NFC 1700 MO)

18 MPa (Nortuff NFA 4400 TO; Petrothene PP 8770-HU)

18.6 MPa (Nortuff NFA 1700 MO)

19 MPa (Petrothene PP 8762-HR)

20 MPa (Petrothene PP 8470-HU)

21 MPa (Petrothene PP 8462-HR, PP 8470-ZU, PP 8755-HK); (yield) (Vestolen P6421

21.4 MPa (Nortuff NFA 1800 TO, NFC 1800 TO)

22 MPa (Petrothene PP 1510-HC, PP 1510-LC); (yield) (Vestolen P5400, P5400 F)

22.8–34.5 MPa (yield) (Tenite Polypropylene)

23 MPa (Petrothene PP 8752-HF)

24 MPa (Petrothene PP 8403-HO, PP 8403-TO, PP 8602-HJ); (yield) (Shell 7129, 7623, 7627; Vestolen P5800)

25 MPa (Petrothene PP 8402-HO, PP 8402-TO); (yield) (Shell 7525, 7635, DX6020; Vestolen P4800)

26 MPa (Petrothene PP 8310-GO, PP 8404-HJ, PP 8404-ZJ, PP 8410-ZR, PP 8502-HK, PP 8815-ZR); (yield) (Shell 7328, DX6016)

27 MPa (Petrothene PP 8310-KO, PP 8412-HK, PP 8412-TK, PP 8420-HK, PP 8802-HO, PP 8820-HU); (yield) (Hostalen PP998)

27.6 MPa (yield) (Marlex CHM-040-01)

28 MPa (Petrothene PP 7300-MF, PP 8411-ZR); (yield) (Shell 7912)

29 MPa (Norchem NPP 7300-AF, NPP 7300-GF; Petrothene PP 7200-AF, PP 7200-GF, PP 7200-MF, PP 7300-KF); (yield) (Hostalen PP996; Shell 7221, 7521, 7522)

29.6 MPa (yield) (Marlex CHM-040-02)

30 MPa (yield) (Vestolen P4700, P6206 S, P6500, P6522); (break) (Vestolen P5272 T)

31 MPa (yield) (Shell 5225; Vestolen P5206 S); (break) (Vestolen P5232 G, P5232 T)

32 MPa (Petrothene PP 2085-GU, PP 8080-AW, PP 8080-GW, PP 8080-ZW); (yield) (Hostalen PP975; Shell 5419, 5431; Vestolen P2300, P2300 F, P6200)

33 MPa (yield) (Shell 5384, 5520, 5550)

34 MPa (Petrothene PP 8000-GK, PP 8001-LK); (yield) (Amoco 4228, CR35, CR35A; Hostalen PP920, PP927, PP936, PP941, PP942, PP989; Shell 5524, 5610; Vestolen P5200, P5200 F)

34.3 MPa (Norchem NPP 1001-LF)

34.4 MPa (Norchem NPP 1010-LC)

34.5 MPa (yield) (Marlex HGH-050, HGN-350, HGX-030, HGV-040-01, HGZ-040, HGZ-050-02, HGZ-080-02, HGZ-120-02, HGZ-120-04, HGZ-350, HNS-080)

35 MPa (Petrothene PP 2004-MR, PP 8004-MR, PP 8005-AR); (yield) (Amoco 4039, 4222, 5219; Hostalen PP933, PP934; Shell 5820, 5840, 5944S; Vestolen P4200, P4200 F)

Polypropylene *(cont'd.)*

35.1 MPa (Norchem NPP 1006-GF, NPP 1008-AK, NPP 2000-GJ, NPP 2003-GJ, NPP 7000-ZF, NPP 8000-GK, NPP 8001-LK, NPP 8002-HK, NPP 8006-GF)

35.8 MPa (yield) (Marlex HGN-120-01, HGN-120-A)

35.9 MPa (yield) (Marlex HLN-020-01)

36 MPa (Norchem NPP 2004-MR, NPP 3007-GO, NPP 3010-SO, NPP 8004-MR, NPP 8005-AR, NPP 8007-GO, NPP 8004-ZR; Petrothene PP 8004-ZR, PP 8020-AU, PP 8020-GU); (yield) (Amoco 10-4017, 4018, 7233; Shell 5824S, 5864; Vestolen P3200, P3200 F)

37 MPa (Petrothene PP 8020-ZU); (yield) (Amoco 1012, 1046, 4036, 6431; Vestolen P2200, P2200 F)

38 MPa (yield) (Amoco 1088)

39 MPa (yield) (Amoco CR22N, CR22NA, CR35N, CR35NA)

39.3 MPa (Nortuff NFC 2200 CO)

40 MPa (yield) (Amoco 7232, 7234; Vestolen P1200, P1200 F)

48.3 MPa (Nortuff NFA 2400 CO)

51.7 MPa (Nortuff NFC 2400 CO)

55.8 MPa (Nortuff NFC 4400 CO)

60.1 MPa (Nortuff NFC 2440 CF)

62 MPa (Nortuff NFA 2600 CO, NFC 2600 CO)

65.5 MPa (Nortuff NFC 4600 CO)

6 kg/cm² (A-Fax 500)

8 kg/cm² (A-Fax 940)

180 kg/cm² (yield) (Modic P-300F, P-300M)

290 kg/cm² (yield) (Modic P-110F)

300 kg/cm² (break) (Modic P-300F)

320 kg/cm² (break) (Modic P-300M)

490 kg/cm² (break) (Modic P-110F)

2400 psi (PP-C6CC-4); (yield) (PP-C3CC-4)

2500 psi (yield) (P04)

2600 psi (Stat-Kon ME)

3000 psi (RTP 100 GB40; Stat-Kon M, M-2); (yield) (Pro-fax SA595)

3100 psi (Stat-Kon MM-3340); (yield) (Pro-fax 8523, 8623; Rexene 9403E)

3150 psi (yield) (P032)

3200 psi (yield) (P0119, P2230)

3300 psi (PP-C5CC-2); (yield) (P012; Rexene 9403)

3400 psi (yield) (Pro-fax SB787, SE191)

3500 psi (EMI-X MA-40; PP-H7CC-4; RTP 100 GB30, 140, 150; Stat-Kon M-1 HI); (yield) (Bapolene 5072; PP-C4IM; Pro-fax 7131, SB751)

3600 psi (RTP 143); (yield) (Fortilene 42X07; P0113, P2140)

3700 psi (yield) (PP-C2TF-4; Pro-fax SA862; Rexene 9500)

3800 psi (PP-H6CC-4; RTP 100 GB20, 141); (yield) (Bapolene 5042, 5052; P2130, P9125; Rexene 9401C)

3850 psi (yield) (Pro-fax 65F5-4, SB661)

3900 psi (yield) (Fortilene 41X04, 4141, 4141F; PP-C2TF-1, -VB-511; Pro-fax 7523; Rexene 9402)

4000 psi (RTP 100 GB10, 142; Stat-Kon MF-15); (yield) (Fortilene 40X05, 41X03; PP-C3TFA-2; Pro-fax 7823, SB782)

4050 psi (yield) (Pro-fax SB786)

4080 psi (yield) (Pro-fax SA868M)

4100 psi (PP-C3TF-2)

4200 psi (PP-H2CC-2; RTP 101 SP Foamed); (yield) (Fortilene 54X02; PP-C2IM, -HIFR; Rexene 6310C25, 9234, 9401)

4338 psi (yield) (Empee PP 402)

4400 psi (yield) (Empee PP 401, PP 401CS; Fortilene 18X02, 18X03; P2120; Pro-fax 6329, PD626)

4500 psi (Bapolene 4114); (yield) (Fortilene 38X01; PP-H2TF-4; Pro-fax SA878)

4600 psi (yield) (Pro-fax 65F4-4; Rexene 9400)

4650 psi (yield) (Fortilene 21X04)

4700 psi (PP-H2MFS-3); (yield) (Fortilene 10X01, 19X02, 19X03)

4750 psi (yield) (Pro-fax 6131, PD701)

4800 psi (PP-H1CC-1; RTP 103 SP Foamed); (yield) (Fortilene 2251, 3907; PP-H6TF-4; Pro-fax PD195; Rexene 9400T2, PP41E4)

4850 psi (yield) (Fortilene 3151; PP-H1TF-4)

4900 psi (PP-H4TF-3; RTP 100T30, 100T40); (yield) (Fortilene 3230, 3251, 9205)

5000 psi (Bapolene 4112; PP-H2LTF-4; RTP 100T10, 100T20, 199x28016; Thermo-comp MF-1004FR); (yield) (Bapolene 4042, 4082; Fortilene 14X01, 16X01, 1401, 1602, 1602A, 9605; Micalite PP-H2MFS-4; PP-H2MFS-4; Rexene PP41E2)

5050 psi (yield) (Pro-fax 6323)

5075 psi (yield) (Pro-fax 6524)

5100 psi (Norchem NPP 2013-UJ, NPP 7000-GF); (yield) (Micalite PP-H2MFS-6; Rexene 6310T4)

5120 psi (yield) (Rexene XO-325)

5150 psi (yield) (Fortilene 3606)

5200 psi (yield) (Bapolene 4062; Rexene 6310, 6310T2)

5250 psi (yield) (Micalite PP-G2MF-4; PP-G2MF-4, -H2MF-4)

5300 psi (PP-H3MFQ-1); (yield) (Bapolene 4072)

5400 psi (yield) (Micalite PP-G2MF-6; Pro-fax 6331NW)

5500 psi (PP-H3MFQ-5; RTP 101FR, 199x23835)

5600 psi (RTP 175X; Thermocomp MF-1002HI)

5800 psi (PP-H3MF-2; RTP 177); (yield) (PP-H2TF-2)

6000 psi (RTP 175)

6100 psi (RTP 101)

6200 psi (yield) (Micalite PP-H2MFS-5; PP-H2MFS-5)

6300 psi (PP-H2MFQ-3; RTP 105 SP Foamed)

6500 psi (PP-H2MFQ-4; Thermocomp MF-1002); (yield) (Micalite PP-G2MF-5; PP-G2MF-5, -H2MF-5, -HFR-2)

Polypropylene (cont'd.)

7000 psi (RTP 178X); (yield) (PP-HFR-3)
7200 psi (RTP 103)
7400 psi (RTP 199x22989)
7700 psi (Thermocomp MF-1004HI)
7800 psi (RTP 105)
8000 psi (RTP 101CC, 178)
8400 psi (RTP 107)
9100 psi (Thermocomp MF-1004)
9700 psi (Thermocomp MF-1006HI)
9800 psi (Thermocomp MF-1006)
10,000 psi (RTP 105CC FR)
10,500 psi (Thermocomp MF-1008, MF-1008HI)
11,500 psi (Thermocomp MFX-1004HS)
12,000 psi (RTP 103CC)
13,500 psi (Thermocomp MFX-1006HS)
14,000 psi (RTP 105CC)
15,000 psi (Thermocomp MFX-1008HS)
16,000 psi (RTP 107CC)

Tens. Elong.:

1% (yield) (PP-HFR-2)
1–2% (Thermocomp MF-1008)
1.1% (yield) (PP-HFR-3)
1.5% (RTP 105 SP Foamed, 107)
1.7% (RTP 199x22898)
1.8% (Nortuff NFC 1700 MO)
< 2% (yield) (PP-H2MF-5)
2% (RTP 101FR, 105CC FR, 177)
2–3% (Thermocomp MF-1004, MF-1006)
2.5% (RTP 103, 103CC, 103 SP Foamed, 105, 105CC, 107CC)
2.7% (RTP 101)
3% (EMI-X MA-40; Nortuff NFA 1700 MO; PP-H2MFQ-4, -H3MFQ-5; RTP 100T40, 101CC, 101 SP Foamed, 175, 175X, 178, 178X)
3–4% (Thermocomp MF-1002, MFX-1008HS)
3.5% (yield) (Pro-fax 65F4-4, 65F5-4)
4% (Nortuff NFC 1800 TO, NFC 4400 CO; RTP 100T30, 199x23835; Stat-Kon MF-15)
4–5% (Thermocomp MF-1008HI, MFX-1004HS, MFX-1006HS)
5% (Bapolene 4114; Nortuff NFC 4600 CO; PP-H2MFS-3; RTP 100T20); (break) (PP-H2MFQ-3, -H2TF-4, -H3MF-2)
5–6% (Thermocomp MF-1006HI)
5.6% (RTP 199x28016)
6% (yield) (Pro-fax SA595); (break) (PP-H6TF-4)
6–7% (Thermocomp MF-1002HI, MF-1004HI)

Polypropylene *(cont'd.)*

6.5% (yield) (Shell 7912)

7% (Nortuff NFA 2400 CO, NFA 2600 CO, NFC 2400 CO, NFC 2440 CF, NFC 2600 CO)

7.5% (PP-H2LTF-4)

8% (Nortuff NFC 2200 CO); (yield) (Amoco 10-4017, 1088, 4018, 7232, 7234, CR22N, CR22NA, CR35N, CR35NA; Hostalen PP989; Pro-fax 8523; Shell 5944S, 7635)

8.5% (yield) (Amoco 7233, CR35, CR35A; Bapolene 5072)

9% (Nortuff NFA 1800 TO); (yield) (Amoco 1012, 1046, 4036, 6431; Bapolene 4062, 4072; Pro-fax 8623, SB661, SB786, SB787; Shell 5524, 5824S, 5864, 7522, 7623)

9.5% (yield) (Bapolene 5042, 5052)

10% (Bapolene 4112; PP-H3MFQ-1); (yield) (Bapolene 4042, 4082; Shell 5610, 5820, 5840, 7221, 7521, 7525); (break) (Vestolen P5272 T)

10–20% (Stat-Kon M-2)

10+% (RTP 100 GB10, 100 GB20, 100 GB30, 100 GB40, 100T10, 140, 141, 142, 143, 150)

10.5% (yield) (Pro-fax 6331NW)

11% (Stat-Kon M); (yield) (Hostalen PP934; Petrothene PP 8820-HU; Pro-fax 6323; Shell 5419, 5431, 5520, 5550, 7129, 7328)

12% (Nortuff NFA 4400 TO; PP-H4TF-3); (yield) (Hostalen PP920, PP933; Petrothene PP 8802-HO; Pro-fax 6131, 6524, 7131, PD701, SB751, SB782; Shell 5225, 5384, 7627, DX6020)

12.5% (yield) (Pro-fax PD626)

13% (yield) (Pro-fax 7523; Shell DX6016)

14% (Nortuff NFC 4000 FR; Stat-Kon MM-3340); (yield) (Pro-fax PD195, SA862, SA878)

14.5% (yield) (Pro-fax 6329)

15% (yield) (Hostalen PP927, PP936, PP941, PP942, PP975; Petrothene PP 8411-ZR, PP 1510-HC, PP 1510-LC; Pro-fax SA868M); (break) (Vestolen P5232 T)

16% (yield) (Hostalen PP996; Norchem NPP 2000-GJ, NPP 2003-GJ, NPP 2004-MR, NPP 3007-GO, NPP 3010-SO, NPP 8004-MR, NPP 8005-AR, NPP 8007-GO, NPP 8004-ZR; Petrothene PP 2004-MR, PP 8004-MR, PP 8004-ZR, PP 8005-AR, PP 8020-AU, PP 8020-GU, PP 8020-ZU, PP 8402-HO, PP 8402-TO, PP 8410-ZR, PP 8815-ZR)

17% (yield) (Norchem NPP 1008-AK, NPP 8000-GK, NPP 8001-LK, NPP 8002-HK; Petrothene PP 8000-GK, PP 8001-LK); (break) (PP-H1TF-4)

18% (Norchem NPP 2013-UJ); (yield) (Petrothene PP 2085-GU, PP 8080-AW, PP 8080-GW, PP 8080-ZW, PP 8412-HK, PP 8412-TK, PP 8420-HK, PP 8462-HR, PP 8502-HK); (break) (PP-H2TF-2)

19% (Norchem NPP 7000-GF); (yield) (Norchem NPP 1006-GF, NPP 7000-ZF, NPP 8006-GF)

20% (yield) (Hostalen PP998; Norchem NPP 1001-LF, NPP 1010-LC; Petrothene PP 7200-AF, PP 7200-GF, PP 7200-MF, PP 8310-GO, PP 8310-KO, PP 8404-HJ, PP

Polypropylene *(cont'd.)*

8404-ZJ, PP 8470-HU, PP 8470-ZU, PP 8602-HJ, PP 8755-HK, PP 8762-HR, PP 8770-HU); (break) (PP-C2TF-4; Vestolen P5232 G)

22% (Stat-Kon M-1 HI)

23% (yield) (Pro-fax 7823)

25% (yield) (Petrothene PP 7300-KF, PP 7300-MF, PP 8403-HO, PP 8403-TO, PP 8752-HF; PP-HIFR); (break) (Rexene 6310T4)

30% (yield) (Norchem NPP 7300-AF, NPP 7300-GF); (break) (Amoco 10-4017, 4018)

40% (break) (A-Fax 940; PP-C3TF-2, -H6CC-4, -H7CC-4)

50% (PP-H2CC-2); (break) (A-Fax 500; Amoco 1012, 1046, 4036)

80% (break) (Amoco 1088; PP-C6CC-4)

90% (break) (Rexene 6310, 6310C25)

100% (yield) (PP-C3CC-4); (break) (Amoco 4222, 4228, 5219; PP-C3TFA-2, -H1CC-1; Rexene 6310T2; Vestolen P1200, P1200 F)

> 100% (yield) (PP-C2TF-1); (break) (PP-C4IM, -C5CC-2, -VB-511; Rexene PP41E2, PP41E4, XO-325)

> 125% (Stat-Kon ME)

180% (break) (Amoco 4039)

190% (break) (Empee PP 401, PP 401CS, PP 402)

200% (break) (Rexene 9400T2)

220% (break) (Rexene 9401C)

250% (break) (Rexene 9400)

350% (yield) (PP-C2IM); (break) (Rexene 9401, 9402)

360% (break) (Rexene 9403)

400% (break) (Fortilene 10X01; Pro-fax SE191; Vestolen P2200, P2200 F, P3200, P3200 F)

440% (break) (Fortilene 41X03)

450% (break) (Fortilene 41X04, 4141, 4141F; Vestolen P4700)

460% (break) (Fortilene 42X07, 54X02)

500% (break) (Fortilene 18X02, 18X03, 38X01)

560% (break) (Fortilene 3907)

570% (break) (Fortilene 16X01, 1602, 1602A, 3606, 9605)

580% (break) (Fortilene 14X01, 1401)

590% (break) (Fortilene 2251, 3151, 3230, 3251, 9205)

600% (break) (Fortilene 19X02, 19X03, 21X04; Vestolen P4200, P4200 F, P4800)

> 600% (break) (Rexene 9500)

605% (Marlex HNS-080 film)

700% (break) (Rexene 9234; Vestolen P2300, P2300 F, P5200, P5200 F, P5800, P6206 S)

> 700% (break) (Rexene 9403E)

750% (break) (Vestolen P5206 S, P6500)

800% (break) (Fortilene 40X05; Vestolen P5400, P5400 F, P6200, P6421, P6522)

830% (Modic P-300F)

860% (Modic P-110F)
900% (Modic P-300M)
Tens. Mod.:
730 MPa (Shell DX6020)
800 MPa (Vestolen P5400, P5400 F, P6421)
850 MPa (Shell 7129)
875 MPa (Shell DX6016)
900 MPa (Shell 7328)
970 MPa (Shell 7627)
1000 MPa (Vestolen P5206 S, P5800, P6206 S, P6522)
1030 MPa (Shell 7525, 7623, 7635)
1055 MPa (Shell 7221)
1100 MPa (Vestolen P4800, P6500)
1150 MPa (Shell 7912)
1200 MPa (Shell 5225, 7521, 7522; Vestolen P2300, P2300 F, P4700, P6200)
1300 MPa (Shell 5384, 5419, 5431, 5520, 5550, 5610, 5944S)
1375 MPa (Shell 5820, 5840)
1450 MPa (Shell 5524)
1500 MPa (Shell 5824S, 5864; Vestolen P5200, P5200 F)
1600 MPa (Vestolen P3200, P3200 F, P4200, P4200 F)
1700 MPa (Vestolen P2200, P2200 F)
1800 MPa (Vestolen P1200, P1200 F)
2400 MPa (Vestolen P5232 T)
2900 MPa (Vestolen P5232 G)
3100 MPa (Vestolen P5272 T)
120,000 psi (P04)
125,000 psi (P032)
145,000 psi (P0113, P0119)
155,000 psi (P012)
190,000 psi (Rexene PP41E2, PP41E4)
220,000 psi (RTP 100 GB10)
250,000 psi (P2230; Rexene XO-325)
270,000 psi (RTP 100 GB20)
280,000 psi (P2130; RTP 150)
300,000 psi (RTP 100 GB30, 142)
310,000 psi (P2120)
320,000 psi (RTP 100 GB40)
330,000 psi (P9125; RTP 100T10, 101 SP Foamed)
350,000 psi (RTP 199x28016)
370,000 psi (RTP 100T20)
380,000 psi (P2140)
390,000 psi (RTP 141)
440,000 psi (RTP 143)

233

Polypropylene *(cont'd.)*

450,000 psi (RTP 100T30)
490,000 psi (RTP 100T40)
500,000 psi (RTP 101CC, 140)
510,000 psi (RTP 101)
550,000 psi (RTP 103 SP Foamed)
600,000 psi (RTP 175, 175X, 199x23835)
700,000 psi (RTP 199x22989)
730,000 psi (RTP 103, 103CC)
750,000 psi (RTP 101FR)
775,000 psi (RTP 105 SP Foamed)
800,000 psi (RTP 177)
850,000 psi (RTP 178, 178X)
10^6 psi (RTP 105, 105CC)
1.2×10^6 psi (RTP 105CC FR)
1.3×10^6 psi (RTP 107CC)
1.5×10^6 psi (RTP 107)

Flex. Str.:

3000 psi (Stat-Kon ME)
3200 psi (P04)
3600 psi (Stat-Kon M)
4000 psi (P032; PP-C6CC-4; Stat-Kon M-2)
4200 psi (P2120)
4500 psi (P012; RTP 100 GB40; Stat-Kon M-1 HI)
4700 psi (P0113, P0119; RTP 100 GB30; Stat-Kon MM-3340)
5000 psi (EMI-X MA-40; PP-C2TFT-1, -C4IM, -C5CC-2, -VB-511; RTP 100 GB10, 100 GB20)
5200 psi (P2230)
5300 psi (P9125; PP-H1TF-4)
5500 psi (Stat-Kon MF-15)
5600 psi (RTP 150)
5800 psi (PP-C3TF-2)
6000 psi (P2130; PP-H2CC-2)
6200 psi (P2140; PP-H7CC-4; RTP 101 SP Foamed)
6500 psi (PP-H6CC-4; Thermocomp MF-1002HI)
6600 psi (PP-H1CC-1; RTP 103 SP Foamed)
6700 psi (RTP 142; Thermocomp MF-1004FR)
6800 psi (RTP 141)
6900 psi (PP-C2TF-4; RTP 143)
7000 psi (PP-C3TFA-2, -H3MFQ-1; RTP 140)
7400 psi (RTP 100T10)
7500 psi (PP-H2MFS-3, -H4TF-3)
7800 psi (RTP 100T20)
7900 psi (Micalite PP-H2MFS-4; PP-H2MFS-4)

8000 psi (RTP 100T30, 101, 177, 199x23835; Thermocomp MF-1002)

8500 psi (PP-H2LTF-4; RTP 100T40, 199x28016)

8600 psi (RTP 103)

8644 psi (Micalite PP-G2MF-4; PP-G2MF-4, -H2MF-4)

8700 psi (PP-H2TF-2, -H2TF-4; RTP 105 SP Foamed)

8800 psi (PP-H6TF-4)

8876 psi (Micalite PP-G2MF-6)

9000 psi (Micalite PP-H2MFS-6; RTP 175X)

9500 psi (RTP 101FR; Thermocomp MF-1004HI)

9600 psi (Micalite PP-H2MFS-5; PP-H2MFS-5)

9900 psi (RTP 105)

10,000 psi (PP-H3MF-2; RTP 175, 178X)

10,500 psi (RTP 107, 199x22989)

10,600 psi (PP-H2MF-5)

10,900 psi (PP-H2MFQ-3)

11,000 psi (Micalite PP-G2MF-5; PP-G2MF-5, -H3MFQ-5; RTP 101CC; Thermo-comp MF-1004, MF-1006HI)

11,500 psi (PP-H2MFQ-4; RTP 178; Thermocomp MF-1008HI)

12,000 psi (Thermocomp MF-1006)

12,500 psi (Thermocomp MF-1008)

14,000 psi (RTP 103CC; Thermocomp MFX-1004HS)

16,000 psi (RTP 105CC FR)

18,000 psi (RTP 105CC; Thermocomp MFX-1006HS)

19,000 psi (RTP 107CC)

22,000 psi (Thermocomp MFX-1008HS)

Flex. Mod.:

760 MPa (Petrothene PP 8462-HR; Shell DX6020)

772 MPa (Nortuff NFC 4000 FR)

795 MPa (Petrothene PP 1510-HC, PP 1510-LC, PP 8470-HU, PP 8755-HK, PP 8762-HR, PP 8770-HU)

860 MPa (Hostalen PP998; Petrothene PP 7300-MF, PP 8403-HO, PP 8403-TO, PP 8470-ZU, PP 8602-HJ, PP 8752-HF)

895 MPa (Petrothene PP 8310-GO)

900 MPa (Hostalen PP996)

930 MPa (Petrothene PP 7300-KF; Shell 7129)

965 MPa (Petrothene PP 8310-KO)

970 MPa (Shell DX6016)

1000 MPa (Hostalen PP920; Norchem NPP 7300-AF, NPP 7300-GF; Petrothene PP 8502-HK; Shell 7328, 7627)

1035 MPa (Petrothene PP 7200-AF, PP 7200-GF, PP 7200-MF; Shell 7221)

1100 MPa (Hostalen PP934, PP975; Marlex CHM-040-01; Petrothene PP 8404-HJ; Shell 7525, 7623, 7635)

1170 MPa (Hostalen PP933, PP936, PP941, PP942, PP989; Petrothene PP 8402-HO,

Polypropylene *(cont'd.)*

PP 8402-TO, PP 8404-ZJ, PP 8411-ZR, PP 8412-HK, PP 8412-TK, PP 8420-HK)

1220 MPa (Norchem NPP 1010-LC)

1240 MPa (Hostalen PP927; Marlex CHM-040-02; Norchem NPP 1001-LF)

1250 MPa (Shell 7521)

1270 MPa (Shell 7912)

1300 MPa (Amoco 4228, CR35, CR35A; Norchem NPP 1006-GF, NPP 7000-ZF, NPP 8006-GF; Shell 5225, 7522)

1340 MPa (Norchem NPP 1008-AK, NPP 2000-GJ, NPP 2003-GJ, NPP 8000-GK, NPP 8001-LK, NPP 8002-HK; Petrothene PP 2085-GU, PP 8080-AW, PP 8080-GW)

1380 MPa (Petrothene PP 8410-ZR, PP 8802-HO, PP 8820-HU)

1400 MPA (Amoco 4036, 4039, 5219, 6431, CR22N, CR22NA, CR35N, CR35NA; Norchem NPP 2004-MR, NPP 3007-GO, NPP 3010-SO, NPP 8004-MR, NPP 8005-AR, NPP 8007-GO)

1450 MPa (Amoco 10-4017, 1012, 4018, 4222; Petrothene PP 8000-GK, PP 8001-LK; Shell 5384, 5419, 5431, 5520, 5550, 5610, 5944S)

1470 MPA (Amoco 7233)

1485 MPa (Petrothene PP 8080-ZW)

1500 MPa (Amoco 1046; Shell 5820, 5840)

1520 MPa (Norchem NPP 8004-ZR)

1540 MPa (Amoco 1088)

1545 MPa (Petrothene PP 8815-ZR)

1550 MPa (Marlex HGH-050, HGX-030, HGV-040-01, HGZ-040, HGZ-050-02)

1575 MPa (Shell 5524)

1590 MPa (Petrothene PP 2004-MR, PP 8004-MR, PP 8005-AR)

1620 MPa (Nortuff NFA 4400 TO)

1640 MPA (Amoco 7232, 7234)

1650 MPa (Shell 5824S, 5864)

1655 MPa (Petrothene PP 8020-AU, PP 8020-GU)

1660 MPa (Marlex HGZ-350)

1725 MPa (Marlex HGZ-080-02, HGZ-120-02, HGZ-120-04, HLN-020-01, HNS-080; Petrothene PP 8004-ZR)

1790 MPa (Marlex HGN-350)

1793 MPa (Nortuff NFC 2200 CO)

1795 MPa (Petrothene PP 8020-ZU)

1930 MPa (Marlex HGN-120-01, HGN-120-A)

2345 MPa (Nortuff NFA 1800 TO)

2483 MPa (Nortuff NFC 1700 MO, NFC 1800 TO)

2758 MPa (Nortuff NFA 1700 MO, NFA 2400 CO)

2931 MPa (Nortuff NFC 2400 CO)

4000 MPa (Nortuff NFA 2600 CO, NFC 4400 CO)

4138 MPa (Nortuff NFC 2440 CF)

4207 MPa (Nortuff NFC 2600 CO)

4483 MPa (Nortuff NFC 4600 CO)
0.135 psi (Pro-fax SA682)
0.150 psi (Pro-fax 7131)
0.155 psi (Pro-fax SA868M)
0.160 psi (Pro-fax 7823)
0.180 psi (Pro-fax SA878)
0.185 psi (Pro-fax 7523)
0.188 psi (Pro-fax PD626)
0.190 psi (Pro-fax SB782)
0.200 psi (Pro-fax SB661, SB786)
0.210 psi (Pro-fax PD701)
0.215 psi (Pro-fax PD195)
0.250 psi (Pro-fax 6524)
0.260 psi (Pro-fax 6323)
0.285 psi (Pro-fax 6331NW)
0.395 psi (Pro-fax 65F5-4)
0.493 psi (Pro-fax 65F4-4)
60,000 psi (PP-C4IM)
80,000 psi (Stat-Kon ME)
95,000 psi (Fortilene 40X05)
110,000 psi (Fortilene 42X07, 4141, 4141F; Rexene 9403E; Stat-Kon M)
120,000 psi (Fortilene 41X04; P04)
125,000 psi (P032; PP-HIFR; Rexene 9403)
140,000 psi (P012; Stat-Kon M-2)
145,000 psi (Bapolene 5072; Rexene 9500)
150,000 psi (P0113, P0119; PP-C2IM, -VB-511; Rexene 9402)
155,000 psi (Bapolene 5052; Fortilene 41X03, 54X02)
158,000 psi (Fortilene 38X01)
160,000 psi (Fortilene 18X02, 18X03, 21X04, 2251, 3151; PP-C2TF-1; Stat-Kon M-1 HI)
165,000 psi (Fortilene 3230, 3251, 9205)
170,000 psi (Bapolene 4082; Fortilene 10X01, 19X02, 19X03)
175,000 psi (Fortilene 14X01, 1401, 3606, 9605; Rexene 9401)
180,000 psi (Fortilene 1602, 1602A, 3907)
185,000 psi (Fortilene 16X01)
190,000 psi (Norchem NPP 7000-GF; Rexene PP41E2, PP41E4, XO-325; RTP 100 GB10)
195,000 psi (Norchem NPP 2013-UJ)
200,000 psi (Bapolene 4042, 4062; PP-C5CC-2, -C6CC-4; Rexene 9400)
215,000 psi (Empee PP 401, PP 401CS, PP 402)
220,000 psi (Bapolene 4072; Rexene 9401C; RTP 100 GB20, 150)
230,000 psi (P2230)
240,000 psi (P2120; Rexene 6310)

Polypropylene *(cont'd.)*

250,000 psi (PP-H1CC-1; RTP 100 GB30, 142; Thermocomp MF-1002HI)
260,000 psi (PP-C3TFA-2)
270,000 psi (PP-H2CC-2)
280,000 psi (P2130; PP-C3TF-2; RTP 100 GB40)
290,000 psi (RTP 100T10)
300,000 psi (RTP 199x28016)
310,000 psi (Rexene 6310C25; RTP 100T20)
315,000 psi (Stat-Kon MM-3340)
320,000 psi (P9125)
325,000 psi (RTP 101 SP Foamed)
330,000 psi (Rexene 9400T2; RTP 141)
350,000 psi (PP-H7CC-4; Thermocomp MF-1002)
370,000 psi (PP-H6CC-4)
375,000 psi (Thermocomp MF-1004HI)
380,000 psi (Bapolene 4112; P2140; RTP 100T30, 143)
400,000 psi (EMI-X MA-40; PP-H1TF-4, -H2TF-2; Rexene 6310T2; RTP 101CC)
420,000 psi (PP-H3MFQ-1; RTP 101)
425,000 psi (RTP 103 SP Foamed)
430,000 psi (RTP 140)
450,000 psi (RTP 100T40, 175, 175X)
460,000 psi (PP-C2TF-4)
470,000 psi (Stat-Kon MF-15)
475,000 psi (PP-H4TF-3)
480,000 psi (RTP 199x23835
500,000 psi (Bapolene 4114; PP-H2TF-4)
510,000 psi (PP-HFR-2)
520,000 psi (PP-H3MF-2, -H6TF-4; RTP 103, 199x22989)
540,000 psi (Rexene 6310T4; RTP 101FR)
550,000 psi (RTP 103CC, 177; Thermocomp MF-1006HI)
590,000 psi (PP-H2MFS-3)
600,000 psi (Thermocomp MF-1004, MFX-1004HS)
610,000 psi (PP-HFR-3)
625,000 psi (PP-H2LTF-4)
650,000 psi (RTP 105 SP Foamed)
660,000 psi (RTP 178, 178X)
700,000 psi (RTP 105CC; Thermocomp MF-1008HI)
720,000 psi (Thermocomp MF-1004FR)
730,000 psi (PP-H2MFQ-3; RTP 105)
800,000 psi (Thermocomp MF-1006, MFX-1006HS)
900,000 psi (RTP 107CC)
950,000 psi (RTP 107)
1×10^6 psi (Micalite PP-H2MFS-4; PP-H2MFQ-4, -H2MFS-4; RTP 105CC FR; Thermocomp MF-1008, MFX-1008HS)

1.2 × 10⁶ psi (Micalite PP-G2MF-4; PP-G2MF-4, -H2MF-4)
1.2×10^6 psi (Micalite PP-G2MF-4; PP-G2MF-4, -H2MF-4)
1.28×10^6 psi (Micalite PP-H2MFS-5; PP-H2MFS-5)
1.3×10^6 psi (Micalite PP-G2MF-5; PP-G2MF-5, -H2MF-5, -H3MFQ-5)
1.875×10^6 psi (Micalite PP-G2MF-6)
1.912×10^6 psi (Micalite PP-H2MFS-6)
1.5×10^8 psi (Rexene 9234)

Compr. Str.:
3900 psi (RTP 100 GB40)
4200 psi (RTP 100 GB30)
4400 psi (RTP 100 GB20)
4500 psi (RTP 100 GB10)
5000 psi (RTP 150)
5200 psi (RTP 199x28016)
5700 psi (RTP 177)
5900 psi (RTP 175X)
6000 psi (RTP 199x23835)
6400 psi (RTP 175)
6800 psi (RTP 100T10, 142)
6900 psi (RTP 100T20, 141)
7000 psi (RTP 101, 101FR, 143)
7200 psi (RTP 100T30, 140)
7300 psi (RTP 178X)
7500 psi (RTP 100T40, 103, 199x22989)
7600 psi (RTP 101CC)
7700 psi (Thermocomp MFX-1004HS)
8400 psi (RTP 105, 178)
8500 psi (Thermocomp MFX-1006HS)
8900 psi (RTP 107)
9700 psi (RTP 103CC)
9800 psi (Thermocomp MFX-1008HS)
11,000 psi (RTP 105CC FR)
12,000 psi (RTP 105CC)
13,000 psi (RTP 107CC)

ESCR:
> 500 h, no failures (Pro-fax SE191)

Impact Str. (Izod):
No break notched (Fortilene 40X05; Stat-Kon M-1 HI, ME)
11 J/m notched to no break (Tenite Polypropylene)
16 J/m notched (Amoco CR22N, CR22NA, CR35, CR35A, CR35N, CR35NA)
20 J/m notched (Shell 5944S; Vestolen P5272 T)
21 J/m notched (Amoco 7232, 7234; Petrothene PP 2004-MR, PP 2085-GU, PP 8004-MR, PP 8005-AR, PP 8020-AU, PP 8020-GU, PP 8080-AW, PP 8080-GW)
21.4 J/m notched (Norchem NPP 2004-MR, NPP 8004-MR, NPP 8005-AR, NPP

Polypropylene *(cont'd.)*

8004-ZR)

24 J/m notched (Amoco 4039)

25 J/m notched (Vestolen P1200, P1200 F, P2200, P2200 F)

26.7 J/m notched (Norchem NPP 3007-GO, NPP 3010-SO, NPP 8007-GO)

27 J/m notched (Amoco 10-4017, 1046, 1088, 4018, 4222, 4228, 5219, 7233; Hostalen PP934, PP975; Marlex HGN-120-01, HGN-120-A, HGN-350, HGZ-080-02, HGZ-120-02, HGZ-120-04, HGZ-350, HNS-080; Petrothene PP 8004-ZR, PP 8020-ZU, PP 8080-ZW; Shell 5820, 5840)

30 J/m notched (Shell 5520, 5550, 5610, 5824S, 5864; Vestolen P3200, P3200 F)

32 J/m notched (Hostalen PP933, PP989; Petrothene PP 8000-GK, PP 8001-LK)

34.7 J/m notched (Norchem NPP 1008-AK, NPP 2000-GJ, NPP 8000-GK, NPP 8001-LK, NPP 8002-HK)

35 J/m notched (Vestolen 5232 T)

36 J/m notched (Shell 5419, 5431, 5524)

37 J/m notched (Amoco 4036; Hostalen PP927, PP936, PP941, PP942; Marlex HGH-050, HGV-040-01, HGX-030, HGZ-040, HGZ-050-02, HLN-020-01)

37.4 J/m notched (Norchem NPP 2003-GJ)

40 J/m notched (Shell 5384; Vestolen P2300, P2300 F, P4200, P4200 F, P5232 G)

45.4 J/m notched (Norchem NPP 1006-GF, NPP 7000-ZF, NPP 8006-GF)

48 J/m notched (Amoco 6431; Petrothene PP 8310-GO)

50 J/m notched (Shell 5225; Vestolen P5200, P5200 F, P5206 S)

53 J/m notched (Amoco 1012; Petrothene PP 7200-AF, PP 7200-GF, PP 7200-MF, PP 8310-KO)

53.4 J/m notched (Norchem NPP 1001-LF)

58.7 J/m notched (Norchem NPP 1010-LC)

60 J/m notched (Vestolen P4700, P6200, P6206 S)

64 J/m notched (Petrothene PP 7300-KF, PP 7300-MF, PP 8820-HU)

69 J/m notched (Nortuff NFC 1700 MO; Petrothene PP 8815-ZR; Shell DX6016)

75 J/m notched (Shell 7912)

80 J/m notched (Hostalen PP996; Norchem NPP 7300-AF, NPP 7300-GF; Petrothene PP 8410-ZR; Shell 7521, DX6020; Vestolen P4800)

85 J/m notched (Petrothene PP 8802-HO)

90.8 J/m notched (Nortuff NFA 1700 MO)

91 J/m notched (Marlex CHM-040-02; Nortuff NFC 2440 CF, NFC 4600 CO; Petrothene PP 8402-HO, PP 8402-TO, PP 8403-HO, PP 8403-TO, PP 8411-ZR)

96 J/m notched (Nortuff NFA 4400 TO; Petrothene PP 8412-HK, PP 8412-TK)

100 J/m notched (Vestolen P6500)

101 J/m notched (Nortuff NFC 1800 TO)

107 J/m notched (Nortuff NFC 4400 CO; Petrothene PP 8420-HK)

110 J/m notched (Shell 7522, 7525, 7627)

112 J/m notched (Nortuff NFC 2200 CO)

117 J/m notched (Hostalen PP998; Petrothene PP 8404-HJ, PP 8404-ZJ; Shell 7623, 7635)

Polypropylene (cont'd.)

118 J/m notched (Marlex CHM-040-01)
123 J/m notched (Hostalen PP920; Petrothene PP 8502-HK)
128 J/m notched (Nortuff NFC 2400 CO, NFC 2600 CO)
130 J/m notched (Vestolen P5800)
133 J/m notched (Nortuff NFC 4000 FR)
150 J/m notched (Vestolen P6522)
160 J/m notched (Nortuff NFA 2400 CO; Petrothene PP 8470-HU, PP 8470-ZU)
187 J/m notched (Nortuff NFA 2600 CO)
200 J/m notched (Vestolen P5400, P5400 F)
203 J/m notched (Nortuff NFA 1800 TO)
230 J/m notched (Vestolen P6421)
267 J/m notched (Shell 7221)
320 J/m notched (Petrothene PP 8462-HR, PP 8770-HU)
427 J/m notched (Petrothene PP 8602-HJ)
534 J/m notched (Petrothene PP 8762-HR)
640 J/m notched (Petrothene PP 8755-HK)
750 J/m notched (Petrothene PP 8752-HF; Shell 7328)
800 J/m notched (Petrothene PP 1510-HC, PP 1510-LC)
910 J/m notched (Shell 7129)
> 30 kg-cm/cm² (Modic P-100F, P-300F, P-300M)
0.3 ft lb/in. notched (Fortilene 38X01; RTP 199x28016)
0.4 ft lb/in. notched (Bapolene 4082; Fortilene 14X01, 16X01, 18X02, 18X03, 19X02, 19X03, 1401, 1602, 1602A, 3606; Pro-fax 65F4-4)
0.45 ft lb/in. notched (PP-H3MFQ-5, -H6TF-4; RTP 100 GB10)
0.5 ft lb/in. (Bapolene 4114; P2120, P2130, P2140); notched (Empee PP 401, PP 401CS, PP 402; Fortilene 41X03, 2251, 3151, 3230, 3251, 9205, 9605; PP-H2LTF-4, -H2TF-4, -H7CC-4; RTP 100T30, 100T40, 199x23835)
0.51 ft lb/in. notched (PP-H1TF-4)
0.52 ft lb/in. notched (Micalite PP-G2MF-5, PP-H2MFS-5; PP-G2MF-5, -H2MF-5, –H2MFS-5)
0.53 ft lb/in. notched (PP-H2MFQ-4, -H2MFS-3)
0.54 ft lb/in. notched (Micalite PP-H2MFS-4; PP-H2MFS-4)
0.55 ft lb/in. notched (Micalite PP-G2MF-4; PP-C3CC-4, -G2MF-4, -H2MF-4, –H2MFQ-3, -H3MF-2, -H4TF-3; RTP 100T20)
0.59 ft lb/in. notched (PP-H2TF-2)
0.60 ft lb/in. notched (Bapolene 4072, 4112; Fortilene 21X04, 41X04, 54X02; PP-H3MFQ-1, -H6CC-4; Pro-fax 6131, PD701; RTP 100 GB20, 100 GB30, 100 GB40, 100T10, 140, 150)
0.61 ft lb/in. notched (Micalite PP-G2MF-6)
0.62 ft lb/in. notched (Micalite PP-H2MFS-6)
0.65 ft lb/in. notched (Norchem NPP 2013-UJ; PP-H2CC-2; RTP 143)
0.7 ft lb/in. (P9125; Rexene 6310T4); notched (Bapolene 4062; Fortilene 10X01, 42X07; Pro-fax 6323; RTP 101FR, 141, 142, 175X, 199x22989)

Polypropylene (cont'd.)

0.8 ft lb/in. notched (Fortilene 4141, 4141F; Pro-fax 65F5-4, 6329, 6524; RTP 101 SP Foamed, 177, 178X; Thermocomp MF-1004FR)

0.85 ft lb/in. (Norchem NPP 7000-GF)

0.9 ft lb/in. notched (Bapolene 4042; PP-C6CC-4, -H1CC-1; RTP 175)

0.95 ft lb/in. notched (PP-C2TF-4)

1.0 ft lb/in. (P2230); notched (PP-HFR-2; Pro-fax 6331NW, PD626; RTP 101, 103 SP Foamed, 105CC FR, 178; Thermocomp MF-1002)

1.1 ft lb/in. notched (PP-C3TF-2, -C5CC-2, -HFR-3; RTP 105 SP Foamed)

1.2 ft lb/in. (P0113); notched (RTP 103)

1.3 ft lb/in. (Rexene 9400); notched (PP-C2IM; RTP 105)

1.4 ft lb/in. (P0119); notched (RTP 107; Thermocomp MF-1004)

1.5 ft lb/in. (Rexene 6310T2); notched (PP-HIFR; Pro-fax 7131, SB751; RTP 101CC)

1.6 ft lb/in. notched (PP-C3TFA-2; Pro-fax SA595, SB661; Thermocomp MF-1006)

1.7 ft lb/in. notched (Bapolene 5072; PP-VB-511; Pro-fax SA862)

1.8 ft lb/in. (Rexene 6310); notched (RTP 103CC; Thermocomp MF-1008, MFX-1004HS)

1.9 ft lb/in. notched (Pro-fax PD195; RTP 105CC; Thermocomp MFX-1006HS)

2.0 ft lb/in. (P012; Rexene 6310C25, 9400T2); notched (PP-C2TF-1, -C4IM; Pro-fax SB782, SB786; RTP 107CC; Thermocomp MFX-1008HS)

2.1 ft lb/in. notched (Pro-fax SB787)

2.3 ft lb/in. notched (EMI-X MA-40)

2.5 ft lb/in. (Rexene 9234); notched (Bapolene 5052; Pro-fax 7523; Stat-Kon MF-15)

2.8 ft lb/in. (Rexene 9401); notched (Thermocomp MF-1002HI)

3.0 ft lb/in. (P032); notched (Bapolene 5042; Thermocomp MF-1004HI, MF-1006HI)

3–4 ft lb/in. notched (Stat-Kon M-2)

3.2 ft lb/in. notched (Stat-Kon M, MM-3340; Thermocomp MF-1008HI)

3.6 ft lb/in. notched (Pro-fax SA868M)

4.0 ft lb/in. notched (Pro-fax 8523)

4.5 ft lb/in. notched (Pro-fax SA878)

4.8 ft lb/in. (Rexene 9401C)

5.8 ft lb/in. (Rexene 9402)

6.0 ft lb/in. notched (Pro-fax 8623)

8.0 ft lb/in. (P04; Rexene 9500)

10 ft lb/in. notched (Pro-fax 7823)

> 12 ft lb/in. (Rexene 9403)

> 15 ft lb/in. (Rexene 9403E)

Tens. Impact:

53 kJ/m² (Amoco 10-4017, 4018, 4039, 5219)

64 kJ/m² (Amoco 1088)

74 kJ/m² (Amoco 1046)

137 kJ/m² (Amoco 1012)

Hardness:

Ball Indentation 40 MPa (Vestolen P5400, P5400 F, P6421)

Ball Indentation 60 MPa (Vestolen P5800)
Ball Indentation 62 MPa (Vestolen P4800, P6206 S)
Ball Indentation 63 MPa (Vestolen P6522)
Ball Indentation 64 MPa (Vestolen P5206 S, P6500)
Ball Indentation 65 MPa (Vestolen P2300, P2300 F)
Ball Indentation 67 MPa (Vestolen P4700)
Ball Indentation 68 MPa (Vestolen P6200)
Ball Indentation 72 MPa (Vestolen P5200, P5200 F)
Ball Indentation 74 MPa (Vestolen P4200, P4200 F)
Ball Indentation 81 MPa (Vestolen P5232G, P5232 T)
Ball Indentation 82 MPa (Vestolen P3200, P3200 F)
Ball Indentation 83 MPa (Vestolen P2200, P2200 F)
Ball Indentation 85 MPa (Vestolen P1200, P1200 F)
Ball Indentation 88 MPa (Vestolen P5272 T)
Rockwell M49 (Thermocomp MFX-1004HS)
Rockwell M57 (Thermocomp MFX-1006HS)
Rockwell M59 (Thermocomp MFX-1008HS)
Rockwell R42 (Rexene 9403E)
Rockwell R58 (Rexene 9403)
Rockwell R60 (Shell 7129)
Rockwell R63–96 (Tenite Polypropylene)
Rockwell R65 (Petrothene PP 8762-HR, PP 8770-HU; Pro-fax 8523, 8623; Shell
    7328)
Rockwell R67 (Shell 7221)
Rockwell R68 (Rexene 9500; Shell DX6020)
Rockwell R70 (Petrothene PP 8470-HU, PP 8470-ZU, PP 8752-HF, PP 8755-HK;
    Shell 7627, 7635, DX6016)
Rockwell R73 (Rexene 9402)
Rockwell R74 (Pro-fax SB787)
Rockwell R75 (Petrothene PP 8462-HR; Pro-fax 7823; Shell 7525, 7623)
Rockwell R78 (Pro-fax 7131, SA862, SB751)
Rockwell R79 (Pro-fax SB786)
Rockwell R80 (Bapolene 5042, 5072; Norchem NPP 7300-AF, NPP 7300-GF;
    Petrothene PP 1510-HC, PP 1510-LC, PP 7200-AF, PP 7200-GF, PP 7200-MF, PP
    7300-KF, PP 7300-MF, PP 8310-GO, PP 8310-KO, PP 8502-HK, PP 8602-HJ; Pro-
    fax 7523; Rexene 9401, 9401C; Shell 7521, 7522)
Rockwell R82 (Bapolene 5052; Pro-fax SB661; RTP 100 GB10, 142)
Rockwell R83 (Pro-fax SA595, SA868M, SB782; RTP 100T10)
Rockwell R84 (Petrothene PP 8411-ZR; RTP 101)
Rockwell R85 (Petrothene PP 8402-HO, PP 8402-TO, PP 8403-HO, PP 8403-TO, PP
    8404-HJ, PP 8404-ZJ, PP 8410-ZR, PP 8412-HK, PP 8412-TK, PP 8420-HK, PP
    8802-HO, PP 8815-ZR, PP 8820-HU; Rexene 9234, PP41E2, PP41E4; RTP
    101CC, 199x28016; Shell 5225, 7912)

Polypropylene *(cont'd.)*

Rockwell R88 (Pro-fax SA878, 141)

Rockwell R90 (Pro-fax 6329; Rexene 6310C25, 6310T4, 9400, 9400T2; RTP 100 GB20, 100T20, 103, 150; Shell 5384, 5419, 5431, 5520, 5550)

Rockwell R92 (Shell 5610)

Rockwell R93 (RTP 103CC)

Rockwell R94 (Petrothene PP 2085-GU, PP 8080-AW, PP 8080-GW, PP 8080-ZW; Rexene 6310T2; RTP 100 GB30, 143)

Rockwell R95 (Bapolene 4082; RTP 100T30, 105, 199x23835; Shell 5524, 5820, 5840)

Rockwell R96 (RTP 100 GB40; Shell 5944S)

Rockwell R97 (Pro-fax 65F4-4, 65F5-4, PD195; Rexene 6310)

Rockwell R98 (Pro-fax 6131, PD701; RTP 105CC)

Rockwell R99 (Pro-fax 6524; RTP 140)

Rockwell R100 (Bapolene 4042, 4062, 4072; Empee PP 401, PP 401CS, PP 402; Pro-fax 6323; RTP 101FR, 175, 175X, 177, 178, 178X; Shell 5824S, 5864)

Rockwell R102 (Amoco 6431, CR22N, CR22NA, CR35, CR35A; RTP 100T40, 107)

Rockwell R103 (Norchem NPP 1001-LF, NPP 1006-GF, NPP 1010-LC, NPP 7000-ZF; Pro-fax 6331NW; Rexene XO-325)

Rockwell R104 (Norchem NPP 1008-AK, NPP 2000-GJ, NPP 2003-GJ, NPP 2004-MR, NPP 3007-GO, NPP 3010-SO, NPP 8000-GK, NPP 8001-LK, NPP 8002-HK, NPP 8004-MR, NPP 8005-AR, NPP 8004-ZR; Petrothene PP 2004-MR, PP 8000-GK, PP 8001-LK, PP 8004-MR, PP 8004-ZR, PP 8005-AR, PP 8020-AU, PP 8020-GU, PP 8020-ZU)

Rockwell R105 (Amoco CR35N, CR35NA; RTP 105CC FR, 107CC, 199x22989)

Rockwell R107 (Amoco 7232, 7234)

Shore D60 (P032, P04)

Shore D63 (P012, P0113, P0119; PP-C5CC-2)

Shore D66 (PP-H3MFQ-1)

Shore D67 (PP-C6CC-4)

Shore D68 (P9125)

Shore D70 (Marlex CHM-040-01; P2130, P2230; PP-C3TF-2)

Shore D71 (Marlex CHM-040-02; P2120)

Shore D72 (Marlex HGH-050, HGV-040-01, HGX-030, HGZ-040, HGZ-050-02; PP-H2CC-2)

Shore D73 (Marlex HGZ-080-02, HGZ-120-02, HGZ-120-04, HGZ-350, HNS-080)

Shore D74 (Marlex LHN-020-01; PP-H1CC-1, -H4TF-3, -H6CC-4)

Shore D75 (Amoco 1012, 4036; Marlex HGN-120-01, HGN-120-A, HGN-350; PP-H2LTF-4, -H2MFS-3, -H7CC-4)

Shore D76 (Amoco 10-4017, 1046, 4018, 4039, 4222, 4228, 5219, 7233; PP-H2MFQ-3, -H2MFQ-4, -H3MFQ-5)

Shore D77 (Micalite PP-G2MF-4; P2140; PP-H3MF-2)

Shore D78 (Micalite PP-G2MF-5, PP-H2MFS-5; PP-G2MF-5, -H2MF-5, -H2MFS-5)

Polypropylene (cont'd.)

Shore D80 (Micalite PP-G2MF-6)
Mold Shrinkage:
  0.010–0.025 mm/mm (Amoco 6431, 7232, 7233, 7234, CR22N, CR22NA, CR35, CR35A, CR35N, CR35NA)
  0.015–0.020 mm/mm (Amoco 10-4017, 1012, 1046, 1088, 4018, 4036, 4039, 4222, 4228, 5219)
  0.001–0.003 in./in. (Nortuff NFA 2600 CO, NFC 2600 CO)
  0.001–0.004 in./in. (Nortuff NFC 4600 CO)
  0.002–0.006 in./in. (PP-H3MFQ-5)
  0.003–0.005 in./in. (Nortuff NFC 2440 CF)
  0.003–0.006 in./in. (Nortuff NFC 4400 CO)
  0.003–0.008 in./in. (PP-H2MFQ-4)
  0.004–0.006 in./in. (Nortuff NFA 2400 CO, NFC 2400 CO)
  0.004–0.010 in./in. (PP-H2LTF-4)
  0.005–0.008 in./in. (PP-H2MFQ-3, -H2MFS-3, -H3MF-2)
  0.005–0.012 in./in. (PP-H4TF-3)
  0.006–0.012 in./in. (PP-H6CC-4, -H7CC-4)
  0.007 in./in. (Stat-Kon MF-15)
  0.007–0.009 in./in. (Nortuff NFC 2200 CO)
  0.008 in./in. (Stat-Kon MM-3340)
  0.008–0.015 in./in. (PP-C6CC-4, -H3MFQ-1)
  0.009–0.012 in./in. (Nortuff NFA 1700 MO, NFA 1800 TO, NFC 1700 MO, NFC 1800 TO)
  0.010–0.014 in./in. (Nortuff NFA 4400 TO, NFC 4000 FR)
  0.010–0.016 in./in. (PP-C5CC-2)
  0.010–0.017 in./in. (PP-H2CC-2)
  0.010–0.018 in./in. (PP-C3TF-2)
  0.012 in./in. (Bapolene 4114; EMI-X MA-40; P2140)
  0.012–0.025 in./in. (Shell 5225, 5384, 5419, 5431, 5520, 5524, 5550, 5610, 5820, 5824S, 5840, 5864, 5944S, 7129, 7221, 7328, 7521, 7522, 7525, 7623, 7627, 7635, 7912, DX6016, DX6020)
  0.014 in./in. (P2130, P2230)
  0.015 in./in. (Bapolene 4112; P2120, P9125)
  0.016 in./in. (Stat-Kon M-1 HI, M-2)
  0.016–0.025 in./in. (PP-H1CC-1)
  0.017 in./in. (Stat-Kon ME)
  0.018 in./in. (P04, P0113, P0119)
  0.020 in./in. (P012, P032)
Water Absorp.:
  0% (Empee PP 402)
  < 0.01% (Amoco 1012, 1046, 1088, 4018, 4036, 4039, 5219)
  0.01% (Empee PP 401, PP 401CS; PP-H3MF-2; Stat-Kon M-1 HI, MF-15, MM-3340)
  0.02% (PP-C3TF-2, -C5CC-2, -C6CC-4, -H1CC-1, -H2CC-2, -H2LTF-4, -H4TF-3,

245

Polypropylene *(cont'd.)*

–H6CC-4, -H7CC-4; Stat-Kon M-2, ME)
< 0.03% (Shell 5225, 5384, 5419, 5431, 5520, 5524, 5550, 5610, 5820, 5824S, 5840, 5864, 5944S, 7129, 7221, 7328, 7521, 7522, 7525, 7623, 7627, 7635, 7912, DX6016, DX6020)
0.03% (PP-H3MFQ-1)
0.04% (PP-H2MFQ-3, -H2MFS-3)
0.05% (PP-H2MFQ-4)
< 0.06% (Amoco 10-4017)
0.06% (EMI-X MA-40; PP-H3MFQ-5)

THERMAL PROPERTIES:

Soften. Pt. (Vicat):
114 C (Modic P-300M)
125 C (Vestolen P5400, P5400 F, P6421)
130 C (Modic P-300F)
130–146 C (Tenite Polypropylene)
131 C (Shell DX6020)
132 C (Shell DX6016)
140 C (Vestolen P2300, P2300 F, P4800, P6500, P6522)
143 C (Shell 7525, 7623, 7627, 7635)
145 C (Modic P-110F; Vestolen P4700, P5206 S, P5800, P6206S)
146 C (Shell 7129, 7221, 7328, 7521)
150 C (Vestolen P1200, P1200 F, P2200, P2200 F, P3200, P3200 F, P4200, P4200 F, P5200, P5200 F, P6200)
152 C (Shell 5225, 5384, 5419, 5431, 5520, 5524, 5550, 5610, 5820, 5824S, 5840, 5864, 5944S, 7522, 7912)
155 C (Vestolen P2200, P2200 F, P5232 G, P5232 T, P5272 T)
300 F (Bapolene 5042, 5072)
303 F (Bapolene 5052)
305 F (Bapolene 4042, 4062, 4072, 4082)

Conduct.:
2.8×10^{-4} Cal/s/cm²/C/cm (Shell 5225, 5384, 5419, 5431, 5520, 5524, 5550, 5610, 5820, 5824S, 5840, 5864, 5944S)
3.0×10^{-4} Cal/s/cm²/C/cm (Shell 7129, 7221, 7328, 7521, 7522, 7525, 7623, 7627, 7635, 7912
0.90 Btu/h/ft²/F/in. (RTP 100 GB10)
1.0 Btu/h/ft²/F/in. (RTP 142)
1.1 Btu/h/ft²/F/in. (RTP 100T10)
1.2 Btu/h/ft²/F/in. (RTP 100 GB20)
1.3 Btu/h/ft²/F/in. (RTP 101, 101CC)
1.6 Btu/h/ft²/F/in. (RTP 100 GB30, 141)
1.7 Btu/h/ft²/F/in. (RTP 143; Stat-Kon ME)
1.8 Btu/h/ft²/F/in. (RTP 100 GB40, 100T20, 199x28016; Stat-Kon M-1 HI, M-2; Thermocomp MF-1002)

2.0 Btu/h/ft²/F/in. (RTP 101FR, 103, 103CC, 140; Stat-Kon MM-3340)

2.1 Btu/h/ft²/F/in. (RTP 100T30, 150, 175, 175X, 199x23835; Thermocomp MF-1004, MFX-1004HS)

2.2 Btu/h/ft²/F/in. (RTP 100T40)

2.3 Btu/h/ft²/F/in. (RTP 105, 105CC, 105CC FR, 178, 178X, 199x22898; Stat-Kon MF-15)

2.35 Btu/h/ft²/F/in. (Thermocomp MF-1006, MFX-1006HS)

2.4 Btu/h/ft²/F/in. (RTP 177)

2.45 Btu/h/ft²/F/in. (RTP 107, 107CC)

2.5 Btu/h/ft²/F/in. (Thermocomp MF-1002HI)

2.55 Btu/h/ft²/F/in. (Thermocomp MF-1008, MFX-1008HS)

2.6 Btu/h/ft²/F/in. (Thermocomp MF-1004HI)

2.7 Btu/h/ft²/F/in. (Thermocomp MF-1006HI)

2.8 Btu/h/ft²/F/in. (Thermocomp MF-1008HI)

5.0 Btu/h/ft²/F/in. (EMI-X MA-40)

Distort. Temp.:

45 C (Vestolen P5400, P5400 F, P6421)

46 C (264 psi) (Hostalen PP998; Nortuff NFC 4000 FR)

48 C (1.82 MPa) (Nortuff NFA 4400 TO)

48–57 C (1.82 MPa) (Tenite Polypropylene)

49 C (264 psi) (Hostalen PP933, PP934, PP936, PP996)

52 C (Vestolen P5800, P6206 S)

53 C (Vestolen P4800, P5206 S, P6500, P6522)

54 C (264 psi) (Hostalen PP920, PP927, PP941, PP942, PP975, PP989; Marlex CHM-040-01, CHM-040-02; P032, P0119)

55 C (Vestolen P4700, P6200); (264 psi) (Amoco 7233; P04)

56 C (264 psi) (Amoco CR35, CR35A)

57 C (Vestolen P5200, P5200 F); (1.82 MPa) (Marlex HGH-050, HGV-040-01, HGX-030, HGZ-040, HGZ-050-02; P012)

58 C (264 psi) (Amoco 6431; P0113)

60 C (Vestolen P2300, P2300 F, P3200, P3200 F, P4200, P4200 F); (264 psi) (Amoco CR22N, CR22NA, CR35N, CR35NA; Marlex HGZ-080-02, HGZ-120-02, HGZ-120-04, HGZ-350, HNS-080)

63 C (264 psi) (Amoco 7232, 7234; Marlex HLN-020-01)

65 C (Vestolen P1200, P1200 F, P2200, P2200 F); (1.82 MPa) (Nortuff NFC 1800 TO)

65.5 C (1.82 MPa) (Nortuff NFA 1700 MO)

66 C (1.82 MPa) (Marlex HGN-120-01, HGN-120-A, HGN-350)

67 C (264 psi) (P2230)

68 C (264 psi) (P2120)

72 C (66 psi) (Pro-fax 8623); (264 psi) (P2130, P2140)

73 C (66 psi) (Pro-fax 8523, SA862)

74 C (1.82 MPa) (Nortuff NFA 1800 TO)

75 C (66 psi) (Pro-fax 7131, SA868M, SB751)

Polypropylene *(cont'd.)*

77 C (264 psi) (P9125)

78 C (66 psi) (Pro-fax 7823, SB787)

80 C (Vestolen P5232 T); (66 psi) (Pro-fax 6329); (264 psi) (Empee PP 401, PP 401CS, PP 402)

81 C (66 psi) (Pro-fax 7523)

82 C (66 psi) (Petrothene PP 7300-MF; Pro-fax SA878; Shell DX6016)

82.2 C (1.82 MPa) (Nortuff NFC 1700 MO)

84 C (66 psi) (Pro-fax PD626, SB661, SB782)

85 C (Vestolen P5232 G); (66 psi) (Petrothene PP 1510-HC, PP 1510-LC, PP 8310-GO, PP 8310-KO, PP 8403-HO, PP 8403-TO, PP 8462-HR, PP 8470-HU, PP 8502-HK, PP 8602-HJ, PP 8752-HF, PP 8755-HK, PP 8762-HR, PP 8770-HU)

86 C (66 psi) (Pro-fax SB786)

87 C (Vestolen P5272 T)

88 C (0.46 MPa) (Norchem NPP 7300-AF, NPP 7300-GF; Petrothene PP 7200-AF, PP 7200-GF, PP 7200-MF, PP 7300-KF)

90 C (66 psi) (Petrothene PP 8402-HO, PP 8402-TO, PP 8404-HJ, PP 8404-ZJ, PP 8410-ZR, PP 8411-ZR, PP 8412-HK, PP 8412-TK, PP 8420-HK, PP 8470-ZU, PP 8815-ZR; Pro-fax 6131, PD701; Shell 7129)

91 C (66 psi) (Pro-fax PD195)

93 C (455 kPa) (Shell 7328, DX6020); (1.82 MPa) (Nortuff NFC 2200 CO)

95 C (66 psi) (Petrothene PP 2085-GU, PP 8020-AU, PP 8020-GU, PP 8080-AW, PP 8080-GW, PP 8080-ZW)

99 C (455 kPa) (Shell 5225)

100 C (66 psi) (Petrothene PP 8020-ZU, PP 8802-HO, PP 8820-HU; Pro-fax 6524; Shell 7221, 7521, 7522, 7525, 7623, 7627, 7635)

102 C (66 psi) (Amoco 4039; Pro-fax 6323; Shell 5419, 5431, 5520, 5550, 5610)

104 C (66 psi) (Amoco 10-4017, 4018, 4222, 4228; Shell 5384, 5820, 5840)

105 C (66 psi) (Petrothene PP 8000-GK, PP 8001-LK)

106 C (66 psi) (Pro-fax SA595)

107 C (66 psi) (Amoco 1012, 5219)

110 C (66 psi) (Amoco 1046, 1088, 4036; Petrothene PP 2004-MR, PP 8004-MR, PP 8005-AR; Shell 5524, 5944S, 7912)

112 C (0.46 MPa) (Norchem NPP 1001-LF, NPP 1006-GF, NPP 1010-LC, NPP 7000-ZF, NPP 8006-GF)

113 C (0.46 MPa) (Norchem NPP 1008-AK, NPP 8000-GK, NPP 8001-LK, NPP 8002-HK; Shell 5824S, 5864)

115 C (0.46 MPa) (Norchem NPP 2000-GJ, NPP 2003-GJ, NPP 2004-MR, NPP 3007-GO, NPP 3010-SO, NPP 8004-MR, NPP 8005-AR, NPP 8007-GO, NPP 8004-ZR; Petrothene PP 8004-ZR)

120 C (66 psi) (Pro-fax 6331NW)

121 C (1.82 MPa) (Nortuff NFA 2400 CO)

129 C (1.82 MPa) (Nortuff NFA 2600 CO, NFC 2400 CO, NFC 2440 CF, NFC 2600 CO)

132 C (1.82 MPa) (Nortuff NFC 4600 CO)
134 C (66 psi) (Pro-fax 65F4-4)
138 C (1.82 MPa) (Nortuff NFC 4400 CO)
100 F (264 psi) (PP-VB-511)
108 F (264 psi) (PP-C4IM)
115 F (264 psi) (PP-C3CC-4)
125 F (264 psi) (PP-C5CC-2; Stat-Kon ME)
130 F (264 psi) (PP-C3TF-2; Stat-Kon M-1 HI, M-2)
135 F (264 psi) (PP-C2TF-1, -C6CC-4; RTP 100 GB10)
140 F (264 psi) (PP-C2TF-4, -H1CC-1; RTP 100T10, 142)
145 F (264 psi) (PP-H2CC-2; RTP 100 GB20)
148 F (264 psi) (PP-H2TF-2)
150 F (264 psi) (PP-H6CC-4, -HIFR; RTP 141)
155 F (264 psi) (RTP 100 GB30)
158 F (264 psi) (PP-H4TF-3, -H6TF-4)
160 F (264 psi) (PP-H7CC-4; RTP 143; Stat-Kon MM-3340)
165 F (264 psi) (PP-C3TFA-2; RTP 100 GB40)
167 F (66 psi) (Rexene 9403E)
170 F (264 psi) (PP-H3MFQ-1; RTP 140)
175 F (66 psi) (Bapolene 5072)
176 F (66 psi) (Rexene 9403)
180 F (66 psi) (Rexene 9500); (264 psi) (PP-H1TF-4, -H2LTF-4, -H2TF-4; Stat-Kon M)
185 F (66 psi) (Bapolene 5042; Rexene 9234)
189 F (66 psi) (Rexene 9401C)
190 F (66 psi) (Bapolene 5052; PP-C2IM; Rexene 9402)
192 F (66 psi) (Rexene 9401)
195 F (264 psi) (PP-H2MFQ-3, -H2MFS-3)
200 F (264 psi) (RTP 178X)
203 F (264 psi) (PP-H3MF-2)
204 F (66 psi) (Rexene XO-325)
205 F (66 psi) (Rexene PP41E2, PP41E4)
210 F (264 psi) (RTP 150, 199x28016)
212 F (66 psi) (Rexene 9400)
217 F (264 psi) (Micalite PP-H2MFS-4, PP-H2MFS-6)
220 F (66 psi) (Bapolene 4042, 4072, 4082); (264 psi) (EMI-X MA-40; Micalite PP-H2MFS-5; PP-H2MFQ-4; RTP 101 SP Foamed)
221 F (66 psi) (Rexene 6310C25)
225 F (66 psi) (Bapolene 4062; Rexene 6310)
230 F (264 psi) (RTP 103 SP Foamed; Stat-Kon MF-15)
239 F (66 psi) (Rexene 9400T2)
240 F (264 psi) (PP-H2MFS-4; RTP 175X)
245 F (264 psi) (Micalite PP-G2MF-4; PP-G2MF-4, -H2MF-4)

Polypropylene *(cont'd.)*

250 F (264 psi) (PP-HFR-2; RTP 105 SP Foamed, 175, 199x23835)

252 F (264 psi) (PP-H2MFS-5)

255 F (264 psi) (PP-HFR-3)

257 F (66 psi) (Rexene 6310T2); (264 psi) (PP-H2MF-5)

260 F (264 psi) (PP-H3MFQ-5; RTP 100T20, 100T30)

265 F (264 psi) (RTP 101)

267 F (264 psi) (Micalite PP-G2MF-6)

268 F (264 psi) (Micalite PP-G2MF-5; PP-G2MF-5)

270 F (264 psi) (RTP 100T40, 101CC)

275 F (66 psi) (Rexene 6310T4); (264 psi) (Thermocomp MF-1002, MF-1002HI)

280 F (264 psi) (RTP 101FR, 177, 199x22898)

285 F (264 psi) (RTP 103; Thermocomp MF-1004, MF-1004HI)

290 F (264 psi) (RTP 103CC, 105CC FR, 178; Thermocomp MF-1004FR)

295 F (264 psi) (RTP 105, 105CC; Thermocomp MF-1006, MF-1006HI, MF-1008HI)

300 F (264 psi) (RTP 107, 107CC; Thermocomp MF-1008)

305 F (264 psi) (Thermocomp MFX-1004HS)

310 F (264 psi) (Thermocomp MFX-1006HS, MFX-1008HS)

Brittle Temp.:

−15 C (Pro-fax SE191)

−10 C (Modic P-110F, P-300F)

−5 C (Modic P-300M)

Coeff. of Linear Exp.:

0.8×10^{-4} K^{-1} (Vestolen P5272 T)

1.0×10^{-4} K^{-1} (Vestolen P5206 S, P5232 G, P6206 S)

1.2×10^{-4} K^{-1} (Vestolen P5232 T)

1.5×10^{-4} K^{-1} (Vestolen P1200, P1200 F, P2200, P2200 F, P2300, P2300 F, P3200, P3200 F, P4200, P4200 F, P4700, P4800, P5200, P5200 F, P5400, P5400 F, P5800, P6200, P6421, P6500, P6522)

6.0–9.0×10^{-5} cm/cm/C (Shell 7129, 7221, 7328, 7521, 7522, 7525, 7623, 7627, 7635, 7912, DX6016, DX6020)

8.0–10.2×10^{-5} cm/cm/C (Shell 5225, 5384, 5419, 5431, 5520, 5524, 5550, 5610, 5820, 5824S, 5840, 5864, 5944S)

1.5×10^{-5} in./in./F (Thermocomp MFX-1008HS)

1.7×10^{-5} in./in./F (Thermocomp MF-1008, MF-1008HI)

2.0×10^{-5} in./in./F (Thermocomp MF-1006, MF-1006HI)

2.1×10^{-5} in./in./F (Thermocomp MFX-1006HS)

2.4×10^{-5} in./in./F (Stat-Kon MF-15; Thermocomp MF-1004, MF-1004FR, MF-1004HI)

2.5×10^{-5} in./in./F (Stat-Kon MM-3340; Thermocomp MFX-1004HS)

2.8×10^{-5} in./in./F (Thermocomp MF-1002HI)

3.0×10^{-5} in./in./F (PP-H2TF-4; Thermocomp MF-1002)

3.6×10^{-5} in./in./F (Stat-Kon M-1 HI)

3.7×10^{-5} in./in./F (Stat-Kon M-2)

3.8 × 10⁻⁵ in./in./F (Stat-Kon ME)

4 × 10⁻⁵ in./in./F (PP-H2TF-2)

Sp. Heat:

0.44–0.46 Cal/g/C (Shell 5225, 5384, 5419, 5431, 5520, 5524, 5550, 5610, 5820, 5824S, 5840, 5864, 5944S)

0.45–0.50 Cal/g/C (Shell 7129, 7221, 7328, 7521, 7522, 7525, 7623, 7627, 7635, 7912, DX6016, DX6020)

0.65 Cal/g/C (A-Fax 500, 600, 940)

0.38 Btu/lb/F (Thermocomp MFX-1008HS)

0.41 Btu/lb/F (Thermocomp MFX-1006HS)

0.44 Btu/lb/F (Thermocomp MFX-1004HS)

Heat of Fusion:

7.6 Cal/g (A-Fax 500, 600, 940)

Flamm.:

V-0 (Nortuff NFA 4400 TO, NFC 4400 CO, NFC 4600 CO; Pro-fax SA595; RTP 101FR, 105CC FR, 150; Thermocomp MF-1004FR)

V-2 (Empee PP 401, PP 401CS, PP 402; Nortuff NFC 4000 FR)

HB (Amoco 1012, 1046; EMI-X MA-40; Norchem NPP 8001-LK; Petrothene PP 8001-LK; RTP 100 GB10, 100 GB20, 100 GB30, 100 GB40, 100T10, 100T20, 100T30, 100T40, 101, 101CC, 101 SP Foamed, 103, 103CC, 103 SP Foamed, 105, 105CC, 105 SP Foamed, 107, 107CC, 140, 141, 142, 143, 175, 175X, 177, 178, 178X, 199x22898, 199x23835, 199x28016; Stat-Kon M-1 HI, M-2, ME, MF-15, MM-3340; Thermocomp MF-1002, MF-1002HI, MF-1004, MF-1004HI, MF-1006, MF-1006HI, MF-1008, MF-1008HI, MFX-1004HS, MFX-1006HS, MFX-1008HS)

ELECTRICAL PROPERTIES:

Dissip. Factor:

0.000001 (50 Hz) (Rexene XO-325)

0.00008 (50 Hz) (Rexene PP41E2, PP41E4)

0.0001 (1 kHz) (A-Fax 600)

0.0002–0.0010 (1 MHz) (Fortilene 10X01, 14X01, 16X01, 18X02, 18X03, 19X02, 19X03, 21X04, 38X01, 40X05, 41X03, 41X04, 42X07, 54X02, 1401, 1602, 1602A, 2251, 3151, 3230, 3251, 3606, 3907, 4141, 4141F, 9205, 9605)

0.0003 (1 kHz) (Pro-fax SE191); (1 MHz) (Shell 5225, 5384, 5419, 5431, 5520, 5524, 5550, 5610, 5820, 5824S, 5840, 5864, 5944S, 7129, 7221, 7328, 7521, 7522, 7525, 7623, 7627, 7635, 7912, DX6016, DX6020)

0.0004 (50 Hz) (Vestolen P5232 G, P5232 T); (1 kHz) (A-Fax 500, 940)

< 0.0005 (50 Hz) (Vestolen P1200, P1200 F, P2200, P2200 F, P2300, P2300 F, P3200, P3200 F, P4200, P4200 F, P4700, P4800, P5200, P5200 F, P5400, P5400 F, P5800, P6200, P6421, P6500, P6522)

0.0008 (1 kHz) (Empee PP 402)

0.001 (1 MHz) (RTP 100 GB10, 100 GB20, 100 GB30, 101, 101CC, 101FR, 103, 103CC, 105, 105CC, 175, 175X)

Polypropylene *(cont'd.)*

0.0013 (50 Hz) (Vestolen P5272 T)

0.002 (1 MHz) (RTP 100 GB40, 107, 107CC, 140)

0.003 (50 Hz) (Vestolen P5206 S, P6206 S); (1 MHz) (RTP 100T10, 105CC FR, 178, 178X)

0.004 (1 MHz) (RTP 100T20, 142)

0.004–0.014 (60 to 10^6 Hz) (Thermocomp MF-1004FR)

0.005 (1 MHz) (RTP 141)

0.006 (1 MHz) (RTP 100T30, 143, 177)

0.008 (1 MHz) (RTP 100T40)

0.010 (1 MHz) (RTP 150)

Dielec. Str.:

45 kV/mm (Vestolen P5206 S, P5232 G, P5232 T, P5272 T, P6206 S)

75 kV/mm (Vestolen P1200, P1200 F, P2200, P2200 F, P2300, P2300 F, P3200, P3200 F, P4200, P4200 F, P4700, P4800, P5200, P5200 F, P5400, P5400 F, P5800, P6200, P6421, P6500, P6522)

450 V/mil (RTP 105CC FR)

490 V/mil (RTP 101FR)

500 V/mil (RTP 100 GB10, 100 GB20, 100 GB30, 100 GB40, 100T10, 100T20, 100T30, 100T40, 105CC, 107CC, 140, 141, 142, 143, 150, 175, 175X, 177, 178, 178X; Thermocomp MF-1004FR)

500–660 V/mil (Fortilene 10X01, 14X01, 16X01, 18X02, 18X03, 19X02, 19X03, 21X04, 38X01, 40X05, 41X03, 41X04, 42X07, 54X02, 1401, 1602, 1602A, 2251, 3151, 3230, 3251, 3606, 3907, 4141, 4141F, 9205, 9605)

510 V/mil (RTP 107)

520 V/mil (RTP 103CC, 105)

530 V/mil (RTP 103)

540 V/mil (RTP 101CC)

550 V/mil (RTP 101)

600 V/mil (A-Fax 600; Shell 5225, 5384, 5419, 5431, 5520, 5524, 5550, 5610, 5820, 5824S, 5840, 5864, 5944S, 7129, 7221, 7328, 7521, 7522, 7525, 7623, 7627, 7635, 7912, DX6016, DX6020)

700 V/mil (A-Fax 500, 940)

900 V/mil (Empee PP 401, PP 401CS)

1175 V/mil (Empee PP 402)

> 16,000 V/mil (Rexene PP41E2, PP41E4, XO-325)

Dielec. Const.:

2.1–2.3 (1 MHz) (Fortilene 10X01, 14X01, 16X01, 18X02, 18X03, 19X02, 19X03, 21X04, 38X01, 40X05, 41X03, 41X04, 42X07, 54X02, 1401, 1602, 1602A, 2251, 3151, 3230, 3251, 3606, 3907, 4141, 4141F, 9205, 9605)

2.11 (1 kHz) (Empee PP 402)

2.2 (1 kHz) (A-Fax 500, 600, 940)

2.24 (1 kHz) (Pro-fax SE191)

2.3 (Vestolen P1200, P1200 F, P2200, P2200 F, P2300, P2300 F, P3200, P3200 F,

P4200, P4200 F, P4700, P4800, P5200, P5200 F, P5400, P5400 F, P5800, P6200, P6421, P6500, P6522); (1 MHz) (RTP 100T10; Shell 5225, 5384, 5419, 5431, 5520, 5524, 5550, 5610, 5820, 5824S, 5840, 5864, 5944S, 7129, 7221, 7328, 7521, 7522, 7525, 7623, 7627, 7635, 7912, DX6016, DX6020)

2.4 (50 Hz) (Vestolen P5206 S, P5232 G, P5232 T, P5272 T, P6206 S); (1 MHz) (RTP 100T20, 142)

2.5 (50 Hz) (Rexene PP41E2, PP41E4, XO-325); (1 MHz) (RTP 100 GB10, 101CC)

2.6 (1 MHz) (RTP 100 GB20, 100T30, 100T40, 103CC)

2.7 (1 MHz) (RTP 100 GB30, 101, 105CC, 150)

2.8 (1 MHz) (RTP 103, 107CC, 141, 175, 175X)

2.86–2.79 (60 to 10^6 Hz) (Thermocomp MF-1004FR)

2.9 (1 MHz) (RTP 100 GB40, 101FR, 105, 178, 178X)

3.0 (1 MHz) (RTP 143, 177)

3.1 (1 MHz) (RTP 140)

3.2 (1 MHz) (RTP 107)

3.3 (1 MHz) (RTP 105CC FR)

Vol. Resist.:

50 ohm-cm (RTP 199x22898, 199x23835, 199x28016)

100 ohm-cm (EMI-X MA-40)

1000 ohm-cm (Stat-Kon M-1 HI)

10,000 ohm-cm (Stat-Kon M-2, MF-15, MM-3340)

100,000 ohm-cm (Stat-Kon ME)

10^{15} ohm-cm (RTP 101FR, 105, 105CC, 105CC FR, 107, 107CC)

4.0×10^{15} ohm-cm (Pro-fax SE191)

5.1×10^{15} ohm-cm (Empee PP 402)

10^{16} ohm-cm (RTP 100 GB10, 100 GB20, 100 GB30, 100 GB40, 100T10, 100T20, 100T30, 100T40, 101, 101CC, 103, 103CC, 140, 141, 142, 143, 150)

> 10^{16} ohm-cm (Fortilene 10X01, 14X01, 16X01, 18X02, 18X03, 19X02, 19X03, 21X04, 38X01, 40X05, 41X03, 41X04, 42X07, 54X02, 1401, 1602, 1602A, 2251, 3151, 3230, 3251, 3606, 3907, 4141, 4141F, 9205, 9605; Vestolen P1200, P1200 F, P2200, P2200 F, P2300, P2300 F, P3200, P3200 F, P4200, P4200 F, P4700, P4800, P5200, P5200 F, P5206 S, P5232 G, P5232 T, P5272 T, P5400, P5400 F, P5800, P6200, P6206 S, P6421, P6500, P6522)

1.5×10^{16} ohm-cm (Empee PP 401, PP 401CS)

10^{17} ohm-cm (Shell 5225, 5384, 5419, 5431, 5520, 5524, 5550, 5610, 5820, 5824S, 5840, 5864, 5944S, 7129, 7221, 7328, 7521, 7522, 7525, 7623, 7627, 7635, 7912, DX6016, DX6020)

> 10^{17} ohm-cm (Rexene PP41E2, PP41E4, XO-325)

3×10^{17} ohm-cm (A-Fax 600)

6×10^{17} ohm-cm (A-Fax 500, 940)

Surf. Resist.:

100 ohm/sq. (EMI-X MA-40)

1000 ohm/sq. (Stat-Kon M-1 HI)

Polypropylene (cont'd.)

10,000 ohm/sq. (Stat-Kon M-2, MF-15, MM-3340)
100,000 ohm/sq. (Stat-Kon M, ME)
> 10^{13} ohm (Vestolen P1200, P1200 F, P2200, P2200 F, P2300, P2300 F, P3200, P3200 F, P4200, P4200 F, P4700, P4800, P5200, P5200 F, P5206 S, P5232 G, P5232 T, P5272 T, P5400, P5400 F, P5800, P6200, P6206 S, P6421, P6500, P6522)

Arc Resist.:

63 s (Empee PP 401, PP 401CS)
80 s (RTP 101FR, 150)
85 s (RTP 105CC FR)
102 s (RTP 107)
115 s (RTP 100 GB40)
120 s (RTP 100 GB30, 100T30, 100T40, 105, 107CC, 177)
123 s (RTP 103)
125 s (RTP 100 GB20, 100T20, 101, 105CC, 175, 175X, 178, 178X)
128 s (RTP 100 GB10, 140)
129 s (RTP 100T10)
130 s (RTP 103CC, 141, 143)
132 s (RTP 101CC, 142)
140 s (Empee PP 402)

STD. PKGS.:

55-gal metal drums (A-Fax 800)
Avail. in bulk tank cars or in metric ton palletized units of 5-kg slabs (A-Fax 500, 940)
Avail. in bulk tank cars, 72-gal fiber or 55-gal metal drums (A-Fax 600)

Polyquaternium-1 (CTFA)

STRUCTURE:

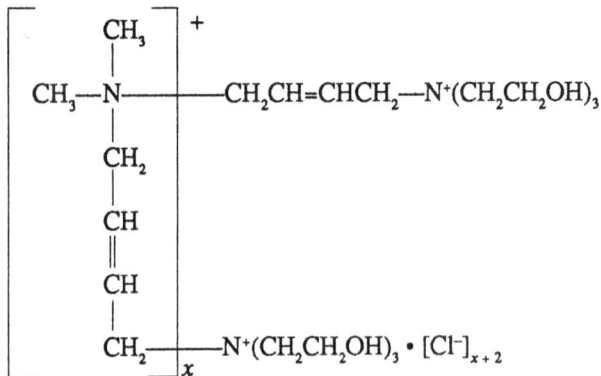

CAS No.:
68518-54-7

TRADENAME EQUIVALENTS:
Onamer M [Millmaster-Onyx]

CATEGORY:
Polymer quat., conditioner

APPLICATIONS:
Cosmetic industry preparations: creams and lotions (Onamer M); hair preparations (Onamer M); shampoos (Onamer M)

PROPERTIES:
Form:
Liquid (Onamer M)
Composition:
30% conc. (Onamer M)

GENERAL PROPERTIES:
Ionic Nature:
Cationic (Onamer M)

Polyquaternium-2 (CTFA)

STRUCTURE:

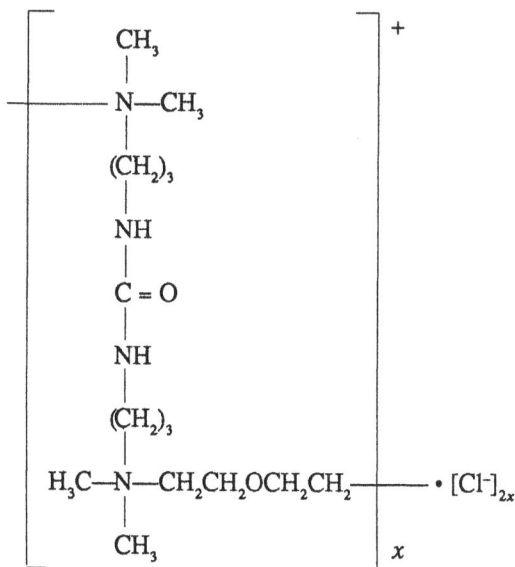

$$\left[\begin{array}{c} CH_3 \\ | \\ -N-CH_3 \\ | \\ (CH_2)_3 \\ | \\ NH \\ | \\ C=O \\ | \\ NH \\ | \\ (CH_2)_3 \\ | \\ H_3C-N-CH_2CH_2OCH_2CH_2- \\ | \\ CH_3 \end{array}\right]^+_x \cdot [Cl^-]_{2x}$$

255

Polyquaternium-2 *(cont'd.)*

CAS No.:
63451-27-4; 68555-36-2
TRADENAME EQUIVALENTS:
Mirapol A-15 [Miranol]
CATEGORY:
Polymeric quat., softener, conditioner, surface modifier, antistat
APPLICATIONS:
Cosmetic industry preparations: conditioners (Mirapol A-15); cream rinses (Mirapol A-15); shampoos (Mirapol A-15)
PROPERTIES:
Form:
Viscous liquid (Mirapol A-15)
Color:
Amber (Mirapol A-15)
Composition:
64% active (Mirapol A-15)
GENERAL PROPERTIES:
Ionic Nature:
Cationic (Mirapol A-15)
Solubility:
Dissolves readily in water in all proportions (Mirapol A-15)
M.W.:
2260 (Mirapol A-15)
pH:
8.5 (Mirapol A-15)

Polyquaternium-5 (CTFA)

SYNONYMS:
Acrylamide/beta-methacrylyloxyethyl trimethyl ammonium methosulfate copolymer
Ethanaminium, N,N,N-trimethyl-2-[(2-methyl-1-oxo-2-propenyl) oxy]-, methyl sulfate, polymer with 2-propenamide
Quaternium-39
CAS No.:
26006-22-4
TRADENAME EQUIVALENTS:
Catamer Q [Richardson]
Emcol Q [Witco]
Reten 210, 220 [Hercules]

CATEGORY:

Copolymer, conditioner, flocculant, retention aid, slip agent, thickener, antistat, film former, suspending agent, cross-linking agent, adhesive

APPLICATIONS:

Cosmetic industry preparations: hair preparations (Catamer Q; Emcol Q); hand and body lotions (Catamer Q); shampoos (Emcol Q)

FDA-approved applications: (Reten 210, 220)

Food-contact applications: (Reten 210, 220)

Industrial applications: pulp and paper industry (Reten 210, 220)

PROPERTIES:

Form:

Liquid (Catamer Q; Emcol Q)

Powder (Reten 210, 220)

Fineness:

30% max. through 200 mesh (Reten 210, 220)

Color:

White (Reten 210, 220)

Composition:

15% max. volatiles (Reten 210, 220)

30% conc. (Emcol Q)

31% solids (Catamer Q)

GENERAL PROPERTIES:

Ionic Nature:

Cationic (Catamer Q; Emcol Q; Reten 210, 220)

Solubility:

Dissolves in warm or cold water (Reten 210, 220)

Density:

9.14 lb/gal (Catamer Q)

42 lb/ft^3 (Reten 210, 220)

Visc.:

300 cps (Catamer Q)

700 cps (1% solids) (Reten 210)

750 cps (1% solids) (Reten 220)

pH:

2.1 (Catamer Q)

5 (1% solids) (Reten 210, 220)

TOXICITY/HANDLING:

Low order of oral toxicity; not a primary skin irritant or sensitizer; spills become slippery (Reten 210, 220)

STORAGE/HANDLING:

Store in a cool, dry place (Reten 210, 220)

Polyquaternium-6 (CTFA)

SYNONYMS:
Dimethyl diallyl ammonium chloride polymer
N,N-Dimethyl-N-2-propenyl-2-propen-1-aminium chloride, homopolymer
Poly (dimethyl diallyl ammonium chloride)
Poly (DMDAAC)
2-Propen-1-aminium, N,N-dimethyl-N-2-propenyl-, chloride, homopolymer
Quaternium-40

EMPIRICAL FORMULA:
$(C_8H_{16}N \cdot Cl)_x$

CAS No.:
26062-79-3; 28301-34-0

TRADENAME EQUIVALENTS:
Jordaquat 40 [PPG-Mazer]
Merquat 100 [Calgon]

CATEGORY:
Polymer, emollient, conditioner

APPLICATIONS:
Bath products: (Merquat 100)
Cosmetic industry preparations: conditioners (Merquat 100); creams and lotions (Merquat 100); hair preparations (Jordaquat 40; Merquat 100); moisturizers (Merquat 100); shampoos (Merquat 100); skin care products (Jordaquat 40; Merquat 100)

PROPERTIES:
Form:
Clear viscous liquid (Jordaquat 40)
Composition:
39–41% active (Jordaquat 40)

GENERAL PROPERTIES:
Ionic Nature:
Cationic (Merquat 100)
Solubility:
Limited sol. in alcohol (Merquat 100)
Sol. in propylene glycol (Merquat 100)
Sol. in water (Merquat 100)

Polyquaternium-7 (CTFA)

SYNONYMS:
Acrylamide/dimethyl diallyl ammonium chloride copolymer

N,N-Dimethyl-N-2-propenyl-2-propen-1-aminium chloride, polymer with 2-prope-
namide
2-Propen-1-aminium, N,N-dimethyl-N-2-propenyl-, chloride, polymer with 2-prope-
namide
Quaternium-41

EMPIRICAL FORMULA:
$(C_8H_{16}N \cdot C_3H_5NO \cdot Cl)_x$

CAS No.:
26590-05-6

TRADENAME EQUIVALENTS:
Jordaquat 41 [PPG-Mazer]
Merquat 550 [Calgon]

CATEGORY:
Polymeric quat., conditioner

APPLICATIONS:
Cosmetic industry preparations: hair preparations (Jordaquat 41; Merquat 550); skin
care products (Jordaquat 41; Merquat 550); shampoos (Merquat 550)

PROPERTIES:
Form:
Clear viscous liquid (Jordaquat 41)
Composition:
8.1–9.1% active (Jordaquat 41)

GENERAL PROPERTIES:
Ionic Nature:
Cationic (Merquat 550)
Solubility:
Limited sol. in alcohol (Merquat 550)
Sol. in propylene glycol (Merquat 550)
Sol. in water (Merquat 550)

Polyquaternium-11 (CTFA)

SYNONYMS:
Quaternium-23
Vinylpyrrolidone/dimethylaminoethylmethacrylate copolymer, reacted with di-
methyl sulfate

CAS No.:
37348-62-2; 37348-63-3

Polyquaternium-11 *(cont'd.)*

TRADENAME EQUIVALENTS:
Gafquat 734, 755, 755N [GAF]
CATEGORY:
Polymeric quat., conditioner, film former
APPLICATIONS:
Cleansers: toilet soaps (Gafquat 734, 755N)

Cosmetic industry preparations: conditioners (Gafquat 734); hair preparations (Gafquat 734, 755, 755N); shampoos (Gafquat 734); shaving preparations (Gafquat 734, 755N); skin care products (Gafquat 734, 755, 755N)

Pharmaceutical applications: antiperspirants/deodorants (Gafquat 734, 755N); antiseptics (Gafquat 734, 755N); sunburn remedies (Gafquat 734, 755N)
PROPERTIES:
Form:

Liquid (Gafquat 734, 755, 755N)

Composition:

20% active in water (Gafquat 755, 755N)

50% active in alcohol (Gafquat 734)

Polyurethane

SYNONYMS:
PU

Urethane polymer
STRUCTURE:

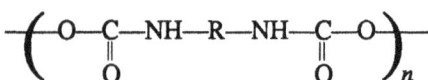

CAS No.:
9009-54-5
TRADENAME EQUIVALENTS:
Thermoplastic:

Q-Thane P-49, P-250, PA-01, PA-05 (linear), PA-06, PA-07 (linear), PA-10, PA-11, PA-20, PA-29, PA-30, PA-40, PA-58, PA-80, PA-93 (linear), PH-56, PH-89 [K.J. Quinn]

Rucothane 2010L (aliphatic), 2030L (aliphatic), 2060L, 3000L, 3104, 3105, 3106, 5000L, 5100 (aromatic), CO-1-620 (in DMF/toluene sol'n.), CO-A-640, CO-A-670/671, CO-A-710, CO-A-819L, CO-A-832L, CO-A-904L (in alcohol/toluene sol'n.), CO-A-2880 FR, CO-A-2885 FR, CO-A-3907, CO-A-3908L (in alcohol/toluene sol'n.), CO-A-3982L, CO-A-3983L, CO-A-4041L (in alcohol/aromatic

Polyurethane *(cont'd.)*

sol'n.), CO-A-4078, CO-A-5002L, CO-A-5054, CO-B-4030L (in alcohol/toluene sol'n.) [Ruco]

Polycaprolactone-based:

CAPA 200 (linear), 205 (linear), 210 (linear), 212 (linear), 215 (linear), 220 (linear), 222 (linear), 223 (linear), 231 (linear), 240 (linear), 304 (branched), 305 (branched), 520 (linear), 600 (linear), 600M (linear), 601 (linear), 601M (linear), Monomer [Solvay]

Polyester-based:

Estane 5701 F1, 5702, 5702 F1, 5702 F2, 5703, 5703, 5703 F1, 5703 F2, 5707 F1, 5708 F1, 5710 F1, 5711, 5712, 5713, 5716, 5715, 58013, 58091, 58092, 58109, 58121, 58130, 58133, 58134, 58136, 58137, 58271, 58277, 58360 [BF Goodrich]

Q-Thane P-279, P-280, P-360, P-440, P-455, PC-58, PI-76, PI-86, PI-95, PI-96, PI-176, PI-186, PI-195, PI-196, PN03-100 (aliphatic), PN3429-100 (aliphatic), PS-16, PS-62, PS-63, PS-65, PS-82, PS-94 [K.J. Quinn]

Rucothane CO-A-610 (aromatic, in MEK/DME sol'n.) [Ruco]

Polyether-based:

Estane 5714 F1, 58300, 58309, 58311, 58370, 58630 [BF Goodrich]

Q-Thane PE-23, PE-36, PE-47, PE-49, PE-50, PE-55, PE74, PE-90, PE103-100 (aliphatic), PE192-100 (aliphatic), PE192-101 (aliphatic) [K.J. Quinn]

Rucothane CO-A-5069 [Ruco]

Carbon powder grade:

Stat-Kon T [LNP]

Glass-reinforced:

RTP 1201-80D (10% glass), 1203-80D (20% glass fibers), 1205-80D (30% glass fibers), 1207-80D (40% glass fibers) [Fiberite]

Thermocomp TF-1004 (20% fiber glass), TF-1006 (30% fiber glass), TF-1008 (40% fiber glass) [LNP]

Thermoset:

Adiprene BL-16, L-42, L-83, L-100, L-167, L-200, L-213, L-300, L-315, L-325, L-367, L-700, L-767, LW-500, LW-510, LW-520, LW-550, LW-570 [Uniroyal]

Calthane ND 1100, 2300, 3200 [Cal Polymers]

Chempol 33-4199/34-5075 [Freeman]

Conathane TU-4010, UC-33 [Conap]

Desmocol 110, 130, 176, 400, 406, 420, 500, 510, 530 [Bayer AG]

Hexcel 164M [Hexcel/Rezolin]

Hysol PC28, PC18 [Hysol/Dexter Corp.]

Lamal 408-40(65)/Lamal C, Lamal HSA/Lamal C [Polymer Industries]

Mira-Glos RT A and C [Polymer Industries]

NeoRez EX-466 (aromatic), R-961 (aliphatic), U-105 (aromatic), U-110 (aromatic), U-760 (aromatic), U-800 (aliphatic prepolymer), U-912 (aliphatic) [Polyvinyl Chem. Industries]

Norcast PR-1020 [R.H. Carlson]

Polyurethane *(cont'd.)*

Poron 4701-01, 4701-05, 4701-09, 4716-16 [Rogers]

P.U.R.E.-CMC [Perma-Flex Mold]

Purelast 204, 207, 208, 209, 220, 221, 223, 224, 226, 228, 242, 243, 245, 245H, 234, 235, 240H, 241, 241H, 247, 249, 251, 253, 255, 254 [Polymer Systems]

Ren:C:O-Thane RP-6400, RP-6401, RP-6402, RP-6403, RP-6405, RP-6410, RP-6413, RP-6414, RP-6422, TDT-178-34, TDT-178-53, TDT-186-1 [Ren Plastics]

RestEasy Foam [BASF]

Rezolin 185N [Hexcel/Rezolin]

Specflex [Dow]

Spectrim 5, 15, 25S, 25W, 35S, 35W, 50S, 50W, MM300, MM353, Polyurea HF, Polyurea HT, SF500, SP400 [Dow]

Stepanfoam A-210, AX-64, AX-66, BX-105 Series, BX-150-5, BX-250 (A-D) Series, BX-289, BX-316-3, BX-316-5, BX-341, Bx-341A. BX-341B, BX-345, BX-350-7, BX-351, BX-352, BX-352P, BX-352M, BX-359, BX-364, BX-369, BX-370, BX-372-2, C-600 Series, F-202, F-302, F-506, FX-250, G Series, H-100 Series, H-102N, H-402N, H-602N, HC-2/30, HC-2/40, HC-3/40, HC-4/60, HC-5/60, HC-9/10 Series, HC-17/40, HW-8/25, HW-8/50, HW-10/25, HW-11/60, HW-12/25, HW-16/60, HW-20/60, MW-20/20, P-502, P-506, PF-15, PR-5, PR-5-O, R-222/R-109, R-223/R-110, R-226/R-112, R-231G/R-110, R-244, R-245, R-246, R-247, SF-3, SF-5/60, SX-159 Series, SX-195A, SX-202, SX-209J, SX-211, SX-214, SX-215, SX-216, SX-217, SX-218A, SX-219 [Stepan]

Stycast CPC-16, CPC-17 [Emerson & Cuming]

Trymer 160, 190, 190-3, 190-4, 210, 9501, 9501-3, 9501-4, 9501-6 [Dow]

Unoflex, 100 [Polymer Industries]

UR 101, 102, 103, 104, 105 [Thermoset Plastics]

Uralite 3111, 3113, 3115, 3121, 3121S, 3122, 3124, 3125, 3128, 3130, 3132, 3139, 3140, 3150, 3152, 3154, 3155, 3156, 3158, 3167, 3175 Fastset, 6108, 7250, 7252 [Hexcel/Rezolin]

Polyester-based:

Conathane RN-3038, RN-3039 [Conap]

Millathane 76, 80, 300, HT [TSE Industries]

Stepanfoam F-403 [Stepan]

Polyether-based:

Adiprene CM [Uniroyal]

Millathane E-34 [TSE Industries]

Presto-Foam 800, 805, 900, 945, 960 [Presto Mfg.]

Rezolin 164, 170 [Hexcel/Rezolin]

TYPES/MODIFICATIONS/SPECIALTY GRADES:

Elastomer:

Adiprene BL-16, CM, L-42, L-83, L-100, L-167, L-200, L-213, L-300, L-315, L-325, L-367, L-700, L-767, LW-500, LW-510, LW-520, LW-550, LW-570; Calthane ND 1100, 2300, 3200; CAPA 200, 205, 210, 212, 220, 222, 223; Conathane RN-3038, RN-3039, TU-4010; Desmocol 110, 130, 176, 400, 406, 420, 500, 510, 530;

Polyurethane *(cont'd.)*

Millathane 76, 80, 300, E-34; P.U.R.E.-CMC; Purelast 204, 207, 208, 209, 242, 243, 228, 242, 243, 245, 245H; Q-Thane P-49, P-250, P-279, P-280, P-360, P-440, P-455, PA-10, PA-20, PA-30, PA-40, PA-80, PE-23, PE-36, PE-47, PE-49, PE-50, PE-55, PE-74, PE-90, PH-56, PH-89, PI-76, PI-86, PI-95, PI-96, PI-176, PI-186, PI-195, PI-196, PS-16, PS-62, PS-63, PS-65, PS-82, PS-94; Ren:C:O-Thane RP-6400, RP-6401, RP-6402, RP-6403, RP-6405, RP-6410, RP-6413, RP-6414, Rp-6422, TDT-178-34, TDT-178-53, TDT-186-1; Rezolin 185N; Rucothane 2060L, CO-1-620, CO-A-610, CO-A-710, CO-A-4078; Stepanfoam MW-20/20; Stycast CPC-16, CPC-17; Uralite 3111, 3113, 3115, 3121, 3121S, 3122, 3124, 3125, 3128, 3130, 3132, 3140, 3150, 3152, 3154, 3155, 3156, 3158, 3167, 3175 Fastset, 7250, 7252

Foams:

CAPA 220, 222, 223; Presto-Foam 800, 805, 900, 945, 960; Spectrim SF 500; Stepanfoam BX-351

Rigid:

· CAPA 304; Chempol 34-4199/34-5075 (with Fluorocarbon 12); Stepanfoam A-210, BX-105 Series, BX-150-5, BX-250 (A–D) Series, BX-289, BX-316-3, BX-316-5, BX-341, BX-341A, BX-241B, BX-345, BX-350-7, BX-352, BX-352P, BX-352M, BX-359, BX-364, BX-369, BX-370, BX-372-2, C-600 Series, G Series, H-100 Series, H-102N, H-402N, H-602N, HC-2/30, HC-2/40, HC-3/40, HC-4/60, HC-17/40, HW-8/25, HW-10/25, HW-12/25, HW-8/50, HW-11/60, HW-16/60, HW-20/60, PR-5, PR-5-O, R-222/R-109, R-223/R-110, R-245, R-247, R-226/R-112, R-231G/R-110, R-244, R-246, SF-3, SF-5/60; Trymer 160, 190, 190-3, 190-4, 210, 9501, 9501-3, 9501-4, 9501-5

Semirigid:

Stepanfoam HC-9/10 Series, P-502, P-506

Semiflexible:

Stepanfoam AX-64, AX-66, SX-195A, SX-202, SX-211, SX-215, SX-216, SX-218A

Flexible:

Stepanfoam F-202, F-302, F-403, F-506, F-250, PF-15, SX-209J, SX-214, SX-217, SX-219

Microcellular:

CAPA 220, 222, 223, 520; Specflex; Stepanfoam MW-20/20

High-density microcellular:

Poron 4701-01, 4701-05, 4701-09, 4716-16

Slabstock:

RestEasy Foam

Molding and extrusion compounds:

Estane 58013, 58091, 58092, 58109, 58121, 58130, 58133, 58134, 58136, 58137, 58271, 58277, 58300, 58309, 58311, 58360, 58370; P.U.R.E.-CMC; Q-Thane P-360, P-440, P-455, PC-58, PE-23, PE-36, PE-47, PE-49, PE-50, PE-55, PE-74, PE-90, PE103-100, PE192-100, PE192-101, PH-56, PH-89, PI-76, PI-86, PI-95, PI-96, PI-176, PI-186, PI-195, PI-196, PN03-100, PN3429-100, PS-16, PS-62, PS-63, PS-

263

Polyurethane *(cont'd.)*

65, PS-82, PS-94; Rezolin 164; RTP 1201-80D, 1203-80D, 1205-80D, 1207-80D; Spectrim 5, 15, 25S, 25W, 35S, 35W, 50S, 50W, MM300, MM353, Polyurea HF, Polyurea HT

Coatings:

CAPA 200, 205, 210, 212, 215, 305; Estane 5701 F1, 5702, 5702 F1, 5702 F2, 5707 F1, 5708 F1, 5710 F1, 5714 F1, 5715, 58630; Hysol PC18, PC28; Lamal 408-40(65)/C; Mira-Glos RT A and C; NeoRez EX-466, R-961, U-105, U-110, U-760, U-800, U-912; Purelast 220, 221, 226, 228, 242, 243, 245, 245H, 234, 235, 247, 249, 251, 253, 255, 254; Q-Thane P-279, PA-80; Rucothane 3000L, 5000L, 5100, CO-1-620, CO-A-610, CO-A-640, CO-A-670/671, CO-A-710, CO-A-819L, CO-A-832L, CO-A-904L, CO-A-2880 FR, CO-A-2885 FR, CO-A-3907, CO-A-3908L, CO-A-3982L, CO-A-3983L, CO-A-4041L, CO-A-4078, CO-A-5002L, CO-A-5054, CO-A-5069, CO-B-4030L; Uralite 3139

Adhesives:

Calthane ND 1100, 2300; CAPA 231, 240, 601M; Desmocoll 110, 130, 176, 400, 406, 420, 500, 510, 530; Estane 5701 F1, 5702, 5702 F1, 5702 F2, 5703, 5703 F1, 5703 F2, 5711, 5712, 5713, 5716; Lamal 408-40(65)/C, HSA; Purelast 223, 224, 240H, 241, 241H; Q-Thane P-279, PA-01, PA-05, PA-06, PA-07, PA-10, Pa-11, PA-20, PA-30, PA-40, PA-58, PA-93, PA-29; Rucothane 2010L, 2030L, 2060L, 3000L, 3104, 3105, 3106, CO-1-620, CO-A-4078; Stepanfoam SX-159 Series; Unoflex, 100, 6108

Latex:

Rucothane 2010L, 2030L, 2060L, 3000L, 3104, 3105, 3106, 5000L, 5100

Statically dissipative:

Stat-Kon T

High-flow:

Stepanfoam BX-341, BX-345, HW-8/50

Flame-retardant:

Poron 4716-16; Ren:C:O-Thane TDT-178-34, TDT-178-53; Rucothane CO-A-2880 FR (halogenated), CO-A-2885 FR (halogenated)

Heat-resistant:

Estane 58130, 58133, 58137; Millathane HT; Norcast PR-1020; Q-Thane PA-01, PA-05, PA-06, PA-58, PA-80; Uralite 3113, 3121S, 3140

Hydrolysis-resistant:

Adiprene LW-520, LW-570; Estane 5714 F1, 58300, 58309, 58311, 58630; Millathane E-34; Poron 4701-01, 4701-05, 4701-09, 4716-16; Purelast 204, 207, 208, 209; Rezolin 185N; Rucothane CO-A-5002L; Uralite 3121S, 3130, 3132, 3139, 3140, 3150, 3152, 7250, 7252

Fungus-resistant:

Estane 5714 F1, 58300, 58311, 58630; Millathane E-34; Rezolin 185N

UV-resistant:

Calthane ND 1100, 2300, 3200; NeoRez R-961, U-800; Rucothane 2060L, CO-A-904L, CO-A-3908L, CO-A-3982L, CO-A-3983L, CO-A-4041L, CO-B-4030L;

Polyurethane *(cont'd.)*

Unoflex
High gloss:
NeoRez U-110, U-760, U-912; Rucothane CO-A-5002L
Plasticized:
Millathane 80
Calcium stearate-dusted:
Estane 5701 F1, 5702, 5702 F1, 5702 F2, 5707 F1, 5708 F1, 5710 F1, 5715
Talc-dusted:
Estane 5703, 5703 F1, 5703 F2, 5711, 5712, 5713, 5716

CATEGORY:
Thermoset or thermoplastic polymer

PROCESSING:
Calendering:
Q-Thane P-279, P-280
Casting:
Adiprene BL-16, L-42, L-83, L-100, L-167, L-200, L-213, L-300, L-315, L-325, L-367, L-700, L-767, LW-500, LW-510, LW-520, LW-550, LW-570; Calthane ND 1100, 2300, 3200; Conathane RN-3038, RN-3039, TU-4010, UC-33; Norcast PR-1020; Purelast 204, 207, 208, 209; Ren:C:O-Thane RP-6400, RP-6401, RP-6413, RP-6414, RP-6422; Rezolin 185N; Stycast CPC-16, CPC-17; Uralite 3113, 3115, 3121, 3121S, 3122, 3124, 3125, 3128, 3130, 3132, 3140, 3150, 3152, 3154, 3155, 3156, 1358, 3167, 3175 Fastset, 7250, 7252
Extrusion:
Estane 58013, 58092, 58271, 58277, 58300, 58309, 58311, 58360, 58370; Q-Thane P-49, P-360, P-440, P-455, PC-58, PE-36, PE-49, PE-50, PE-55, PE-74, PE-90, PE103-100, PE192-100, PE192-101, PH-56, PH-89, PN03-100, PN3429-100, PS-16, PS-62, PS-63, PS-65, PS-82, PS-94
Film casting:
Estane 5707 F1; Rucothane CO-A-4078
Injection molding:
Adiprene CM; Estane 58091, 58092, 58109, 58121, 58130, 58133, 58134, 58136, 58137, 58300, 58309, 58311, 58370; Q-Thane P-440, PE-23, PE-36, PE-47, PE-74, PI-76, PI-86, PI-95, PI-96, PI-176, PI-186, PI-195, PI-196; Ren:C:O-Thane RP-6405, TDT-186-1; Rezolin 164, 170; RTP 1201-80D, 1203-80D, 1205-80D, 1207-80D; Spectrim MM300, MM353; Stepanfoam BX-352M, BX-352P
Lamination:
Estane 5703, 5703 F1, 5703 F2, 5711, 5712, 5713, 5716
Plastic-gum processing:
Adiprene L-213, LW-520
Potting/encapsulating:
Purelast 204, 207, 208, 209; Rezolin 185N; Stepanfoam A-210, BX-105 Series, C-600 Series, G Series; UR 101, 102, 103, 104, 105; Uralite 3121S, 3125, 3130, 3132, 3150, 3152, 3154, 3155, 3156, 3158, 7250, 7252

Polyurethane (cont'd.)

Reaction-injection molding (RIM):
Spectrim 5, 15, 25S, 25W, 35S, 35W, 50S, 50W, Polyurea HF, Polyurea HT

Structural foam molding:
Stepanfoam HC-9/10 Series, HW-8/25, HW-10/25, HW-12/25, HW-8/50, HW-11/60, HW-16/60, HW-20/60

Thermoforming:
Ren:C:O-Thane RP-6405, TDT-186-1

APPLICATIONS:

Agriculture industry: cattle tags (Estane 58136); components (Q-Thane Series)

Architectural applications: beam systems (Stepanfoam HC-17/40); construction applications (Purelast 220, 221, 226, 228, 234, 235, 240H, 241, 241H, 247, 249, 251, 253, 255); deck coating (Purelast 234, 235, 247, 249); roof coating (Purelast 228, 245, 245H, 251, 253, 255, 254); roof insulation (Trymer 210); windows (Conathane UC-33)

Automotive applications: (Spectrim 5, 15, 25S, 25W, 35S, 35W, 50S, 50W, Polyurea HF, Polyurea HT); body panels (Spectrim Polyurea HF, Polyurea HT); exterior trim (Spectrim 15, 25S, 25W, 35S, 35W, 50S, 50W, Polyurea HF, Polyurea HT); padding (Poron 4701-01, 4701-05, 4701-09, 4716-16); parts (Estane 58091, 58130, 58133, 58137; Q-Thane Series); structural composites (Spectrim MM300, MM353, SF500, SP400); windows (Spectrim 5)

Aviation industry: coatings (NeoRez U-800)

Consumer products: ; appliance gears (Estane 58091); appliance parts (Q-Thane Series); art objects (Uralite 3113, 3140); athletic equipment padding (Presto-Foam 945, 960); crafts (Presto-Foam 800); footwear (Adiprene CM; Desmocoll 110; Estane 5703, 5703 F1, 5703 F2, 5711, 5712, 5713, 5716, 58121; Presto-Foam 945; Rucothane 3104, 3105, 3106, CO-A-3907; Stepanfoam HC-9/10 Series, HW-8/25, HW-10/25, HW-12/25, MW-20/20); furniture/bedding (RestEasy Foam; Stepanfoam HC-2/30, HC-2/40, HC-3/40, HC-4/60, HC-5/60, SX-209J, SX-214, SX-219; Uralite 3113, 3140); golf ball covers (CAPA 601M); ice skates (Estane 58133); novelties (Presto-Foam 800); picnic coolers (Stepanfoam BX-359, BX-369); recreational equipment (RTP 1201-80D, 1203-80D, 1205-80D, 1207-80D); ski boots (Estane 58137); sporting equipment (Q-Thane Series); toys (Q-Thane Series)

Electrical/electronic industry: (Q-Thane Series); business machine parts (Q-Thane Series); cable and wire jacketing (Adiprene CM; Estane 58300, 58309, 58311, 58360, 58370; Q-Thane Series); electrical components (Stat-Kon T); electrical enclosures/packaging (Stat-Kon T); insulated telecommunications cable (Hexcel 164M; Rezolin 164, 170, 185N); junctions (Estane 58300); magnetic tape binder (Q-Thane P-280); padding (Poron 4701-01, 4701-05, 4701-09, 4716-16); potting/encapsulation (Norcast PR-1020; Rezolin 185N; Stepanfoam A-210, BX-105 Series, C-600 Series, G Series; UR 101, 102, 103, 104, 105; Uralite 3121S, 3125, 3130, 3132, 3154, 3155, 3156, 3158, 7250, 7252); printed circuit coatings (Hysol PC18, PC28); RF-shielding applications (Poron 4701-01, 4701-05, 4701-09, 4716-16); UL-listed applications (Stepanfoam BX-350-7)

Polyurethane (cont'd.)

FDA-approved applications: (Lamal HSA; Unoflex)

Food-contact applications: (Estane 58271, 58277); USDA-approved applications: (Estane 58271, 58277; Unoflex)

Functional additives: binder (Rucothane 2010L, 3000L); blending resin (Estane 5701 F1, 5702, 5702 F1, 5702 F2, 5710 F1; Q-Thane P-279, P-280; Rucothane 2010L); modifier (CAPA Monomer; Q-Thane P-279; Rucothane CO-1-620, CO-A-610, CO-A-832L); saturant (Rucothane 2010L); vehicle (Adiprene BL-16; CAPA 200, 304, 305, 600M, Monomer)

Industrial applications: adhesives (Adiprene BL-16; CAPA 231, 240, 600M, 601M; Calthane ND 1100, 2300; Desmocol 110, 130, 176, 400, 406, 420, 500, 510, 530; Estane 5701 F1, 5702, 5702 F1, 5702 F2, 5703, 5703 F1, 5703 F2, 5711, 5712, 5713, 5716; Lamal 408-40(65)/C, HSA; Purelast 223, 224, 240H, 241, 241H; Q-Thane Series, P-250, P-279, P-280, PA-01, PA-05, PA-06, PA-07, PA-10, PA-11, PA-20, PA-29, PA-30, PA-40, PA-58, PA-93, PH-56, PH-89; Rucothane 2010L, 2030L, 2060L, 3000L, 3104, 3105, 3106, CO-1-620, CO-A-610, CO-A-670/671, CO-A-4078; Unoflex, 100; Uralite 3139, 6108); bearing applications (Adiprene L-100, L-213, L-315, L-325, LW-570); bearings and bushings (Q-Thane Series; Thermocomp TF-1004, TF-1006, TF-1008); belting (Rucothane CO-A-710); binders (Estane 5701 F1, 5703, 5703 F1, 5703 F2, 5707 F1, 5710 F1, 5715); castings (Calthane ND 1100, 2300, 3200; Conathane RN-3038, RN-3039, TU-4010; Ren:C:O-Thane RP-6400); chemical industry parts (Q-Thane Series); closures (Q-Thane Series); coatings (Adiprene BL-16; CAPA 305; Estane 5715; Lamal 408-40(65)/C; NeoRez EX-466, R-961, U-105, U-912; Purelast 220, 221; Q-Thane Series, P-279, PA-80; Rucothane 3000L); compounding (Q-Thane Series); concrete/cement coatings (Purelast 220, 221, 223, 224, 242, 243); conveyor belts (Adiprene CM; Q-Thane Series); core boxes (Ren:C:O-Thane RP-6413, RP-6414, 3121S, 3122, 3124, 3125, 3175 Fastset, 7250, 7252); couplings (Calthane ND 1100, 2300); decorative parts (Stepanfoam BX-289, BX-316-3, BX-316-5, HC-2/30, HC-2/40, HC-3/40, HC-4/60, HC-5/60, HC-17/40, SF-3, SF-5/60); displays (Conathane UC-33); dunnage (Presto-Foam 805; Stepanfoam MW-20/20); elastomers (Spectrim 5, 15, 25S, 25W, 35S, 35W, 50S, 50W, Polyurea HF, Polyurea HT); engine mounts (Uralite 3121, 3150, 3152, 3154, 3155, 3156, 3158, 7250, 7252); extrusion coating (Lamal 408-40(65)/C); films (Adiprene BL-16; Estane 58092, 58309, 58311, 58360, 58630; Q-Thane Series); foams (CAPA 220, 222, 223, 520; Rucothane 3000L); foundry patterns (Calthane ND 3200; Ren:C:O-Thane RP-6413, RP-6414, 3121S, 3122, 3124, 3125, 3175 Fastset, 7250, 7252); frothable adhesive (Rucothane 2010L, 2030L, 2060L, 3000L); gaskets/seals/rings (Adiprene CM; Estane 58109; Poron 4701-01, 4701-05, 4701-09, 4716-16; Presto-Foam 800, 805, 900, 945, 960; RTP 1201-80D, 1203-80D, 1205-80D, 1207-80D; Thermocomp TF-1004, TF-1006, TF-1008; Uralite 3111, 3113, 3115, 3121, 3121S, 3128, 3130, 3132, 3140, 3150, 3152, 3154, 3155, 3156, 3158, 3167, 7250, 7252); gears (Ren:C:O-Thane RP-6403; RTP 1201-80D, 1203-80D, 1205-80D, 1207-80D; Thermocomp TF-1004, TF-1006, TF-1008; Uralite 3121S, 3125); graphic arts

Polyurethane *(cont'd.)*

applications (Mira-Glos RT A and C); heat-seal applications (Estane 5711, 5712, 5713, 5716; Rucothane CO-A-610, CO-A-640, CO-A-670/671, CO-A-710, CO-A-819L, CO-A-904L, CO-A-4041L, CO-A-5054, CO-A-5069); impact parts (Uralite 3125); impellers (Calthane ND 1100, 2300; Ren:C:O-Thane RP-6403; Uralite 3121S, 3125); insulation (Stepanfoam BX-150-5, BX-250 (A–D) Series, BX-350-7, BX-352, BX-352P, BX-352M, BX-369, H-100 Series, H-402N, H-602N, R-222/R-109, R-223/R-110, R-245, R-247, R-226/R-112, R-231G/R-110, R-244, R-246; Trymer 160, 190, 190-3, 190-4, 210, 9501, 9501-3, 9501-4, 9501-6); laminates (Estane 5703, 5703 F1, 5703 F2, 5711, 5712, 5713, 5716; Lamal 408-40(65)/C, HSA; Q-Thane P-280, PA-11; Rucothane 3000L; Unoflex, 100; Uralite 3139, 6108); large parts (Uralite 3128); leather finishes (Estane 5707 F1, 5715; Rucothane CO-A-3982L, CO-A-3983L); lens materials (Conathane UC-33); mechanical goods (Adiprene L-83; Q-Thane Series; Ren:C:O-Thane RP-6402); metal coatings (NeoRez U-110; Purelast 254); metal forming (Calthane ND 1100, 2300, 3200; Ren:C:O-Thane RP-6402, RP-6403; Uralite 3122, 3124, 3125, 3130, 3132, 3150, 3152, 3154, 3155, 3156, 3158, 3167, 7250, 7252); moisture blocks (Hexcel 164M; Rezolin 164, 170, 185N); molded goods (Adiprene CM; Ren:C:O-Thane RP-6405); molds and dies (Calthane ND 1100, 2300; Conathane TU-4010; P.U.R.E.-CMC; Ren:C:O-Thane RP-6400, RP-6401, RP-6410, TDT-178-34; Uralite 3111, 3113, 3115, 3121S, 3122, 3125, 3130, 3132, 3140, 3167, 3175 Fastset); nozzles (RTP 1201-80D, 1203-80D, 1205-80D, 1207-80D); outdoor applications (Estane 58136); packaging (Desmocoll 110, 130, 176; Mira-Glos RT A and C; Presto-Foam 800, 805, 900, 945, 960; Q-Thane Series; Stepanfoam BX-369, BX-370, BX-372-2, F-403, F-506, H-102N, PF-15, PR-5, PR-5-O); pads/protective padding (Poron 4701-01, 4701-05, 4701-09, 4716-16; Presto-Foam 800, 805, 900, 945; Stepanfoam AX-64, F-202, F-302; Uralite 3111, 3113, 3115, 3121, 3140, 3150, 3152, 3154, 3155, 3156, 3158, 3167, 7250, 7252); paper coatings (Estane 5715; Lamal HSA); paper/pulp industry (Hexcel 164M; Q-Thane Series; Rezolin 164); petroleum industry (Adiprene CM); pipe covering (Trymer 190, 190-3, 190-4, 9501, 9501-3, 9501-4, 9501-6); pipe fittings (Q-Thane Series); pipes (Q-Thane Series); plaster/gypsum/cement molds (P.U.R.E.-CMC); plastics (Q-Thane Series); printing rollers (Uralite 3121S, 3124); production parts (Ren:C:O-Thane RP-6402, RP-6422; Uralite 3111, 3121, 3125); reinforcement applications (Trymer 9501-3, 9501-4, 9501-6); resilient parts (Ren:C:O-Thane RP-6400, RP-6410, TDT-178-34); rolls/rollers (Adiprene CM, L-83; Calthane ND 1100, 2300, 3200; Ren:C:O-Thane RP-6403; Uralite 3111, 3121, 3125, 3150, 3152, 3154, 3155, 3156, 3158, 7250, 7252); rubber (CAPA 200, 205, 210, 212, 220, 222, 223); rubber coatings (Adiprene BL-16; Q-Thane Series; Rucothane CO-B-4030L); sealants/caulking (NeoRez U-110; Purelast 220, 221, 226, 242, 243; Rucothane CO-1-620; Uralite 3139, 6108); sheet (Estane 58013, 58092, 58309, 58311, 58360; Q-Thane Series); solution coatings (Q-Thane P-280; Rucothane CO-1-620, CO-A-610, CO-A-640, CO-A-670/671, CO-A-710, CO-A-819L, CO-A-832L, CO-A-904L, CO-A-2880 FR, CO-A-2885 FR, CO-A-3907, CO-A-3908L, CO-A-3982L, CO-A-3983L, CO-A-4041L, CO-

A-4078, CO-A-5002L, CO-A-5054, CO-A-5069, CO-B-4030L); sound dampening (Ren:C:O-Thane RP-6402; Stepanfoam FX-250; Uralite 3139); structural foam applications (Stepanfoam BX-341, BX-341A, BX-341B, BX-345, BX-364, HC-9/10 Series, HW-8/25, HW-10/25, HW-12/25, HW-8/50, HW-11/60, HW-16/60, HW-20/60); textile applications (Estane 5703, 5703 F1, 5703 F2; Rucothane 2010L); textile coatings (Adiprene BL-16, CM; CAPA 205, 210, 212, 215; Estane 5701 F1, 5702, 5702 F1, 5702 F2, 5707 F1, 5708 F1, 5710 F1, 5714 F1, 58630; Q-Thane P-280; Rucothane 2030L, 2060L, 5000L, 5100, CO-A-710); thin sections (Adiprene BL-16); tires (Adiprene CM; Q-Thane Series, 3121S, 3124); tooling (Conathane TU-4010; Uralite 3121, 3125, 3175 Fastset); tools (Ren:C:O-Thane RP-6413, RP-6414); tubing/hoses (Adiprene CM; Estane 58013, 58092, 58309, 58311); vibration mounts/damping (Calthane ND 1100, ND 2300; Poron 4701-01, 4701-05, 4701-09, 4716-16; Ren:C:O-Thane RP-6422; Stepanfoam F-403, F-506, PF-15); void filling (Stepanfoam BX-369, H-102N, P-502, P-506; Trymer 160, 190, 190-3, 190-4, 210, 9501, 9501-3, 9501-4, 9501-6); waterproofing (Purelast 220, 221, 223, 224, 226, 228, 234, 235, 240H, 241, 241H, 247, 249, 251, 253, 255); wear pads (Calthane ND 1100, 2300); weatherstripping (Presto-Foam 800); wheels (Adiprene L-83; Calthane ND 1100, ND 2300; Ren:C:O-Thane RP-6403); wood coatings (NeoRez U-105, U-110)

Marine applications: coatings (NeoRez U-800); flotation/buoyancy (Stepanfoam BX-289, BX-369, H-402N, HC-2/40; Trymer 160)

Medical applications: biomedical tubing (Q-Thane Series); orthopedic applications (CAPA Monomer)

Military applications: (Stepanfoam A-210, BX-105 Series, BX-289, C-600 Series, G Series, H-402N, PF-15; Uralite 3150, 3152, 3154, 3155, 3156, 3158, 7250, 7252)

PROPERTIES:

Form:

Liquid (Adiprene L-42, L-83, L-100, L-167, L-200, L-213, L-300, L-315, L-325, L-367, L-700, L-767, LW-500, LW-510, LW-520, LW-550, LW-570; Calthane ND 1100, 2300, 3200; CAPA 200, 304, 305, Monomer; Conathane TU-4010, UC-33; Desmocoll 400; Purelast 223, 224, 226, 228, 234, 235; Ren:C:O-Thane RP-6400, RP-6401, RP-6402, RP-6403, RP-6405, RP-6410, RP-6413, RP-6414, RP-6422, TDT-178-34, TDT-178-53, TDT-186-1; Stycast CPC-16, CPC-17; Uralite 3111, 3113, 3115, 3121, 3121S, 3122, 3124, 3125, 3128, 3130, 3132, 3140, 3150, 3152, 3154, 3155, 3156, 3158, 3167, 3175 Fastset, 6108, 7250, 7252)

Viscous liquid (Adiprene BL-16)

Thixotropic (Uralite 3139)

Paste (CAPA 210, 212, 520)

Soft (Purelast 204)

Solid (Conathane RN-3038, RN-3039; Millathane 76, 80, 300, E-34, HT)

Granular (Estane 5701 F1, 5702, 5702 F1, 5702 F2, 5703, 5703 F1, 5703 F2, 5707 F1, 5708 F1, 5710 F1, 5711, 5712, 5713, 5714 F1, 5715, 5716; Q-Thane P-280, PE103-100, PE192-100, PE192-101)

Polyurethane *(cont'd.)*

Solid slabs (Adiprene CM)
Wax-like (CAPA 215, 220, 222, 223, 231, 240)
Tough polymer (CAPA 600, 600M, 601, 601M)
Foam (Specflex; Stepanfoam Series)
Foam slabstock (RestEasy Foam)
Foam avail. 0.1–0.5 in. (15 pcf), 0.062–0.5 in. (20 pcf) (Poron 4701-01, 4701-05)
Foam avail. 0.062–0.45 in. (Poron 4716-16)
Foam avail. 0.062–0.1 in. (20 pcf), 0.035–0.062 in. (25 pcf) (Poron 4701-09)
Foam avail. in thicknesses from 3/32 in., in sheets, strips, shapes, etc. (Presto-Foam 800, 805, 900, 945, 960)

Color:

Clear (Conathane UC-33 Parts A and B; Ren:C:O-Thane TDT-186-1); (cured) (Purelast 207)
Water clear (cured) (Conathane UC-33)
Clear transparent (Calthane ND 1100, 2300, 3200)
Transparent (Rezolin 185)
Translucent (Purelast 204, 208, 209)
Natural (Poron 4716-16; Presto-Foam 805, 900)
Natural, black (RTP 1201-80D, 1203-80D, 1205-80D, 1207-80D)
White (Conathane TU-4010 Part B; Ren:C:O-Thane RP-6405, TDT-178-34 (hardener), TDT-178-53); (cured) (Conathane TU-4010)
White to cream (Millathane 300)
Clear light straw (Unoflex)
Pale amber (Estane 5701 F1, 5702, 5702 F1, 5702 F2, 5703, 5703 F1, 5703 F2, 5707 F1, 5708 F1, 5710 F1, 5711, 5712, 5713, 5714 F1, 5715, 5716; Ren:C:O-Thane RP-6413, RP-6414, RP-6422, TDT-178-34 (resin))
Clear amber (Uralite 3111, 3121, 3122, 3124)
Amber (Adiprene BL-16; Conathane TU-4010 Part A; Millathane 80; Norcast PR-1020; UR 104; Uralite 3113, 3121S, 3130, 3150, 3154, 3156, 3167, 7250)
Semitransparent amber (Uralite 3140)
Dark to light amber (Millathane 76, E-34, HT)
Yellow (Presto-Foam 900; Ren:C:O-Thane RP-6403, RP-6410)
Azure (Poron 4701-09)
Honey (Adiprene L-42, L-83, L-100, L-167, L-200, L-213, L-300, L-315, L-325, L-367, L-700, L-767, LW-520, LW-570)
Orange (Adiprene CM; Uralite 3139)
Red (Ren:C:O-Thane RP-6402)
Green (Ren:C:O-Thane RP-6400)
Blue (Ren:C:O-Thane RP-6401)
Blue tint (UR 103)
Opaque brown (Uralite 3175 Fastset)
Gray (Poron 4701-05; Presto-Foam 960)
Gray, yellow, blue, sand, custom colors (Presto-Foam 800)

Black (Poron 4701-01; UR 101, 102, 105; Uralite 3115, 3125, 3132, 3152, 3155, 3158, 7252)
Odor:
  Low (NeoRez R-961)
  Faint, characteristic (Adiprene CM; Millathane 76, 80, 300, E-34, HT)
Composition:
  2.8% isocyanate content (Adiprene L-42)
  3.15% isocyanate content (Adiprene LW-500)
  3.2% isocyanate content (Conathane RN-3038)
  3.25% isocyanate content (Adiprene L-83)
  4.1% isocyanate content (Adiprene L-100)
  4.25% isocyanate content (Adiprene LW-510)
  4.3% isocyanate content (Conathane RN-3039)
  4.75% isocyanate content (Adiprene LW-520)
  5.3–5.8% blocked isocyanate content (Adiprene BL-16)
  5.5% isocyanate content (Adiprene L-700)
  5.55% isocyanate content (Adiprene LW-550)
  6.3% isocyanate content (Adiprene L-167)
  7.5% isocyanate content (Adiprene L-200, LW-570)
  7.8% isocyanate content (Adiprene L-767)
  9.15% isocyanate content (Adiprene L-325)
  9.3% isocyanate content (Adiprene L-213)
  9.45% isocyanate content (Adiprene L-315)
  20% nonvolatiles (NeoRez U-912)
  31% nonvolatiles (NeoRez EX-466)
  34% nonvolatiles (NeoRez R-961)
  40% nonvolatiles (NeoRez U-105, U-110, U-800)
  60% nonvolatiles (NeoRez U-760)
  57 ± 1% solids (Mira-Glos RT C)
  65 ± 1% solids (Lamal 408-40(65))
  70% solids in denatured ethanol (Lamal HSA)
  70 ± 1% solids (Mira-Glos RT A)
  75 ± 2% solids in MEK (Unoflex)
  90+% closed cell content (Trymer 9501-3, 9501-4, 9501-6)
  92% closed cell content (Trymer 160, 190, 190-3, 1901-4, 210, 9501)
  100% solids (Lamal C; Purelast 204, 207, 208, 209, 220, 221, 223, 224, 226, 228, 242, 243, 245, 245H, 234, 235, 240H, 241, 241H, 247, 249; Unoflex 100; Uralite 3139)

GENERAL PROPERTIES:
Solubility:
  Sol. in acetone (Q-Thane P-250, P-279, PA-11)
  Limited sol. in alcohols (Adiprene BL-16); insol. (Q-Thane PA-93); dilutable in alcohol/water mixtures (Lamal HSA)
  Limited sol. in aliphatic solvents (Adiprene BL-16, L-42, L-83, L-100, L-167, L-200,

Polyurethane (cont'd.)

L-213, L-315, L-325, L-700, L-767, LW-500, LW-510, LW-520, LW-550, LW-570); insol. (Q-Thane PA-93)

Sol. in aromatic hydrocarbons (Adiprene L-42, L-83, L-100, L-167, L-200, L-213, L-315, L-325, L-700, L-767, LW-500, LW-510, LW-520, LW-550, LW-570); swells readily (Q-Thane PA-93)

Sol. in benzene (Q-Thane P-250, P-279, PA-11)

Sol. in butyl acetate (Q-Thane P-250, PA-11)

Sol. in Cellosolve acetate (Q-Thane P-279)

Sol. in chlorinated hydrocarbons (Adiprene L-42, L-83, L-100, L-167, L-200, L-213, L-315, L-325, L-700, L-767, LW-500, LW-510, LW-520, LW-550, LW-570; Q-Thane PA-93); swollen by chlorinated solvents (Adiprene CM)

Sol. in cyclohexanone (Q-Thane P-250, P-279, PA-11, PA-29)

Sol. in diacetone alcohol (Q-Thane P-279)

Sol. in dimethylformamide (Adiprene CM; Millathane 76, 80, 300, E-34, HT; Q-Thane P-250, P-279, PA-29)

Sol. in 1,4-dioxane (Q-Thane P-250, P-279)

Sol. in esters (Adiprene L-42, L-83, L-100, L-167, L-200, L-213, L-315, L-325, L-700, L-767, LW-500, LW-510, LW-520, LW-550, LW-570; Q-Thane PA-93)

Dilutable in ethanol (Lamal HSA)

Sol. in ethyl acetate (Adiprene BL-16; Mira-Glos RT A and C; Q-Thane P-250, P-279, PA-11, PA-29); dilutable (Lamal HSA)

Sol. in ethylene dichloride (Q-Thane PA-11, PA-29)

Dilutable in isopropanol (Lamal HSA)

Sol. in ketones (Adiprene L-42, L-83, L-100, L-167, L-200, L-213, L-315, L-325, L-700, L-767, LW-500, LW-510, LW-520, LW-550, LW-570; Q-Thane PA-93)

Sol. in MEK (Adiprene BL-16, CM; Q-Thane P-250, P-279, PA-11)

Dilutable in methanol (Lamal HSA)

Sol. in methylene chloride (Millathane 76, 80, 300, E-34, HT; Q-Thane P-250, P-279, PA-11)

Sol. in MIBK (Q-Thane PA-11, PA-29)

Swollen by petroleum fractions (Adiprene CM)

Dilutable in normal propanol (Lamal HSA)

Sol. in tetrahydrofuran (Adiprene CM; Millathane 76, 80, 300, E-34, HT; Q-Thane P-250, P-279, PA-11, PA-29)

Sol. in toluene (Adiprene BL-16; Q-Thane P-250, PA-11); partly sol. (Q-Thane P-279)

Sol. in trichloroethylene (Adiprene BL-16)

Insol. in water (Millathane 76, 80, 300, E-34, HT)

Sol. in xylene (Q-Thane P-250, PA-11); partly sol. (Q-Thane P-279)

M.W.:

114 (CAPA Monomer)

250 (CAPA 304)

540 (CAPA 305)

550 (CAPA 200)

830 (CAPA 205)
1000 (CAPA 210, 212)
1250 (CAPA 215)
2000 (CAPA 220, 222, 223, 520)
3000 (CAPA 231)
4000 (CAPA 240)
20,000 (CAPA 600, 600M)
45,000 (CAPA 601, 601M)
Sp. Gr.:
0.040 (cured) (Uralite 3111, 3113)
0.60 (Spectrim SP400)
0.94 (UR 104, 105)
0.95 (UR 101, 103)
1.00 (P.U.R.E.-CMC)
1.01 (Adiprene LW-500)
1.02 (Hexcel 164M; UR 102)
1.029 Uralite 3140 Part A)
1.03 (Adiprene L-42, LW-510, LW-520, LW-550; Uralite 3130 Part A, 3132 Part A)
1.03 ± 0.02 (Conathane UC-33 Part B)
1.035 (Rezolin 170)
1.04 (Calthane ND 3200; Uralite 3154 Part A, 3155 Part A); (cured) (Ren:C:O-Thane RP-6400, RP-6410, RP-6422)
1.048 (Uralite 3139 Part A)
1.05 (Adiprene L-83, LW-570; Ren:C:O-Thane RP-6413; Uralite 3125 Part B)
1.056 (Uralite 7250 Part A, 7252 Part A)
1.1056 (Uralite 3156 Part A, 3158 Part A)
1.06 (Adiprene CM, L-100, L-300; Uralite 3115 Part A, 3167 Part A)
1.06 ± 0.02 (Adiprene BL-16)
1.068 (cured) (Uralite 3139)
1.07 (Adiprene L-167, L-367, L-700; Rezolin 185); (cured) (Conathane UC-33; Ren:C:O-Thane RP-6401; Uralite 3130, 3132)
1.073 (cured) (Uralite 3140)
1.074 (Uralite 3140 Part B)
1.08 (Millathane E-34; Ren:C:O-Thane RP-6402); (cured) (Uralite 3125, 3115, 3154, 3155, 3167)
1.09 (Adiprene L-200, L-767; Ren:C:O-Thane RP-6414)
1.09 ± 0.02 (Conathane UC-33 Part A)
1.10 (Ren:C:O-Thane RP-6403, TDT-186-1; Uralite 3121S, 3130 Part B, 3132 Part B); (cured) (Stycast CPC-16; Uralite 7250, 7252)
1.11 (Adiprene L-213, L-315, L-325; Millathane HT; Uralite 3154 Part B, 3155 Part B, 3156 Part B, 3158 Part B, 7250 Part B, 7252 Part B)
1.114 (cured) (Uralite 3156, 3158)
1.12 (Rezolin 164)

Polyurethane (cont'd.)

1.13 (Estane 58300; Millathane 80; Q-Thane PE-49); (cured) (Ren:C:O-Thane RP-6405; Stycast CPC-17)

1.135 (Uralite 3121, 3122)

1.14 (Conathane TU-4010 Part A; Estane 58309, 58311, 58630; Q-Thane PE-74; Spectrim SF500; Uralite 3139 Part B)

1.143 (Uralite 3125 Part A)

1.15 (Estane 58091; Millathane 300; Q-Thane PE-55)

1.16 (Q-Thane PE-23, PE-50)

1.17 (Q-Thane P-49, P-280, P-455, PE-36)

1.18 (Estane 58370; Q-Thane PA-01, PA-06, PA-58, PA-80, PC-58, PE-47, PS-82; Uralite 3115 Part B, 3167 Part B)

1.19 (Millathane 76; Q-Thane PA-10, PA-11, PA-20, PA-30, PA-40, PS-16, PS-62, PS-63, PS-65, PS-94; Uralite 3124)

1.20 (Conathane RN-3039; Q-Thane P-440, PE-90, PI-86, PI-95, PI-186, PI-195)

1.2–1.6 (Spectrim MM300)

1.21 (Estane 58013, 58121, 58130, 58133, 58134, 58136, 58137, 58271, 58277; Q-Thane P-360, PI-76, PI-96, PI-176, PI-196)

1.22 (Estane 58092, 58109; Ren:C:O-Thane TDT-178-34)

1.23 (Ren:C:O-Thane TDT-178-53)

1.24 (Conathane RN-3038)

1.26 (Stat-Kon T)

1.28 (Estane 58360)

1.29 (RTP 1201-80D)

1.3–1.8 (Spectrim MM353)

1.37 (RTP 1203-80D; Thermocomp TF-1004)

1.40 (cured) (Conathane TU-4010)

1.44 (Conathane TU-4010 Part B)

1.45 (RTP 1205-80D)

1.46 (Thermocomp TF-1006)

1.54 (RTP 1207-80D)

1.55 (Thermocomp TF-1008)

1.74 (cured) (Uralite 3175 Fastset)

Sp. Vol.:

16.4 g/in.^{3} (P.U.R.E.-CMC)

17.9 in.^{3}/lb (Thermocomp TF-1008)

19.0 in.^{3}/lb (Thermocomp TF-1006)

20.2 in.^{3}/lb (Thermocomp TF-1004)

22.5 in.^{3}/lb (Ren:C:O-Thane TDT-178-53)

22.7 in.^{3}/lb (Ren:C:O-Thane TDT-178-34)

24 in.^{3}/lb (Purelast 208, 209)

24.5 in.^{3}/lb (Ren:C:O-Thane RP-6405)

25 in.^{3}/lb (Purelast 207)

25.2 in.^{3}/lb (Ren:C:O-Thane RP-6403)

25.4 in.^{3}/lb (Ren:C:O-Thane TDT-186-1)
25.6 in.^{3}/lb (Ren:C:O-Thane RP-6402)
25.9 in.^{3}/lb (Ren:C:O-Thane RP-6401)
26 in.^{3}/lb (Purelast 204)
26.4 in.^{3}/lb (Ren:C:O-Thane RP-6414)
26.6 in.^{3}/lb (Ren:C:O-Thane RP-6400, RP-6410, RP-6413, RP-6422)

Density:

8.76 g/cc (Uralite 3111 Part A, 3113 Part A)
8.84 g/cc (Uralite 3111 Part B, 3113 Part B)
7.9 lb/gal (NeoRez U-800, U-912)
8.1 lb/gal (NeoRez U-105, U-110)
8.5 lb/gal (mixed) (Purelast 223, 224, 226, 228, 245, 245H, 240H, 241, 241H)
8.6 lb/gal (Mira-Glos RT A and C; NeoRez U-760); (mixed) (Purelast 220, 221, 242, 243, 251)
8.64 lb/gal (Uralite 3150 Part A, 3152 Part A)
8.7 lb/gal (NeoRez EX-466)
8.8 lb/gal (NeoRez R-961); (mixed) (Purelast 254)
8.9 lb/gal (mixed) (Purelast 234, 247)
9.0 lb/gal (mixed) (Purelast 235, 249, 253)
9.1 lb/gal (mixed) (Purelast 255)
9.1 ± 0.1 lb/gal (Unoflex 100)
9.27 lb/gal (Uralite 3150 Part B, 3152 Part B)
9.8 lb/gal (Chempol 34-4199)
10.3 lb/gal (Chempol 34-5075)
0.039 lb/in.^{3} (cured) (Conathane UC-33; Uralite 3150, 3152)
0.0505 lb/in.^{3} (cured) (Conathane TU-4010)
0.5 pcf (Stepanfoam PR-5, PR-5-O)
1.2 pcf (Presto-Foam 800, 805)
1.5 pcf (Stepanfoam BX-150-5, BX-359, BX-369, PF-15)
1.6 pcf (Trymer 160)
1.7 pcf (Stepanfoam HC-17/40)
1.7–1.8 (Stepanfoam R-231G/R-110, R-244, R-246)
1.9 pcf (Stepanfoam HC-2/40; Trymer 190)
1.9–2.0 pcf (Stepanfoam R-222/R-109, R-223/R-110, R-226/R-112, R-245, R-247)
2.0 pcf (Stepanfoam BX-289, BX-351, BX-352, BX-352P, BX-352M, BX-370, BX-372-2, H-102N, H-402N, H-602N, P-502, SX-219; Trymer 9501)
2–9 pcf (Stepanfoam H-100 Series)
2–10 pcf (Stepanfoam G Series)
2–20 pcf (Stepanfoam BX-105 Series)
2.1 pcf (Trymer 210)
2.2 pcf (Presto-Foam 900)
2.3 pcf (Stepanfoam SX-217, SX-218A)
2.5 pcf (Stepanfoam FX-250, SX-209J)

Polyurethane *(cont'd.)*

2.5+ pcf (Stepanfoam F-202, F-302)
2.7 pcf (Stepanfoam SX-214)
3.0 pcf (Stepanfoam BX-316-3, BX-350-7, SF-3; Trymer 190-3, 9501-3)
3.5 pcf (RestEasy Foam)
3.5+ pcf (Stepanfoam F-403)
4.0 pcf (Stepanfoam AX-64; Trymer 190-4, 9501-4)
4–7 pcf (Stepanfoam F-506)
4.5 pcf (Presto-Foam 945; Stepanfoam SX-195A, SX-202)
5.0 pcf (Stepanfoam BX-316-5, SF-5/60, SX-215, SX-216)
5.0–20 pcf (Stepanfoam C-600 Series)
5.5 pcf (Stepanfoam SX-211)
6.0 pcf (Presto-Foam 960; Stepanfoam P-506; Trymer 9501-6)
6.7–34 pcf (Specflex)
7+ pcf (Stepanfoam BX-364)
8.0 pcf (Stepanfoam AX-66, HW-8/25, HW-8/50)
10 pcf (Stepanfoam A-210, BX-341B, BX-345, HC-9/10 Series, HW-10/25)
11 pcf (Stepanfoam HW-11/60)
12 pcf (Stepanfoam BX-341, BX-341A, HW-12/25)
15 and 20 pcf (Poron 4701-01)
15, 20, 30 pcf (Poron 4701-05)
16 pcf (Stepanfoam HW-16/60)
20 pcf (Poron 4716-16; Stepanfoam HW-20/60)
20+ pcf (Stepanfoam MW-20/20)
20, 25 pcf (Poron 4701-09)

Visc.:

0.2 ± 0.1 Pa•s (15% sol'n. in MEK) (Desmocoll 110, 130)
0.6 ± 0.2 Pa•s (15% sol'n. in MEK) (Desmocoll 176, 400, 406, 500)
1.2 ± 0.2 Pa•s (15% sol'n. in MEK) (Desmocoll 420, 510, 530)
50 cps (Ren:C:O-Thane TDT-178-34 (resin))
80–135 cps (Mira-Glos RT C)
85 cps (Conathane TU-4010 Part A)
100 cps (NeoRez U-105, U-110, U-760, U-800)
100–200 cps (20% in MEK) (Estane 5715)
140–300 cps (15% in MEK) (Q-Thane P-279)
150–250 cps (Rezolin 164 Part B)
200 cps (Chempol 34-5075)
250 ± 50 cps (Conathane UC-33 Part B)
300 cps (Chempol 34-4199; NeoRez EX-466)
300–500 cps (15% in THF) (Estane 5702, 5703; Q-Thane P-250-1)
300–700 cps (15% in THF) (Estane 5701 F1); (15% in MEK) (Q-Thane P-280)
300–800 cps (15% in THF) (Q-Thane PA-10)
386 cps (Uralite 3115 Part B)
400 cps (NeoRez R-961)

400 ± 50 cps (Conathane UC-33 Part A)
400–800 cps (15% in THF) (Estane 5710 F1)
420–780 cps (15% in DMF) (Estane 5708 F1)
500–800 cps (15% in THF) (Estane 5702 F1, 5703 F1)
500–1000 cps (15% in MEK) (Q-Thane PA-29)
510 cps (Uralite 3111 Part B)
600 cps (NeoRez U-912)
600–800 cps (mixed) (Rezolin 164)
600–1000 cps (20% in MEK) (Q-Thane PA-11)
600–1200 cps (15% in THF) (Estane 5714 F1; Q-Thane P-250-2)
670–1290 cps (15% in DMF) (Estane 5707 F1)
725 cps (Hexcel 164M Part B)
750 cps (100 C) (Conathane RN-3039)
800–1200 cps (15% in THF) (Estane 5702 F2, 5703 F2; Q-Thane PA-20); (15% in MEK) (Q-Thane PA-01, PA-06, PA-58)
800–1300 cps (15% in THF) (Estane 5713)
900 cps (100 C) (Adiprene LW-520, LW-550, LW-570)
1000–1500 cps (Rezolin 164 Part A)
1200 cps (100 C) (Conathane RN-3038)
1200–2000 cps (15% in THF) (Q-Thane PA-30)
1600 cps (Hexcel 164M mixed)
1700 cps (100 C) (Adiprene LW-510)
1750 cps (mixed) (P.U.R.E.-CMC)
1800–3000 cps (20% in MEK) (Estane 5711)
2000 cps (Uralite 3113 Part B); (mixed) (Rezolin 170, 185N; Uralite 3121S, 3130, 3132)
2000–2400 cps (Ren:C:O-Thane TDT-178-34 (hardener))
2000–3000 cps (15% in THF) (Q-Thane PA-40)
2100–3800 cps (20% in MEK) (Estane 5712)
2200–4000 cps (20% in MEK) (Estane 5716)
2350 cps (Uralite 3175 Fastset Parts A and B)
2500 cps (UR 103, 104)
2500–3500 cps (20% in MEK) (Q-Thane PA-80)
2500–4000 cps (Mira-Glos RT A)
2600 cps (mixed) (Uralite 7250, 7252)
2700 cps (Hexcel 164M Part A; UR 105)
3000 cps (mixed) (Purelast 204; Uralite 6108)
3100 cps (mixed) (Uralite 3111)
3300 cps (100 C) (Adiprene LW-500); (mixed) (Uralite 3115, 3167)
3800 cps (mixed) (Uralite 3140, 3156, 3158)
3900 cps (Uralite 3115 Part A)
4000–8000 cps (Lamal 408-40(65); Unoflex)
4400 cps (Uralite 3111 Part A, 3113 Part A)

Polyurethane *(cont'd.)*

4800 cps (mixed) (Uralite 3113)
5000 cps (UR 101)
5200 cps (mixed) (Uralite 3154, 3155)
5500 cps (mixed) (Uralite 3150, 3152)
5700 cps (mixed) (Uralite 3121)
5800 cps (mixed) (Uralite 3125)
6000 cps (Adiprene L-167, L-367); (mixed) (Uralite 3122)
6000–10,000 cps (Lamal HSA)
6400 cps (mixed) (Uralite 3124)
6500 cps (UR 102)
7000 cps (Adiprene L-767); (mixed) (Stycast CPC-16)
7500–15,000 cps (50 C) (Adiprene BL-16)
8000 cps (Adiprene L-200)
8500 cps (Adiprene L-700)
10,000 cps (Adiprene L-213)
10,000–16,000 cps (Lamal C)
12,000 cps (Conathane TU-4010 Part B)
13,000 cps (mixed) (Stycast CPC-17)
15,000 cps (Adiprene L-315)
17,000 cps (Adiprene L-42)
18,000 cps (Adiprene L-100)
20,000 cps (Adiprene L-300, L-325)
28,000 cps (Adiprene L-83)
40,000–80,000 cps (Unoflex 100)
45,000 cps (mixed) (Uralite 3139)
70,000 cps (mixed) (Purelast 207)
95,000 cps (mixed) (Purelast 208)
125,000 cps (mixed) (Purelast 209)
Mooney 5–15 (ML1+4, 100 C) (Millathane 80)
Mooney 20–55 (ML1+4, 100 C) (Millathane 76)
Mooney 25–55 (ML1+4, 100 C) (Millathane HT)
Mooney 35–55 (ML1+4, 100 C) (Millathane E-34)
Mooney 35–65 (ML1+4, 100 C) (Millathane 300)
Mooney ≈ 60 (MS10, 100 C) (Adiprene CM)
M.P.:
1 C (CAPA Monomer)
15 C (CAPA 200)
25 C (CAPA 205)
26 C (CAPA 520)
37 C (CAPA 210, 212)
40 C (CAPA 215)
47 C (CAPA 220, 222, 223)
52 C (CAPA 231)

55 C (CAPA 240)
58 C (CAPA 600, 600M, 601, 601M)
Flash Pt.:
> 204 C (OC) (Millathane 76, 80, 300, E-34, HT)
81 F (PM) (NeoRez U-110)
83 F (PM) (NeoRez U-800)
84 F (PM) (NeoRez U-105)
87 F (PM) (NeoRez U-760)
95 F (PM) (NeoRez U-912)
106 F (PM) (NeoRez EX-466)
> 212 F (PM) (NeoRez R-961)
Hydroxyl No.:
2 (CAPA 601, 601M)
5 (CAPA 600, 600M)
28 (CAPA 240)
36 (CAPA 231)
56 (CAPA 220, 222, 223, 520)
90 (CAPA 215)
112 (CAPA 210, 212)
135 (CAPA 205)
204 (CAPA 200)
310 (CAPA 305)
540 (CAPA 304)
Stability:
Outstanding resistance to abrasion, good resistance to heat deterioration, ozone cracking, weathering, and swelling in oils or solvents (Adiprene CM)
Unaffected by high humidities or high ambient temps. (Lamal 408-40(65)/Lamal C; Lamal HSA/Lamal C)
Excellent ozone, weathering, oil and fuel resistance (Millathane HT)
Very good ozone, weathering, oil, and fuel resistance (Millathane 80, 300, E-34)
Good ozone and weathering resistance; very good oil and fuel resistance (Millathane 76)
Excellent abrasion and scuff resistance (NeoRez U-105)
Good chemical resistance (NeoRez U-110)
Good uv resistance (NeoRez U-760)
Good resistance to dry heat aging (Rezolin 185N)
Resistant to aq. sol'ns. at ambient temps. of 20% acetic acid, 20% sulfuric acid, 20% hydrochloric acid, 20% sodium hydroxide, 17% ammonia, 1% sodium hypochlorite, and 20% EDTA tetrasodium salt (Purelast 204, 207, 208, 209)
Resistant to water to 220 F; resistant at ambient temps. to acids (20% aq. acetic, sulfuric, and hydrochloric), alkalis (20% caustic and 17% ammonia), and 1% bleach; resistant to aliphatic hydrocarbons (gasoline, jet fuel, oil, and grease); aromatic hydrocarbons cause swelling and softening (Purelast 220, 221, 223, 224,

Polyurethane *(cont'd.)*

226, 228, 242, 243, 245, 245H, 234, 235, 240H, 241, 241H, 247, 249, 251, 253, 254, 255)

Resistant to aliphatic hydrocarbons and drycleaning solvents (Q-Thane P-280)

Unaffected by uv; abrasion resistant (NeoRez U-800, U-912)

Unaffected by mild inorganic acids and bases; modest swelling with oils, greases, and other linear hydrocarbons; strongly polar solvents will greatly swell product; good resistance to bacteria; antifungicidal; excellent ozone resistance (Poron 4701-01, 4701-05, 4701-09, 4716-16)

Cured rubber has good heat resistance to 100 C, negligible shrinkage in storage and use, and good cold flow resistance (P.U.R.E.-CMC)

Storage Stability:

Excellent raw polymer storage stability @ R.T. (Adiprene BL-16)

Excellent storage stability (Adiprene CM; Millathane 76, 80, 300, E-34, HT; NeoRez U-760)

3 mos shelf life when stored in original sealed containers below 75 F (Unoflex 100)

6 mos storage life (Unoflex)

12 mos shelf life (Conathane TU-4010)

12 mos unopened shelf life @ R.T. (Conathane UC-33 Parts A and B)

Ref. Index:

1.452 (Conathane UC-33 Part B)

1.485 (cured) (Conathane UC-33)

1.497 (Conathane UC-33 Part A)

pH:

7.5–8.5 (NeoRez R-961)

8.5–9.5 (NeoRez EX-466)

MECHANICAL PROPERTIES:

Tens. Str.:

20.7 MPa (cured) (Adiprene L-42)

27.6 MPa (cured) (Adiprene L-767)

30.3 MPa (cured) (Adiprene L-83)

31.0 MPa (Millathane HT); (cured) (Adiprene L-100)

32.4 MPa (cured) (Adiprene L-700, LW-520)

34.5 MPa (cured) (Adiprene L-167)

37.2 MPa (cured) (Adiprene BL-16)

37.9 MPa (cured) (Adiprene LW-500, LW-510)

40.0 MPa (cured) (Adiprene LW-550)

43.3 MPa (cured) (Adiprene LW-570)

57.2 MPa (cured) (Adiprene L-200)

60.7 MPa (cured) (Adiprene L-213)

62.0 MPa (cured) (Adiprene L-315)

64.3 MPa (cured) (Adiprene L-325)

193 kg/cm^2 (Uralite 3130, 3132)

281–352 kg/cm^2 (Adiprene CM)

299.6 kg/cm^2 (Uralite 3124)
315 kg/cm^2 (Uralite 3121)
336 kg/cm^2 (Uralite 3122)
350 kg/cm^2 (Uralite 3121S)
8.8 psi (3.5 pcf) (RestEasy Foam)
40 psi (20 pcf) (Poron 4701-09, 4716-16)
43 psi (2.0 pcf) (Chempol 34-4199/34-5075)
55 psi (15 pcf) (Poron 4701-01)
60 psi (25 pcf) (Poron 4701-09)
90 psi (20 pcf) (Poron 4701-01)
95 psi (15 pcf) (Poron 4701-05)
120 psi (UR 105)
130 psi (UR 103, 104)
150 psi (20 pcf) (Poron 4701-05)
200 psi (Uralite 3128)
225 psi (Purelast 228, 245, 245H, 251)
230 psi (30 pcf) (Poron 4701-05)
280 psi (Purelast 204, 223, 224, 226, 240H, 241, 241H)
305 psi (Purelast 220, 221, 242, 243)
400 psi (P.U.R.E.-CMC)
510 psi (cured) (Ren:C:O-Thane RP-6410)
550 psi (Uralite 6108)
700 psi (Uralite 3111)
700–1000 psi (cured) (Conathane TU-4010)
750 psi (Purelast 207); (cured) (Ren:C:O-Thane RP-6400)
770 psi (UR 101)
800 psi (Calthane ND 2300; Uralite 3113)
855 psi (Purelast 208, 234, 247, 253, 254)
1025 psi (Uralite 3140)
1285 psi (Purelast 209, 235, 249, 255)
1450 psi (cured) (Ren:C:O-Thane TDT-178-34)
1500 psi (Spectrim 15)
1520 psi (UR 102)
1600 psi (cured) (Ren:C:O-Thane RP-6422)
1800 psi (Uralite 3115, 3167)
2000 psi (cured) (Ren:C:O-Thane RP-6401)
2100 psi (Spectrim 5)
2200 psi (Uralite 3150, 3512); (cured) (Ren:C:O-Thane RP-6402)
2300 psi (Calthane ND 1100)
2500 psi (cured) (Ren:C:O-Thane RP-6403, RP-6413)
2600 psi (Spectrim SP 400; Uralite 7250, 7252); (cured) (Ren:C:O-Thane RP-6414)
2700–3300 psi (Q-Thane PA-10)
2800 psi (Spectrim 25S); (break) (Q-Thane P-250)

Polyurethane *(cont'd.)*

2900 psi (Spectrim 25W; Uralite 3156, 3158)
3000 psi (Spectrim 35W; Stat-Kon T; Uralite 3139)
3000–3500 psi (Q-Thane PA-11)
3100 psi (Spectrim 35S, 50S; Uralite 3154, 3155)
3200 psi (Spectrim 50W)
3200–3500 psi (Q-Thane PA-80)
3300–3600 psi (Q-Thane PA-20)
3600–4000 psi (Q-Thane PA-30)
3910 psi (Uralite 3175 Fastset)
4000 psi (Millathane 80, 300, E-34)
4000–4500 psi (Q-Thane PA-40)
4500 psi (Spectrim Polyurea HT)
4700 psi (cured) (Ren:C:O-Thane TDT-186-1)
4800 psi (Spectrim Polyurea HF)
5000 psi (Millathane 76; Q-Thane PE103-100)
5300 psi (cured) (Ren:C:O-Thane RP-6405)
6000 psi (Q-Thane PE-49, PE-55, PE-74, PE192-100, PE-192-101)
6100 psi (Spectrim SF 500)
6300 psi (Calthane ND 3200)
6500 psi (Q-Thane PE-23, PE-50)
6800 psi (Thermocomp TF-1004)
7000 psi (Q-Thane P-360, PE-47, PE-90, PN03-100)
7200 psi (Q-Thane PI-196); (cast film) (Q-Thane P-279)
7500 psi (Q-Thane PI-76, PI-86, PI-96, PI-176, PI-186)
7600 psi (cast film) (Q-Thane P-280)
8000 psi (Q-Thane PS-82; Uralite 3125)
8200 psi (Thermocomp TF-1006)
8500 psi (Q-Thane P-49, PC-58, PE-36, PS-63); (break) (cured) (Conathane RN-3039)
8600 psi (break) (cured) (Conathane RN-3038)
9000 psi (Q-Thane P-440, P-455, PN3429-100, PS-16, PS-62, PS-65)
9500 psi (RTP 1201-80D); (cured) (Conathane UC-33)
9600 psi (Thermocomp TF-1008)
10,000 psi (Q-Thane PI-195, PS-94)
10,100 psi (cured) (Ren:C:O-Thane TDT-178-53)
10,000–20,000 psi (Spectrim MM300)
10,000–30,000 psi (Spectrim MM353)
11,000 psi (Q-Thane PI-95)
14,000 psi (RTP 1023-80D)
17,000 psi (RTP 1205-80D)
19,000 psi (RTP 1207-80D)
Tens. Elong.:
3.3% (RTP 1203-80D)
3.5% (RTP 1205-80D)

3.8% (RTP 1207-80D)
6.7% (cured) (Ren:C:O-Thane TDT-178-53)
7.28% (cured) (Ren:C:O-Thane TDT-178-34)
8% (RTP 1201-80D)
10% (cured) (Calthane ND 3200)
13.3% (cured) (Ren:C:O-Thane RP-6405)
15% (Uralite 3125); (yield) (Q-Thane P-250); (break) (cured) (Conathane UC-33)
20% (Thermocomp TF-1008); (break) (Spectrim Polyurea HT)
25% (Thermocomp TF-1006)
28% (break) (Spectrim Polyurea HF)
30% (Thermocomp TF-1004)
50% (cured) (Ren:C:O-Thane TDT-186-1)
85% (Uralite 6108)
100% min. (Poron 4701-01, 4701-05, 4701-09, 4716-16)
100–110% (3.5 pcf) (RestEasy Foam)
130% (UR 105; Uralite 3156, 3158); (cured) (Calthane ND 1100)
140% (UR 103, 104)
150% (Stat-Kon T)
180% (cured) (Adiprene LW-570; Calthane ND 2300)
190% (break) (Spectrim 50S)
200% (Q-Thane PI-196; Uralite 3124, 3139); (break) (Spectrim 15)
210% (Uralite 3150, 3152); (break) (Spectrim 50W); (cured) (Adiprene L-315)
220% (Uralite 3154, 3155)
225% (Q-Thane PI-96; UR 101; Uralite 7250, 7252)
235% (break) (Spectrim 35W)
240% (break) (Spectrim 35S); (cured) (Adiprene L-213; Ren:C:O-Thane RP-6403)
250% (Uralite 3121, 3122, 3130, 3132)
260% (break) (Spectrim 25S, 25W); (cured) (Adiprene L-325)
290% (UR 102)
295% (cast film) (Q-Thane P-280)
300% (Uralite 3121S); (cured) (Ren:C:O-Thane RP-6414)
320% (cured) (Adiprene L-200, L-767; Ren:C:O-Thane RP-6402)
330% (cured) (Ren:C:O-Thane RP-6410)
340% (cured) (Adiprene LW-550)
350% (Purelast 209, 235, 249, 255); (break) (Spectrim 5); (cured) (Adiprene LW-520)
360% (Ren:C:O-Thane RP-6401)
365% (Purelast 208, 234, 247, 253, 254)
375% (Q-Thane P-360)
400% (Purelast 207; Q-Thane PI-76, PN3429-100); (cured) (Adiprene L-167, LW-510; Ren:C:O-Thane RP-6414); (cast film) (Q-Thane P-279)
425% (Q-Thane PE-47, PI-176)
430% (cured) (Adiprene L-700, LW-500)
450% (Q-Thane P-440, PE-36, PE103-100, PI-186, PN03-100); (cured) (Adiprene L-

Polyurethane *(cont'd.)*

    100; Ren:C:O-Thane RP-6400, RP-6422)
   455% (break) (cured) (Adiprene BL-16)
   475% (Q-Thane PI-86, PS-63)
   500% (Millathane 76; Q-Thane P-49, PE192-100, PE192-101, PI-195, PS-62, PS-65, PS-94; Uralite 3115, 3167)
   500–600% (Uralite 3128)
   525% (Q-Thane PS-16)
   550% (Millathane 300; Q-Thane PE-50, PE-90, PI-95)
   580% (cured) (Adiprene L-83)
   600% (Purelast 204, 228, 245, 245H, 251; Q-Thane PC-58, PE-23, PE-49, PE-55; Uralite 3113); (break) (cured) (Conathane TU-4010)
   615% (break) (Q-Thane P-250)
   650% (Q-Thane P-455, PE-74, PS-82); (break) (cured) (Conathane RN-3039)
   690% (break) (Q-Thane PA-11)
   700% (P.U.R.E.-CMC; Uralite 3140); (break) (cured) (Conathane RN-3038)
   750% (break) (Q-Thane PA-30)
   800% (break) (Q-Thane PA-20); (cured) (Adiprene L-42)
   850% (Uralite 3111); (break) (Q-Thane PA-10)
   950% (Purelast 220, 221, 242, 243)
   1025% (Purelast 223, 224, 226, 240H, 241, 241H)

Tens. Mod.:
   600,000 psi (RTP 1201-80D)
   800,000 psi (RTP 1203-80D)
   1.05×10^6 psi (RTP 1205-80D)
   1.30×10^6 psi (RTP 1207-80D)

100% Mod.:
   2.8 MPa (cured) (Adiprene L-42)
   4.8 MPa (cured) (Adiprene L-83, LW-500)
   7.6 MPa (cured) (Adiprene L-100)
   7.8 MPa (cured) (Adiprene LW-520)
   8.1 MPa (cured) (Adiprene LW-510)
   9.0 MPa (cured) (Adiprene L-700)
   11.4 MPa (cured) (Adiprene BL-16)
   12.4 MPa (cured) (Adiprene L-167, L-767)
   12.8 MPa (cured) (Adiprene LW-550)
   20.7 MPa (cured) (Adiprene L-200)
   25.8 MPa (cured) (Adiprene L-325)
   26.9 MPa (cured) (Adiprene L-213)
   29.5 MPa (cured) (Adiprene LW-570)
   33.4 MPa (cured) (Adiprene L-315)
   80 psi (cast film) (Q-Thane P-279)
   100–150 psi (cured) (Conathane TU-4010)
   500 psi (Q-Thane PE-49)

550 psi (Q-Thane P-49, PS-82)
575 psi (Q-Thane P-455)
600 psi (Q-Thane PC-58)
700 psi (Q-Thane PE-74, PS-16)
760 psi (cast film) (Q-Thane P-280)
800 psi (Q-Thane PE-23, PE-55); (cured) (Conathane RN-3038)
900 psi (Q-Thane PE192-100, PE192-101, PN3429-100)
1000 psi (Q-Thane PE-50)
1050 psi (Q-Thane PE103-100)
1200 psi (Q-Thane PE-90, PI-86, PI-95, PI-186, PI-195, PN03-100, PS-62)
1300 psi (Q-Thane PS-63)
1400 psi (Q-Thane PS-65); (cured) (Conathane RN-3039)
1500 psi (Q-Thane PE-36, PS-94)
1800 psi (Q-Thane P-440)
2500 psi (Q-Thane PI-76, PI-96, PI-196)
2600 psi (Q-Thane PE-47)
2700 psi (Q-Thane PI-176)
3000 psi (Q-Thane P-360)

Flex. Str.:
4300 psi (Thermocomp TF-1004)
4840 psi (cured) (Conathane UC-33)
5600 psi (Thermocomp TF-1006)
6854 psi (Uralite 3175 Fastset)
6900 psi (Thermocomp TF-1008)
15,000 psi (RTP 1201-80D)
19,000 psi (RTP 1203-80D)
23,000 psi (RTP 1205-80D)
25,000 psi (RTP 1207-80D)

Flex. Mod.:
586 MPa (cured) (Adiprene L-325)
662 MPa (cured) (Adiprene L-315)
689 MPa (cured) (Adiprene LW-570)
800 MPa (cured) (Adiprene L-213)
500–1,000,000 psi (Spectrim MM300)
500–2,000,000 psi (Spectrim MM353)
4700 psi (Spectrim 5)
15,000 psi (Spectrim 15)
25,000 psi (Spectrim 25S, 25W)
35,000 psi (Spectrim 35S, 35W)
50,000 psi (Spectrim 50S, 50W)
125,000 psi (Thermocomp TF-1004)
140,000 psi (Spectrim SP 400)
190,000 psi (Thermocomp TF-1006)

Polyurethane *(cont'd.)*

200,000 psi (Stat-Kon T)
240,000 psi (Spectrim Polyurea HF, Polyurea HT)
260,000 psi (Spectrim SF 500; Thermocomp TF-1008)
357,324 psi (cured) (Conathane UC-33)
450,000 psi (RTP 1201-80D)
620,000 psi (RTP 1203-80D)
800,000 psi (RTP 1205-80D)
10^6 psi (RTP 1207-80D)

Compr. Str.:
22 psi (Trymer 160)
25 psi (Trymer 190)
28 psi (Trymer 9501)
30 psi (Trymer 210)
35 psi (2.0 pcf, yield) (Chempol 34-4199/34-5075)
50 psi (Trymer 9501-3)
52 psi (Trymer 190-3)
83 psi (Trymer 190-4)
90 psi (Trymer 9501-4)
160 psi (Trymer 9501-6)
7000 psi (RTP 1201-80D)
8003 psi (Uralite 3175 Fastset)
10,000 psi (RTP 1203-80D)
12,000 psi (RTP 1205-80D)
13,000 psi (RTP 1207-80D)

Tear Str.:
45 kg/cm (Uralite 31301, 3132)
53.7 kg/cm (Uralite 3121)
62.6 kg/cm (Uralite 3122)
71.2–89 kg/cm (Adiprene CM)
71.5 kg/cm (Uralite 3121S)
134.2 kg/cm (Uralite 3124)
9.6 kN/m (cured) (Adiprene LW-500, LW-520)
12.2 kN/m (cured) (Adiprene L-42)
12.3 kN/m (cured) (Adiprene LW-510)
13.1 kN/m (cured) (Adiprene L-100)
14.9 kN/m (cured) (Adiprene L-83)
16.6 kN/m (cured) (Adiprene LW-550)
17.5 kN/m (cured) (Adiprene L-700)
18.4 kN/m (cured) (Adiprene LW-570)
19.2 kN/m (cured) (Adiprene L-315)
22.7 kN/m (cured) (Adiprene L-767)
23.6 kN/m (cured) (Adiprene L-200)
25.4 kN/m (cured) (Adiprene L-213)

26.2 kN/m (cured) (Adiprene L-167)
1.0–1.3 pli (3.5 pcf) (RestEasy Foam)
30 pli (cured) (Calthane ND 2300)
35 pli (Purelast 223, 224, 226, 240H, 241, 241H)
40 pli (P.U.R.E.-CMC; Purelast 204)
42 pli (Purelast 220, 221, 242, 243)
47 pli (Purelast 228, 245, 245H, 251)
50 pli (Uralite 3128, 6108)
65 pli (Purelast 207)
70 pli (Purelast 208); (cured) (Ren:C:O-Thane RP-6400)
71 pli (Purelast 234, 247, 253, 254)
75 pli (cured) (Ren:C:O-Thane RP-6410)
98 pli (Uralite 3111)
100–150 pli (cured) (Conathane TU-4010)
110 pli (Uralite 3113)
115 pli (Purelast 209, 235, 249, 255); (cured) (Ren:C:O-Thane TDT-178-34)
145 pli (cured) (Ren:C:O-Thane RP-6401)
150 pli (cured) (Ren:C:O-Thane RP-6405)
180 pli (cured) (Ren:C:O-Thane RP-6402)
200 pli (Uralite 3115, 3167)
220 pli (cured) (Ren:C:O-Thane RP-6422)
230 pli (cured) (Calthane ND 1100)
250 pli (Spectrim 15)
260 pli (Uralite 3140)
280 pli (Uralite 3139)
295 pli (Uralite 3154, 3155)
300 pli (Spectrim 5)
302 pli (Uralite 3150, 3152)
325 pli (Q-Thane PE-49)
350 pli (cured) (Ren:C:O-Thane RP-6413)
385 pli (Uralite 7250, 7252)
400 pli (Q-Thane P-49)
410 pli (Uralite 3156, 3158)
425 pli (Q-Thane PE-55, PE-74)
450 pli (Q-Thane PE-23, PE-50, PS-82)
470 pli (cured) (Ren:C:O-Thane RP-6403)
475 pli (Spectrim 25S)
490 pli (Spectrim 25W)
500 pli (Q-Thane PE-90; Spectrim 35S); (cured) (Conathane RN-3038; Ren:C:O-Thane RP-6414, TDT-186-1)
520 pli (Spectrim 35W)
540 pli (Spectrim 50W)
550 pli (Q-Thane PE-36)

Polyurethane *(cont'd.)*

600 pli (cured) (Conathane RN-3039)
620 pli (Spectrim 50S)
650 pli (Q-Thane PE-47)
700 pli (Q-Thane PI-86, PI-95, PI-186, PI-195, PS-16, PS-62, PS-65)
725 pli (Q-Thane PS-63)
750 pli (Q-Thane PS-94)
850 pli (Q-Thane PI-76, PI-176, PI-196)
900 pli (Q-Thane PI-96; Uralite 3125)
290 pit (cast film) (Q-Thane P-279)
400 pit (Q-Thane P-250)
630 pit (cast film) (Q-Thane P-280)

Shear Str.:
4300 psi (Thermocomp TF-1004)
4800 psi (Thermocomp TF-1006)
5400 psi (Thermocomp TF-1008)

Impact Str. (Izod):
640 J/m notched (cured) (Adiprene L-315)
750 J/m notched (cured) (Adiprene LW-570)
800 J/m notched (cured) (Adiprene L-213)
1.3 ft lb/in. notched (RTP 1201-80D)
1.5 ft lb/in. notched (RTP 1203-80D)
2.0 ft lb/in. notched (RTP 1205-80D)
2.5 ft lb/in. notched (Spectrim SP 400)
3.0 ft lb/in. notched (RTP 1207-80D)
4.2 ft lb/in. notched (Spectrim SF 500)
5–12 ft lb/in. notched (Spectrim MM300)
5–14 ft lb/in. notched (Spectrim MM353)
7.5 ft lb/in. notched (Stat-Kon T)
9.0 ft lb/in. notched (Thermocomp TF-1008)
9.5 ft lb/in. notched (Thermocomp TF-1006)
10 ft lb/in. notched (Thermocomp TF-1004)
> 32 ft lb/in. notched (UR 101, 102, 103, 104, 105)
Very high (Stycast CPC-16, CPC-17)
No break (Q-Thane P-49, P-360, P-440, P-455, PC-58, PE-23, PE-36, PE-47, PE-49, PE-50, PE-55, PE-74, PE-90, PI-76, PI-86, PI-95, PI-96, PI-176, PI-186, PI-195, PI-196, PS-16, PS-62, PS-63, PS-65, PS-82, PS-94); (cured) (Adiprene L-325)

Hardness:
Rockwell R114 (RTP 1201-80D)
Rockwell R115 (RTP 1203-80D)
Rockwell R117 (RTP 1205-80D)
Rockwell R119 (RTP 1207-80D)
Shore A7–11 (15 pcf) (Poron 4701-01)
Shore A8–12 (Poron 4716-16)

Polyurethane *(cont'd.)*

Shore A9–14 (20 pcf) (Poron 4701-01)
Shore A10–40 (cured) (Ren:C:O-Thane TDT-178-34)
Shore A11–16 (15 pcf) (Poron 4701-05)
Shore A14–24 (20 pcf) (Poron 4701-05)
Shore A15–95 (Millathane 76)
Shore A25 ± 5 (Uralite 3128)
Shore > A26 (30 pcf) (Poron 4701-05)
Shore A30 (Purelast 223, 224, 226, 240H, 241, 241H; UR 105)
Shore A30–80 (Millathane 80)
Shore A30–95 (Millathane HT)
Shore A35–40 (cured) (Ren:C:O-Thane RP-6410)
Shore A38 (P.U.R.E.-CMC)
Shore A40 (Purelast 220, 221, 242, 243; UR 103, 104); (cured) (Conathane TU-4010)
Shore A40–45 (cured) (Ren:C:O-Thane RP-6400)
Shore A45 (Purelast 204, 228, 245, 245H, 251)
Shore A45–90 (Millathane E-34)
Shore A45–95 (Millathane 300)
Shore A50–60 (Rezolin 164)
Shore A50–90 (Adiprene CM)
Shore A55 (Uralite 3111)
Shore A60 (UR 101; Uralite 3140); (cast film) (Q-Thane P-279)
Shore A60–65 (cured) (Ren:C:O-Thane RP-6401)
Shore A65 (Uralite 3113)
Shore A65 ± 5 (cured) (Calthane ND 2300)
Shore A65–70 (Uralite 6108)
Shore A67 (Hexcel 164M)
Shore A68 (Q-Thane P-280)
Shore A70 (Estane 5702, 5702 F1, 5702 F2, 5703, 5703 F1, 5703 F2; Purelast 207)
Shore A70–75 (Q-Thane P-49, PE-49, PE-74, PH-56)
Shore A75 (Estane 58121; Stycast CPC-16)
Shore A75–80 (Q-Thane P-455, PE-55, PE192-100, PE192-101, PS-82; Rezolin 170); (cured) (Ren:C:O-Thane RP-6402, RP-6422)
Shore A78 (Estane 5710 F1)
Shore A80 (Estane 5714 F1, 58300; Purelast 208, 234, 247, 253, 254; Uralite 3115, 3167); (cured) (Adiprene L-42)
Shore A80–85 (Q-Thane PC-58, PE-23, PE-50; Uralite 3130, 3132)
Shore A82 (Estane 58630)
Shore A83 (cured) (Adiprene L-83, LW-500)
Shore A84 (cured) (Conathane RN-3038)
Shore A85 (Estane 58309, 58311; Q-Thane PA-80; Rezolin 185N)
Shore A85–90 (Q-Thane PE-90, PN3429-100, PS-16, PS-62, PS-63, PS-65; Uralite 3121S); (cured) (Ren:C:O-Thane RP-6403)
Shore A86 (Estane 58013, 58370)

Polyurethane (cont'd.)

Shore A87 (Estane 5708 F1, 58271, 58277; Spectrim 5); (cured) (Adiprene L-300)

Shore A88 (Estane 5701 F1)

Shore A90 (Estane 58109; Purelast 209, 235, 249, 255; UR 102; Uralite 3121, 3139, 3150, 3152, 3154, 3155); (cured) (Adiprene L-100, LW-510, LW-520; Conathane RN-3039)

Shore A90–95 (Q-Thane P-440, PE-36, PH-89, PI-95, PI-195, PN03-100, PS-94); (cured) (Ren:C:O-Thane RP-6413)

Shore A92 (cured) (Adiprene BL-16, L-700)

Shore A92–95 (Q-Thane PA-11)

Shore A93 (Estane 58360)

Shore A93 ± 5 (cured) (Calthane ND 1100)

Shore A95 (Q-Thane PA-10, PA-20, PA-30, PA-40); (cured) (Adiprene L-167, L-367)

Shore A95–100 (Q-Thane P-360, PE-47, PE103-100, PI-86, PI-186)

Shore A96 (Spectrim 15)

Shore D30 (Estane 5716)

Shore D35 (Stycast CPC-17)

Shore D42 (Estane 5711, 5712, 5713)

Shore D45 (Estane 58092, 58134, 58136)

Shore D48 (cured) (Adiprene LW-550)

Shore D50 (Estane 5707 F1, 58130)

Shore D53 (cured) (Adiprene L-767)

Shore D55 (Estane 58133)

Shore D55–60 (cured) (Ren:C:O-Thane RP-6414)

Shore D58 (cured) (Adiprene L-200)

Shore D60 (Thermocomp TF-1004; Uralite 3122, 3156, 3158, 7250, 7252)

Shore D60–65 (Q-Thane PI-76, PI-176)

Shore D63 (Estane 5715)

Shore D65 (Estane 58137; Thermocomp TF-1006)

Shore D65–70 (Q-Thane PI-96, PI-196)

Shore D70 (Estane 58091; Thermocomp TF-1008)

Shore D70–75 (cured) (Ren:C:O-Thane RP-6405)

Shore D72 (cured) (Adiprene L-325)

Shore D73 (cured) (Adiprene L-213, L-315)

Shore D75 (Uralite 3124); (cured) (Adiprene LW-570)

Shore D75–80 (cured) (Ren:C:O-Thane TDT-186-1)

Shore D80 (Uralite 3125, 3175 Fastset); (cured) (Calthane ND 3200)

Shore D85 (cured) (Ren:C:O-Thane TDT-178-53)

Shore D85 ± 5 (cured) (Conathane UC-33)

Sward 28 (NeoRez U-912)

Sward 40 (NeoRez EX-466, U-760)

Sward 44 (NeoRez R-961)

Sward 45 (NeoRez U-105, U-110, U-800)

Mold Shrinkage:
0.70% (cured) (Conathane UC-33)
0.0001 in./in. (cured) (Conathane TU-4010)
0.0008 in./in. (cured) (Ren:C:O-Thane TDT-178-34)
0.001 in./in. (cured) (Ren:C:O-Thane RP-6400, RP-6401, RP-6402, RP-6403, RP-6413, RP-6414, RP-6422)
0.002 in./in. (cured) (Ren:C:O-Thane TDT-186-1)
0.0023 in./in. (cured) (Ren:C:O-Thane TDT-178-53)
0.003 in./in. (cured) (Ren:C:O-Thane RP-6410)
0.0045 in./in. (cured) (Ren:C:O-Thane RP-6405)
0.008 in./in. (Stat-Kon T)

Water Absorp.:
0.39% (cured) (Conathane UC-33)
0.4% (Stat-Kon T)

THERMAL PROPERTIES:

Glass Transition Temp.:
105 C (cured) (Conathane UC-33)

Conduct.:
5.0×10^{-4} Cal/s/cm²/C/cm (cured) (Conathane UC-33)
6.1×10^{-4} Cal/s/cm²/C/cm (Stycast CPC-16, CPC-17)
0.5 Btu/h/ft²/F/in. (Poron 4701-01, 4701-05, 4716-16)
1.8 Btu/h/ft²/F/in. (RTP 1201-80D; Stat-Kon T)
2.5 Btu/h/ft²/F/in. (RTP 1203-80D)
2.6 Btu/h/ft²/F/in. (Thermocomp TF-1004)
2.8 Btu/h/ft²/F/in. (RTP 1205-80D)
2.9 Btu/h/ft²/F/in. (Thermocomp TF-1006)
3.0 Btu/h/ft²/F/in. (RTP 1207-80D)
3.1 Btu/h/ft²/F/in. (Thermocomp TF-1008)

Distort. Temp.:
57 C (264 psi) (Spectrim SP 400)
70 C (264 psi) (Spectrim SF 500)
84 C (cured) (Ren:C:O-Thane TDT-178-53)
85 C (cured) (Adiprene L-213)
110–150 C (264 psi) (Spectrim MM300)
140 C (cured) (Adiprene L-315)
145 C (cured) (Adiprene L-325)
150 F (264 psi) (RTP 1201-80D)
150–250 C (264 psi) (Spectrim MM353)
152 F (Uralite 3175 Fastset)
170 F (264 psi) (RTP 1203-80D)
185 F (264 psi) (RTP 1205-80D)
200 F (264 psi) (RTP 1207-80D)
335 F (264 psi) (Thermocomp TF-1004)

Polyurethane *(cont'd.)*

340 F (264 psi) (Thermocomp TF-1006)
345 F (264 psi) (Thermocomp TF-1008)
Brittle Temp.:
–55 C (Poron 4701-09)
–54 C (Poron 4716-16)
–40 C (Poron 4701-01, 4701-05)
–95 F (Q-Thane P-440, PI-86, PI-95, PS-65, PS-82)
–90 F (Q-Thane P-49, P-455, PC-58, PI-186, PI-195, PS-16, PS-62, PS-63, PS-94)
–80 F (Millathane E-34)
–70 F (Q-Thane P-360, PI-76, PI-96, PI-176, PI-196)
–65 F (Q-Thane PE-49)
–60 F (Millathane 300; Q-Thane PE-23, PE-50, PE-55, PE-74, PE-90)
–59 F (Millathane HT)
–50 F (Q-Thane PE-36, PE-47)
–40 F (Millathane 76)
Coeff. of Linear Exp.:
$1.3–1.8 \times 10^{-4}$/C (Poron 4701-01, 4701-05, 4716-16)
165×10^{-6} cm/cm/C (Stycast CPC-17)
170×10^{-6} cm/cm/C (Stycast CPC-16)
7.75×10^{-5} in./in./C (cured) (Conathane UC-33)
1.4×10^{-5} in./in./F (Thermocomp TF-1008)
2.5×10^{-5} in./in./F (Thermocomp TF-1006)
4.5×10^{-5} in./in./F (Thermocomp TF-1004)
8.5×10^{-5} in./in./F (Stat-Kon T)
Flamm.:
Burns readily unless specifically compounded for flame resistance (Adiprene CM)
Self-extinguishing (Rezolin 164, 170, 185N)
V-0 (Ren:C:O-Thane TDT-178-53)
HB (Poron 4716-16; RestEasy Foam; RTP 1201-80D, 1203-80D, 1205-80D, 1207-80D; Stat-Kon T; Thermocomp TF-1004, TF-1006, TF-1008)
Fire Class I (Trymer 9501, 9501-3, 9501-4)
Fire Class II (Trymer 160, 190, 190-3, 190-4, 210, 9501-6)

ELECTRICAL PROPERTIES:
Dissip. Factor:
0.007 (100 Hz) (cured) (Conathane UC-33)
0.011 (1 MHz) (RTP 1201-80D)
0.012 (1 MHz) (RTP 1203-80D)
0.014 (1 MHz) (RTP 1205-80D)
0.016 (1 MHz) (UR 103, 104)
0.017 (1 MHz) (RTP 1207-80D)
0.022 (1 MHz) (UR 102)
0.028 (1 MHz) (UR 101)
0.029 (1 MHz) (UR 105)

0.04 (1 kHz) (Stycast CPC-16)
0.061 (1 kHz) (Hexcel 164M)
0.08 (100 cps) (Adiprene CM)
0.16 (1 MHz) (Stycast CPC-17)
Dielec. Str.:
52.6 (Poron 4701-01, 4716-16)
58.1 (Poron 4701-05)
206 V/mil (Rezolin 164)
240 V/mil (Rezolin 185N; Uralite 3130, 3132)
355 V/mil (Rezolin 170)
450 V/mil (RTP 1201-80D, 1203-80D, 1205-80D, 1207-80D)
550 V/mil (Stycast CPC-16, CPC-17)
740 V/mil (cured) (Conathane UC-33)
Dielec. Const.:
1.25 ± 0.15 (1 MHz) (Poron 4701-01, 4701-05, 4716-16)
2.90 (1 MHz) (UR 101)
3.04 (1 MHz) (UR 104)
3.06 (1 MHz) (UR 103)
3.12 (1 MHz) (UR 102)
3.15 (1 MHz) (UR 105)
3.37 (100 Hz) (cured) (Conathane UC-33)
4.2 (100 Hz) (Rezolin 170); (1 MHz) (RTP 1207-80D)
4.5 (1 MHz) (RTP 1205-80D)
4.8 (1 MHz) (RTP 1203-80D)
4.99 (1 kHz) (Hexcel 164M)
5.0 (1 MHz) (RTP 1201-80D)
5.6 (1 kHz) (Stycast CPC-16)
6.7 (100 Hz) (Rezolin 164)
7.2 (1 kHz) (Uralite 3130, 3132); (1 MHz) (Stycast CPC-17)
7.4 (1 kHz) (Rezolin 185N)
Vol. Resist.:
1000 ohm-cm (Stat-Kon T)
1.0×10^{11} ohm-cm (RTP 1201-80D, 1203-80D, 1205-80D, 1207-80D)
1.0×10^{12} ohm-cm (Stycast CPC-17; Uralite 3140)
8.0×10^{12} ohm-cm (Uralite 3154, 3155)
1.0×10^{13} ohm-cm (Rezolin 185N; Stycast CPC-16; Uralite 3130, 3132, 3150, 3152)
2.0×10^{13} ohm-cm (Rezolin 164)
3.0×10^{14} ohm-cm (Rezolin 170)
4.0×10^{14} ohm-cm (UR 104)
1.0×10^{15} ohm-cm (Uralite 3121S)
3.0×10^{15} ohm-cm (UR 101)
4.0×10^{16} ohm-cm (UR 105)
5.0×10^{16} ohm-cm (UR 103)

Polyurethane *(cont'd.)*

6.0×10^{16} ohm-cm (cured) (Conathane UC-33)

7.0×10^{16} ohm-cm (UR 102)

Surf. Resist.:

1000 ohm/sq. (Stat-Kon T)

10^{11} ohm (Poron 4701-01, 4701-05, 4716-16)

1.0×10^{13} ohm (Uralite 3154, 3155)

2.0×10^{13} ohm (Rezolin 185N; Uralite 3130, 3132)

2.8×10^{13} ohm (Uralite 3150, 3152)

4.0×10^{14} ohm (Rezolin 164)

5.0×10^{15} ohm (Rezolin 170)

4.1×10^{17} ohm (cured) (Conathane UC-33)

Arc Resist.:

90 s (RTP 1201-80D, 1203-80D, 1205-80D, 1207-80D)

> 120 s (cured) (Conathane UC-33)

CURING CHARACTERISTICS:

Mix Ratio:

1/1 (Mira-Glos RT A and C)

1/1 by volume (Norcast PR-1020)

1A/1B (Calthane ND 1100); by volume (Purelast 254)

2A/1B by volume (Purelast 220, 221, 242, 243, 253, 255)

2A/3B (Calthane ND 2300)

3A/1B by volume (Purelast 223, 224, 226, 228, 240H, 241, 241H, 245, 245H, 251)

3A/2B (Calthane ND 3200)

4A/1B (Purelast 204. 207, 208, 209); by volume (Purelast 234, 235, 247)

10/100 (Ren:C:O-Thane RP-6400 resin/hardener, TDT-178-34 resin/hardener)

10A/100B (Conathane TU-4010)

25/100 (Ren:C:O-Thane RP-6401 resin/hardener)

35/100 (Ren:C:O-Thane RP-6402 resin/hardener)

50/100 (Ren:C:O-Thane RP-6403 resin/hardener)

95A/100B (Chempol 34-4199/34-5075)

100/7.5 (Lamal 408-40(65)/Lamal C)

100A/9B (UR 105)

100A/10B (UR 103, 104)

100A/22B (Uralite 3140)

100A/25B (UR 101)

100A/26B (Uralite 3139, 3150, 3152)

100A/28B (Uralite 3154, 3155)

100A/30B (Uralite 31301, 3132)

100A/32B (Uralite 3124, 3125)

100A/34B (Uralite 3115, 3167)

100A/36B (Uralite 3113, 3156, 3158)

100A/38B (Uralite 7250, 7252)

100A/40B (Uralite 3121S)

100A/46B (Uralite 3111)
100A/46.5B (Conathane UC-33)
100A/50B (UR 102)
100/58 (Ren:C:O-Thane RP-6413 resin/hardener)
100/60 (Ren:C:O-Thane RP-6414 resin/hardener)
100A/60B (Uralite 3122)
100/80 (Ren:C:O-Thane TDT-178-53 resin/hardener)
100A/80B (Uralite 3121)
100A/82B (Uralite 6108)
100/100 (Ren:C:O-Thane RP-6405 resin/hardener, RP-6410 resin/hardener, RP-6422 resin/hardener, TDT-186-1 resin/hardener)
100A/100B (Uralite 3175 Fastset)
100A/200B (Uralite 3128)

Typical/Suggested Curing Agent:
Caytur 21 (Adiprene L-300, L-367)
27.2 phr Caytur 21 (Adiprene L-700)
38.6 phr Caytur 21 (Adiprene L-767)
Dicumyl peroxide or Varox curatives (Millathane 300)
9.5 phr MBOCA (Conathane RN-3038)
12.4 phr MBOCA (Conathane RN-3039)
7.6 phr MDA (Adiprene LW-500)
10.1 phr MDA (Adiprene LW-510)
10.5 phr MDA (Adiprene LW-520)
11.8 phr MDA (Adiprene BL-16)
13.0 phr MDA (Adiprene LW-550)
16.8 phr MDA (Adiprene LW-570)
8.8 phr MOCA (Adiprene L-42)
10.3 phr MOCA (Adiprene L-83)
12.5 phr MOCA (Adiprene L-100)
19.5 phr MOCA (Adiprene L-167)
23.2 phr MOCA (Adiprene L-200)
25.8 phr MOCA (Adiprene L-325)
26.0 phr MOCA (Adiprene L-315)
26.6 phr MOCA (Adiprene L-213)
Sulfur (Adiprene CM)
Sulfur or peroxide-based curatives (Millathane 76, 80, E-34, HT)

Gel Time:
95 s @ 75–80 F (Chempol 34-4199/34-5075)
5 min (Uralite 3175 Fastset)
10 min (Rezolin 185N)
15 min (Rezolin 170)
17 min (Uralite 3156, 3158)
20 min (Uralite 3115, 3167)

Polyurethane *(cont'd.)*

22 min (Hexcel 164M; Uralite 3139)
25 min (Uralite 3150, 3152)
25–30 min (Rezolin 164)
28 min (Uralite 7250, 7252)
30 min (Uralite 3140)
32 min (Uralite 3111)
35–40 min (Uralite 6108)
43 min (Uralite 3154, 3155)
50 min (Uralite 3113)
75 min (Uralite 3125)

Pour Time:
12 min (P.U.R.E.-CMC)

Set Time:
14 min (P.U.R.E.-CMC)

Pot Life:
10 min (Uralite 3122)
14 min (Uralite 3130, 3132)
15 min (Conathane UC-33; Uralite 3121, 3121S)
20-30 min (Conathane TU-4010)
35 min (Uralite 3124)
35 ± 5 min (Uralite 3128)

Working Life:
1 min @ 77 C (Adiprene L-315)
1.5 min @ 85 C (Adiprene L-213)
1.5 min @ 100 C (Adiprene LW-570)
2–3 min @ 100 C (Adiprene LW-550)
2.5 min @ 70 C (Adiprene L-325)
3 min (Uralite 3175 Fastset); @ 85 C (Adiprene L-200)
3–4 min @ 100 C (Adiprene LW-510)
4 min @ 100 C (Adiprene LW-500)
4–6 min @ 100 C (Adiprene LW-520)
5 min @ 85 C (Adiprene L-167)
5 min @ 100 C (Adiprene L-83)
6 min (Conathane RN-3038, RN-3039)
7 min @ 100 C (Adiprene L-42)
10 min @ 100 C (Adiprene L-100)
10–12 min (Uralite 3156, 3158)
10–20 min @ R.T. (UR 104)
15 min (Uralite 3115, 3167)
15 min to gelled visc. (Ren:C:O-Thane RP-6403)
15–20 min (Uralite 3139)
18 min (Uralite 3150, 3152)
20 min to gelled visc. (Ren:C:O-Thane RP-6401, RP-6402, RP-6405, RP-6413, RP-

6414, TDT-178-34, TDT-178-53)
20 min to 20,000 cps (Uralite 7250, 7252)
20–30 min @ R.T. (UR 102)
20+ min @ 70 C (Adiprene L-700)
25 min (Uralite 3140)
25–30 min (Uralite 6108)
25+ min @ 70 C (Adiprene L-767)
27 min (Uralite 3111)
30 min (Purelast 221, 224, 240H, 242, 247, 249; Uralite 3113)
30 min to gelled visc. (Ren:C:O-Thane RP-6400, RP-6410, TDT-186-1)
30 min to 20,000 cps visc. (Ren:C:O-Thane RP-6422)
30–40 min @ R.T. (UR 101, 105)
33 min (Uralite 3154, 3155)
40–60 min @ R.T. (UR 103)
45 min @ 75 F (Purelast 204, 207, 208, 209)
50 min (Uralite 3125)
1 h (Purelast 220, 223, 226, 228, 234, 235)
2 h (Purelast 241, 241H, 243, 245, 245H, 251, 253, 254, 255)
Set Time:
1 h @ 75 F (Purelast 204, 207, 208, 209)
Typical/Suggested Cure:
Curable @ R.T. (Calthane ND 1100, 2300, 3200; Mira-Glos RT A and C; Ren:C:O-Thane RP-6400, RP-6401, RP-6402, RP-6403, RP-6405, RP-6410, RP-6413, RP-6414, RP-6422, TDT-186-1)
Curable @ R.T. or elevated temps. (Norcast PR-1020)
Cure 0.5 h @ 100 C plus 7 days R.T. (Conathane RN-3038)
Cure 1 h @ 100 C (Adiprene L-167, L-200, L-213)
Cure 1 h @ 100 C plus 7 days R.T. (Conathane RN-3039)
Cure 1 h @ 130 C (Adiprene BL-16)
Cure 1–2 h @ 100 C (Adiprene L-315)
Cure 1–3 h @ 100 C (Adiprene L-100)
Cure 2 h @ 60 C or 24 h @ R.T. (Hysol PC28)
Cure 2 h @ 60 C or 5–7 days @ R.T. (Hysol PC18)
Cure 2 h @ 121 C (Adiprene L-700, L-767)
Cure 3 h @ 100 C (Adiprene L-42, L-83)
Cure 6 h @ 80 C (Conathane TU-4010)
Cure 6 h @ 100 C (Adiprene L-325)
Cure 8 h @ 60 C (Conathane UC-33)
Cure 16 h @ 100 C (Adiprene LW-520)
Cure 20 h @ 100 C (Adiprene LW-500, LW-510, LW-550, LW-570)
Cure 6 days @ R.T. (Ren:C:O-Thane TDT-178-34)
Cure overnight @ 25 C (UR 101, 102, 103, 104, 105)
Postcure required (Ren:C:O-Thane TDT-178-53)

Polyurethane *(cont'd.)*

Postcure @ 150 F (Stepanfoam F-506)

Postcure @ 200 F (Stepanfoam F-403)

80–90% of the final bond strength is obtained within 24 h; complete cure: 5 days (Lamal 408-40(65)/Lamal C)

89–90% of bond strength is obtained within 24 h; complete cure: 5–10 days (Lamal HSA/Lamal C)

Demold Time:

3+ min (Stepanfoam SX-159 Series)

< 5 min (Stepanfoam MW-20/20)

5 mi (Stepanfoam BX-359, BX-369)

5–10 min (Stepanfoam HC-17/40)

6 min (Stepanfoam HC-9/10 Series, HW-8/25, HW-10/25, HW-12/25)

6–8 min (Stepanfoam BX-341A, BX-341B, SX-202)

8–10 min (Stepanfoam AX-64, AX-66, BX-341, BX-345, HC-2/40, HW-8/50, HW-11/60, HW-16/60, HW-20/60, SX-195A, SX-211, SX-214, SX-215, SX-216, SX-218A, SX-219)

10 min (Stepanfoam F-202, F-302, F-403, H-102N)

10–12 min (Stepanfoam HC-2/30, HC-3/40, HC-4/60, HC-50/60, SF-3, SF-5/60)

10–15 min (Stepanfoam BX-289, SX-217)

12–15 min (Stepanfoam BX-316-3, BX-315-5)

20 min @ 85 C (Conathane UC-33)

30 min (Stepanfoam F-506, P-502, P-506; Uralite 3175 Fastset)

1 h (Stepanfoam BX-105 Series)

2 h (Stepanfoam A-210, C-600 Series)

4 h (Conathane TU-4010; Uralite 3130, 3132)

4–6 h (Uralite 3150, 3152)

5 h (Uralite 3121S, 3121S)

6–12 h (Uralite 3156, 3158)

10–16 h (Uralite 3115, 3115, 3167)

12 h (Purelast 208, 209; Uralite 3121, 3121, 3122)

12–24 h (Uralite 3111, 3113, 3113, 3139, 3140)

< 16 h (Ren:C:O-Thane TDT-178-34)

16 h (P.U.R.E.-CMC; Purelast 207; Uralite 3124, 3128)

18–36 h (Uralite 3154, 3155, 7250, 7252)

24 h (Purelast 204; Uralite 3125)

Initial Cure:

$^1/_2$ day (Purelast 221, 224, 240H, 242)

1 day (Purelast 220, 223, 226, 228, 234, 235, 247, 249, 251, 253, 354, 255)

$1^1/_2$ days (Purelast 241, 241H, 243, 245, 245H)

Complete Cure:

12–24 h (Uralite 3175 Fastset)

2 days (Uralite 3121S, 3121S)

2–3 days (Purelast 207, 208, 209)

2–4 days (Uralite 3130, 3132, 3150, 3152, 3156, 3158)

3 days (Uralite 3121, 3121, 3122, 3128, 3140)

3–4 days (Purelast 204)

3–5 days (Uralite 3111, 3113, 3113, 3139, 3154, 3155, 7250, 7252)

4 days (Uralite 3125)

4–6 days (Uralite 3115, 3115, 3167, 6108)

5 days (Uralite 3124)

Peak Exotherm:

126 F (Uralite 3154, 3155)

147 F (Uralite 3150, 3152)

164 F (Hexcel 164M)

TOXICITY/HANDLING:

During heat cure, the blocking agent (MEK), a flammable liquid, is evolved; tolylene diisocyanate may also be released; avoid inhalation of these vapors (Adiprene BL-16)

Avoid skin contact and inhalation of vapors (Chempol 34-4199/34-5075)

Do not take internally; use in well ventilated areas; avoid breathing vapors; protect skin and eyes from contact (Conathane TU-4010)

Avoid contact; use protective clothing; use in well ventilated areas; avoid breathing of vapors (Conathane UC-33)

Use with sufficient ventilation (Millathane 76, 80, 300, E-34, HT)

Use with good ventilation; protect skin and eyes against contact (Purelast 204, 207, 208, 209, 220, 221)

Good ventilation and protective clothing, gloves, and goggles are recommended when handling isocyanate compounds (P.U.R.E.-CMC)

Resin causes irritation; may cause allergic skin or respiratory reaction; harmful if inhaled or swallowed; hardener is harmful if swallowed; may cause irritation; avoid prolonged/repeated skin contact (Ren:C:O-Thane TDT-178-34)

Part A contains free monomeric TDI—vapors are very irritating to mucous membranes and eyes; can cause chest pain, strangling cough, and shortness of breath; strong eye irritant; may cause asthma and dermatitis; Part B contains MOCA, a carcinogen for some animals; overexposure may cause temporary liver damage cyanosis, mild eye irritation, irritation to mucous membranes, delayed chemical burns; ingestion may cause poisoning; use with adequate ventilation and protective clothing (Uralite 3111, 3113, 3115, 3167)

Part A contains free monomeric TDI—may cause skin irritation, strong eye irritation; vapors may cause breathlessness, chest discomfort; ingestion is harmful; Part B contains heavy metal catalyst and amines—may cause skin and eye irritation; vapors may cause nausea; ingestion is very harmful; use with adequate ventilation and protective clothing (Uralite 6108)

Part A contains free monomeric isocyanate—may cause breathlessness, chest discomfort, and reduced pulmonary functions; may cause skin and eye irritation; Part B may cause skin irritation; use with adequate ventilation and protective clothing

Polyurethane *(cont'd.)*

(Uralite 3125)

Part A contains free monomeric isocyanate—may cause skin irritation, strong eye irritation, temporary breathlessness, chest discomfort, reduced pulmonary function; Part B contains amines and a heavy metal catalyst—can cause temporary liver damage, cyanosis, mild eye irritation, delayed chemical burns; ingestion may cause poisoning; use with ventilation and protective clothing (Uralite 3139, 3140, 3150, 3152, 3154, 3155, 3156, 3158, 7250, 7252)

Part A contains free monomeric isocyanate and aromatic hydrocarbons—may cause skin irritation, strong eye irritation; vapors may cause breathlessness and chest discomfort; ingestion is harmful; Part B contains aromatic hydrocarbons—may cause skin irritation and strong eye irritation; vapors may cause nausea; use with adequate ventilation and protective clothing (Uralite 3175 Fastset)

Contains an isocyanate-based prepolymer, amines, and heavy metal catalysts—harmful if swallowed; may cause burns or skin irritation; use with ventilation and protective clothing; avoid inhalation of vapors (Uralite 3130, 3132)

STORAGE/HANDLING:

May present a serious fire hazard; avoid mostire contamination of components; maintain component temps. @ 75–80 F (Chempol 34-4199/34-5075)

Moisture sensitive (all parts); flush contnainers with dry nitrogen after opening; store in original, unopened container at 65–85 F for max. storage life (Conathane TU-4010)

Flush with dry nitrogen to prevent moisture contamination of opened containers; Part A may slightly crystallize or become highly viscous during storage—place container in oven @ 70–80 C until liquid, agitate, and cool to R.T. (Conathane UC-33)

High temps. will release cyanates and hydrocarbons, oxides of carbon, nitrogen, and small amonts of HCN under burning conditions (Millathane 76, 80, 300, E-34, HT)

Do not use in contact with nitric acid in any conc.; protect materials from moisture (Purelast 220, 221, 223, 224, 226, 228, 234, 235, 240H, 241, 241H, 242, 243, 245, 245H, 247, 249, 252, 253, 254, 255)

Store in a cool place; flammable liquid; 6 mo. storage life for each part (Lamal 408-40(65)/Lamal C)

Store in a cool place; 6 mo. storage life (Lamal HSA/Lamal C)

Store in a dry room ≥ 40 F; both components are Red Label materials; 6 mos. storage life (Mira-Glos RT A and C)

Store in a cool, dry place; Red Label material (Unoflex)

Store in a cool, dry place; ≥ 1 yr shelf life when stored at 65–80 F in unopened containers (Uralite 3111, 3113, 3115, 3125, 3167, 3175 Fastset, 6108)

Store in a cool, dry place; ≥ 1 yr shelf life when stored at 65–80 F; blanket with dry nitrogen before resealing (Uralite 3139, 3140, 3150, 3152, 3154, 3155, 3156, 3158, 7250, 7252)

Store in a cool, dry area at 65–90 F; blanket with dry nitrogen before resealing (Uralite 3130, 3132)

Store resin and hardener components at 70–100 F; avoid moisture contamination by

Polyurethane *(cont'd.)*

blanketing with dry nitrogen (Ren:C:O-Thane RP-6400, RP-6401, RP-6402, RP-6403, RP-6405, RP-6410, RP-6413, RP-6414, RP-6422, TDT-178-34, TDT-178-53, TDT-186-1)

Store Part A at 70–90 F; crystallization may occur below 70 F; heat to 120–150 F to dissolve crystals; store part B above 75 F to prevent cloudiness and crystallization (Uralite 3121S)

All parts of the system are somewhat sensitive to moisture or humidity (P.U.R.E.-CMC)

STD. PKGS.:

8-, 15-, 31-, and 60-oz kits (in cans); 8-, 15-, and 23-oz kits (in bags) (Rezolin 185N)

24 $2^1/_2$ or 6 fl oz cartridges per case (Rezolin 164, 170)

24 15-oz (350-g) mixing bags/case; 20 23-oz (650-g) mixing bags/case (Hexcel 164M)

$1^1/_4$ qt and $1^1/^4$ gal two-package units; 5 gal single package unit; 5×55 gal drum unit (Purelast 204, 207, 208, 209)

Qt pack, pail pack, drum pack (Uralite 3139)

12-qt pack, 5-gal pack, 55-gal drum pack (Uralite 3111, 3128, 3140, 6108)

12-qt pack, 5-gal pail packs (Uralite 3121, 3121)

12-qt pack, gal kit, pail pack, drum pack (Uralite 3121S)

12-qt pack, pail pack (Uralite 3122, 3124)

12-qt pack, pail pack, drum pack (Uralite 3125, 3130, 3132, 3150, 3152, 3154, 3155, 3156, 3158, 7250, 7252)

1-gal and 5-gal units (Conathane UC-33)

1-gal, 5-gal, 55-gal units (Conathane TU-4010)

55-gal (400 lb net) steel drums (Lamal 408-40(65)); 5-gal (49 lb net) cans (Lamal C)

55-gal (425 lb net) steel drums (Lamal HSA); 5-gal (49 lb net) cans (Lamal C)

55-gal (430 lb net)) drums (Unoflex)

55-gal (430 lb net) steel drums (Mira-Glos RT A); 5-gal (40 lb net) steel pails (Mira-Glos RT C)

40 lb cans and 450 lb drums (Adiprene L-42, L-83, L-100, L-167, L-200, L-213, L-300, L-315, L-325, L-367, L-700, L-767, LW-520, LW-570)

450 lb drums (Adiprene BL-16)

Polyvinyl chloride

SYNONYMS:
PVC

EMPIRICAL FORMULA:
$(—H_2CCHCl—)_x$

Polyvinyl choride *(cont'd.)*

STRUCTURE:

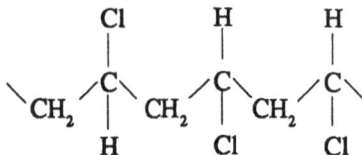

CAS No.:
9002-86-2

TRADENAME EQUIVALENTS:
Conoco 5385, 80172, 80273, 90171 [Conoco]

Conoco Compound No. 14331, 15041, 15071, 16081, 17461, 18161, 18961, 26831, 27121, 28131, 28831, 34431, 34551, 34641, 35441, 35851, 38551, 38601, 38631, 38671, 38931, 44171, 44331, 45541, 48561, 48562 [Conoco]

FPC 1238, 1239, 1240, 1265C, 1359, 1363A, 1379, 1380, 1380A, 1381A, 1415, 1418, 1419, 1430, 1433, 5000B, 5002, 5004, 5005, 5006A, 5008, 5009, 5060, 5105, 5125A, 5160, 5175, 5264, 5290, 5305, 5360, 5390, 5439A, 5440, 5460, 5605, 5760, 5760B, 5790, 5805, 9275 [Firestone Plastics]

Geon 85721, 85858, 87001 [B.F. Goodrich]

Harwal AOB, DBP-C, DBP-F, DBP-G, DBP-I, DBP-NT, GBM, GI, GMW, GP-C, GP-G, GP-I, GP-NT, GPR, GSB, GSB-F, GSF 1, GSG [U.S. Polymers]

Harwal Purge Compound [U.S. Polymers]

Kohinor 311 FR, 434, 509, 510, 702, 906, 940, 941, 947, 948 [Pantasote]

Rucoblend 3-6, 8-289, 8-436C, 8-451B, 8-1241B, 8-1251, 8-1314, 8-1311K, 9-17, 55B, 55E, 103, 105, 161J, 172A, 179A, 207, 247, 300, 309A, 500, 501, 508F, 600, 601A, 2117G, 2160, 2785A, 2831, 3500 [Ruco Polymer]

Rucodur 1505 [Ruco Polymer]

Ultralite [Herbert Lushan Plastics]

PVC graft polymer:

Rucodur 1500, 1503, 1504, 1505, 1706, 1707, 1803 [Ruco Polymer]

PVC homopolymer:

Conoco 5305, 5425, 5465 [Conoco]

FPC 965, 9269A, 9282, 9290, 9300, 9326, 9339, 9418 [Firestone Plastics]

Rucon B-221, B-253, B-282, B-341 [Ruco Polymer]

SCC-608, -614, -616, -676, -676P, -686 [Stauffer]

Vinyl chloride/vinyl acetate copolymer:

SCC-133 (14.48% VA), -156 (17.2% VA), -421 (2% VA) [Stauffer]

Suspension PVC:

Vestolit S 6058, S 6059, S 6554, S 6555, S 6558, S 6858, S 7054, S 7055, S 7554, S 8054 [Hüls AG]

PVC/elastomer copolymer:

Vestolit HI S 6882 (6% elastomer), HI S 6883 (10% elastomer), HI S 7587 (50% elastomer), P 1976 K (10% elastomer), P 1982 K (6% elastomer) [Hüls AG]

Microsuspension PVC:
Vestolit B 7021, B 7022, B 7521 [Hüls AG]
Vinyl chloride/vinyl acetate copolymer:
Vestolit B 7090 (5% VA), B 7092 (5% VA) [Hüls AG]
Mass PVC:
Vestolit M 5867, M 6067, M 6267, M 6567, M 6867 [Hüls AG]
Dispersion/emulsion PVC:
FPC 605, 654, 654H, 6337, 6411, 6422, 6854, 6866 [Firestone Plastics]
Vestolit E 6003, E 6007, E6017, E 6503, E 6507, E7001, E 7003, E 7004, E 7006, E
7007, E 7008, E 7012, E 7031, E 7033, E 7037, E 8001, E 8003, E 8019, P 1331 K,
P 1333 K, P 1341 K, P 1342 K, P 1344 K, P 1345 K 70, P 1345 K 80 [Hüls AG]
PVC homopolymer:
SCC-24, -28, -54, -58 [Stauffer]
Vinyl chloride/vinyl acetate copolymer:
FPC 6338 [Firestone Plastics]
SCC-40, -52 [Stauffer]
Vestolit E 7091 (5% VA) [Hüls AG]
PVC/elastomer copolymer:
Vestolit HI E 6077 (20% elastomer), HI E 6577 (20% elastomer), HI E 7077 (20%
elastomer) [Hüls AG]
Chlorinated PVC:
Geon 3007, 3010, 88933, 88934, CPVC [B.F. Goodrich]

MODIFICATIONS/SPECIALTY GRADES:
Antistatic:
Vestolit E 6007, E 6507, E 7007
High-impact:
Harwal AOB, GBM, GMW; Rucoblend 3-6, 9-17, 105, 161J, 172A, 179A, 309A,
508F, 2117G, 2160, 2831; Rucodur 1500, 1503, 1505, 1706, 1707, 1803
High-strength:
Rucodur 1504
Flame-retardant:
FPC 5000B, 5002, 5006A, 5009; Geon CPVC; Rucodur 1500, 1503, 1504, 1505, 1706
Heat-stabilized:
Harwal GMW (calcium-zinc); Harwal Purge Compound; Rucoblend 3-6 (dioctyl tin),
55B (dioctyl tin), 55E (dioctyl tin), 247 (calcium-zinc), 2117G (dioctyl tin), 2160
(dioctyl tin), 3500 (tin)
Plasticized:
Harwal GSB-F, GSF 1, GSG
Lubricated:
Harwal Pruge Compound

CATEGORY:
Thermoplastic resin

Polyvinyl choride *(cont'd.)*

PROCESSING:

Blow molding:
FPC 9290; Harwal AOB, GMB; Rucoblend 3-6; Rucon B-253; SCC-608

Blown film:
Rucon B-341; SCC-608, -614, -616, -676

Calendering:
Conoco 5305, 5465; FPC 9269A, 9282, 9290, 9300, 9339; Rucon B-221, B-253, B-282; SCC-421, -608, -614, -616, -676, -676P, -686; Vestolit E 6003, E 6503, E 7003, E 7004, E 7008, E 7033, E 8003, E 8019, M 6067, M 6267, S 6058, S 6059, S 6554, S 7054, S 7554, S 8054

Cast cellular processing:
FPC 605, 6366

Casting:
Vestolit B 7021

Coating:
FPC 605, 654, 654H, 6337, 6338, 6366, 6411, 6422, 6854, 6866; SCC-28, -58; Vestolit B 7021, E 7001, E 8001

Cold blending:
FPC 9275

Compression molding:
Geon 3007, 3010, 88933, 88934

Custom compounding:
SCC-421

Dip coating:
FPC 605, 654, 654H, 6337, 6338, 6366, 6854, 6866; SCC-28, -58; Vestolit B 7021

Dry processing:
FPC 9269A, 9282, 9290, 9300, 9326, 9339, 9418

Extrusion:
Conoco 5305, 5465, 80172, 80273; Conoco Compound No. 14331, 15041, 15071, 16081, 17461, 18161, 18961, 34431, 34551, 34641, 35441, 35851, 38551, 38601, 38631, 38671, 38931, 44331, 45541; FPC 1238, 1239, 1240, 1265C, 1359, 1379, 1380, 1380A, 1381A, 1418, 5000B, 5002, 5004, 5005, 5006A, 5008, 5009, 5060, 5105, 5125A, 5160, 5175, 5264, 5290, 5305, 5360, 5390, 5439A, 5440, 5460, 5605, 5760, 5760B, 5790, 5805, 9282, 9300, 9326, 9339, 9418; Geon 3007, 85721, 85858, 87001, 88933, 88934, CPVC; Harwal DBP-C, DBP-F, DBP-G, DBP-I, DBP-NT, GP-C, GP-G, GP-I, GP-NT, GPR, GSG, Purge Compound; Kohinor 311 FR, 434, 509, 510, 702, 906, 940, 941, 947, 948; Rucoblend 3-6, 8-289, 8-436C, 8-451B, 8-1241B, 8-1251, 8-1314, 8-1311K, 103, 105, 300, 309A, 500, 501, 508F, 600, 601A; Rucodur 1706, 1707, 1803; Rucon B-221, B-253, B-282, B-341; SCC-421, -676P, -686; Vestolit E 6503, E 7003, E 7008, E 7007, E 7033, E 8003, HI E 6077, HI E 6577, HI E 7077, M 6067, M 6267, P 1333 K, P 1341 K, S 6058, S 6554, S 6555, S7055, S 6558, S 6858, S 7054, S 7554, S 8054

Extrusion blow molding:
 Rucoblend 9-17, 161J, 172A, 179A, 247, 2117G, 2160, 2785A, 2831, 3500; Rucodur
 1803; Vestolit M 5867, M 6567, M 6867
Foamed plastisols:
 SCC-24, -40, -52, -54
Foam/expansion processing:
 FPC 6338, 6854; SCC-614, -616; Vestolit B 7022, B 7092, E 7012, M 5867, M 6567,
 M 6867, P 1341 K
Injection blow molding:
 Harwal GSB; Rucoblend 55B, 55E
Injection molding:
 Conoco 5305, 5465, 90171; Conoco Compound No. 26831, 27121, 28131, 28831,
 38671, 44171, 48561, 48562; FPC 1363A, 1415, 1419, 1430, 1433, 9290; Harwal
 GI, GSF 1; Rucoblend 207; Rucodur 1500, 1503, 1504, 1505; Rucon B-221;
 Vestolit M 5867, M 6567, M 6867, S 6058, S 6554, S 6555, S 7055, S 7054, S 7554,
 S 8054
Paste processing:
 Vestolit B 7021, B 7022, B 7090, B 7092, B 7521, E 7001, E 7012, E 7031, E 7091,
 E 8001, P 1331 K, P 1344 K, P 1345 K 70, P 1345 K 80
Powder processing:
 Vestolit S 6555, S 7055
Rotational casting:
 FPC 605, 654, 654H, 6337, 6338, 6366, 6854, 6866; SCC-28, -58
Sintering:
 Vestolit E 7006
Slush molding:
 FPC 605, 654, 654H, 6337, 6338, 6366, 6854, 6866; SCC-28, -58
Spraying:
 FPC 605, 654H, 6337, 6338, 6366, 6854, 6866
Vacuum forming:
 Rucoblend 103, 500, 501, 508F, 600, 601A

APPLICATIONS:

 Agriculture industry: irrigation conduit (Conoco 80172, 80273; Harwal DBP-I, GP-I)
 Architectural applications: construction applications (Vestolit HI E 6077, HI E 6577,
 HI E 7077, M 5867); gutters (Rucoblend 309A); siding (Conoco 5385; FPC 9418);
 windows (Rucoblend 309A; Rucodur 1706, 1707; Vestolit HI S 6882, HI S 6883,
 HI S 7587, P 1982 K)
 Automotive applications: (Conoco 5385; Rucodur 1504); battery separators (Vestolit
 E 7006); jumper cables (FPC 1418, 5439A); under-the-hood wiring (Conoco 5465;
 FPC 5125A); windows (Ultralite)
 Consumer products: appliance parts (Conoco 5305; Rucodur 1503); appliances
 (Rucodur 1504); cosmetics/shampoos containers (Harwal GBM; Rucoblend 55B,
 55E, 161J, 172A, 179A, 2785A, 2831); fencing (Conoco Compound No. 44331,

Polyvinyl choride *(cont'd.)*

45541); footwear (Conoco 5385; FPC 1363A, 1415, 1419, 1430, 1433; Harwal GSF 1); furniture (Vestolit HI E 6077, HI E 6577, HI E 7077, M 5867); garden hose (Harwal GSG); housewares (Conoco 5305); lawn furniture (Conoco Compound No. 44331); outdoor applications (Vestolit HI S 6882, HI S 6883, P 1982 K); phonograph records (Rucoblend 8-289, 8-436C, 8-451B, 8-1241B, 8-1251, 8-1314, 8-1311K; SCC-133, -156)

Electrical/electronic industry: (Rucodur 1504); appliance wire (Conoco Compound No. 34431, 34551, 34641, 35441, 35851, 38551, 38601, 38631, 38931; FPC 5005, 5105, 5290, 5360, 5460, 5805; Kohinor 509, 510, 906); building wire (Conoco Compound No. 35851); cable and wire (Conoco 5385, 5425, 5465; Conoco Compound No. 34431, 34551, 34641, 35441, 35851, 38551, 38601, 38631, 38671, 38931; FPC 5160, 5175, 5790, 9269A; Kohinor 311 FR, 940, 941, 947, 948; SCC-676; Vestolit S 6554, S 7054, S 7554, S 8054); cable coatings (Conoco Compound No. 14331, 15041); communications/telephone conduit (Conoco 80172, 80273; Kohinor 434, 702); conduits (Harwal DBP-C, GP-C; SCC-676P; Vestolit S 6058); connectors (Rucodur 1503); insulation (Conoco Compound No. 34431, 34551, 35441, 35851, 38551, 38631, 38931; FPC 5005, 5060, 5105, 5305, 5390, 5460, 5605, 5760, 5760B, 5805; Kohinor 509, 510, 702); jacketing (Conoco Compound No. 34431, 38551, 38601, 38631, 44331, 45541; FPC 5000B, 5002, 5004, 5006A, 5008, 5009, 5264, 5439A, 5440, 5760, 5760B; Kohinor 434; Vestolit S 6555, S 7055); outdoor applications (FPC 5004, 5006A); protective covers for conductors (Geon CPVC)

FDA-approved applications: (Harwal GSB; Rucoblend 3-6, 600, 601A, 2117G, 2160)

Food-contact applications: beverage containers/tubing (Harwal GMW, GSB; Rucoblend 2160, 300); drinking straws (Vestolit S 6058); food containers (Harwal GBM; Rucoblend 3-6, 247, 300, 60, 601A, 2117G, 3500)

Functional additives: cleaning extruders (Harwal Purge Compound); modifier (FPC 965, 9282); processing aid (Vestolit P 1342 K)

Industrial applications: blister packaging (Rucoblend 103, 500, 501, 508F); bottles (Harwal AOB, GBM, GSB; Rucoblend 3-6, 9-17, 55B, 55E, 161J, 172A, 179A, 2117G, 2160, 2785A, 2831, 3500; Rucodur 1803; SCC-614); carpet backing/runners (Conoco Compound No. 14331, 15041, 15071; SCC-40, -52); cellular products (FPC 605, 654H, 6337, 6366); coatings (FPC 6338; Vestolit E 7001, E 8001, P 1345 K 80); coil coatings (FPC 6411, 6422); containers (Rucoblend 247; SCC-608); detergent/bleach, 3785A, 2831 containers (Harwal GBM; Rucoblend 161J, 172A, 179A); ductwork (Geon CPVC); expanded profile (SCC-614, -616-676P); exterior applications (Geon 85858; Rucoblend 309A); fabric coating (FPC 6338, 6411, 6422; SCC-40); fence coating (FPC 1265C); films (Conoco 5305; Rucoblend 103, 300; Rucon B-341; SCC-421, 608, -614, -616, -676, -686; Vestolit E 6503, E 6007, E 6507, E 6017, E 7004, E 7008, E 7033, E 7037, E 8003, E 8019, HI S 7587, M 6067, M 6267, P 1976 K, S 6058, S 6059, S 6554, S 7054, S 7554, S 8054); fittings (Geon 3010; Vestolit S 6058); flexible applications (Conoco 5385, 5425, 5465; Rucoblend 300; Rucon B-221, B-253, B-282; SCC-421, -616, -676,

Polyvinyl chloride *(cont' d.)*

–676P, -686; Vestolit E 7033, E 7037); flocking adhesives (SCC-40); flooring (FPC 6411, 6422; Vestolit E 6003, E 6503, E 7003, E 7004, E 7008, E 7033, E 8003, S 7054, S 7554, S 8054); foamed products (FPC 9282; SCC-24, -40, -52, -54; Vestolit B 7022, B 7090, B 7092, E 7012, P 1341 K, P 1345 K 70); graphic arts (Rucoblend 500, 501, 508F; heat seal applications (SCC-421); hollow articles (Vestolit M 5867); inks (FPC 6338); interior applications (Geon 85721, 87001; Harwal GPR); interior decorative items (Rucoblend 103, 105, 309A); intricate/hard-to-fill moldings (Rucodur 1500); laminates (SCC-40); molded parts (Conoco Compound No. 38671); NSF-approved applications (Conoco 80172, 80273, 90171; Rucoblend 207; Rucodur 1500); packaging (Rucon B-221, B-253, B-282); pipe fittings (Conoco 5305, 90171; Harwal GI; Rucoblend 207; Rucodur 1500); pipe/conduit (Conoco 5385, 80172, 80273; FPC 9418; Geon 3007, 88934, CPVC; Harwal DBP-F, DBP-G, DBP-NT, GP-G, GP-NT; Rucodur 1706, 1707; SCC-676P; Vestolit E 6503, E 7003, E 7004, E 7008, E 7033, E 8003, S 6058, S 6554, S 6558, S 6858, S 7054, S 7554, S 8054); plastisols/organosols (FPC 605, 654, 654H, 6337; SCC-24, -28, -40, -52, -54, -58); plugs (Conoco Compound No. 38671); potable water pipe/pipe fittings (Conoco 90171; Harwal DBP-NT, GP-NT; Rucoblend 207; Rucodur 1500); profiles (Conoco 5305, 5385, 5425, 5465; Conoco Compound No. 15041, 17461, 18161, 18961; FPC 1238, 1239, 1240, 1359, 1379, 1380, 1380A, 1381A, 9282, 9418; Geon 88934; Harwal GPR; Rucoblend 105, 300, 309A; Rucodur 1706, 1707; SCC-421, -676P); rigid applications (Conoco 5305, 5385, 5425; FPC 9282, 9326, 9418; Geon 85721, 85858, 87001; Rucoblend 9-17, 55B, 55E, 103, 105, 247, 309A, 500, 501, 508F, 600, 601A, 2785A, 2831; Rucodur 1500, 1503, 1505, 1706, 1707, 1803; Rucon B-221, B-253, B-282; SCC-608, -614, -616, -676, -676P; Vestolit E 6017, E 8003, E 8019, P 1333 K, S 6058, S 6059); sealants (FPC 605, 654, 654H, 6337); sections (Vestolit E 6503, E 6507, E 7003, E 7004, E 7008, E 7007, E 7033, E 8003, HI E 6077, HI E 6577, HI E 7077, HI S 6882, HI S 6883, HI S 7587, M 5867, M 6067, M 6267, P 1333 K, P 1982 K, S 6058, S 6554, S 6558, S 6858, S 7054, S 7554, S 8054); semirigid applications (Vestolit S 6554); sheet (Conoco 5425, 5465; Rucoblend 103, 500, 501, 600, 601A; Rucodur 1803; SCC-421, -608, -686; Ultralite; Vestolit E 6503, E 6007, E 6507, E 6017, E 7004, E 7008, E 7033, E 7037, E 8003, E 8019, HI S 6882, HI S 6883, HI S 7587, M 6067, M 6267, P 1976 K, P 1982 K, S 6058, S 6059, S 6554, S 7054, S 7554, S 8054); structural foam (Vestolit M 5867); sun-sensitive packaging (Rucoblend 9-17); tanks (Geon CPVC); thick sheet (Vestolit M 5867, M 6067, M 6267); tubing/hoses (Conoco Compound No. 38601, 38631; Geon CPVC); valves (Geon CPVC); waterstops (Harwal GSB-F); weatherstripping (Conoco Compound No. 15071, 16081, 17461, 18161, 18961; FPC 1238, 1239, 1240); welting (Conoco Compound No. 15071)
Marine equipment: windows (Ultralite)
Medical applications: tubing (Rucoblend 300)
Military applications: naval jacket (FPC 5000B)

Polyvinyl choride *(cont'd.)*

PROPERTIES:
Form:
Solid (Conoco 5305, 5385, 5425, 5465; Conoco Compound No. 14331, 15041, 15071, 16081, 17461, 18161, 18961, 26831, 27121, 28131, 28831, 34431, 34551, 34641, 35441, 35851, 38551, 38601, 38631, 38671, 38931, 44171, 44331, 45541, 48561, 48562)

Solid dryblend (Conoco 80172, 80273, 90171; Harwal DBP-C, DBP-G, DBP-I, DBP-NT)

Opaque, filled solid (FPC 1238, 1239, 1265C, 1359, 1379, 1380, 1380A, 1381A, 1418, 1419, 1430, 5439A)

Filled solid (FPC 1363A, 1415)

Pellets (Rucoblend 8-1241B, 8-1251, 8-1314, 8-1311K, 309A, 2117G, 2831, 3500; Rucodur 1500, 1503, 1505, 1706)

Cube (Geon 3007, 3010, 88933, 88934)

Granular (Harwal AOB, GP-C, GP-G, GP-I, GP-NT, GI)

Powder (Rucoblend 8-289, 8-436C, 8-451B; SCC-28, -40)

Free-flowing powder (SCC-52, -54, -58

Fine powder (SCC-24)

Particulate (FPC 965)

Particulate (80–200 mesh) (FPC 9269A, 9339)

Particulate (100–200 mesh) (FPC 9282, 9290, 9300, 9326, 9418)

Particulate (140–200 mesh) (FPC 9275)

Sheets in various sizes (Ultralite)

Color:
Translucent (Rucoblend 2831)

Transparent; other colors avail. (Harwal GSF 1)

Transparent, yellow (Rucoblend 3-6)

White (Conoco 90171; Rucoblend 309A; SCC-24, -28, -40, -52, -54, -58)

Neutral (Rucoblend 105)

Natural (FPC 1265C)

Natural, black (FPC 5439A)

Natural, black, gray (FPC 1359, 1379, 1380, 1381A)

Natural, black, white, gray (FPC 1238, 1239)

Avail. in different colors (Harwal GSG; Rucodur 1500, 1503, 1505, 1706)

Avail. in different colors incl. terra-cotta brown (Harwal GI)

Avail. in different colors incl. metallic and wood-grain (Harwal GPR)

Black (Harwal GSB-F)

GENERAL PROPERTIES:
Sp. Gr.:
1.18 (Conoco Compound No. 27121; FPC 1433)

1.20 ± 0.02 (Kohinor 509, 906)

1.21 (FPC 1415)

1.21 ± 0.02 (Kohinor 510)

1.23 (Conoco Compound No. 14331, 26831, 28131, 28831)
1.24 (Conoco Compound No. 44331; FPC 1363A, 1419, 1430, 5439A)
1.25 (Conoco Compound No. 45541; FPC 1418)
1.26 (Conoco Compound No. 15041)
1.29 (FPC 5000B, 5760)
1.31 (FPC 1265C)
1.32 (FPC 5760B)
1.32 ± 0.02 (Kohinor 311 FR)
1.33 ± 0.02 (Conoco Compound No. 38551)
1.34 (FPC 5005, 5105)
1.34 ± 0.01 (Rucodur 1500, 1503, 1505)
1.34 ± 0.02 (Kohinor 941)
1.35 (FPC 1359, 5008)
1.35 ± 0.02 (Rucodur 1803)
1.36 (Conoco Compound No. 48561, 48562; FPC 1379, 5290, 5440, 5460)
1.36 ± 0.02 (Conoco Compound No. 34451)
1.37 (FPC 1238, 1380, 1380A, 5004, 5175, 5305, 5360)
1.37 ± 0.02 (Conoco Compound No. 34431, 34641, 38601, 38931; Kohinor 947)
1.38 (Conoco Compound No. 17461, 18161, 18961; FPC 5060, 5160, 5790, 6388)
1.38 ± 0.02 (Conoco Compound No. 35851)
1.39 (FPC 1239, 1381A, 5006A, 5009, 5805; SCC-40, -52)
1.40 (FPC 605, 654, 654H, 965, 1240, 6337, 6366, 6411, 6422, 6854, 6866, 9269A,
    9275, 9282, 9290, 9300, 9326, 9339, 9418; SCC-24, -28, -54, -58)
1.40 ± 0.02 (Conoco Compound No. 35441, 38631)
1.41 (Conoco 80172, 80273, 90171; Conoco Compound No. 15071)
1.41 ± 0.02 (Rucodur 1706)
1.42 (FPC 5002)
1.42 ± 0.02 (Kohinor 940)
1.43 (FPC 5390)
1.44 (Conoco Compound No. 44171; FPC 5264)
1.44 ± 0.02 (Kohinor 434, 702)
1.45 (Conoco Compound No. 16081)
1.45 ± 0.02 (Conoco Compound No. 38671)
1.46 (FPC 5605)
1.50 ± 0.02 (Kohinor 948)
1.52 ± 0.02 (Geon 3010, 88933)
1.55 ± 0.02 (Geon 3007)
1.57 ± 0.02 (Geon 88934)
1.63 (FPC 5125A)

Density:
1.21 kg/m³ (Harwal GSF 1, GSG)
1.30–1.32 kg/m³ (Harwal AOB)
1.31–1.33 kg/m³ (Harwal GBM, GMW, GSB-F)

Polyvinyl choride *(cont'd.)*

1.33–1.35 kg/m³ (Harwal GSB)
1.40 kg/m³ (Harwal GPR)
1.40–1.41 kg/m³ (Harwal GI)
1.65 kg/m³ (Harwal Purge Compound)
0.38 g/cc (FPC 9275)
0.41 g/cc (SCC-52)
0.45 g/cc (SCC-54)
0.48 g/cc (FPC 9269A, 9339; SCC-58)
0.50 g/cc (FPC 9300; SCC-421)
0.50–0.60 g/cc (expanded) (Geon 85721)
0.52 g/cc (SCC-616)
0.53 g/cc (SCC-676P)
0.56 g/cc (FPC 9326, 9418)
0.58 g/cc (Conoco 5305)
0.60 g/cc (FPC 965, 9282)
0.70 g/cc (SCC-133)
0.70–0.80 g/cc (expanded) (Geon 85858, 87001)
0.72 g/cc (SCC-156)
0.30 g/ml (Vestolit B 7021)
0.32 g/ml (Vestolit B 7521)
0.35 g/ml (Vestolit B 7090, E 6503, E 7003, E 8003)
0.40 g/ml (Vestolit E 6003, E 7001, E 7008, E 8001, P 1345 K 70, P 1345 K 80)
0.42 g/ml (Vestolit B 7092)
0.43 g/ml (Vestolit E 7004)
0.44 g/ml (Vestolit E 7031)
0.45 g/ml (Vestolit B 7022, B 7033)
0.47 g/ml (Vestolit S 7554, S 8054)
0.48 g/ml (SCC-686; Vestolit S 7054)
0.50 g/ml (SCC-676; Vestolit HI E 6077, HI E 6577, HI E 7077, P 1331 K, P 1341 K, P 1344 K)
0.51 g/ml (Vestolit S 6554)
0.53 g/ml (Vestolit P 1342 K, S 6555, S 7055)
0.55 g/ml (Vestolit E 7012, E 7091, S 6858)
0.56 g/ml (Vestolit E 7037, E 8019)
0.57 g/ml (SCC-608, -614; Vestolit P 1333 K, S 6558)
0.58 g/ml (Vestolit E 6017)
0.59 g/ml (Vestolit S 6058)
0.60 g/ml (Vestolit E 6007, E 6507, E 7007, M 6567, S 6059)
0.62 g/ml (Vestolit E 7006, P 1982 K)
0.63 g/ml (Vestolit M 6867)
0.65 g/ml (Vestolit HI S 6882, M 5867, M 6067, M 6267)
0.66 g/ml (Vestolit HI S 6883, P 1976 K)
0.70 g/ml (Vestolit HI S 7587)

14 lb/ft³ (SCC-28, -40)
32 lb/ft³ (Conoco 5425, 5465)
33.5 lb/ft³ (Conoco 5385)

Visc.:
0.49 (inherent) (SCC-133, -156)
0.52 (inherent) (FPC 965)
0.65 (inherent) (FPC 9282)
0.66 (inherent) (SCC-608)
0.78 (inherent) (SCC-614)
0.80 (inherent) (FPC 9290)
0.82 (inherent) (SCC-54)
0.83 (inherent) (SCC-616)
0.86 (inherent) (SCC-24)
0.91 (inherent) (FPC 9418; SCC-676P)
0.95 (inherent) 9SCC-676)
0.96 (inherent) (FPC 9300)
0.97 (inherent) (FPC 9326)
1.02 (inherent) (SCC-52)
1.04 (inherent) (SCC-40)
1.05 (inherent) (SCC-421)
1.07 (inherent) (FPC 9269A)
1.09 (inherent) (SCC-686)
1.17 (inherent) (SCC-58)
1.20 (inherent) (SCC-28)
1.24 (inherent) (FPC 9275)
1.33 (inherent) (FPC 9339)
82 ml/g (Vestolit M 5867)
88 ml/g (Vestolit E 6003, E 6007, E 6017, HI E 6077, M 6067, P 1976 K, S 6058, S 6059)
95 ml/g (Vestolit M 6267)
105 ml/g (Vestolit E 6503, E 6507, HI E 6577, M 6567, S 6554, S 6555, S 6558)
112 ml/g (Vestolit P 1982 K)
116 ml/g (Vestolit HI S 6882, HI S 6883, M 6867, S 6858)
125 ml/g (Vestolit B 7021, B 7022, B 7090, B 7092, E 7003, E 7001, E 7004, E 7006, E 7007, E 7008, E 7012, E 7031, E 7033, E 7037, E 7091, HI E 7077, P 1331 K, P 1333 K, P 1341 K, P 1344 K, P 1345 K 70, S 7055, S 7054)
145 ml/g (Vestolit B 7521, S 7554)
155 ml/g (Vestolit HI S 7587)
169 ml/g (Vestolit E 8001, P 1345 K 80)
170 ml/g (Vestolit E 8003, E 8019, S 8054)
> 230 ml/g (Vestolit P 1342 K)
30–50 poise (FPC 6337, 6338, 6422)
30–60 poise (FPC 605, 654, 654H, 6366)

Polyvinyl choride *(cont'd.)*

40–60 poise (FPC 6411)

50–80 poise (FPC 6854)

Stability:

Excellent heat stability (Conoco 5305, 5385, 5425, 5465; Conoco Compound No. 34431, 34451, 34641, 35441, 35851, 38551, 38601, 38631, 38671, 38931)

Excellent high heat and uv light stability; good weather and sulfur staining resistance (Conoco Compound No. 26831, 27121, 28131, 28831, 44171, 48561, 48562)

Excellent viscosity stability (FPC 6337)

Good chemical resistance (Geon CPVC)

Good heat stability (Conoco 80172, 80273)

Good heat stability during processing and aging (FPC 9300)

Good fusion and heat stability (FPC 9326)

Good uv light stability, heat stability, and weather resistance (Conoco Compound No. 14331, 15041, 15071, 16081, 17461, 18161, 18961, 44331, 45541)

100% retention of tensile and 90% retention of elongation after heat aging 7 days @ 136 C (FPC 5000B)

97% retention of elongation on heat aging 7 days @ 100 C (FPC 5760B)

95% retention of tensile and 90% retention of elongation after sunlight; 95% retention of elongation on heat aging 10 days @ 100 C (FPC 5009)

90% retention of elongation on heat aging 7 days @ 100 C (FPC 5160, 5360, 5460)

89% retention of elongation on heat aging 7 days @ 100 C (FPC 5004)

88% retention of elongation on heat aging 7 days @ 100 C (FPC 5006A)

85% retention of elongation on heat aging 7 days @ 136 C (FPC 5305, 5605)

85% retention of elongation on heat aging 7 days @ 121 C (FPC 5390)

84% retention of elongation on heat aging 7 days @ 100 C (FPC 5060, 5440)

84% retention of elongation on heat aging 7 days @ 121 C (FPC 5175)

80% retention of elongation on heat aging 60 days @ 113 C (FPC 5005)

80% retention of elongation on heat aging 7 days @ 136 C (FPC 5002)

80% retention of elongation on heat aging 7 days @ 121 C (FPC 5290)

80% retention of elongation on heat aging 7 days @ 113 C (FPC 5008)

80% retention of elongation on heat aging 7 days @ 100 C (FPC 5264, 5760)

77% retention of elongation on heat aging 7 days @ 121 C (FPC 5790)

75% retention of elongation on heat aging 7 days @ 136 C (FPC 5105)

74% retention of elongation on heat aging 7 days @ 136 C (FPC 5805)

66% retention of elongation on heat aging 7 days @ 158 C (FPC 5125A)

MECHANICAL PROPERTIES:

Tens. Str.:

10–12 MPa (Geon 85721)

13 MPa (Harwal GSF 1)

14–15 MPa (yield) (Harwal GSB-F)

14–17 MPa (Geon 87001)

17–21 MPa (Geon 85858)

19.5 MPa (Harwal GSG)

45 MPa (Harwal GBM, GMW, GPR)
50 MPa (Harwal GI)
57 MPa (Geon 3010)
58 MPa (Geon 3007)
45 kg/cm² (Harwal AOB)
350 psi (break, fused, 120 C) (FPC 605)
400 psi (break, fused, 120 C) (FPC 6422, 6866)
450 psi (break, fused, 120 C) (FPC 654, 654H, 6366, 6411)
470 psi (break, fused, 120 C) (FPC 6337)
500 psi (break, fused, 120 C) (FPC 6854)
600 psi (break, fused, 120 C) (FPC 6338)
1070 psi (FPC 1359)
1270 psi (FPC 1380A)
1400 psi (FPC 1415)
1490 psi (FPC 1379)
1500 psi (Conoco Compound No. 26831)
1520 psi (FPC 1433)
1600 psi (Conoco Compound No. 17461; FPC 1238)
1700 psi (Conoco Compound No. 18161)
1760 psi (FPC 1380)
1800 psi (Conoco Compound No. 27121; FPC 1430)
1850 psi (FPC 5008)
1870 psi (FPC 1239)
1900 psi (Conoco Compound No. 48561, 48562; FPC 1363A)
2000 psi (FPC 1418; Kohinor 906)
2100 psi (Conoco Compound No. 38671; FPC 1419, 5006A, 5460; Kohinor 948)
2140 psi (FPC 5264)
2150 psi (FPC 5439A)
2175 psi (SCC-54 film)
2200 psi (Conoco Compound No. 28131, 38601, 38931; FPC 5004, 5440)
2225 psi (Kohinor 941)
2250 psi (SCC-52 film)
2300 psi (Conoco Compound No. 18961, 38551, 38631; FPC 5390; Kohinor 509)
2350 psi (FPC 5060; SCC-24 film)
2400 psi (Conoco Compound No. 28831, 34641; FPC 1240, 5290; Kohinor 311 FR)
2500 psi (Conoco Compound No. 34431; FPC 5605)
2550 psi (Kohinor 940)
2565 psi (FPC 5005)
2600 psi (Conoco Compound No. 15071; FPC 5175, 5360; Kohinor 510)
2700 psi (Conoco Compound No. 34451, 35441; FPC 5760B; Kohinor 434, 947)
2750 psi (FPC 5009)
2800 psi (Conoco Compound No. 16081; FPC 5000B, 5760)
2890 psi (FPC 1381A)

Polyvinyl choride *(cont'd.)*

2900 psi (Conoco Compound No. 35851, 44331; FPC 5305; Kohinor 702)
2920 psi (FPC 5790)
2916 psi (SCC-58 film)
2975 psi (SCC-28 film)
3000 psi (Conoco Compound No. 14331; FPC 5160, 5805; SCC-40 film)
3100 psi (Conoco Compound No. 15041)
3125 psi (FPC 5105)
3140 psi (FPC 1265C)
3200 psi (Conoco Compound No. 45541; FPC 5002, 5125A)
6700 psi (Rucodur 1706, 1803)
7000 psi (Rucodur 1500, 1503, 1505)
7800 psi (Conoco 80172, 90171)
7900 psi (Conoco 80273)
8200 psi (60 C) (Geon 88933)
8400 psi (Geon 88934)
Tens. Elong.:
38–40% (Geon 85721)
40–45% (Geon 87001)
60–65% (Geon 85858)
140% (break) (Harwal GI)
160–210% (break) (Harwal AOB)
180% (break) (FPC 5005)
190% (break) (FPC 5790)
220% (break) (FPC 5264, 5290)
250% (Conoco Compound No. 16081)
240% (FPC 1265C); (break) (FPC 1380A, 5125A)
250% (Kohinor 702); (break) (FPC 1359, 1380, 5060)
260% (break) (FPC 5175)
270% (break) (FPC 5002)
275% (break) (FPC 5004, 5160, 5305)
280% (break) (FPC 1240, 5360, 5460, 5605)
290% (break) (FPC 1379, 5008, 5105, 5760, 5805)
300% (Conoco Compound No. 15071, 18961, 34431, 34451, 34641, 35441, 35851, 48561, 48562; Kohinor 510, 947, 948); (break) (FPC 1381A, 1418, 1430, 5390)
300–350% (break) (Harwal GSB-F)
310% (Kohinor 434); (break) (FPC 1239, 5009, 5760B; Harwal GSG)
315% (break) (FPC 5439A)
320% (Kohinor 311 FR); (break) (FPC 1415, 5000B, 5006A, 5440)
325% (Kohinor 940)
330% (break) (FPC 1363A)
350% (Conoco Compound No. 15041, 17461, 18161, 28831, 38601, 38631, 38671, 38931, 45541; Kohinor 906, 941)
360% (break) (FPC 1238)

370% (break) (FPC 1419, 1433)
375% (Kohinor 509)
390% (break) (Harwal GSF 1)
400% (Conoco Compound No. 14331, 38551, 44331)
430% (break) (SCC-54 film)
450% (Conoco Compound No. 26831, 27121, 28131)
460% (break) (SCC-58 film)
470% (break) (SCC-28 film)
500% (break) (SCC-24 film, -52 film)

Tens. Mod.:

2919 MPa (Geon 3007)
3312 MPa (Geon 3010)
401,000 psi (Rucodur 1500, 1503, 1505)
405,000 psi (Geon 88933, 88934)
407,000 psi (Rucodur 1803)
410,000 psi (Rucodur 1706)
443,000 psi (Conoco 80172)
447,000 psi (Conoco 80273)
501,000 psi (Conoco 90171)

100% Mod.:

500 psi (Conoco Compound No. 26831)
510 psi (FPC 1433)
520 psi (FPC 1359)
600 psi (Conoco Compound No. 27121; FPC 1415)
670 psi (FPC 1379)
680 psi (FPC 1380A)
700 psi (Conoco Compound No. 17461, 18161)
730 psi (FPC 1238)
790 psi (FPC 1419)
840 psi (FPC 1239)
850 psi (SCC-24 film, -40 film, -54 film)
880 psi (FPC 1363A)
900 psi (Conoco Compound No. 28131; FPC 5008; SCC-52 film)
950 psi (FPC 1430)
960 psi (FPC 1380, 1418)
980 psi (SCC-58 film)
1000 psi (Conoco Compound No. 38551, 38601, 48561, 48562; SCC-28 film)
1020 psi (FPC 5439A)
1075 psi (FPC 5440)
1100 psi (Conoco Compound No. 38631, 38671)
1200 psi (Conoco Compound No. 38931)
1250 psi (FPC 5460)
1300 psi (Conoco Compound No. 18961, 28831; FPC 5390)

Polyvinyl choride *(cont'd.)*

    1325 psi (FPC 5006A)
    1330 psi (FPC 5264)
    1350 psi (FPC 5060)
    1370 psi (FPC 1240)
    1400 psi (FPC 5760)
    1500 psi (Conoco Compound No. 34431; FPC 5605)
    1550 psi (FPC 5004)
    1600 psi (Conoco Compound No. 34641; FPC 5000B)
    1610 psi (FPC 5760B)
    1650 psi (FPC 5009)
    1700 psi (Conoco Compound No. 14331, 15071, 34451, 44331)
    1750 psi (FPC 5105; FPC 5175)
    1770 psi (FPC 1381A)
    1800 psi (FPC 5290, 5360, 5805)
    1900 psi (Conoco Compound No. 35441)
    2000 psi (Conoco Compound No. 15041, 16081; FPC 5305)
    2100 psi (FPC 5005, 5160)
    2200 psi (Conoco Compound No. 45541; FPC 5125A)
    2300 psi (Conoco Compound No. 35851)
    2370 psi (FPC 5790)
    2420 psi (FPC 1265C)
    2600 psi (FPC 5002)
Flex. Str.:
    23–32 MPa (Geon 85721)
    31–35 MPa (Geon 87001)
    36–39 MPa (Geon 85858)
    108 MPa (Geon 3007)
    110 MPa (Geon 3010)
    11,000 psi (Rucodur 1803)
    11,300 psi (Rucodur 1706)
    14,000 psi (Rucodur 1500, 1503, 1505)
    14,500 psi (Geon 88933)
    15,000 psi (Geon 88934)
Flex. Mod.:
    1035–1380 MPa (Geon 85721)
    1378–1723 MPa (Geon 85858)
    1380–1725 MPa (Geon 87001)
    2760 MPa (Geon 3010)
    2939 MPa (Geon 3007)
    386,000 psi (Rucodur 1803)
    387,000 psi (Geon 88933)
    395,000 psi (Geon 88934)
    410,000 psi (Rucodur 1706)

Polyvinyl chloride (cont'd.)

Impact Str. (Izod):
0.6–0.7 J/cm notched (Geon 87001)
0.7 J/cm notched (Geon 85858)
1.1 J/cm notched (Geon 3010)
1.6 J/cm notched (Geon 3007)
8–12 J/cm (Harwal GBM, GMW)
9 J/cm (Harwal GPR)
12–30 J/cm (Harwal AOB)
0.6 J/m notched (Geon 85721)
0.7 ft lb/in. notched (Conoco 90171)
0.9 ft lb/in. notched (Conoco 80172)
1.0 ft lb/in. notched (Conoco 80273)
2.0 ft lb/in. notched (Geon 88934)
2.3 ft lb/in. notched (Geon 88933)
15 ft lb/in. notched (Rucodur 1500, 1503)
16 ft lb/in. notched (Rucodur 1803)
17 ft lb/in. notched (Rucodur 1505)
20 ft lb/in. notched (Rucodur 1706)
Tens. Impact:
330+ ft lb/in.^{2} (Rucodur 1706, 1803)
Hardness:
Rockwell R101 (Rucodur 1706, 1803)
Rockwell R107 (Rucodur 1500, 1503, 1505)
Rockwell R118 (Geon 3010)
Rockwell R119 (Geon 88933, 88934)
Rockwell R120 (Geon 3007)
Shore A62 (FPC 1433)
Shore A65 (FPC 1359, 1415)
Shore A68 (Conoco Compound No. 26831)
Shore A68 ± 3 (Kohinor 509)
Shore A70 (FPC 1379)
Shore A71 (Conoco Compound No. 27121; FPC 1238)
Shore A73 (FPC 1419)
Shore A74 (Conoco Compound No. 17461)
Shore A74–77 (Harwal GSB-F)
Shore A75 (FPC 1363A, 1380, 1380A, 5439A)
Shore A77 (FPC 1239)
Shore A80 (FPC 1430, 5008)
Shore A81 (Conoco Compound No. 18161, 28131; FPC 5440)
Shore A82 (FPC 5004)
Shore A83 (FPC 5006A)
Shore A84 (FPC 1418)
Shore A85 (Conoco Compound No. 48561, 48562; FPC 5460)

Polyvinyl choride *(cont'd.)*

    Shore A85 ± 3 (Kohinor 510)
    Shore A86 (FPC 5264)
    Shore A86 ± 3 (Conoco Compound No. 38671)
    Shore A87 (FPC 5390)
    Shore A87 ± 3 (Kohinor 434)
    Shore A88 (Conoco Compound No. 28831; FPC 5060, 5760)
    Shore A89 (Conoco Compound No. 18961; FPC 1240)
    Shore A89 ± 3 (Conoco Compound No. 38931)
    Shore A90 (FPC 5605; Harwal GSG)
    Shore A90 ± 3 (Kohinor 311 FR)
    Shore A91 (FPC 5305)
    Shore A92 (FPC 5290, 5360, 5760B)
    Shore A93 (FPC 1381A, 5009, 5105, 5805)
    Shore A94 (FPC 1265C, 5005)
    Shore A95 (FPC 5000B)
    Shore A96 (FPC 5160, 5175)
    Shore C80 ± 3 (Kohinor 702)
    Shore C90 (FPC 5002)
    Shore D40–50 (Geon 85721)
    Shore D41 (Conoco Compound No. 44171)
    Shore D44 (Conoco Compound No. 44331)
    Shore D44 ± 3 (Conoco Compound No. 34431)
    Shore D45 (Conoco Compound No. 14331)
    Shore D45 ± 3 (Conoco Compound No. 34551)
    Shore D46 ± 3 (Conoco Compound No. 34641)
    Shore D50 (Conoco Compound No. 15041, 15071)
    Shore D54 ± 3 (Conoco Compound No. 35441)
    Shore D55 (Conoco Compound No. 45541; Geon 85858, 87001)
    Shore D58 (Harwal GSF 1)
    Shore D58 ± 3 (Conoco Compound No. 35851)
    Shore D60 (Conoco Compound No. 16081)
    Shore D61 (FPC 5790)
    Shore D62 (FPC 5125A)
    Shore D78 (Conoco 80172)
    Shore D78–80 (Harwal AOB)
    Shore D80 (Harwal GBM, GI, GMW, GSB)
    Shore D81 (Conoco 80273, 90171)
    Shore D82 (Harwal GPR, GPR)
Transmittance:
    85% (Harwal AOB, GBM, GMW)
THERMAL PROPERTIES:
Soften. Pt. (Vicat):
    74 C (Harwal GI)

77 C (Harwal GSB)
79 C (Harwal AOB, GBM, GMW)
Conduct.:
0.137 W/mK (Geon 3007)
Distort. Temp.:
57–58 C (1.8 MPa) (Geon 85721)
65 C (1.8 MPa) (Geon 87001)
66 C (1.8 MPa) (Geon 85858)
70 C (264 psi) (Rucodur 1706)
72 C (264 psi) (Rucodur 1500, 1503, 1505, 1803)
73 C (264 psi) (Conoco 80172, 80273, 90171)
100 C (1.8 MPa) (Geon 3010)
105 C (1.8 MPa) (Geon 3007)
212 F (1.8 MPa) (Geon 88933)
216 F (1.8 MPa) (Geon 88934)
Coeff. of Linear Exp.:
6.0×10^{-5} cm/cm/C (Harwal GPR)
6.3×10^5 cm/cm/C (Geon 85721)
6.8×10^5 cm/cm/C (Geon 3007, 3010)
3.6×10^5 in./in./F (Geon 88934)
3.9×10^5 in./in./F (Geon 88933)
Flamm.:
V-0 (Rucodur 1500, 1503, 1504, 1505, 1706)
Meets UL 1072/1277 requirements (Kohinor 311 FR, 940, 941, 947, 948)

ELECTRICAL PROPERTIES:
Dissip. Factor:
0.019 (60 Hz) (Geon 3007, 3010)
Dielec. Str.:
51,220 V/mm (Geon 3007, 3010)
Dielec. Const.:
2.9 (60 Hz) (Harwal GPR)
3.4 (1 kHz) (Geon 3007, 3010)
5.24 (FPC 5440)
5.75 (FPC 5004, 5006A)
Vol. Resist.:
6.6×10^9 ohm-cm (FPC 605)
7.6×10^9 ohm-cm (FPC 654, 654H, 6854)
8.1×10^9 ohm-cm (FPC 6366, 6866)
3×10^{10} ohm-cm (stabilized, 50 C) (Conoco Compound No. 38631)
6×10^{10} ohm-cm (stabilized, 50 C) (Conoco Compound No. 38601)
6.83×10^{10} ohm-cm (FPC 6411)
1.57×10^{11} ohm-cm (FPC 6337)
2×10^{11} ohm-cm (60 C) (FPC 5125A, 5760B)

Polyvinyl chloride *(cont'd.)*

2.19×10^{11} ohm-cm (FPC 6338)

3×10^{11} ohm-cm (stabilized, 50 C) (Conoco Compound No. 38671); (70 C) (Kohinor 509, 906)

$6\,2 \times 10^{11}$ ohm-cm (stabilized, 50 C) (Conoco Compound No. 38551)

7×10^{11} ohm-cm (70 C) (Kohinor 510)

1×10^{12} ohm-cm (60 C) (FPC 5264, 5605)

3×10^{12} ohm-cm (stabilized, 50 C) (Conoco Compound No. 34641); (60 C) (FPC 5760)

5×10^{12} ohm-cm (stabilized, 50 C) (Conoco Compound No. 34551); (60 C) (FPC 5305, 5460)

7×10^{12} ohm-cm (60 C) (FPC 5390)

1×10^{13} ohm-cm (stabilized, 50 C) (Conoco Compound No. 38931); (60 C) (FPC 5105, 5790); (70 C) (Kohinor 702)

1.5×10^{13} ohm-cm (60 C) (FPC 5005, 5060, 5175, 5290, 5360)

3×10^{13} ohm-cm (stabilized, 50 C) (Conoco Compound No. 35441, 35851)

3.5×10^{13} ohm-cm (60 C) (FPC 5805)

5×10^{13} ohm-cm (60 C) (FPC 5160)

7×10^{13} ohm-cm (stabilized, 50 C) (Conoco Compound No. 34431)

STD. PKGS.:

50-lb bags and bulk (Conoco 5305, 5385, 5425, 5465)

50-lb bags, hopper cars, hopper trucks, gaylords (Conoco Compound No. 14331, 15041, 15071, 16081, 17461, 18161, 18961, 26831, 27121, 28131, 28831, 34431, 34551, 34641, 35441, 35851, 38551, 38601, 38631, 38671, 38931, 44171, 44331, 45541, 48561, 48562)

1000-lb gaylords, hopper cars, and trucks (Conoco 80172, 80273, 90171)

Polyvinyl isobutyl ether

SYNONYMS:

Polyvinyl ether
PVI

STRUCTURE:

$[-CHOCH_2CH(CH_3)_2CH_2-]_n$

TRADENAME EQUIVALENTS:

Gantrez B-773 [GAF]
Perenol EI [Henkel Canada]

CATEGORY:

Polymer, plasticizer, leveling agent, defoamer, adhesive, tackifier

APPLICATIONS:

Industrial applications: coatings (Gantrez B-773); paints (Perenol EI)

PROPERTIES:
Form:
   Solution (Gantrez B-773)
Composition:
   70% active in hexane (Gantrez B-773)
GENERAL PROPERTIES:
Solubility:
   Sol. in aliphatic hydrocarbons (Gantrez B-773)
   Sol. in aromatic hydrocarbons (Gantrez B-773)
   Sol. in chlorinated hydrocarbons (Gantrez B-773)

Polyvinylpyrrolidone

SYNONYMS:
   1-Ethenyl-2-pyrrolidinone, homopolymer
   Povidone
   PVP (CTFA)
   2-Pyrrolidinone, 1-ethenyl-, homopolymer
EMPIRICAL FORMULA:
   $(C_6H_9NO)_x$
STRUCTURE:

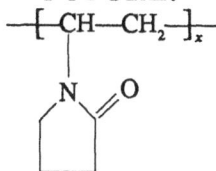

CAS No.:
   9003-39-8
TRADENAME EQUIVALENTS:
   Luviskol K12, K17, K30, K60, K80, K90 [BASF AG]
   Peregal ST [GAF]
   Plasdone, C-30 [GAF]
   Polyclar 10, AT [GAF]
   Polyplasdone XL, XL-10 [GAF]
   PVP K-15, K-26/28, K-29/32, K-30, K-60, K-90 [GAF]
   Sokalan HP-50 [BASF AG]

Polyvinylpyrrolidone (cont'd.)

MODIFICATIONS/SPECIALTY GRADES:
USP grade:
Plasdone C-30, K-26/28, K-29/32
Pharmaceutical grade:
Polyplasdone XL, XL-10

CATEGORY:
Linear polymer, thickener, protective colloid, film former, dispersant, stripping assistant, detergent, suspending agent, binder, disintegrant, stabilizer, solubilizer, cohesive agent, detoxicant, anticaking agent, lubricant, absorbent, complexing agent, viscosity modifier, anti-redeposition agent, leveling agent, vehicle

APPLICATIONS:
Agriculture industry: (PVP K-90)

Cosmetic industry applications: (Luviskol K12, K17, K30, K60, K80, K90); aerosols (PVP K-30); creams and lotions (PVP K-30); dyes (PVP K-30); hair preparations (PVP K-30); makeup (PVP K-30); shampoos (PVP K-30)

Food-contact applications: wines/vinegar (Polyclar 10, AT)

Industrial applications: (Luviskol K12, K17, K30, K60, K80, K90); adhesives (PVP K-60); antifreeze (PVP K-90); chromatography (Polyclar 10, AT; Polyplasdone XL); coatings (PVP K-90); detergents (Peregal ST; PVP K-90; Sokalan HP-50); dyes and pigments (Peregal ST; PVP K-90); inks (PVP K-90); paints (PVP K-90); paper coatings (PVP K-90); paper industry (Peregal ST; PVP K-90); printing industry (PVP K-90); printing pastes (PVP K-90); sizing (PVP K-90); textile applications (Peregal ST; PVP K-90)

Medical applications: (Plasdone C-30)

Pharmaceutical applications: aerosols (Plasdone K-26/28, K-29/32); drugs (Plasdone K-26/28, K-29/32); ophthalmic preparations (Plasdone K-26/28, K-29/32); tablet mfg. (Plasdone, K-26/28, K-29/32; Polyplasdone XL, XL-10); topical preparations (Plasdone K-26/28, K-29/32)

PROPERTIES:
Form:
Liquid (Luviskol K12, K17, K30, K60, K80, K90; Peregal ST; PVP K-60, K-90)
Powder (Luviskol K17, K30, K80, K90; Plasdone, C-30, K-26/28, K-29/32; Polyplasdone XL; PVP K-15, K-30, K-90; Sokalan HP-50)
Finely ground powder (Polyclar 10, AT; Polyplasdone XL-10)
Color:
Clear (PVP K-60)
Off-white (PVP K-15, K-30)
VCS 5 max. (Peregal ST)
Composition:
20% in water (Luviskol K80 liquid, K90 liquid; PVP K-90 liquid)
30% in water (Luviskol K30 liquid)
30-32% active (Peregal ST)

45% in water (Luviskol K60 liquid; PVP K-60)
50% in water (Luviskol K12 liquid, K17 liquid)
95% active (Sokalan HP-50); 95% min. active (PVP K-15, K-30, K-90 powder)

GENERAL PROPERTIES:

Solubility:

Sol. in alcohols (PVP K-30)
Sol. in amines (PVP K-30)
Sol. in some chlorinated compounds (PVP K-30)
Sol. in nitroparaffins (PVP K-30)
Sol. in many organic solvents (PVP K-30)
Sol. in water (Peregal ST; PVP K-15, K-30)

M.W.:

10,000 (PVP K-15)
40,000 (PVP K-30)
160,000 (PVP K-60)
360,000 (PVP K-90)

Density:

9.3 lb/gal (PVP K-60, K-90 sol'n.)
20 lb/ft³ (PVP K-90 powder)
28 lb/ft³ (PVP K-30)
36 lb/ft³ (PVP K-15)

pH:

5–7 (5% aq.) (Peregal ST)

PPG-9 (CTFA)

SYNONYMS:

Polyoxypropylene (9)
Polypropylene glycol (9)
POP (9)
PPG (9)

STRUCTURE:

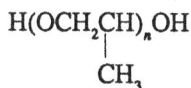

$$H(OCH_2CH)_n OH$$
$$CH_3$$

where avg. $n = 9$

CAS No.:

25322-69-4 (generic)
RD No.: 977055-51-8

PPG-9 *(cont'd.)*

TRADENAME EQUIVALENTS:
Alkapol PPG-425 [Alkaril]
Jeffox PPG-400 [Texaco]
Monolan PPG-440 [Harcros UK]
Pluracol P-410 [BASF]

CATEGORY:
Polymer, intermediate, solvent, cosolvent, lubricant, antifoam, plasticizer, antiblooming agent, binder, antistat

APPLICATIONS:
Cosmetic industry preparations: (Alkapol PPG-425; Monolan PPG440)

Industrial applications: (Alkapol PPG-425); ceramics (Pluracol P-410); dyes and pigments (Monolan PPG440); fermentation (Pluracol P-410); grinding fluids (Pluracol P-410); hydraulic systems (Alkapol PPG-425; Monolan PPG440; Pluracol P-410); inks (Monolan PPG440); lubricants (Jeffox PPG-400; Monolan PPG440; Pluracol P-410); metalworking (Pluracol P-410); ore flotation (Pluracol P-410); paints (Pluracol P-410); paper industry (Pluracol P-410); plastics/resins (Monolan PPG440); polyurethane foams (Pluracol P-410); rubber (Monolan PPG440); surfactants processing (Alkapol PPG-425; Monolan PPG440); wood treatment (Pluracol P-410)

PROPERTIES:
Form:
Liquid (Alkapol PPG-425; Monolan PPG440; Pluracol P-410)
Color:
Colorless (Monolan PPG440)
Water-white (Pluracol P-410)
Odor:
Faint (Monolan PPG440)
Slight ether-like (Pluracol P-410)
Composition:
100% active (Monolan PPG440)

GENERAL PROPERTIES:
Solubility:
Sol. in perchloroethylene (Alkapol PPG-425)
Sol. in water (Alkapol PPG-425; Jeffox PPG-400; Monolan PPG440; Pluracol P-410)
M.W.:
400 (Jeffox PPG-400); (avg.) (Monolan PPG440)
400–450 (Alkapol PPG-425)
425 (Pluracol P-410)
Sp. Gr.:
1.005 (Pluracol P-410)
1.010 (Monolan PPG440)

Density:
  1.01 g/ml (Alkapol PPG-425)
  8.36 lb/gal (Pluracol P-410)
Visc.:
  35 cps (100 F) (Pluracol P-410)
  80 cs (Monolan PPG440)
Pour Pt.:
  < 0 C (Monolan PPG440)
  −35 F (Pluracol P-410)
Flash Pt.:
  > 400 F (PMCC) (Pluracol P-410)
  > 450 F (COC) (Monolan PPG440)
Stability:
  Chemically inert, good thermal stability (Monolan PPG440)
pH:
  6–7 (Pluracol P-410)
  6–8 (5% in 10:1 methanol/water) (Alkapol PPG-425)

TOXICITY/HANDLING:
  Skin irritant on prolonged contact with concentrated state; spillages may be slippery (Monolan PPG440)
  Slight eye irritant (Pluracol P-410)

STORAGE/HANDLING:
  Do not mix with strong oxidizing or reducing agents—potentially explosive (Monolan PPG440)

STD. PKGS.:
  200 kg net mild steel drums or bulk (Monolan PPG440)
  55 gal drums, tank cars (Pluracol P-410)

PPG-12 (CTFA)

SYNONYMS:
  Polyoxypropylene (12)
  Polypropylene glycol (12)
  POP (12)
  PPG (12)

STRUCTURE:
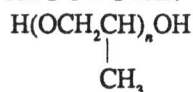

  $H(OCH_2CH)_nOH$
  CH_3

  where avg. $n = 12$

PPG-12 *(cont'd.)*

CAS No.:
25322-69-4 (generic)
RD No.: 977066-97-9

TRADENAME EQUIVALENTS:
Pluracol P-710 [BASF]

CATEGORY:
Polymer, lubricant, binder, antifoam, antiblooming agent, plasticizer, intermediate

APPLICATIONS:
Industrial applications: ceramics (Pluracol P-710); fermentation (Pluracol P-710); paint (Pluracol P-710); wood treatment (Pluracol P-710); ceramics (Pluracol P-710); metalworking (Pluracol P-710); paints (Pluracol P-710); paper industry (Pluracol P-710); polyurethane foams (Pluracol P-710); hydraulic systems (Pluracol P-710); grinding and cutting fluids (Pluracol P-710); lubricants (Pluracol P-710); ore flotation (Pluracol P-710)

PROPERTIES:

Form:
Liquid (Pluracol P-710)

Color:
Water-white (Pluracol P-710)

Odor:
Slight, ether-like (Pluracol P-710)

Composition:
100% active (Pluracol P-710)

GENERAL PROPERTIES:

Ionic Nature:
Nonionic (Pluracol P-710)

Solubility:
Moderately sol. in water (Pluracol P-710)

M.W.:
775 (Pluracol P-710)

Sp. Gr.:
1.004 (Pluracol P-710)

Density:
8.35 lb/gal (Pluracol P-710)

Visc.:
65 cps (100 F) (Pluracol P-710)

Pour Pt.:
–35 F (Pluracol P-710)

Flash Pt.:
> 400 F (PMCC) (Pluracol P-710)

Acid No.:
0.02 max. (Pluracol P-710)

PPG-12 *(cont'd.)*

Hydroxyl No.:
  145 (Pluracol P-710)
pH:
  6–7 (Pluracol P-710)
TOXICITY/HANDLING:
  Relatively nontoxic orally; nonirritating to skin; will cause slight, temporary eye irritation on contact (Pluracol P-710)
STORAGE/HANDLING:
  Storage at 70–100 F is recommended (Pluracol P-710)

PPG-17 (CTFA)

SYNONYMS:
  Polyoxypropylene (17)
  Polypropylene glycol (17)
  POP (17)
  PPG (17)
STRUCTURE:

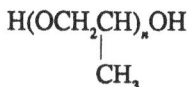

  where avg. $n = 17$
CAS No.:
  25322-69-4 (generic)
  RD No.: 977066-99-1
TRADENAME EQUIVALENTS:
  Monolan PPG1100 [Harcros UK]
  Pluracol P-1010 [BASF]
CATEGORY:
  Polymer, lubricant, intermediate, antifoam, plasticizer, antiblooming agent, binder, antistat, cosolvent
APPLICATIONS:
  Cosmetic industry preparations: (Monolan PPG1100)
  Industrial applications: ceramics (Pluracol P-1010); dyes and pigments (Monolan PPG1100); fermentation (Pluracol P-1010); grinding fluids (Pluracol P-1010); hydraulic systems (Monolan PPG1100; Pluracol P-1010); inks (Monolan PPG1100); metalworking (Pluracol P-1010); ore flotation (Pluracol P-1010); paints (Pluracol P-1010); paper industry (Pluracol P-1010); plastics/resins (Monolan PPG1100); polyurethane foams (Pluracol P-1010); rubber (Monolan PPG1100);

327

PPG-17 *(cont'd.)*

surfactants processing (Monolan PPG1100); wood treatment (Pluracol P-1010)

PROPERTIES:

Form:
Liquid (Monolan PPG1100; Pluracol P-1010)

Color:
Colorless (Monolan PPG1100)
Water-white (Pluracol P-1010)

Odor:
Faint (Monolan PPG1100)
Slight ether-like (Pluracol P-1010)

Composition:
100% active (Monolan PPG1100)

GENERAL PROPERTIES:

Ionic Nature:
Nonionic (Monolan PPG1100)

Solubility:
Insol. in water (Monolan PPG1100); practically insol. in water @ R.T. (Pluracol P-1010)

M.W.:
1000 (avg.) (Monolan PPG1100)
1050 (Pluracol P-1010)

Sp. Gr.:
1.005 (Monolan PPG1100)
1.007 (Pluracol P-1010)

Density:
8.38 lb/gal (Pluracol P-1010)

Visc.:
80 cps (100 F) (Pluracol P-1010)
180 cs (Monolan PPG1100)

Pour Pt.:
< 0 C (Monolan PPG1100)
−35 F (Pluracol P-1010)

Flash Pt.:
> 400 F (PMCC) (Pluracol P-1010)
> 450 F (COC) (Monolan PPG1100)

Stability:
Chemically inert, good thermal stability (Monolan PPG1100)

pH:
6–7 (Pluracol P-1010)

TOXICITY/HANDLING:

Skin irritant on prolonged contact with concentrated state; spillages may be slippery (Monolan PPG1100)

Slight eye irritant (Pluracol P-1010)
STORAGE/HANDLING:
Do not mix with strong oxidizing or reducing agents—potentially explosive (Monolan PPG1100)
STD. PKGS.:
55 gal drums, tank cars (Pluracol P-1010)

PPG-26 (CTFA)

SYNONYMS:
Polyoxypropylene (26)
Polypropylene glycol (26)
POP (26)
PPG (26)
STRUCTURE:

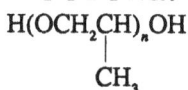

where avg. $n = 26$
CAS No.:
25322-69-4 (generic)
RD No.: 977055-52-9
TRADENAME EQUIVALENTS:
Alkapol PPG-2000 [Alkaril]
Jeffox PPG-2000 [Texaco]
Monolan PPG2200 [Harcros UK]
Pluracol P-2010 [BASF]
Pluriol P 2000 [BASF AG]
Polyglycol P-200 [Dow Europe]
CATEGORY:
Polymer, antifoam, carrier, cosolvent, lubricant, intermediate, antistat, plasticizer, release agent, solubilizer, binder, antiblooming agent
APPLICATIONS:
Cosmetic industry preparations: (Alkapol PPG-2000; Monolan PPG2200)
Industrial applications: (Alkapol PPG-2000); ceramics (Pluracol P-2010); chemical synthesis (Jeffox PPG-2000; Pluriol P 2000); cutting and grinding fluids (Pluracol P-2010; Pluriol P 2000); detergents (Jeffox PPG-2000); dyes and pigments (Monolan PPG2200); fermentation (Pluracol P-2010); hydraulic systems (Monolan PPG2200; Pluracol P-2010; Pluriol P 2000); inks (Monolan PPG2200); latexes

PPG-26 *(cont'd.)*

(Pluriol P 2000); lubricants (Pluriol P 2000); metalworking (Pluracol P-2010); ore flotation (Pluracol P-2010); paper industry (Pluracol P-2010); paints (Pluracol P-2010; Pluriol P 2000; Polyglycol P-200); plastics/resins (Monolan PPG2200); polyurethane elastomers/foams (Jeffox PPG-2000; Pluracol P-2010); refrigerants/coolants (Pluriol P 2000); rubber (Monolan PPG2200; Pluriol P 2000); surfactants processing (Monolan PPG2200); tire lubricants (Alkapol PPG-2000); wood treatment (Pluracol P-2010)

PROPERTIES:
Form:
Liquid (Alkapol PPG-2000; Monolan PPG2200; Pluracol P-2010; Polyglycol P-200)
Clear liquid (Pluriol P 2000)
Color:
Colorless (Monolan PPG2200; Pluriol P 2000)
Water-white (Pluracol P-2010)
Odor:
Faint (Monolan PPG2200)
Slight ether-like (Pluracol P-2010)
Composition:
100% active (Monolan PPG2200; Pluriol P 2000)

GENERAL PROPERTIES:
Ionic Nature:
Nonionic (Monolan PPG2200)
Solubility:
Sol. in aromatic solvents (Alkapol PPG-2000)
Miscible with ethanol (Pluriol P 2000)
Miscible with most oils (Pluriol P 2000)
Sol. in perchloroethylene (Alkapol PPG-2000)
Miscible with toluene (Pluriol P 2000)
Miscible with trichloroethylene (Pluriol P 2000)
Insol. in water (Monolan PPG2200); practically insol. in water @ R.T. (Pluracol P-2010)
M.W.:
1900–2100 (Alkapol PPG-2000)
2000 (Jeffox PPG-2000; Monolan PPG2200; Pluracol P-2010; Pluriol P 2000)
Sp. Gr.:
1.0 (Pluriol P 2000)
1.002 (Pluracol P-2010)
1.004 (Monolan PPG2200)
Density:
1.01 g/ml (Alkapol PPG-2000)
8.34 lb/gal (Pluracol P-2010)
Visc.:
175 cps (100 F) (Pluracol P-2010)

440 cs (Pluriol P 2000)
450 cs (Monolan PPG2200)

Pour Pt.:
< 0 C (Monolan PPG2200)
−35 F (Pluracol P-2010; Pluriol P 2000)

Flash Pt.:
222 C (Pluriol P 2000)
> 400 F (PMCC) (Pluracol P-2010)
> 450 F (COC) (Monolan PPG2200)

Stability:
Chemically inert, good thermal stability (Monolan PPG2200)

pH:
6–7 (Pluracol P-2010)
6–8 (5% in 10:1 methanol/water) (Alkapol PPG-2000)
6.5–7.5 (1% aq.) (Pluriol P 2000)

TOXICITY/HANDLING:
Slight eye irritant (Pluracol P-2010)
Skin irritant on prolonged contact with concentrated state; spillages may be slippery (Monolan PPG2200)

STORAGE/HANDLING:
Do not mix with strong oxidizing or reducing agents—potentially explosive (Monolan PPG2200)

STD. PKGS.:
200 kg net mild steel drums or bulk (Monolan PPG2200)
55 gal drums, tank cars (Pluracol P-2010)

PPG-30 (CTFA)

SYNONYMS:
Polyoxypropylene (30)
Polypropylene glycol (30)
POP (30)
PPG (30)

STRUCTURE:

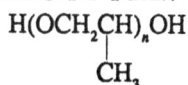

where avg. *n* = 30

PPG-30 *(cont'd.)*

CAS No.:
25322-69-4 (generic)
RD No.: 977055-53-0

TRADENAME EQUIVALENTS:
Alkapol PPG-4000 [Alkaril]
Pluracol P-4010 [BASF]

CATEGORY:
Polymer, antifoam, carrier, cosolvent, lubricant, intermediate, plasticizer

APPLICATIONS:
Cosmetic industry preparations: (Alkapol PPG-4000)
Industrial applications: (Alkapol PPG-4000); ceramics (Pluracol P-4010); fermentation (Pluracol P-4010); grinding fluids (Pluracol P-4010); hydraualic systems (Pluracol P-4010); metalworking (Pluracol P-4010); ore flotation (Pluracol P-4010); paints (Pluracol P-4010); paper industry (Pluracol P-4010); polyurethane foams (Pluracol P-4010); tire lubricants (Alkapol PPG-4000); wood treatments (Pluracol P-4010)

PROPERTIES:
Form:
Liquid (Alkapol PPG-4000; Pluracol P-4010)
Color:
Water-white (Pluracol P-4010)
Odor:
Slight ether-like (Pluracol P-4010)
Composition:
100% conc. (Alkapol PPG-4000)

GENERAL PROPERTIES:
Solubility:
Sol. in perchloroethylene (Alkapol PPG-4000)
Practically insol. in water @ R.T. (Pluracol P-4010)
M.W.:
3900–4100 (Alkapol PPG-4000)
4000 (Pluracol P-4010)
Sp. Gr.:
1.00 (Pluracol P-4010)
Density:
1.01 g/ml (Alkapol PPG-4000)
8.33 lb/gal (Pluracol P-4010)
Visc.:
550 cps (100 F) (Pluracol P-4010)
Pour Pt.:
–20 F (Pluracol P-4010)

Flash Pt.:
> 400 F (PMCC) (Pluracol P-4010)
pH:
6–7 (Pluracol P-4010)
6–8 (Alkapol PPG-4000)
TOXICITY/HANDLING:
Slight eye irritant (Pluracol P-4010)
STD. PKGS.:
55 gal drums, tank cars (Pluracol P-4010)

PVM/MA copolymer (CTFA)

SYNONYMS:
2,5-Furandione, polymer with methoxyethylene
Methyl vinyl ether/maleic anhydride copolymer
Poly(methyl vinyl ether/maleic anhydride)
EMPIRICAL FORMULA:
$(C_4H_2O_3 \cdot C_3H_6O)_x$
CAS No.:
9011-16-9
TRADENAME EQUIVALENTS:
Gaftex PT [GAF]
Gantrez AN, AN-119, AN-139, AN-149, AN-169, AN-179 [GAF]
Thickener L, LN [GAF]
CATEGORY:
Copolymer, thickener, dispersant, stabilizer, protective colloid, gelling agent, corrosion inhibitor, suspending aid, coupling agent, anti-redeposition agent, solubilizer, film former, antistat
APPLICATIONS:
Agriculture industry: pesticides/herbicides (Gantrez AN); soil conditioners (Gantrez AN)

Cleansers: detergent bars (Gantrez AN); hand cleaners (Gantrez AN)

Cosmetic industry applications: (Gantrez AN); hair preparations (Gantrez AN); toiletries (Gantrez AN)

Industrial applications: acid bowl cleaners (Gantrez AN); adhesives (Gantrez AN, AN-119, -139, -149, -169, -179; Thickener L, LN); asbestos (Gantrez AN); carpet and upholstery backing (Thickener L); ceramics (Gantrez AN); chemical processing (Gantrez AN); coatings (Gantrez AN, AN-119, -139, -149, -169, -179); detergents (Gantrez AN, AN-119, -139, -149, -169, -179); dyes and pigments (Gantrez AN,

PVM/MA copolymer (cont'd.)

AN-119, -139, -149, -169, -179); films (Gantrez AN-119, -139, -149, -169, -179; Thickener L); fire fighting foams (Gantrez AN); inks (Gantrez AN); latexes (Gantrez AN; Thickener L, LN); leather applications (Gantrez AN); paints (Gantrez AN; Thickener L, LN); paper industry (Gantrez AN); photoreproduction (Gantrez AN); plastics (Gantrez AN-119, -139, -149, -169, -179); polishes and waxes (Gantrez AN); polymerization (Gantrez AN); printing pastes (Gantrez AN; Thickener L, LN); steel processing (Gantrez AN); textile applications (Gaftex PT; Gantrez AN, AN-119, -139, -149, -169, -179); water treatment (Gantrez AN)

Pharmaceutical applications: (Gantrez AN); tablet coating (Gantrez AN)

PROPERTIES:

Form:
Liquid (Thickener L, LN)
Powder (Gaftex PT)
Fluffy powder (Gantrez AN)
Low-viscosity powder (Gantrez AN-119)
Medium-viscosity powder (Gantrez AN-139, -149)
High-viscosity powder (Gantrez AN-169, -179)

Color:
White (Gantrez AN)

Composition:
15% active (Thickener L, LN)
100% conc. (Gantrez AN)

GENERAL PROPERTIES:

Solubility:
Sol. in acid (Gantrez AN-119, -139, -149, -169, -179)
Sol. in alcohols (Gantrez AN)
Sol. in lower aliphatic esters (Gantrez AN)
Sol. in aldehydes (Gantrez AN)
Sol. in caustic (Gantrez AN-119, -139, -149, -169, -179)
Sol. in ketones (Gantrez AN)
Sol. in lactams (Gantrez AN)
Sol. in certain organic solvents (Gantrez AN, AN-119, -139, -149, -169, -179)
Sol. in phenols (Gantrez AN)
Sol. in pyridine (Gantrez AN)
Sol. in water (Gantrez AN-119, -139, -149, -169, -179); sol. in water over entire pH range (Gantrez AN)

Sp. Gr.:
1.37 (Gantrez AN)

Density:
0.32 g/cc (Gantrez AN)

Softening Pt.:
200–225 C (Gantrez AN)

Stability:
Stable to acid and alkalis (Gantrez AN)
pH:
≈ 2 (free acid—5% aq. sol'n.) (Gantrez AN)
9.0 ± 0.5 (Thickener L, LN)

Sodium polyacrylate

CAS No.:
9003-04-7

TRADENAME EQUIVALENTS:
Acrysol GS, HV-1 [Rohm & Haas]
Alcogum [Alco]
Alcosperse 107, 124, 149, 157, 602-N [Alco]
Colloid 202, 207, 208, 211, 218D, 223, 223D, 230, 233, 245D, 350 [Colloids]
Daxad 37LN7, 37LN10, 37NS [W.R. Grace]
Densol 1010 [Graden]
Drewsperse 611 [Drew]
Good-rite K-7028, K-7058, K-7200, K-7600 [B.F. Goodrich]
Hostacerin PN 73 [Hoechst-Celanese]
Serpol QPA 160 [Servo]
Sokalan CP 5, CP 5 Powder, CP 7, CP 8, CP 10, CP 10S, CP 13S, CP 45, PA 15, PA 20, PA 20 PN, PA 25 PN, PA 30, PA 40, PA 40 Powder, PA 50, PA 70 PN, PA 80 [BASF AG]

CATEGORY:
Polymer, thickener, viscosity control agent, dispersant, stabilizer, anti-redeposition aid, detergent, processing aid, rheology control agent, sequestrant, chelating agent, scale inhibitor, grinding aid, deflocculant, fluidifier

APPLICATIONS:
Consumer products: footwear (Densol 1010)
Cosmetic industry preparations: (Hostacerin PN 73)
Household detergents: (Colloid 208, 233, 245D; Sokalan CP 7, PA 15, PA 20, Pa 80); car wash (Colloid 207, 245D); dishwashing (Colloid 218D, 245D; Good-rite K-7028, K-7058, K-7200, K-7600); laundry detergents (Colloid 202, 207; Good-rite K-7028, K-7058, K-7200, K-7600); spray-dried detergents (Colloid 208; Good-rite K-7028, K-7058, K-7200, K-7600)
Industrial applications: adhesives (Acrysol GS; Densol 1010); carpet and upholstery backing (Acrysol HV-1; Densol 1010); ceramics (Alcosperse 107, 124, 149, 157, 602-N); clays (Colloid 211, 230; Daxad 37LN7, 37LN10, 37NS); coatings (Acrysol GS); detergents (Colloid 245D; Daxad 37LN7, 37LN10); fillers (Colloid 230); films (Acrysol GS); institutional detergents (Colloid 207); latexes (Acrysol GS; Alcosperse 149, 157; Densol 1010); mining industry (Alcosperse 107, 124, 149, 157, 602-N); paints (Acrysol GS; Alcosperse 107, 124, 149, 157, 602-N; Daxad 37LN7, 37LN10; Drewsperse 611); paper coating (Alcosperse 107, 124, 149, 157,

Sodium polyacrylate *(cont'd.)*

602-N; Colloid 230; Daxad 37LN7, 37LN10); paper industry (Serpol QPA 160); petroleum industry (Daxad 37LN7, 37LN10); pigments (Alcosperse 107, 124, 149, 157, 602-N; Colloid 211, 230; Daxad 37LN7, 37LN10, 37NS; Drewsperse 611); rubber (Alcosperse 149, 157); slurries (Alcosperse 107, 124, 149, 157, 602-N; Colloid 211, 230, 350; Daxad 37LN7, 37LN10, 37NS; Sokalan PA 15, PA 20); textile applications (Alcosperse 107, 124, 149, 157, 602-N)

PROPERTIES:

Form:
  Liquid (Alcosperse 107, 124, 149; Colloid 202, 207, 208, 211, 223, 230, 233; Sokalan CP 5, CP 7, CP 8, CP 10, CP 10S, CP 13S, CP 45, PA 15, PA 20, PA 20 PN, PA 25 PN, PA 30, PA 40, PA 50, PA 70 PN, PA 80)
  Clear to slightly hazy liquid (Daxad 37LN7, 37LN10)
  Aq. solution (Alcogum; Alcosperse 157, 602-N; Good-rite K-7028, K-7058, K-7200, K-7600)
  Gel (Densol 1010)
  Powder (Colloid 218D, 245D; Good-rite K-7028, K-7058, K-7200, K-7600; Hostacerin PN 73; Sokalan CP 5 Powder, PA 40 Powder)
  Dry form (Alcosperse 107, 124, 149; Colloid 223D)
Color:
  Colorless (Good-rite K-7028, K-7058, K-7200, K-7600)
  Pale amber (Daxad 37LN7, 37LN10; Densol 1010)
Odor:
  Odorless (Good-rite K-7028, K-7058, K-7200, K-7600)
Composition:
  10% solids (Densol 1010)
  25% min. active in water (Alcosperse 157)
  30% conc. (Sokalan PA 70 PN)
  35% active (Sokalan CP 8, CP 13S, PA 40, PA 80)
  40% active (Sokalan CP 5, CP 7, PA 50)
  40% active in water (Alcosperse 149)
  45% active (Sokalan CP 10, CP 45, PA 15, PA 20, PA 20 N, PA 30)
  45% total solids (Daxad 37LN7, 37LN10)
  50% active (Sokalan CP 10S)
  54% conc. (Sokalan PA 25 PN)
  91% active (Sokalan PA 40 Powder)
  92% active (Sokalan CP 5 Powder)

GENERAL PROPERTIES:

Ionic Nature:
  Anionic (Alcosperse 107; Alcosperse 124, 149, 157, 602-N; Colloid 230)
Solubility:
  Sol. in water (Alcosperse 107, 124, 149, 157, 602-N; Colloid 207, 208, 218D, 230, 245D; Daxad 37LN7, 37LN10; Densol 1010; Hostacerin PN 73; Serpol QPA 160)

Sodium polyacrylate *(cont'd.)*

Sp. Gr.:
1.30 (Daxad 37LN7, 37LN10)
Density:
8.5 lb/gal (Densol 1010)
10.9 lb/gal (Daxad 37LN7, 37LN10)
Visc.:
500 cps (Daxad 37LN7, 37LN10)
Stability:
Stable over wide pH range; hydrolytically stable (Daxad 37LN7, 37LN10)
Storage Stability:
Good storage stability (Alcosperse 149, 157)
pH:
7.0 (Daxad 37LN7)
9 (Alcosperse 149, 157)
10 (Daxad 37LN10); (2% sol'n.) (Densol 1010)
Surface Tension:
71 dynes/cm (1% sol'n.) (Daxad 37LN10)
75 dynes/cm (1% sol'n.) (Daxad 37LN7)
STORAGE/HANDLING:
Store at R.T.; if frozen, warm to 100–120 F before using (Densol 1010)
STD. PKGS.:
55-gal fiber drums (Densol 1010)

Sodium polymethacrylate (CTFA)

SYNONYMS:
2-Propenoic acid, 2-methyl-, homopolymer, sodium salt
EMPIRICAL FORMULA:
$(C_4H_6O_2)_x \bullet XNa$
STRUCTURE:

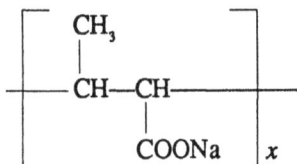

CAS No.:
25086-62-8; 54193-36-1

338

Sodium polymethacrylate (cont'd.)

TRADENAME EQUIVALENTS:
Darvan #7 [R.T. Vanderbilt]
Daxad 30, 30-30, 30S, 34N10, 35, 41 [W.R. Grace]

CATEGORY:
Polymer, surfactant, stabilizer, dispersant, scale inhibitor, viscosity control agent

APPLICATIONS:
Agriculture industry: (Daxad 30, 30S)

Cosmetic industry preparations: (Daxad 30)

Industrial applications: fillers (Daxad 35); latexes (Darvan #7; Daxad 34N10); rubber (Darvan #7); paints (Daxad 30, 30-30, 30S, 35); paper coating (Daxad 34N10); pigments (Daxad 30, 30-30, 30S, 35); polymerization (Daxad 30, 30-30, 34N10); water treatment (Daxad 30, 30-30, 30S, 35, 41); clay slurries (Daxad 30, 30-30, 35, 34N10)

Industrial cleaners: (Daxad 30, 34N10)

PROPERTIES:
Form:
Clear liquid (Daxad 30, 30-30, 34N10, 35)
Clear to slightly opalescent liquid (Darvan #7)
Fine powder (Daxad 30S, 41)

Color:
Water-white (Darvan #7; Daxad 30, 30-30)
White (Daxad 30S, 41)
Pale amber (Daxad 34N10, 35)

Composition:
25% solids (Daxad 30)
$25 \pm 1\%$ total solids in water (Darvan #7)
30% solids (Daxad 30-30, 34N10)
40% total solids (Daxad 35)
90% active, 10% moisture (Daxad 30S, 41)

GENERAL PROPERTIES:
Solubility:
Sol. in water systems (Daxad 30, 30-30, 30S, 34N10, 35, 41)

Sp. Gr.:
1.15 (Daxad 30)
1.21 (Daxad 30-30, 34N10)
1.28 (Daxad 35)

Density:
1.16 ± 0.02 mg/m³ (Darvan #7)
32–35 lb/ft³ (tamped) (Daxad 30S)
9.6 lb/gal (Daxad 30)
10.0 lb/gal (Daxad 34N10)
10.1 lb/gal (Daxad 30-30)

Sodium polymethacrylate *(cont'd.)*

10.6 lb/gal (Daxad 35)

Visc.:

75 cps max. (Darvan #7; Daxad 30)

150 cps max. (Daxad 30-30)

Stability:

Stable over wide pH range (Daxad 30, 30-30, 34N10, 35)

Stable over wide pH range; medium to high hygroscopicity (Daxad 30S)

Stable over wide pH range; stable at high temps. and pressures; medium to high hygroscopicity (Daxad 41)

pH:

7.5 (1% sol'n.) (Daxad 41)

9.5–10.5 (Darvan #7)

10.0 (Daxad 30, 30-30, 34N10); (1% sol'n.) (Daxad 30S)

Surface Tension:

70 dynes/cm (1% sol'n.) (Daxad 30, 30-30, 34N10, 35, 41)

STORAGE/HANDLING:

Should not be stored below 5 C for long periods of time as freezing may cause phase separation; warm and agitate to correct (Daxad 30, 30-30)

Sodium polynaphthalene sulfonate *(CTFA)*

SYNONYMS:

Sodium naphthalene-formaldehyde sulfonate

EMPIRICAL FORMULA:

$(C_{10}H_8O_3S \cdot CH_2O)_x \cdot x$ Na

CAS No.:

9084-06-4

TRADENAME EQUIVALENTS:

Darvan No. 1 [R.T. Vanderbilt]

Intraphor AC [Henkel-Nopco]

Lomar DL, LS, PL, PW, ST [Henkel/Process]

Nopcosant [Henkel/Process]

Protodye QNF [Proctor]

Tamol 819, 819L-43, L Conc., SD, SN [Rohm & Haas]

Tamol NH 3901, NH 7519, NH 9103, NN, NN 2406, NN 2901, NN 4501, NNO, NNOK [BASF AG]

Tek Tan ND, NL [Hamblet & Hayes]

Vultamol, SA Liq. [BASF AG]

Wettol D2 [BASF AG]

Sodium polynaphthalene sulfonate *(cont'd.)*

CATEGORY:

Polyelectrolyte, dispersant, stabilizer, leveling agent, suspending agent, wetting agent, plasticizer, grinding aid, precipitant, tanning agent

APPLICATIONS:

Agriculture industry: EPA-approved applications (Lomar PL, PW); pesticides (Intraphor AC); wettable powders (Wettol D 2)

FDA-approved applications: (Lomar PL, PW)

Food-contact applications: food packaging (Lomar PL, PW)

Industrial applications: carbon black (Tamol 819, 819L-43, NH 3901, NH 7519, NN 2406, NN 2901, NN 4501); cement/concrete (Nopcosant; Tamol 819, 819L-43, NH 9103, NN, NNO, NNOK); ceramics (Lomar PL, PW); dyes and pigments (Intraphor AC; Lomar LS, PL, PW; Nopcosant; Protodye QNF; Tamol 819, 819L-43, L Conc., NH 3901, NH 7519, NH 9103, NN, NN 2406, NN 2901, NN 4501, NNO, NNOK); extenders/fillers (Lomar PL, PW); gypsum (Lomar PL, PW; Tamol NH 9103); latexes (Darvan No. 1; Lomar PL, PW; Nopcosant; Protodye QNF; Tamol NN 4501; Vultamol, SA Liq.); leather processing (Tamol L Conc., SD, SN, ND); mortar (Tamol NH 9103); paints (Lomar PL, PW; Nopcosant); polymerization (Lomar PL, PW); printing industry (Lomar PL, PW); rubber (Darvan No. 1; Lomar PL, PW; Nopcosant; Tamol NN 4501; Vultamol, SA Liq.); sealants (Nopcosant); textile applications (Lomar LS, PL, PW; Protodye QNF); wood carbonizing (Protodye QNF)

Industrial cleaning: metal cleaning (Tamol NN, NNO, NNOK)

PROPERTIES:

Form:

Liquid (Lomar DL, PL; Protodye QNF; Tamol 819L-43, L Conc., NH 3901, NN 2406, NN 2901, NN 4501; Vultamol SA Liq.)

Clear liquid (Tamol NN)

Aq. sol'n. (Tek Tan NL)

Solid (Tamol SD, SN)

Dry form (Tek Tan ND)

Powder (Intraphor AC; Lomar LS; Nopcosant; Tamol NH 7519, NH 9103, NNO, NNOK; Vultamol; Wettol D 2)

Fine powder (Lomar PW)

Granules (Darvan No. 1)

Spray-dried solid (Tamol 819)

Color:

Light (Intraphor AC)

Buff (Darvan No. 1)

Light yellow (Protodye QNF)

Amber (Lomar DL, PL)

Tan (Lomar LS, PW; Nopcosant)

Pale brown (Tamol NN, NNO, NNOK)

Sodium polynaphthalene sulfonate *(cont'd.)*

Composition:
16% active (Protodye QNF)
24% conc. (Tamol 2406)
30% conc. (Tamol NN 2901)
30% active, 34% solids (Lomar DL)
33% active (Tamol NN)
38% conc. (Tamol NH 3901)
41% active, 47% solids (Lomar PL)
43% solids (Tamol 819L-43)
45% conc. (Tamol NN 4501; Vultamol SA Liq.)
47% active (Tamol L Conc.)
75% conc. (Tamol NH 7519)
87% active (Lomar PW); 87% min. active (Darvan No. 1)
90% conc. (Tamol NH 9103)
91% conc. (Vultamol)
95% active (Lomar LS; Tamol NNO, NNOK; Wettol D 2)

GENERAL PROPERTIES:
Ionic Nature:
Anionic (Tamol L Conc., NNOK)
Solubility:
Very sol. in water (Darvan No. 1); sol. in water (Lomar ST; Nopcosant; Protodye QNF; Tamol NH 7519, NH 9103; Tek Tan ND; Vultamol; Wettol D 2); freely sol. in hard or soft water; forms clear 10% aq. sol'ns. (Lomar PW); forms slightly hazy aq. sol'ns. (Lomar LS); miscible with water (Tamol NH 3901, NN 2406, NN 2901, NN 4501; Tek Tan NL; Vultamol SA Liq.)
Sp. Gr.:
1.02 (Nopcosant)
1.20 (Lomar DL)
1.25 (Lomar PL)
Density:
0.66 g/cc (Lomar PW)
pH:
6.5–7.5 (10% aq.) (Tamol NNO)
7.5–8.0 (2%) (Nopcosant)
8.0–10.5 (1% sol'n.) (Darvan No. 1)
9.0 (Lomar DL)
9.5 (10% aq.) (Lomar LS, PW); (20% sol'n.) (Lomar PL)
9.5–10.5 (10% aq.) (Tamol NN, NNOK)

STORAGE/HANDLING:
Store in a cool, dry place (Lomar LS)
Store in a dry place and keep tightly covered (Lomar PW)

Sodium polynaphthalene sulfonate *(cont'd.)*

STD. PKGS.:

55-gal lined fiber drums (Lomar DL, PL)
50-lb multiwall polyethylene-lined paper bags (Lomar LS)
50-lb net multiwall paper bags or 250-lb net drums (Lomar PW)

Styrene/PVP copolymer (CTFA)

SYNONYMS:

1-Ethenyl-2-pyrrolidinone, polymer with ethenylbenzene
2-Pyrrolidinone, 1-ethenyl-, polymer with ethenylbenzene
Styrene/vinylpyrrolidone copolymer

EMPIRICAL FORMULA:

$(C_8H_8 \cdot C_6H_9NO)_x$

CAS No.:

25086-29-7

TRADENAME EQUIVALENTS:

Polectron 430 [GAF]

CATEGORY:

Polymer, binder, stabilizer, opacifier

APPLICATIONS:

Cosmetric industry preparations: (Polectron 430)

Industrial applications: adhesives (Polectron 430); aerosol starch resins (Polectron 430); carpet backings (Polectron 430); coatings (Polectron 430); concrete (Polectron 430); detergents (Polectron 430); floor waxes (Polectron 430); glass fibers (Polectron 430); paper coatings (Polectron 430); textile applications (Polectron 430)

PROPERTIES:

Form:
Emulsion (Polectron 430)

Composition:
40% active in water (Polectron 430)

Vinylidene chloride copolymer

SYNONYMS:
Polyvinylidene choride
VC copolymer
VDC copolymer

EMPIRICAL FORMULA:
——$CH_2=CCl_2$——

TRADENAME EQUIVALENTS:
Daran 220, 229, 820, SL143 [W.R. Grace]
Polidene 33-001, 33-004, 33-021, 33-031, 33-038, 33-041, 33-075 [Scott Bader]

MODIFICATIONS/SPECIALTY GRADES:
Phosphate-plasticized:
Polidene 33-031

CATEGORY:
Thermoplastic resin

APPLICATIONS:
Functional additives: binder (Polidene 33-001, 33-021, 33-031, 33-038, 33-075; Versaflex 6); pigment binder (Polidene 33-004)
Industrial applications: adhesives (Polidene 33-004, 33-031); barrier coating (Daran 220, 229, 820, SL143); coatings (Daran 220, 229, SL143; Polidene 33-004, 33-041, 33-075; Versaflex 6, 7); fiber impregnants (Polidene 33-004); films (Polidene 33-001, 33-004, 33-031, 33-041); fire-retardant applications (Polidene 33-004, 33-021, 33-031, 33-075; Versaflex 6, 7); heat sealing applications (Daran 220, 229, 820); masking tape (Polidene 33-001); paints (Polidene 33-075; Versaflex 6, 7); paper coatings (Daran 229, SL143; Polidene 33-001, 33-031; Versaflex 6); textile applications (Polidene 33-031, 33-038; Versaflex 7); wallpaper (Polidene 33-041)

PROPERTIES:
Form:
Emulsion; 0.08–0.12 µ particle size (Daran 820)
Emulsion; 0.10–0.14 µ particle size (Daran 220)
Emulsion; 0.15 µ particle size (Versaflex 6, 7)
Emulsion; 0.15–0.19 µ particle size (Daran 229)
Emulsion; 0.25 µ particle size (Polidene 33-001, 33-004, 33-021, 33-038, 33-075)
Emulsion; 0.3 µ particle size (Polidene 33-031, 33-041)
Latex; 0.11 µ particle size (Daran SL143)

Color:
White (Versaflex 6, 7)
Odor:
Slight, characteristic (Versaflex 6, 7)
Composition:
45 ± 1% solids (Polidene 33-021, 33-038)
48–50% solids (Veresaflex 7)
49–51% solids (Daran 820; Versaflex 6)
50 ± 1% solids in water (Polidene 33-004, 33-031, 33-041)
54% total solids (Daran SL143)
55 ± 1% solids in water (Polidene 33-001, 33-075)
60–62% solids (Daran 220, 229)

GENERAL PROPERTIES:

Ionic Nature:
Anionic (Polidene 33-021, 33-031, 33-038)
Nonionic/anionic (Polidene 33-001, 33-004, 33-041, 33-075)
Sp. Gr.:
1.10 (Polidene 33-031)
1.16 (Polidene 33-038)
1.17 (Polidene 33-075)
1.20 (Polidene 33-041)
1.22 (Polidene 33-004, 33-021)
1.252–1.265 (Daran 820)
1.27 (Polidene 33-001)
1.326–1.340 (Daran 220, 229)
Density:
10.4–10.5 lb/gal (Daran 820)
10.85 lb/gal (Daran SL143)
11.0–11.2 lb/gal (Daran 220, 229)
11.3 lb/gal (Versaflex 6, 7)
Visc.:
0.05–0.30 poise (Polidene 33-001, 33-004, 33-021, 33-038)
0.05–0.50 poise (Polidene 33-031)
0.25–1.0 poise (Polidene 33-075)
0.4–0.9 poise (Polidene 33-041)
20–60 cps (Versaflex 7)
30 cps max. (Daran 820)
< 50 cps (Daran SL143)
50 cps max. (Daran 229)
50–150 cps (Versaflex 6)
75 cps max. (Daran 220)
F.P.:
36 F (Daran 220, 229, 820, SL143)

Vinylidene chloride copolymer *(cont'd.)*

Stability:

Excellent mechanical stability; alcohol tolerant (Daran SL143)

Unstable freeze-thaw (Polidene 33-001, 33-004, 33-031, 33-041, 33-075)

Unstable freeze-thaw; the film has excellent water, solvent, and oil resistance, and is nonflammable (Polidene 33-021)

No freeze-thaw stability; emulsions exhibit excellent mechanical and storage stability (Daran 220, 229, 820)

Emulsions exhibit good mechanical stability (20+ min); films exhibit good water resistance (Versaflex 6, 7)

Storage Stability:

Recommended use within 6 mos. (Daran 220, 229, 820)

9 mo shelf life (Daran SL143)

pH:

1.5–2.0 (Versaflex 7)

1.5–2.5 (Daran 820)

1.5–3.5 (Daran 229)

2.0 (Daran SL143)

2.0–3.0 (Versaflex 6)

2.5–4.5 (Daran 220)

3.0–4.5 (Polidene 33-001, 33-004, 33-021, 33-038, 33-041, 33-075)

7.0–8.0 (Polidene 33-031)

Surface Tension:

60–70 dynes/cm (Daran SL143)

STORAGE/HANDLING:

Avoid contact with mild steel, copper, zinc, and aluminum; protect against frost (Polidene 33-001, 33-004, 33-021, 33-031, 33-041, 33-075)

STD. PKGS.:

225-kg polyethylene-lined drums; bulk road tanker (Polidene 33-001, 33-004, 33-021, 33-031, 33-041, 33-075)

TRADENAME PRODUCTS AND GENERIC EQUIVALENTS

Abil 10–10,000 Series [Th. Goldschmidt AG]—Dimethicone
Abil AV 20-1000, 8853 [Goldschmidt AG]—Phenylmethyl polysiloxane
A-C Copolymer 540A [Allied-Signal]—Ethylene/acrylate copolymer
A-C Polyethylene 6, 6A, 8, 8A, 9, 9A, 617, 617A, 629A [Allied-Signal]—Polyethylene A-
    critamer 940 [RITA]—Carbomer 940
Acritamer 941 [RITA]—Carbomer 941
Acrysol A-1, A-3, A-5, ASE-60, ASE-75, LMW-Series [Rohm & Haas]—Polyacrylic
    acid
Acrysol G-100 [Rohm & Haas]—Ammonium polyacrylate
Acrysol GS, HV-1 [Rohm & Haas]—Sodium polyacrylate
Acrysol ICS-1 [Rohm & Haas]—Acrylates/steareth-20 methacrylate copolymer
Adiprene BL-16, L-42, L-83, L-100, L-167, L-200, L-213, L-300, L-315, L-325, L-367,
    L-700, L-767, LW-500, LW-510, LW-520, LW-550, LW-570 [Uniroyal]—Poly-
    urethane, thermoset
Adiprene CM [Uniroyal]—Polyurethane, thermoset, polyether-based
AF 60, 66, 72, 75, 9020 [General Electric]—Dimethicone
A-Fax 500, 600, 800, 940 [Hercules]—Polypropylene
Akrochem P-478 [Akron]—Phenol-formaldehyde resin
Akrochem Silicone Emulsion 350, 350 Conc., 1M, 10M, 60M [Akron]—Dimethicone
Akrochem Silicone Fluids 350, 1M, 5M, 10M, 30M, 60M [Akron]—Dimethicone
Akrochem SWS-201 [Akron]—Dimethicone
Alathon 7030, 7040, 7050, 7835, 7840 [DuPont]—Polyethylene (HDPE)
Alathon 7140, 7220, 7230, 7240, 7245, 7320, 7340, 7810, 7815, 7820, 7860, 7910, 7915,
    7960, 7965, 7970, PE-5510 [DuPont]—Polyethylene (HDPE)
Alathon 7440, 7970, PE-5510 [DuPont]—Polyethylene (HDPE)
Alcogum [Alco]—Sodium polyacrylate
Alcogum 9639 [Alco]—Ammonium polyacrylate
Alcogum L [Alco]—Polyacrylic acid
Alcosperse 107, 124, 149, 157, 602-N [Alco]—Sodium polyacrylate
Alcosperse 249 [Alco]—Ammonium polyacrylate
Alcosperse 404, 409 [Alco]—Polyacrylic acid
Alkapol PEG 200 [Alkaril]—PEG-4
Alkapol PEG 300 [Alkaril]—PEG-6

347

Alkapol PEG 400 [Alkaril]—PEG-8
Alkapol PEG 600 [Alkaril]—PEG-12
Alkapol PEG 1500 [Alkaril]—PEG-6-32
Alkapol PEG 3350 [Alkaril]—PEG-75
Alkapol PPG-425 [Alkaril]—PPG-9
Alkapol PPG-2000 [Alkaril]—PPG-26
Alkapol PPG-4000 [Alkaril]—PPG-30
Amoco 10-4017, 1012, 1046, 1088, 4018, 4036, 4039, 4222, 4228, 5219, 6200P, 6400P,
    6420P, 6431, 6800P, 7000P, 7200P, 7220P, 7232, 7233, 7234, CR22N, CR22NA,
    CR35, CR35A, CR35N, CR35NA [Amoco]—Polypropylene
Antifoam Compound SWS-201, SWS-202, SWS-203 [Wacker Silicones]—Dimethicone
Antiprex 461 [Allied Colloids]—Polyacrylic acid
Arcel Moldable Polyethylene Copolymers [Arco]—Polyethylene
Bapolene 1030, 1052, 1072, 1082 [Bamberger Polymers]—Polyethylene (LDPE)
Bapolene 2001, 2011, 2062, 2072, 2082, 2101 [Bamberger Polymers]—Polyethylene
    (HDPE)
Bapolene 3092, 3092L [Bamberger Polymers]—Polyethylene (LLDPE)
Bapolene 4042, 4062, 4072, 4082 [Bamberger Polymers]—Polypropylene
Bapolene 4112, 4114 [Bamberger Polymers]—Polypropylene, talc filled
Bapolene 5042, 5052, 5072 [Bamberger Polymers]—Polypropylene
Bayer CM 2552, CM 3610, CM 3630, CM 3631, CM 3632, CM 4230 [Bayer AG]—
    Polyethylene, chlorinated
Baypren 110, 112, 115, 124, 130, 210, 211, 213, 214, 215, 220, 230, 233, 235, 236, 243,
    320, 320 GR, 321, 321 GR, 330, 331, , 610, 710 [Bayer AG]—Polychloroprene
Baypren Latex Series, B, GK, MKB, SK, T [Bayer AG]—Polychloroprene latex
BCI Nylon #808-809, #818-819, #819-S-10%, #829 [Belding]—Nylon 6/6
BCI Nylon LX 3249, LX 3250, No. 651, No. 653 [Belding]—Nylon
Calthane ND 1100, 2300, 3200 [Cal Polymers]—Polyurethane, thermoset
CAPA 200, 205, 210, 212, 215, 220, 222, 223, 231, 240, 304, 305, 520, 600, 600M, 601,
    601M, Monomer [Solvay]—Polyurethane, thermoplastic, polycaprolactone-based
Capron 8200, 8202, 8202C, 8202CL, 8202L, 8203C, 8207F, 8209F, 8220 HS, 8221 HS,
    8253, 8254, 8255, 8259, 8270 HS, 8280 HS, 8350, 8352 [Allied-Signal]—Nylon 6
Capron 8230G HS, 8231G HS, 8232G HS FR, 8233G HS, 8234G HS [Allied-Signal]—
    Nylon 6, glass reinforced
Capron 8266G HS, 8267G HS [Allied-Signal]—Nylon 6, glass/mineral reinforced
Capron 8360 [Allied-Signal]—Nylon 6, mineral reinforced
Carbopol 907 [B.F. Goodrich]—Polyacrylic acid
Carbopol 940 [B.F. Goodrich]—Carbomer 940
Carbopol 941 [B.F. Goodrich]—Carbomer 941
Carbowax Sentry PEG 200 [Union Carbide]—PEG-4
Carbowax Sentry PEG 300 [Union Carbide]—PEG-6
Carbowax Sentry PEG 400 [Union Carbide]—PEG-8

TRADENAME PRODUCTS AND GENERIC EQUIVALENTS

Carbowax Sentry PEG 600 [Union Carbide]—PEG-12
Carbowax Sentry PEG 1000 [Union Carbide]—PEG-20
Carbowax Sentry PEG 1450 [Union Carbide]—PEG-6-32
Carbowax Sentry PEG 3350 [Union Carbide]—PEG-75
Carbowax Sentry PEG 8000 [Union Carbide]—PEG-150
Cartaretin F-4, F-8 [Sandoz]—Adipic acid/dimethylaminohydroxypropyl diethylene-triamine copolymer
Catamer Q [Richardson]—Polyquaternium-5
Celanese Nylon 1000-1, 1000-2, 1000-4, 1003-1, 1003-2, 1200-1, 1310-1, 1310-2, 1310-4 [Hoechst Celanese]—Nylon 6/6
Celanese Nylon 1500-1, 1500-2, 1503-1, 1503-2, 1600-1, 1600-2, 1603-1, 1603-2 [Hoechst Celanese]—Nylon 6/6, glass reinforced
Chemigum NBR Polymers [Goodyear]—Acrylonitrile-butadiene copolymer
Chemplex 1005, 1007, 1008, 1013, 1014, 1015, 1016, 1017 [Chemplex]—Polyethylene
Chemplex 1005, 1007, 1008, 1013, 1014, 1015, 1016, 1017, 1040, 3401, 3402, 3404, 3405 [Chemplex]—Polyethylene (LDPE)
Chemplex 1044, 1045, 1050, 1054, 1057, 1060, 3040, 3043, 3044, 3104, 3105, 3311 [Chemplex]—Polyethylene (LDPE/VA copolymer)
Chemplex 1101, 3015, 3024, 3052 [Chemplex]—Polyethylene
Chemplex 5003, 5402, 5602, 5604 [Chemplex]—Polyethylene (HDPE)
Chemplex 5701, 5704, 5705 [Chemplex]—Polyethylene (HMW-HDPE)
Chemplex 6001, 6004, 6006, 6008, 6009, 6109 [Chemplex]—Polyethylene (HDPE)
Chempol 33-4199/34-5075 [Freeman]—Polyurethane, thermoset
Clysar ECL, EH [DuPont]—Polyethylene
Colloid 102, 118 [Colloids]—Ammonium polyacrylate
Colloid 117/50, 119/50, 204, 274 [Colloids]—Polyacrylic acid
Colloid 202, 207, 208, 211, 218D, 223, 223D, 230, 233, 245D, 350 [Colloids]—Sodium polyacrylate
Conathane RN-3038, RN-3039 [Conap]—Polyurethane, thermoset, polyester-based
Conathane TU-4010, UC-33 [Conap]—Polyurethane, thermoset
Conoco 5305, 5425, 5465 [Conoco]—Polyvinyl chloride
Conoco 5385, 80172, 80273, 90171 [Conoco]—Polyvinyl chloride
Conoco Compound No. 14331, 15041, 15071, 16081, 17461, 18161, 18961, 26831, 27121, 28131, 28831, 34431, 34551, 34641, 35441, 35851, 38551, 38601, 38631, 38671, 38931, 44171, 44331, 45541, 48561, 48562 [Conoco]—Polyvinyl chloride
Cyanamer A-370, P-26, P-35, P-35 Solution, P-70, P-250 [Amer. Cyanamid; Cyanamid BV]—Polyacrylamide
Daran 220, 229, 820, SL143 [W.R. Grace]—Vinylidene chloride copolymer
Darex 110L [W.R. Grace]—Acrylonitrile-butadiene copolymer
Darvan #1 [R.T. Vanderbilt]—Sodium polynaphthalene sulfonate
Darvan #7 [R.T. Vanderbilt]—Sodium polymethacrylate
Daxad 30, 30-30, 30S, 34N10, 35, 41 [W.R. Grace]—Sodium polymethacrylate

349

Daxad 37L [W.R. Grace]—Polyacrylic acid

Daxad 37LA7 [W.R. Grace]—Ammonium polyacrylate

Daxad 37LN7, 37LN10, 37NS [W.R. Grace]—Sodium polyacrylate

Densol 1010 [Graden]—Sodium polyacrylate

Desmocol 110, 130, 176, 400, 406, 420, 500, 510, 530 [Bayer AG]—Polyurethane, thermoset

Dispersol 130 [Aquatec Quimica]—Ammonium polyacrylate

Dow Corning 190, 193, 196, 198, 1315, 5043, 5098 Surfactant [Dow Corning]—Dimethicone copolyol

Dow Corning 200 Fluid, 1500 Silicone Antifoam, 1520 Silicone Antifoam, FI-1630 [Dow Corning]—Dimethicone

Dow Corning 470A Fluid [Dow Corning]—Dimethicone copolyol

Dow Corning 3225C Formulation Aid [Dow Corning]—Dimethicone copolyol

Drewsperse 611 [Drew]—Sodium polyacrylate

Droxol 400 [Henkel/Process]—PEG-8

Durethan A30S, B25T, B30P, B30S, B31SK, B31F, B40E, B40F, B40SK, B50E, BKV35, BKV50H [Bayer AG]—Nylon 6 and 6/6

Durethan AKV30, AKV30H, BG30X, BKV30, BKV30H, BKV30N, BKV30N1, BKV35, BKV50H [Bayer AG]—Nylon 6 and 6/6, glass reinforced

Durethan BC30, BC40 [Bayer AG]—Nylon 6 and 6/6 with olefin polymer

Durez 115, 157, 16038, 21206, 30959 [Occidental]—Phenol-formaldehyde resin

Durez 17702, 30417, 30698, 30806 [Occidental]—Phenol-formaldehyde resin

Durez 31219 [Occidental]—Phenol-formaldehyde resin, glass-reinforced

Durez 31735 [Occidental]—Phenol-formaldehyde resin, fiber-reinforced

Elfacos ST9 [Akzo Chemie]—PEG-45/dodecyl glycol copolymer

EMA [Monsanto]—Ethylene/maleic anhydride copolymer

Emcol Q [Witco]—Polyquaternium-5

Emery 6686 [Henkel/Emery]—PEG-12

Emery 6687 [Henkel/Emery]—PEG-6

Emery 6709 [Henkel/Emery]—PEG-8

EMI-X MA-40 [LNP]—Polypropylene, aluminum filled

EMI-X PC-1008, PC-100-10 [LNP]—Nylon 6, carbon reinforced

EMI-X RA-30, RA-35, RA-40 [LNP]—Nylon 6/6, aluminum filled

EMI-X RC-1008, RC-100-10, RC-100-12 [LNP]—Nylon 6/6, carbon reinforced

EmPee PP 401, PP 401CS, PP 402 [Monmouth Plastics]—Polypropylene

Epolene C-10, C-15, N-10, N-10-P, N-11, N-11-P, N-12, N-12-P, N-14, N-14-P, N-34, N-34-P, N-45, N-45-P [Eastman]—Polyethylene

Ertalon 6SAU [Chemplast]—Nylon 6

Ertalon 11SA [Chemplast]—Nylon 11

Ertalon 66SA, 66SAM [Chemplast]—Nylon 6/6

Estane 5701 F1, 5702, 5702 F1, 5702 F2, 5703, 5703, 5703 F1, 5703 F2, 5707 F1, 5708 F1, 5710 F1, 5711, 5712, 5713, 5716, 5715, 58013, 58091, 58092, 58109, 58121,

58130, 58133, 58134, 58136, 58137, 58271, 58277, 58360 [B.F. Goodrich]—
Polyurethane, thermoplastic, polyester-based
Estane 5714 F1, 58300, 58309, 58311, 58370, 58630 [B.F. Goodrich]—Polyurethane,
thermoplastic, polyether-based
EXI-X QC-1008, QC-100-10 [LNP]—Nylon 6/10, carbon reinforced
F-751 [Wacker Silicones]—Dimethicone
Firestone Nylon 200-001, 210-001L, 213-001, 228-001 [Firestone]—Nylon 6
Firestone Nylon 415-001HS, 430-001HS [Firestone]—Nylon 6, glass reinforced
Firestone Nylon 540-001HS [Firestone]—Nylon 6, mineral reinforced
Firestone Nylon 640-001HS [Firestone]—Nylon 6, glass/mineral reinforced
FM 510 [Fiberite]—Phenol-formaldehyde resin
FM 1132, 1132P, 3510, 7700, 9294 [Fiberite]—Phenol-formaldehyde resin, fabric-
reinforced
FM 1303 [Fiberite]—Phenol-formaldehyde resin, nylon-reinforced
FM 1389, 14111 [Fiberite]—Phenol-formaldehyde resin, cellulose-filled
FM 3000, 3001, 3002, 4201, 7676, 10365 [Fiberite]—Phenol-formaldehyde resin,
cellulose-filled
FM 4004, 4004F, 4005, 4005F, 4007, 4007F, 4030-190, 404464, 8130 [Fiberite]—
Phenol-formaldehyde resin, glass-reinforced
FM 5367 [Fiberite]—Phenol-formaldehyde resin, cellulose/glass-filled
FM 6101, 11547, 17610 [Fiberite]—Phenol-formaldehyde resin, asbestos-filled
FM 16771, 21288 [Fiberite]—Phenol-formaldehyde resin, glass-reinforced
Foamkill 810F, 830F [Crucible]—Dimethicone
Fortilene 10X01, 21X04, 38X01, 40X05, 41X03, 2251, 3151, 3250, 3251, 3606, 3907,
9205, 9605 [Soltex Polymer]—Polypropylene
Fortilene 14X01, 16X01, 18X02, 18X03, 19X02, 19X03, 1401, 1602, 1602A [Soltex
Polymer]—Polypropylene
Fortilene 41X04, 42X07, 54X02, 4141, 4141F [Soltex Polymer]—Polypropylene
Fosta Nylon 446, 471, 512, 523, 567, 589, 714, 870, 1047, 1210, 1379, 1417, 1525, 1693,
1722 [Hoechst Celanese]—Nylon 6
FPC 605, 654, 654H, 6337, 6411, 6422, 6854, 6866 [Firestone Plastics]—Polyvinyl
chloride
FPC 965, 9269A, 9282, 9290, 9300, 9326, 9339, 9418 [Firestone Plastics]—Polyvinyl
chloride
FPC 1238, 1239, 1240, 1265C, 1359, 1363A, 1379, 1380, 1380A, 1381A, 1415, 1418,
1419, 1430, 1433, 5000B, 5002, 5004, 5005, 5006A, 5008, 5009, 5060, 5105,
5125A, 5160, 5175, 5264, 5290, 5305, 5360, 5390, 5439A, 5440, 5460, 5605, 5760,
5760B, 5790, 5805, 9275 [Firestone Plastics]—Polyvinyl chloride
FPC 6338 [Firestone Plastics]—Polyvinyl chloride (vinyl chloride/vinyl acetate copolymer)
Gafquat 734, 755, 755N [GAF]—Polyquaternium-11
Gaftex PT [GAF]—PVM/MA copolymer

Gantrez AN, AN-119, AN-139, AN-149, AN-169, AN-179 [GAF]—PVM/MA copolymer

Gantrez B-773 [GAF]—Polyvinyl isobutyl ether

Geon 3007, 3010, 88933, 88934, CPVC [B.F. Goodrich]—Polyvinyl chloride, chlorinated

Geon 85721, 85858, 87001 [B.F. Goodrich]—Polyvinyl chloride

Glissoviscal B [BASF AG]—Polyisobutene

Good-rite K-7028, K-7058, K-7200, K-7600 [B.F. Goodrich]—Sodium polyacrylate

Grilamid ELY20NZ, ELY60 [Emser Industries]—Nylon 12 elastomer

Grilamid L16G, L16GM, L20FR, L20G, L20W20, L20W40, L25, L25 natural 6112, L25W10, L25W20, L25W40, L25W40NZ, TR55LX, TR55UV [Emser Industries]—Nylon 12

Grilamid LKN-5H, LV-3H, LV-43H [Emser Industries]—Nylon 12, glass reinforced

Grilon A23, A23GM, A23G natural 6165, A28, A28GM, A28NX, A28NY, A28NZ, A28VO, A28W10, BT40, F40, F47, R40, R40GM, R40N38, R47, R47HW, R50, W3941, W5744 [Emser Industries]—Nylon 6

Grilon CA6, CA6EH, CF35, CR9, CR9 natural 6361 [Emser Industries]—Nylon 6/12

Grilon ELX23NZ [Emser Industries]—Nylon 6/elastomer

Grilon PV-3H, PV-5H, PV-15H, PVN-3H , PVN-15H, PVZ-3H [Emser Industries]—Nylon 6, glass reinforced

Harwal AOB, DBP-C, DBP-F, DBP-G, DBP-I, DBP-NT, GBM, GI, GMW, GP-C, GP-G, GP-I, GP-NT, GPR, GSB, GSB-F, GSF 1, GSG [U.S. Polymers]—Polyvinyl chloride

Harwal Purge Compound [U.S. Polymers]—Polyvinyl chloride

Herox 0200HA, 0200MA, 0200SA [DuPont]—Nylon

Hexcel 164M [Hexcel/Rezolin]—Polyurethane, thermoset

HiD 9300, 9301, 9327, 9346, 9347 [Chevron]—Polyethylene (HDPE)

HiD 9602, 9634, 9660 [Chevron]—Polyethylene (HDPE)

HiD 9606, 9632, 9640, 9642, 9650 [Chevron]—Polyethylene (HDPE)

HiD 9690 [Chevron]—Polyethylene (HMW-HDPE)

Hostacerin PN 73 [Hoechst-Celanese]—Sodium polyacrylate

Hostalen GA 7960, GC 7560 [Hoechst Celanese]—Polyethylene (HDPE)

Hostalen GB 6950, GF 7740 F2, GM 5010 T2, GM 5010 T2N [Hoechst Celanese]—Polyethylene (HDPE)

Hostalen GM 7255, GM 7746 [Hoechst Celanese]—Polyethylene (HDPE)

Hostalen GM 9255 F [Hoechst AG]—Polyethylene (HMW-HDPE)

Hostalen GUR [Hoechst AG]—Polyethylene (UHMW-PE)

Hostalen PP920, PP933, PP934, PP936, PP941, PP942, PP975 [Hoechst Celanese]—Polypropylene

Hostalen PP927, PP989 [Hoechst Celanese]—Polypropylene

Hostalen PP996, PP998 [Hoechst Celanese]—Polypropylene

Hycar 1312, 1422, 1422X8, 1452P50, Butadiene-Acrylonitrile Rubber, Nitrile, Nitrile

Latices [Goodrich]—Acrylonitrile-butadiene copolymer

Hypalon 20, 30, 40, 40S, 4085, LD-999, 45, 48 [DuPont]—Polyethylene, chlorosulfonated

Hysol PC28, PC18 [Hysol/Dexter Corp.]—Polyurethane, thermoset

Interflo Porous Plastic [Chromex]—Polyethylene

Intraphor AC [Henkel-Nopco]—Sodium polynaphthalene sulfonate

Jeffox PPG-400 [Texaco]—PPG-9

Jeffox PPG-2000 [Texaco]—PPG-26

Jonwax 26 [S.C. Johnson]—Polyethylene

Jordaquat 40 [PPG-Mazer]—Polyquaternium-6

Jordaquat 41 [PPG-Mazer]—Polyquaternium-7

Kohinor 311 FR, 434, 509, 510, 702, 906, 940, 941, 947, 948 [Pantasote]—Polyvinyl chloride

Krynac 19.65, 21.65, 25.65, 27.50, 29.60, 32.55, 34.35. 34.50, 34.80, 34.140, 34.E50, 34.E80, 38.50, 40.65, 45,55, 50.75, 800, 803, 810, 822, 825, 826, 826, 843 [Polysar]—Acrylonitrile-butadiene copolymer

Lamal 408-40(65)/Lamal C, Lamal HSA/Lamal C [Polymer Industries]—Polyurethane, thermoset

LBA-133 Polyethylene [Mobil]—Polyethylene (LDPE)

LGA-563 Polyethylene [Mobil]—Polyethylene (LDPE)

LKA-753 Polyethylene [Mobil]—Polyethylene (LDPE)

LLA-533 Polyethylene [Mobil]—Polyethylene (LDPE)

LMA-000 Polyethylene [Mobil]—Polyethylene (LDPE)

Lomar DL, LS, PL, PW, ST [Henkel/Process]—Sodium polynaphthalene sulfonate

Lutrol E 400 [BASF AG]—PEG-8

Lutrol E 4000 [BASF AG]—PEG-75

Lutrol E 6000 [BASF AG]—PEG-150

Lutrol E-1500 [BASF AG]—PEG-6-32

Lutrol E300 [BASF AG]—PEG-6

Luviskol K12, K17, K30, K60, K80, K90 [BASF AG]—Polyvinylpyrrolidone

Marlex BMN 53120, BMN 55200, BMN TR-880, EHM 6006, HHM 4520, HHM 5202, HHM 5502, HHM TR140 [Phillips]—Polyethylene (HDPE)

Marlex BMN TR955, EHM TR160, HHM 4903, HHM TR401, HHM TR418, HHM TR460, HHM TR480, HXM 50100 [Phillips]—Polyethylene (HDPE)

Marlex CHM-040-01, CHM-040-02 [Phillips]—Polypropylene

Marlex CL-50-35, CL-100, CL-100B [Phillips]—Polyethylene

Marlex EHM 6001, EHM 6003, EMN 6030, EMN TR885 [Phillips]—Polyethylene (HDPE)

Marlex HGH-050, HGN-120-01, HGN-120-A, HGN-350, HGV-040-01, HGX-030, HGZ-040, HGZ-050-02, HGZ-080-02, HGZ-120-02, HGZ-120-04, HGZ-350, HLN-020-01, HNS-080 [Phillips]—Polypropylene

Marlex HHM TR130, HHM TR400 [Phillips]—Polyethylene

Masil EM-14, -62, -100, -100 Conc., -100D, -100P, -250 Conc., -350, -350X, -350X
   Conc., -1000, -1000 Conc., -1000P, -10,000, -10,000 Conc., -60,000, -100,000, -N
   [PPG-Mazer]—Dimethicone
Masil SF 5, 20, 50, 100, 200, 350, 500, 1000, 5000, 10,000, 12,500, 30,000, 60,000,
   100,000, 300,000, 600,000 [PPG-Mazer]—Dimethicone
Merpoxen PEG 200 [Kempen]—PEG-4
Merpoxen PEG 300 [Kempen]—PEG-6
Merpoxen PEG 400 [Kempen]—PEG-8
Merpoxen PEG 600 [Kempen]—PEG-12
Merpoxen PEG 1000 [Kempen]—PEG-20
Merpoxen PEG 1500S [Kempen]—PEG-6-32
Merpoxen PEG 6000S [Kempen]—PEG-150
Merquat 100 [Calgon]—Polyquaternium-6
Merquat 550 [Calgon]—Polyquaternium-7
Micalite PP-G2MF-4, PP-G2MF-5, PP-G2MF-6, PP-H2MFS-4, PP-H2MFS-5, PP-
   H2MFS-6 [Washington Penn]—Polypropylene, mica filled
Migralube Q-1000 [LNP]—Nylon 6/10
Migralube QFL-4036, QFL-4536 [LNP]—Nylon 6/10, glass reinforced
Millathane 76, 80, 300, HT [TSE Industries]—Polyurethane, thermoset, polyester-based
Millathane E-34 [TSE Industries]—Polyurethane, thermoset, polyether-based
Minlon 10B [DuPont]—Nylon 6/6
Minlon 11C, 12T, 20B, 21C [DuPont]—Nylon, mineral or mineral/glass reinforced
Mira-Glos RT A and C [Polymer Industries]—Polyurethane, thermoset
Mirapol A-15 [Miranol]—Polyquaternium-2
Modic H-100E, H-100F, H-400C, H-400F [Mitsubishi]—Polyethylene (HDPE)
Modic L-100F, L-400F, L-400H [Mitsubishi]—Polyethylene (LDPE)
Modic P-110F, P-300F, P-300M [Mitsubishi]—Polypropylene
Monocast MC901 [Polymer Corp.]—Nylon
Monolan PPG-440 [Harcros UK]—PPG-9
Monolan PPG1100 [Harcros UK]—PPG-17
Monolan PPG2200 [Harcros UK]—PPG-26
MWB Film [Hercules]—Polypropylene
Neoprene Latex 571, 572 [DuPont]—Polychloroprene latex
Neoprene Latex 601A, 622, 635, 654, 671, 735-A, 842-A [DuPont]—Polychloroprene
   latex
Neoprene Latex 650, 750 [DuPont]—Polychloroprene latex
Neoprene Latex Series, 101, 102, 950 [DuPont]—Polychloroprene latex
Neoprene Series, AC, AD, AF, AG, AH, CG, FB, GN, GNA, GRT, GW, TRT, TW, TW-
   100, W, WB, WD, WHV, WHV-A, WK, WRT [DuPont]—Polychloroprene
NeoRez EX-466, R-961, U-105, U-110, U-760, U-800, U-912 [Polyvinyl Chem. Indus-
   tries]—Polyurethane, thermoset
Nobestos/D-7102 [Rogers]—Polychloroprene

Nobestos/D-7280 [Rogers]—Acrylonitrile-butadiene copolymer

Nopalcol 200 [Henkel/Process]—PEG-4

Nopcosant [Henkel/Process]—Sodium polynaphthalene sulfonate

Norcast 6368 [R.H. Carlson]—Polyethylene

Norcast PR-1020 [R.H. Carlson]—Polyurethane, thermoset

Norchem NHD 6908 [Norchem]—Polyethylene (HDPE)

Norchem NPE 130, NPE 190, NPE 320, NPE 330, NPE 333, NPE 334, NPE 336, NPE 350, NPE 353, NPE 510, NPE 810, NPE 820, NPE 831, NPE 840, NPE 853, NPE 860, NPE 861, NPE 930, NPE 931, NPE 940, NPE 941, NPE 950, NPE 952, NPE 953, NPE 954 [Norchem]—Polyethylene (LDPE)

Norchem NPE 420, NPE 480, NPE 481 [Norchem]—Polyethylene (LDPE/VA copolymer)

Norchem NPP 1001-LF, NPP 1006-GF, NPP 1008-AK, NPP 1010-LC, NPP 2000-GJ, NPP 2003-GJ, NPP 2004-MR, NPP 2013-UJ, NPP 3007-GO, NPP 3010-SO, NPP 7000-GF, NPP 7000-ZF, NPP 8000-GK, NPP 8001-LK, NPP 8002-HK, NPP 8004-MR, NPP 8005-AR, NPP 8006-GF, NPP 8007-GO, NPP 8004-ZR [Norchem]—Polypropylene

Norchem NPP 7300-GF, NPP 7300-AF [Norchem]—Polypropylene

Nortuff NFA 1700 MO, NFA 1800 TO, NFA 2400 CO, NFA 2600 CO, NFA 4400 TO, NFC 1700 MO, NFC 1800 TO, NFC 2200 CO, NFC 2400 CO, NFC 2440 CF, NFC 2600 CO, NFC 4000 FR, NFC 4400 CO, NFC 4600 CO [Norchem]—Polypropylene

NY-30CF [Compounding Technology]—Nylon 6, carbon reinforced

Nydur B 31 SK, B 40 SK [Mobay]—Nylon 6

Nydur BKV 15 H, BKV 30 H, BKV 35 H, BKV 35 Z, BKV 40 H, BKV 50 H, GKV 115, BKV 120, BKV 125, BKV 130, BKV 135, RM KU 2-22501/30, RM KU 2-2521/20, RM KU 2-25121/25, RM KU 2-2521/30, RM KU 2-2521/35 [Mobay]—Nylon 6, glass reinforced

Nylaflow H, T [Polymer Corp.]—Nylon 6/6

Nylaflow LP [Polymer Corp.]—Nylon 6

Nylatron 1018, 1027, GS-51, GS-51-13 [Polymer Corp.]—Nylon 6/6, glass reinforced

Nylatron 1022, 1024, GS, GS-21, GS-HS, NSB-90 [Polymer Corp.]—Nylon 6/6

Nylatron 1025 [Polymer Corp.]—Nylon 6/6, mineral reinforced

Nylatron 1026 [Polymer Corp.]—Nylon 6/6, glass/mineral reinforced

Nylatron 1028 [Polymer Corp.]—Nylon 6/6 alloy

Nylatron 2001, 2005, 2010, 2011, 2015, 2029, 2038, 2039, GS-60, GS-63 [Polymer Corp.]—Nylon 6

Nylatron 2007, 2033 [Polymer Corp.]—Nylon 6, mineral reinforced

Nylatron 2008, 2040, GS-60-9, GS-60-13 [Polymer Corp.]—Nylon 6, glass reinforced

Nylatron GS-71, GS-73 [Polymer Corp.]—Nylon 6/12

Nylatron GS-71-9 [Polymer Corp.]—Nylon 6/12, glass reinforced

Nylatron GSM, NSB [Polymer Corp.]—Nylon

Nyrim P-1 1000, P-1 2000, P-1 3000, P-1 4000 [Custom Urethane Elastomers]—Nylon 6/elastomer

Olict C [Alox]—Polyisobutene

Onamer M [Millmaster-Onyx]—Polyquaternium-1

P.U.R.E.-CMC [Perma-Flex Mold]—Polyurethane, thermoset

P012, P032, P04, P0113, P0119 [M.A. Industries]—Polypropylene

P2120, P2130, P2140, P2230, P9125 [M.A. Industries]—Polypropylene, mineral filled

PA-52-333, -52-334 [Polymer Applications]—Phenol-formaldehyde resin

Parlon P [Hercules]—Polypropylene, chlorinated

Paxon AA55-003, AB50-003, AF40-003, BA50-100 [Allied-Signal]—Polyethylene (HDPE)

Paxon AA60-003, AA60-007, AD60-007 [Allied-Signal]—Polyethylene (HDPE)

Paxon FD60-018 [Allied-Signal]—Polyethylene (HMW-HDPE)

Paxon SS55-100, SS55-180, SS55-250, SS55-400 [Allied-Signal]—Polyethylene (HDPE)

PDX-82428 [LNP]—Nylon 6/10, nickel reinforced

PDX-83392 [LNP]—Nylon 6/6, nickel reinforced

PDX-85384, -85382, -85473 [LNP]—Nylon, glass reinforced

PE 1017, 1018, 1019, 1028, 1117, 2130, 2151, 4517, 4560, 5325, 5554-H, 5555, 5561, 5565, 5613, 5619, 5622, 5625, 5754, 5755, 5861 [Chevron]—Polyethylene (LDPE)

PE 2205, 2207, 2255, 2260 [Chevron]—Ethylene/methyl acrylate copolymer

PE 5220, 5222, 5240, 5254, 5272, 5280, 5290 [Chevron]—Polyethylene (LDPE/VA copolymer)

PE 5961 [Chevron]—Polyethylene

PE-25 [Washington Penn]—Polyethylene

PE005, PE012, PE012/C [M.A. Industries]—Polyethylene

Peem 122 Conc., 397, 410 [GAF]—Polyethylene

PEG 400 [ICI PLC]—PEG-8

PEG 1000 [ICI PLC]—PEG-20

Perbunan N 1807 NS, N 2802 NS, N 2807 NS, N 2810, N 2818 NS, N 3302 NS, N 3307 NS, N 3310, N 3312 NS, N 3807 NS, N 3810 [Bayer AG]—Acrylonitrile-butadiene copolymer

Perbunan N Latex 1590, 3090, 2818, 3310, 3310 HD, 3810, 3415 M, T, VT [Bayer AG]—Acrylonitrile-butadiene copolymer

Peregal ST [GAF]—Polyvinylpyrrolidone

Perenol EI [Henkel Canada]—Polyvinyl isobutyl ether

Petrothene 320, 334, 336, 350, 353, 940, 941, 952, 953, 954, 954-1, 955, 955-0, 957, 957-0, 957-1, 962, 962L, 963, 964, 965, 980, 983, 983-6, 3401, 3401E, 3404, 3404B, 3404D, 3404H, 3407, 3407B, 3407G, 3407H, 3408, 3408A, NA 140, NA 141, NA 142, NA 143, NA 145, NA 147, NA 148, NA 152, NA 153, NA 154, NA 271, NA 272, NA 273, NA 279, NA 284, NA 301, NA 344 [Quantum/USI]—Polyethylene (LDPE)

Petrothene 420, 425, 440, 440-1, 441, 442, 445, 480, 481, 1060, 1060K, 3004A, 3004C, 3004D, 3040B, 3040G, 3040L, 3043, 3043B, 3043G, 3043H, 3104A, 3104D, 3350, NA 233, NA 234, NA 235, NA 238, NA 239, NA 289, NA 290, NA 295, NA 386, NA 387, NA 388 [Quantum/USI]—Polyethylene (LDPE/VA copolymer)

Petrothene 3503, 3503A, 3503C, 3503E, 3503G, 3505, 3505A, 3507, 3507A [Quantum/USI]—Polyethylene (HMW-LDPE)

Petrothene GA 601-030, 601-031, 601-032, 601-033, 601-130, 603-035, 604-040, 605-030, 605-031, 605-033, 605-133, 605-150 [Quantum/USI]—Polyethylene (LLDPE)

Petrothene HD 5002, HD 5003, HD 5601, HD 5602, HD 5604, HD 5703, HD 5704, HD 5705, HD 5711, HD 5712D, HD 5713, HD 6004, HD 6007, HD 6009, HD 6085, LA 203F3, LA 303F3, LA 404F3, LA 408F3, LB 748, LB 830, LB 832, LB 833, LC 732, LR 723, LR 732, LR 734, LR 920, LS 901, NA 225 [Quantum/USI]—Polyethylene (HDPE)

Petrothene LB 924 [Quantum/USI]—Polyethylene (HDPE)

Petrothene LR 923, LY 520, LY 600, LY 660, LY 955 [Quantum/USI]—Polyethylene (HDPE)

Petrothene PA 161, 162 [Quantum/USI]—Polyethylene (LLDPE)

Petrothene PP 1510-HC, PP 1510-LC, PP 8402-HO, PP 8402-TO, PP 8403-HO, PP 8403-TO, PP 8404-HJ, PP 8404-ZJ, PP 8410-ZR, PP 8411-ZR, PP 8412-HK, PP 8412-TK, PP 8420-HK, PP 8462-HR, PP 8470-HU, PP 8470-ZU, PP 8502-HK, PP 8602-HJ, PP 8752-HF, PP 8755-HK, PP 8762-HR, PP 8770-HU, PP 8802-HO, PP 8815-ZR, PP 8820-HU [Quantum/USI]—Polypropylene

Petrothene PP 2004-MR, PP 8085-GU, PP 8000-GK, PP 8001-LK, PP 8004-MR, PP 8004-ZR, PP 8005-AR, PP 8020-AU, PP 8020-GU, PP 8020-ZU, PP 8080-AW, PP 8080-GW, PP 8080-ZW [Quantum/USI]—Polypropylene

Petrothene PP 7200-AF, PP 7200-GF, PP 7200-MF, PP 7300-KF, PP 7300-MF, PP 8310-GO, PP 8310-KO [Quantum/USI]—Polypropylene

PGE-300 [Hefti]—PEG-6

PGE-400 [Hefti]—PEG-8

PGE-600 [Hefti]—PEG-12

PGE-1000 [Hefti]—PEG-20

PGE-1500 [Hefti]—PEG-6-32

Phenolic Molding Compound 868 [Shenango Phenolics]—Phenol-formaldehyde resin

Plasdone, C-30 [GAF]—Polyvinylpyrrolidone

Plenco 118-AF Dark Grey, 118 Dark Grey [Plastics Engineering]—Phenol-formaldehyde resin, mineral/graphite-filled

Plenco 201, 203 [Plastics Engineering]—Phenol-formaldehyde resin, aluminum-filled

Plenco 300 Black, 300 Brown, 300 Red, 308 Black, 308 Brown, 317 Red, 320 Green, 411 Walnut Mottle, 482 Black, 512 Black, 535 Black, 571 Black, 579 Black [Plastics Engineering]—Phenol-formaldehyde resin, woodflour-filled

Plenco 307 Black, 500 Black, 500 Brown, 523 Black [Plastics Engineering]—Phenol-formaldehyde resin, woodflour/flock-filled

Plenco 321 Black, 554 Black [Plastics Engineering]—Phenol-formaldehyde resin, fabric-reinforced

Plenco 343 Natural [Plastics Engineering]—Phenol-formaldehyde resin, mica-filled

Plenco 349 Black, 4443 [Plastics Engineering]—Phenol-formaldehyde resin, mineral-filled

Plenco 368 Black, 417 Black, 476 Black [Plastics Engineering]—Phenol-formaldehyde resin, cotton flock-filled

Plenco 369 Black, 480 Black, 557 Black, 567 Black, 4543 Red [Plastics Engineering]—Phenol-formaldehyde resin, woodflour-filled

Plenco 400 Black, 400 Brown, 485 Black, 485 Brown, 548 Black [Plastics Engineering]—Phenol-formaldehyde resin, mineral/woodflour-filled

Plenco 414 Black, 466 Black, 466 Brown, 467 Black, 509 Black, 509 Brown [Plastics Engineering]—Phenol-formaldehyde resin, mineral/flock-filled

Plenco 507 Black [Plastics Engineering]—Phenol-formaldehyde resin, filled

Plenco 527 Black, 586 Brown [Plastics Engineering]—Phenol-formaldehyde resin, mineral/flock-filled

Pluracol E-200 [BASF; BASF AG]—PEG-4

Pluracol E-300 [BASF]—PEG-6

Pluracol E-400, E-400 NF [BASF]—PEG-8

Pluracol E-600, E-600 NF [BASF; BASF AG]—PEG-12

Pluracol E-1500 [BASF]—PEG-6-32

Pluracol E-4000, E-4000 NF [BASF]—PEG-75

Pluracol E-6000 [BASF]—PEG-150

Pluracol P-410 [BASF]—PPG-9

Pluracol P-710 [BASF]—PPG-12

Pluracol P-1010 [BASF]—PPG-17

Pluracol P-2010 [BASF]—PPG-26

Pluracol P-4010 [BASF]—PPG-30

Pluriol E400 [BASF AG]—PEG-8

Pluriol P 2000 [BASF AG]—PPG-26

Pogol 400 [Hart Chem. Ltd.]—PEG-8

Polectron 430 [GAF]—Styrene/PVP copolymer

Polidene 33-001, 33-004, 33-021, 33-031, 33-038, 33-041, 33-075 [Scott Bader]—Vinylidene chloride copolymer

Polyclar 10, AT [GAF]—Polyvinylpyrrolidone

Poly-G 400 [Olin]—PEG-8

Polyglycol P-200 [Dow Europe]—PPG-26

Polypenco Nylon 101 [Polymer Corp.]—Nylon 6/6

Polyplasdone XL, XL-10 [GAF]—Polyvinylpyrrolidone

Poron 4701-01, 4701-05, 4701-09, 4716-16 [Rogers]—Polyurethane, thermoset

PP-C2IM, PP-C4IM, PP-HIFR [Washington Penn]—Polypropylene

TRADENAME PRODUCTS AND GENERIC EQUIVALENTS

PP-C2TF-1, -C2TF-4, -C3TF-2, -C3TFA-2, -H1TF-4, -H2LTF-4, -H2TF-2, -H2TF-4, –H4TF-3, -H6TF-4, -VB-511 [Washington Penn]—Polypropylene, talc filled

PP-C3CC-4, C5CC-2, -C6CC-4, H1CC-1, -H2CC-2, -H6CC-4, H7CC-4 [Washington Penn]—Polypropylene, calcium carbonate filled

PP-G2MF-4, -G2MF-5, -H2MF-4, -H2MF-5, -H3MF-2, -H2MFQ-3, -H2MFQ-4, –H3MFQ-1, -H3MFQ-5, -H2MFS-3, -H2MFS-4, -H2MFS-5 [Washington Penn]—Polypropylene, mica filled

PP-HFR-2, -HFR-3 [Washington Penn]—Polypropylene, glass reinforced

Presto-Foam 800, 805, 900, 945, 960 [Presto Mfg.]—Polyurethane, thermoset, polyether-based

Primacor 4990 Dispersion [Dow]—Ethylene/acrylate copolymer

Pro-fax 65F4-4 [Hercules]—Polypropylene, talc filled

Pro-fax 65F5-4 [Hercules]—Polypropylene, calcium carbonate filled

Pro-fax 6131, 6323, 6323F, 6329, 6331NW, 6523F, 6524, 6532F, 7531, PC-072, PC942, PC968, PD064, PD195, PD401, PD626, PD701, PF101, PF151 [Hercules]—Polypropylene

Pro-fax 7131, 7523, 7823, 8523, 8623, SA-595, SA-752, SA-861, SA-862, SA-868M, SA-878, SB-661, SB-751, SB-782, SB-786, SB-787, SD-062, SD-101, SE-191 [Hercules]—Polypropylene

Protodye QNF [Proctor]—Sodium polynaphthalene sulfonate

Purelast 204, 207, 208, 209, 220, 221, 223, 224, 226, 228, 242, 243, 245, 245H, 234, 235, 240H, 241, 241H, 247, 249, 251, 253, 255, 254 [Polymer Systems]—Polyurethane, thermoset

PVP K-15, K-26/28, K-29/32, K-30, K-60, K-90 [GAF]—Polyvinylpyrrolidone

Q-Thane P-49, P-250, PA-01, PA-05, PA-06, PA-07, PA-10, PA-11, PA-20, PA-29, PA-30, PA-40, PA-58, PA-80, PA-93, PH-56, PH-89 [K.J. Quinn]—Polyurethane, thermoplastic

Q-Thane P-279, P-280, P-360, P-440, P-455, PC-58, PI-76, PI-86, PI-95, PI-96, PI-176, PI-186, PI-195, PI-196, PN03-100, PN3429-100, PS-16, PS-62, PS-63, PS-65, PS-82, PS-94 [K.J. Quinn]—Polyurethane, thermoplastic, polyester-based

Q-Thane PE-23, PE-36, PE-47, PE-49, PE-50, PE-55, PE74, PE-90, PE103-100, PE192-100, PE192-101 [K.J. Quinn]—Polyurethane, thermoplastic, polyether-based

Ren:C:O-Thane RP-6400, RP-6401, RP-6402, RP-6403, RP-6405, RP-6410, RP-6413, RP-6414, RP-6422, TDT-178-34, TDT-178-53, TDT-186-1 [Ren Plastics]—Polyurethane, thermoset

RestEasy Foam [BASF]—Polyurethane, thermoset

Reten 210, 220 [Hercules]—Polyquaternium-5

Reten 420 [Hercules]—Polyacrylamide

Reten 421, 423, 425 [Hercules]—Acrylamide/sodium acrylate copolymer

Reten 521 [Hercules]—Acrylamides copolymer

Rexene 6310, PP41E2, PP41E4, XO-325 [Rexene Products]—Polypropylene

Rexene 6310C25, 9401C2 [Rexene Products]—Polypropylene, calcium carbonate filled

Rexene 6310T2, 6310T4, 9400T2 [Rexene Products]—Polypropylene, talc filled

Rexene 9234, 9400, 9401, 9402, 9403, 4903E, 9500 [Rexene Products]—Polypropylene

Rezolin 164, 170 [Hexcel/Rezolin]—Polyurethane, thermoset, polyether-based

Rezolin 185N [Hexcel/Rezolin]—Polyurethane, thermoset

Rheolate 1 [NL Treating]—Acrylamides copolymer

Rilsan BECNO, BECVO P40 TL, BESHVO, BESN Black T, BESN F15, BESNO,
BESNO P20, BESNO P40, BESNO P40 TL, BESNO TL, BESVO, BMN F15,
BMN F25, BMNO, BMNO P20, BMNO P40, BMNY BZ TL, BMV Black T,
KMVOTL, RDP15-10 Natural, RDP-17-1 Dispersion Coating, RDP Pigmented
[Rilsan]—Nylon 11

Rilsan BZM23G9, BZM43G9 [Rilsan]—Nylon 11, glass/graphite reinforced

Rilsan BZM300 [Rilsan]—Nylon 11, glass reinforced

RTP 100GB10, 100GB20, 100GB30, 100GB40, 101, 101CC, 101FR, 101 SP Foamed,
103, 103CC, 103 SP Foamed, 105, 105CC, 105CC FR, 105 SP Foamed, 107, 107CC
[Fiberite]—Polypropylene, glass reinforced

RTP 100T10 or 131, 100T20 or 128, 100T30 or 132, 100T40 or 127 [Fiberite]—
Polypropylene, talc filled

RTP 140, 141, 142, 143 [Fiberite]—Polypropylene, calcium carbonate filled

RTP 150, 199x22898, 199x23835, 199x28016 [Fiberite]—Polypropylene

RTP 175, 175X, 177, 178, 178X [Fiberite]—Polypropylene, glass/mineral reinforced

RTP 200FR [Fiberite]—Nylon 6/6

RTP 200H-FR, 204H-FR, 205H-FR [Fiberite]—Nylon

RTP 201, 201FR, 202FR, 203, 203FR, 204FR, 204GB FR, 205, 205FR, 205TFE5,
205TFE15, 205TFE20, 207, 209 [Fiberite]—Nylon 6/6, glass reinforced

RTP 201A, 203A, 205A, 207A, 209A, 299x27567 [Fiberite]—Nylon 6, glass reinforced

RTP 201B, 203B, 205B, 207B [Fiberite]—Nylon 6/10, glass reinforced

RTP 201C, 203C, 205C, 207C [Fiberite]—Nylon 11, glass reinforced

RTP 201D, 203D, 205D, 207D [Fiberite]—Nylon 6/12, glass reinforced

RTP 201H, 203H, 205H, 207H [Fiberite]—Nylon, glass reinforced

RTP 201P25, 203P25 [Fiberite]—Nylon 6/6, glass/carbon reinforced

RTP 202M, 203GB20, 203M GB20 [Fiberite]—Nylon 6/6, glass/mineral reinforced

RTP 203H-CF15, 204H-CF10 [Fiberite]—Nylon, carbon reinforced

RTP 225, 227 [Fiberite]—Nylon 6/6, mineral/talc reinforced

RTP 227FR [Fiberite]—Nylon 6/6, mineral reinforced

RTP 283, 285, 285P, 287 PAN, 287P [Fiberite]—Nylon 6/6, carbon reinforced

RTP 700FR [Fiberite]—Polyethylene

RTP 701, 703, 705, 707 [Fiberite]—Polyethylene (HDPE, glass reinforced)

RTP 1201-80D, 1203-80D, 1205-80D, 1207-80D [Fiberite]—Polyurethane, thermoplastic, glass-reinforced

Rucoblend 3-6, 8-289, 8-436C, 8-451B, 8-1241B, 8-1251, 8-1314, 8-1311K, 9-17, 55B,
55E, 103, 105, 161J, 172A, 179A, 207, 247, 300, 309A, 500, 501, 508F, 600, 601A,

TRADENAME PRODUCTS AND GENERIC EQUIVALENTS

2117G, 2160, 2785A, 2831, 3500 [Ruco Polymer]—Polyvinyl chloride
Rucodur 1500, 1503, 1504, 1505, 1706, 1707, 1803 [Ruco Polymer]—Polyvinyl chloride
Rucodur 1505 [Ruco Polymer]—Polyvinyl chloride
Rucon B-221, B-253, B-282, B-341 [Ruco Polymer]—Polyvinyl chloride
Rucothane 2010L, 2030L, 2060L, 3000L, 3104, 3105, 3106, 5000L, 5100, CO-1-620, CO-A-640, CO-A-670/671, CO-A-710, CO-A-819L, CO-A-832L, CO-A-904L, CO-A-2880 FR, CO-A-2885 FR, CO-A-3907, CO-A-3908L, CO-A-3982L, CO-A-3983L, CO-A-4041L, CO-A-4078, CO-A-5002L, CO-A-5054, CO-B-4030L [Occidental]—Polyurethane, thermoplastic
Rucothane CO-A-610 [Occidental]—Polyurethane, thermoplastic, polyester-based
Rucothane CO-A-5069 [Occidental]—Polyurethane, thermoplastic, polyether-based
RX 340, 350, 352 [Rogers]—Phenol-formaldehyde resin, cellulose/graphite-filled
RX 342 [Rogers]—Phenol-formaldehyde resin, cellulose/TFE-filled
RX 363 [Rogers]—Phenol-formaldehyde resin, asbestos/TFE-filled
RX 429, 448, 525, 950 [Rogers]—Phenol-formaldehyde resin, cellulose-filled
RX 431, 475 [Rogers]—Phenol-formaldehyde resin, fabric-reinforced
RX 462, 465, 466, 467, 468, 490, 495 [Rogers]—Phenol-formaldehyde resin, asbestos-filled
RX 468A [Rogers]—Phenol-formaldehyde resin, asbestos-filled
RX 610N, 611, 625, 630, 650, 655, 660, 842, 850, 862, 865, 866D, 867, 867D [Rogers]—Phenol-formaldehyde resin, glass-reinforced
RX 640, 655A, 867A [Rogers]—Phenol-formaldehyde resin, glass-reinforced
Sag Silicone Antifoam 10, 30, 100, 720 [Union Carbide]—Dimethicone
SB522/1S Film [Hercules]—Polypropylene
SCC-24, -28, -54, -58 [Stauffer]—Polyvinyl chloride
SCC-40, -52 [Stauffer]—Polyvinyl chloride (vinyl chloride/vinyl acetate copolymer)
SCC-133, -156, -421 [Stauffer]—Polyvinyl chloride (vinyl chloride/vinyl acetate copolymer)
SCC-608, -614, -616, -676, -676P, -686 [Stauffer]—Polyvinyl chloride
Serpol QPA 150 [Servo]—Ammonium polyacrylate
Serpol QPA 160 [Servo]—Sodium polyacrylate
SF 18, 69, 81, 96, 97, 1093 [General Electric]—Dimethicone
SF 1173, 1202, 1204 [General Electric]—Cyclomethicone
SF 1188 [General Electric]—Dimethicone copolyol
Shell 5225, 5384, 5419, 5431, 5520, 5524, 5550, 5610, 5820, 5824S, 5840, 5864, 5944S [Shell]—Polypropylene
Shell 7129, 7221, 7328, 7521, 7522, 7525, 7623, 7627, 7635, 7912 [Shell]—Polypropylene
Shell DX6016, DX6020 [Shell]—Polypropylene
Shell Polybutylene 0200, 0300, 0400, 0700, 1600A, 4101, 4103, 4110, 4121, 4127, 4128, 8240, 8640 [Shell]—Polyisobutene
Silicone AF-10FG, AF-10 IND, AF-30 FG, AF-30 IND, AF-600M [Harcros]—Dimeth-

361

icone

Silicone C111 [ICI Specialty]—Dimethicone

Silicone Fluid 200, 203, 225, 230 [Dow Corning]—Dimethicone

Silicone Mold Release SEM-35 [Harcros]—Dimethicone

Silicone Release Agent #5038 [Polymer Research Corp. of Amer.]—Dimethicone

SM 2061, 2140, 2155, 2162 [General Electric]—Dimethicone

Sokalan CP 5, CP 5 Powder, CP 7, CP 8, CP 10, CP 10S, CP 13S, CP 45, PA 15, PA 20, PA 20 PN, PA 25 PN, PA 30, PA 40, PA 40 Powder, PA 50, PA 70 PN, PA 80 [BASF AG]—Sodium polyacrylate

Sokalan HP-50 [BASF AG]—Polyvinylpyrrolidone

Solidur 10 100, 10 100 IV 20, 10 802 AST, 10 Color Series, 10 DS Series, 25, Ceram P, Marble [Solidur Plastics] Polyethylene (UHMW-PE)

SP-8014 [Schenectady]—Phenol-formaldehyde resin

Specflex [Dow]—Polyurethane, thermoset

Spectrim 5, 15, 25S, 25W, 35S, 35W, 50S, 50W, MM300, MM353, Polyurea HF, Polyurea HT, SF500, SP400 [Dow]—Polyurethane, thermoset

SR 80, SR 100, SR 120 [Rilsan]—Nylon 11, carbon or carbon/glass reinforced

Stat-Kon AS-F, AS-FE, BLM, FE [LNP]—Polyethylene (HDPE)

Stat-Kon H [LNP]—Nylon 11, carbon reinforced

Stat-Kon IC-1006 [LNP]—Nylon 6/12, carbon reinforced

Stat-Kon M [LNP]—Polypropylene

Stat-Kon M-1 HI, M-2, ME [LNP]—Polypropylene, carbon reinforced

Stat-Kon MF-15 [LNP]—Polypropylene, carbon/glass reinforced

Stat-Kon MM-3340 [LNP]—Polypropylene, carbon/mica reinforced

Stat-Kon PC-1006 [LNP]—Nylon 6, carbon reinforced

Stat-Kon QC-1002, QC-1006 [LNP]—Nylon 6/10, carbon reinforced

Stat-Kon R, RC-1002, RC-1004, RC-1004 FR, RC-1006, RC-1006 HI, RCL-4036, RCL-4042, RCL-4536 [LNP]—Nylon 6/6, carbon reinforced

Stat-Kon R-15 [LNP]—Nylon 6/6, glass reinforced

Stat-Kon RCF-1006, RF-15 [LNP]—Nylon 6/6, glass/carbon reinforced

Stat-Kon T [LNP]—Polyurethane, thermoplastic, carbon powder grade

Stat-Kon VC-1003 [LNP]—Nylon, carbon reinforced

Stepanfoam A-210, AX-64, AX-66, BX-105 Series, BX-150-5, BX-250 (A-D) Series, BX-289, BX-316-3, BX-316-5, BX-341, Bx-341A, BX-341B, BX-345, BX-350-7, BX-351, BX-352, BX-352P, BX-352M, BX-359, BX-364, BX-369, BX-370, BX-372-2, C-600 Series, F-202, F-302, F-506, FX-250, G Series, H-100 Series, H-102N, H-402N, H-602N, HC-2/30, HC-2/40, HC-3/40, HC-4/60, HC-5/60, HC-9/10 Series, HC-17/40, HW-8/25, HW-8/50, HW-10/25, HW-11/60, HW-12/25, HW-16/60, HW-20/60, MW-20/20, P-502, P-506, PF-15, PR-5, PR-5-O, R-222/R-109, R-223/R-110, R-226/R-112, R-231G/R-110, R-244, R-245, R-246, R-247, SF-3, SF-5/60, SX-159 Series, SX-195A, SX-202, SX-209J, SX-211, SX-214, SX-215, SX-216, SX-217, SX-218A, SX-219 [Stepan]—Polyurethane, thermoset

Stepanfoam F-403 [Stepan]—Polyurethane, thermoset, polyester-based

STX Nylon [Allied-Signal]—Nylon, glass reinforced

Stycast CPC-16, CPC-17 [Emerson & Cuming]—Polyurethane, thermoset

SWS-101, -230 Series, -231, -232, -235 [Wacker Silicones]—Dimethicone

SWS-03314 [Wacker Silicones]—Cyclomethicone

Tamol 819, 819L-43, L Conc., SD, SN [Rohm & Haas]—Sodium polynaphthalene sulfonate

Tamol NH 3901, NH 7519, NH 9103, NN, NN 2406, NN 2901, NN 4501, NNO, NNOK [BASF AG]—Sodium polynaphthalene sulfonate

Tek Tan ND, NL [Hamblet & Hayes]—Sodium polynaphthalene sulfonate

Tenite Polyethylene [Eastman]—Polyethylene

Tenite Polypropylene [Eastman]—Polypropylene

Texalon 600A, 600A HS, 600A NU, 600A PL-2, 600A Zip 1, 600A Zip 3, 670A, 680A, 1106, 1108, 1110A PL, 1110A PL HS, 1203D, XP-1296 [Texapol]—Nylon 6

Texalon 1200A, 1200A Black 11, 1200A HS, 1200AXL, 1200A ZIP-1, 1200A ZIP-6, 1308A, 1310, XP-1317 [Texapol]—Nylon 6/6

Texalon GF 600 A(6-33), GF 604 (15 and 35) [Texapol]—Nylon 6, glass reinforced

Texalon GF 1200A (13-40), GF 1308A (13-33), GF 1310 (13 and 33) [Texapol]—Nylon 6/6, glass reinforced

Texalon GMF 600-40 [Texapol]—Nylon 6/6, glass/mineral reinforced

Texalon MF 1200A-40, MF 1200A1-40 [Texapol]—Nylon 6/6, mineral reinforced

Thermocomp FF-1004, FF-1006, FF-1008 [LNP]—Polyethylene (HDPE), glass reinforced

Thermocomp HF-1006 [LNP]—Nylon 11, glass reinforced

Thermocomp IF-1002, IF-1004, IF-1006, IF-1008, IF-100-10, IF-100-12 [LNP]—Nylon 6/12, glass reinforced

Thermocomp MF-1002, MF-1002HI, MF-1004, MF-1004FR, MF-1004HI, MF-1006, MF-1006HI, MF-1008, MF-1008HI, MFX-1004HS, MFX-1006HS, MFX-1008HS [LNP]—Polypropylene, glass reinforced

Thermocomp PC-1006 [LNP]—Nylon 6, carbon reinforced

Thermocomp PF-1002, PF-1002HI, PF-1004, PF-1004HI, PF-1006, PF-1006FR, PF-1006HI, PF-1008, PF-1008HI, PF-100-10, PF-100-12, PFL-4216, PFL-4218 [LNP]—Nylon 6, glass reinforced

Thermocomp QC-1006 [LNP]—Nylon 6/10, carbon reinforced

Thermocomp QF-1002, QF-1004, QF-1006, QF-1006FR, QF-1008, QF-100-10, QF-100-12 [LNP]—Nylon 6/10, glass reinforced

Thermocomp R-1000, R-1000FR-HS, RL-4040, RL-4310, RL-4540 (Migralube Resin) [LNP]—Nylon 6/6

Thermocomp RC-1002, RC-1004, RC-1006, RC-1006HI, RC-1006PC Pitch, RC-1008, RCL-4036, RCL-4536 [LNP]—Nylon 6/6, carbon reinforced

Thermocomp RF-1002, RF-1002HI, RF-1004, RF-1004FR-HS, RF-1004HI, RF-1006, RF-1006FR-HS, RF-1006HI, RF-1008, RF-1008HI, RF-100-10, RF-100-12, RFL-

4216, RFL-4218 [LNP]—Nylon 6/6, glass reinforced

Thermocomp SF-1006, SF-100-10 [LNP]—Nylon 12, glass reinforced

Thermocomp TF-1004, TF-1006, TF-1008 [LNP]—Polyurethane, thermoplastic, glass-reinforced

Thermocomp XF-1004, XF-1006, XF-1008, XFL-4036 [LNP]—Nylon, glass reinforced

Thickener L, LN [GAF]—PVM/MA copolymer

Trymer 160, 190, 190-3, 190-4, 210, 9501, 9501-3, 9501-4, 9501-6 [Dow]—Polyurethane, thermoset

1900 UHMW Polymers [Hercules]—Polyethylene (UHMW-PE)

Ultralite [Herbert Lushan Plastics]—Polyvinyl chloride

Ultramid A3, A3K, A3KN, A3 Pellets 24, A3R, A3SK, A3W, A4, A4H, A4K, A5 B3, B3K, B3L, B3S, B4, B4K, B5, B5W, B35, B35M, B35MF01, B35SK, B35W, KR 4205, KR 4405, KR 4406, KR 4407, KR 4409, KR 4411, KR 4412, S3, S3K, S4 [Badische]—Nylon

Ultramid A3EG5, A3EG6, A3EG7, A3EG10, A3G5, A3G6, A3G7, A3HG5, A3WG5, A3WG6, A3WG7, A3WG10, A3X1G7, A3X1G10, A3XG5, A3XG7, B3EG5, B3EG6, B3EG7, B3EG10, B3G5, B3G5HS, B3G6, B3G6HS, B3G7, B3G7HS, B3G10, B3WG5, B3WG6, B3WG7, B3WG10, B35G3, KR 4445/1, KR 4447/1 [Badische]—Nylon, glass reinforced

Ultramid B3WM601, B3WM602, KR 4250, KR 4446, KR-B3X2V6 [Badische]—Nylon, mineral reinforced

Ultramid KR 4448, KR 4450 [Badische]—Nylon, glass/mineral reinforced

Ultramid KR 4449 [Badische]—Nylon, glass/ballotini reinforced

Ultramid KR 4609 [Badische]—Nylon 6/9

Ultramid KR 4645 [Badische]—Nylon 6 and 6/6

Ultramid KR 4650, KR 4652 [Badische]—Nylon 6 and 6/6, mineral reinforced

Ultramid KR 4651 [Bacische]—Nylon 6 and 6/6, glass/mineral reinforced

Ultramid KR 4653 [Badische]—Nylon 6 and 6/6, glass reinforced

Ultra-Wear UHMWPE [Polymer Corp.]—Polyethylene (UHMW-PE)

Union Carbide Dimethyl Fluid L-45 [Union Carbide]—Dimethicone

Union Carbide LE-45, LE-46, LE-420, LE-453HS, LE-458HS, LE-461, LE-462, LE-462HS, LE-467, LE-467HS [Union Carbide]—Dimethicone

Unoflex, 100 [Polymer Industries]—Polyurethane, thermoset

UR 101, 102, 103, 104, 105 [Thermoset Plastics]—Polyurethane, thermoset

Uralite 3111, 3113, 3115, 3121, 3121S, 3122, 3124, 3125, 3128, 3130, 3132, 3139, 3140, 3150, 3152, 3154, 3155, 3156, 3158, 3167, 3175 Fastset, 6108, 7250, 7252 [Hexcel/Rezolin]—Polyurethane, thermoset

Varcum 647, 1359, 1364, 4326, 4631, 4727, 9836A, 9874, C-86, DR-406 [Reichhold]—Phenol-formaldehyde resin

Varcum 1417, 1481, 1494, 5145, 5160, 5302, 5485, 6820, 8366 [Reichhold]—Phenol-formaldehyde resin

Vekton 6PA, 6PAG, 6PAM, 6PB, 6XAU [Chemplast]—Nylon 6

TRADENAME PRODUCTS AND GENERIC EQUIVALENTS

Vestolen A3512, A3512 R, A3513, A3515, A4516, A5016 F, A5017, A5018, A5041 R, A5543, A5561, A5561 P, A6012, A6013, A6014, A6016, A6017, A6042 [Huls AG]—Polyethylene (HDPE)

Vestolen P1200, P1200F, P2200, P2200F, P3200, P3200F, P3230 F, P4200, P4200F, P5200, P5200F, P5202, P5204, P5206S, P5212 LF, P6200, P6202, P6206S [Huls AG]—Polypropylene

Vestolen P2300, P2300F, P2330 F, P5400, P5400F, P6421 [Huls AG]—Polypropylene

Vestolen P4700, P4702L, P4800, P4802L, P5800, P5802 L, P6500, P6502, P6503, P6522 [Huls AG]—Polypropylene

Vestolen P5232G [Huls AG]—Polypropylene, glass reinforced

Vestolen P5232T, P5272T [Fiberite]—Polypropylene, talc filled

Vestolit B 7021, B 7022, B 7521 [Hüls AG]—Polyvinyl chloride

Vestolit B 7090, B 7092 [Hüls AG]—Polyvinyl chloride (vinyl chloride/vinyl acetate copolymer)

Vestolit E 6003, E 6007, E6017, E 6503, E 6507, E7001, E 7003, E 7004, E 7006, E 7007, E 7008, E 7012, E 7031, E 7033, E 7037, E 8001, E 8003, E 8019, P 1331 K, P 1333 K, P 1341 K, P 1342 K, P 1344 K, P 1345 K 70, P 1345 K 80 [Hüls AG]—Polyvinyl chloride

Vestolit E 7091 [Hüls AG]—Polyvinyl chloride (vinyl chloride/vinyl acetate copolymer)

Vestolit HI E 6077, HI E 6577, HI E 7077 [Hüls AG]—Polyvinyl chloride/elastomer copolymer

Vestolit HI S 6882, HI S 6883, HI S 7587, P 1976 K, P 1982 K [Hüls AG]—Polyvinyl chloride/elastomer copolymer

Vestolit M 5867, M 6067, M 6267, M 6567, M 6867 [Hüls AG]—Polyvinyl chloride

Vestolit S 6058, S 6059, S 6554, S 6555, S 6558, S 6858, S 7054, S 7055, S 7554, S 8054 [Hüls AG]—Polyvinyl chloride

Viscasil 5M, 10M, 60M [General Electric]—Dimethicone

Volatile Silicones 7158, 7207, 7349 [Union Carbide]—Cyclomethicone

Vultamol, SA Liq. [BASF AG]—Sodium polynaphthalene sulfonate

Vydyne 21, 21SP, 21X, 22H, 22HSP, 24 NSL, 24NSP, 25W, 25X, 65B, 66B [Monsanto]—Nylon 6/6

Vydyne 60H [Monsanto]—Nylon 6/9

Vydyne 909 [Monsanto]—Nylon, glass reinforced

Vydyne M-340, M-344, M-345 [Monsanto]—Nylon

Vydyne R-100, R-200, R-220, R-250 [Monsanto]—Nylon, mineral reinforced

Vydyne R-400G [Monsanto]—Nylon 6/6, glass/mineral reinforced

Vydyne R-513, R-513H, R-533, R-533H, R-538H-02, R-543, R-543H [Monsanto]—Nylon 6/6, glass reinforced

Vylor 7264, ME-1791 [DuPont]—Nylon 6/6

Wellamid 2BRH-NW, 21LN2-NNT, 22LH-N, 22LH13-N, 22L-N, 22LN2-N, 220-N, FR22F-N [Wellman]—Nylon 6/6

Wellamid 42LH-N, 42L-N, 42LN2-N [Wellman]—Nylon 6

Wellamid FRGF25-66, FRGS25-66, GS25-66, GS40-66, GSF25/15-66 [Wellman]—
 Nylon 6/6, glass reinforced

Wellamid GS40-60 [Wellman]—Nylon 6, glass reinforced

Wellamid MR329HS, MR409HS, MR-410HS [Wellman]—Nylon 6/6, mineral reinforced

Wellamid MRG30/10 [Wellman]—Nylon 6, glass/mineral reinforced

Wellamid MRG30/10 [Wellman]—Nylon 6/6, glass/mineral reinforced

Wettol D2 [BASF AG]—Sodium polynaphthalene sulfonate

Zytel FR-50 [DuPont]—Nylon 6/6, glass reinforced

Zytel ST FR-80 [DuPont]—Nylon

GENERIC CHEMICAL SYNONYMS
AND CROSS REFERENCES

Acrylamide/beta-methacrylyloxyethyl trimethyl ammonium methosulfate copolymer. See Polyquaternium-5

Acrylamide/dimethyl diallyl ammonium chloride copolymer. See Polyquaternium-7

Acrylonitrile-butadiene rubber. See Acrylonitrile-butadiene copolymer

Acrylonitrile rubber. See Acrylonitrile-butadiene copolymer

Butadiene-acrylonitrile copolymer. See Acrylonitrile-butadiene copolymer

Chlorinated polyethylene. See under Polyethylene

Chlorinated PVC. See under Polyvinyl chloride

Chloroprene rubber. See Polychloroprene

Chlorosulfonated polyethylene. See under Polyethylene

CR. See Polychloroprene

Cyclic dimethylpolysiloxane. See Cyclomethicone

Cyclic dimethylsiloxane. See Cyclomethicone

Dimethyl diallyl ammonium chloride polymer. See Polyquaternium-6

Dimethylpolysiloxane. See Dimethicone

N,N-Dimethyl-N-2-propenyl-2-propen-1-aminium chloride, homopolymer. See Polyquaternium-6

N,N-Dimethyl-N-2-propenyl-2-propen-1-aminium chloride, polymer with 2-propenamide. See Polyquaternium-7

Dimethyl silicone. See Dimethicone

Dimethylsiloxane. See Dimethicone

Dimethylsiloxane-glycol copolymer. See Dimethicone copolyol

EMA. See Ethylene/methyl acrylate copolymer

Ethanaminium, N,N,N-trimethyl-2-[(2-methyl-1-oxo-2-propenyl) oxy]-, methyl sulfate, polymer with 2-propenamide. See Polyquaternium-5

Ethene, homopolymer. See Polyethylene

1-Ethenyl-2-pyrrolidinone, homopolymer. See Polyvinylpyrrolidone

1-Ethenyl-2-pyrrolidinone, polymer with ethenylbenzene. See Styrene/PVP copolymer

Ethylene/acrylic acid copolymer (EAA). See Ethylene/acrylate copolymer

Ethylene/methacrylate copolymer. See Ethylene/acrylate copolymer

Ethylene/methacrylic acid copolymer (EMAA). See Ethylene/acrylate copolymer

2,5-Furandione, polymer with ethene. See Ethylene/maleic anhydride copolymer

2,5-Furandione, polymer with methoxyethylene. See PVM/MA copolymer

HDPE. See Polyethylene
HMW-HDPE. See Polyethylene
HMW-LDPE. See Polyethylene
LDPE. See Polyethylene
LDPE/vinyl acetate copolymer. See under Polyethylene
LLDPE. See Polyethylene
LMWPE. See Polyethylene
Macrogol 200. See PEG-4
Macrogol 300. See PEG-6
Macrogol 400. See PEG-8
Macrogol 600. See PEG-12
Macrogol 1500. See PEG-6-32
Macrogol 4000. See PEG-75
Macrogol 6000. See PEG-150
MDPE. See Polyethylene
Methyl vinyl ether/maleic anhydride copolymer. See PVM/MA copolymer
NBR. See Acrylonitrile-butadiene copolymer
Neoprene. See Polychloroprene
Nitrile-butadiene rubber. See Acrylonitrile-butadiene copolymer
Nitrile rubber. See Acrylonitrile-butadiene copolymer
Novalac resin. See Phenol-formaldehyde resin
Novalak resin. See Phenol-formaldehyde resin
Nylon 5. See Nylon
Nylon 6/6. See Nylon
Nylon 6/10. See Nylon
Nylon 9. See Nylon
Nylon 11. See Nylon
Nylon 12. See Nylon
PDMS. See Dimethicone
PE. See Polyethylene
PEG 200. See PEG-4
PEG 300. See PEG-6
PEG 400. See PEG-8
PEG 600. See PEG-12
PEG 1000. See PEG-20
PEG 1500. See PEG-6-32
PEG 4000. See PEG-75
PEG 6000. See PEG-150
Phenol-formaldehyde novolak resin. See Phenol-formaldehyde resin
Phenolic resin. See Phenol-formaldehyde resin
POE (4). See PEG-4
POE (6). See PEG-6

POE (8). See PEG-8
POE (12). See PEG-12
POE (20). See PEG-20
POE (75). See PEG-75
POE (150). See PEG-150
POE 1500. See PEG-6-32
Poly (11-aminoundecanoic acid). See Nylon 11 under Nylon
Polybutene (CTFA). See Polyisobutene
Polybutylene. See Polyisobutene
Polycaprolactam. See Nylon 6 under Nylon
Poly (dimethyl diallyl ammonium chloride). See Polyquaternium-6
Polydimethylsiloxane. See Dimethicone
Poly (DMDAAC). See Polyquaternium-6
Polyethylene glycol 200. See PEG-4
Polyethylene glycol 300. See PEG-6
Polyethylene glycol 400. See PEG-8
Polyethylene glycol 600. See PEG-12
Polyethylene glycol 1000. See PEG-20
Polyethylene glycol 1500. See PEG-6-32
Polyethylene glycol 4000. See PEG-75
Polyethylene glycol 6000. See PEG-150
Poly(hexamethylene adipamide). See Nylon 6/6 under Nylon
Polyisobutylene. See Polyisobutene
Poly(methyl vinyl ether/maleic anhydride). See PVM/MA copolymer
Poly [oxy (dimethylsilylene)], α-(trimethylsilyl)-ω-methyl-. See Dimethicone
Polyoxyethylene (4). See PEG-4
Polyoxyethylene (6). See PEG-6
Polyoxyethylene (8). See PEG-8
Polyoxyethylene (12). See PEG-12
Polyoxyethylene (12). See PEG-20
Polyoxyethylene (75). See PEG-75
Polyoxyethylene (150). See PEG-150
Polyoxyethylene 1500. See PEG-6-32
Polyoxypropylene (9). See PPG-9
Polyoxypropylene (12). See PPG-12
Polyoxypropylene (17). See PPG-17
Polyoxypropylene (26). See PPG-26
Polyoxypropylene (30). See PPG-30
Polypropylene glycol (9). See PPG-9
Polypropylene glycol (12). See PPG-12
Polypropylene glycol (17). See PPG-17
Polypropylene glycol (26). See PPG-26

Polypropylene glycol (30). See PPG-30

Polyvinyl ether. See Polyvinyl isobutyl ether

Polyvinylidene choride. See Vinylidene chloride copolymer

POP (9). See PPG-9

POP (12). See PPG-12

POP (17). See PPG-17

POP (26). See PPG-26

POP (30). See PPG-30

Povidone. See Polyvinylpyrrolidone

PP. See Polypropylene

PPG (9). See PPG-9

PPG (12). See PPG-12

PPG (17). See PPG-17

PPG (26). See PPG-26

PPG (30). See PPG-30

2-Propenamide, homopolymer. See Polyacrylamide

2-Propenamide, polymer with 2-propenoic acid, sodium salt. See Acrylamide/sodium
    acrylate copolymer

2-Propen-1-aminium, N,N-dimethyl-N-2-propenyl-, chloride, homopolymer.   See
    Polyquaternium-6

2-Propen-1-aminium, N,N-dimethyl-N-2-propenyl-, chloride, polymer with 2-prope-
    namide. See Polyquaternium-7

2-Propenoic acid, homopolymer. See Polyacrylic acid

2-Propenoic acid, 2-methyl-, homopolymer, sodium salt. See Sodium polymethacrylate

2-Propenoic acid, sodium salt, polymer with 2-propenamide. See Acrylamide/sodium
    acrylate copolymer

Propylene polymer. See Polypropylene

PU. See Polyurethane

PVC. See Polyvinyl chloride

PVI. See Polyvinyl isobutyl ether

PVP (CTFA). See Polyvinylpyrrolidone

2-Pyrrolidinone, 1-ethenyl-, homopolymer. See Polyvinylpyrrolidone

2-Pyrrolidinone, 1-ethenyl-, polymer with ethenylbenzene. See Styrene/PVP copolymer

Quaternium-23. See Polyquaternium-11

Quaternium-39. See Polyquaternium-5

Quaternium-40. See Polyquaternium-6

Quaternium-41. See Polyquaternium-7

Resole. See Phenol-formaldehyde resin

Sebacic acid/hexamethylenediamine condensation product. See Nylon 6/10 under Nylon

Sodium naphthalene-formaldehyde sulfonate. See Sodium polynaphthalene sulfonate

Styrene/vinylpyrrolidone copolymer. See Styrene/PVP copolymer

UHMW-PE. See Polyethylene

Urethane polymer. See Polyurethane

VC copolymer. See Vinylidene chloride copolymer

VDC copolymer. See Vinylidene chloride copolymer

Vinyl chloride/vinyl acetate copolymer. See under Polyvinyl chloride

Vinylpyrrolidone/dimethylaminoethylmethacrylate copolymer, reacted with dimethyl sulfate. See Polyquaternium-11

TRADENAME PRODUCT MANUFACTURERS

Akron Chemical Co.
255 Fountain St.
Akron, OH 44304

Akzo Chemie America
Akzo Chemie America/Armak Chemical
300 S. Riverside Plaza
Chicago, IL 60606

Akzo Chemie UK Ltd.
1-5 Queens Rd., Hersham
Waltham-on-Thames
Surrey KT12 5NL UK

Akzo Chemie
PO Box 186
LS Arnhem Netherlands 6800

Akzo Chemie B.V.
POB 975
3800 AZ Amersfoort, Netherlands

Akzo Chemie Italia SpA.
Via Vismara, 20020 Arese
Milano, Italy

Alco Chemical Corp.
909 Mueller Dr.
Chattanooga, TN 37406

Alkaril Chemicals Inc.
Industrial Pkwy., PO Box 1010
Winder, GA 30680

Alkaril Chemicals Ltd.
3265 Wolfedale Road
Mississauga, Ontario L5C 1V8, Canada

Allied Colloids Inc.
One Robinson Lane
Ridgewood, NJ 07450

Allied Colloids Ltd.
PO Box 38, Low Moor, Bradford
Yorkshire BD12 OJZ, UK

Allied-Signal
PO Box 2332R, Columbia Rd. & Park Ave.
Morristown, NJ 07960

Allied Corporation International NV-SA
Haasrode Research Park
B-3030 Heverlee, Belgium

Allied Corporation International NV-SA
International House
Bickenhill Lane
Birmingham B37 7HQ
England

Allied Chemical International Corporation
P.O. Box 99067, Tsimshatsui Post Office
Hong Kong

Alox Corp.
3943 Buffalo Ave., PO Box 517
Niagara Falls, NY 14302

American Cyanamid Co.
American Cyanamid Co./Industrial Chem.
Div.
Berdan Ave.
Wayne, NJ 07470

American Cyanamid Co./
Polymer & Chem. Dept.
Berdan Ave.
Wayne, NJ 07470

Cyanamid B.V.
Postbus 1523, 3000 BM
Rotterdam, The Netherlands

Cyanamid India Ltd.
Nyloc House, 254-D2 Dr. Annie Besant Rd.
Bombay 400 025 India

Cyanamid Quimica do Brasil Ltda.
Av. Imperatriz Leopoldina, 86
Sao Paulo, Brazil

Cyanamid Taiwan Corp.
8/F Union Commercial Bldg.
137, Nanking E. Rd., Sec. 2
Taipei, Taiwan, R.O.C.

Amoco Chemicals Corp.
200 East Randolph Dr., PO Box 8640A
Chicago, IL 60680

Amoco Chemical Europe S.A.
15, Rue Rothschild
CH-1211 Geneva 21,
Switzerland

Amoco Performance Products,
Japan Ltd.
10th Floor, Tonichi Building
2-31 Roppongi 6-Chome
Minato Ku, Tokyo 106, Japan

Aquatec Quimica S/A
Rua Sampaio Viana, 425, CX. Postal 4885
04004 Sao Paulo Sp., Brazil

Arco Chemical Co./Div. Atlantic Richfield Co.—*See Sartomer Co. Inc.*

Arco Chemical Company
Toranomon 37 Mori Bldg., 5th fl.
5-1 Toranomon 3-chome
Minato-ku
Tokyo, 105 Japan

Arco Chemical Company
Windsor Bridge House
1 Brocas Street
Eton, Berkshire
SL4 6BW England

Atochem U.S.A. Inc.
PO Box 159, 13 Sunflower Ave.
Paramus, NJ 07652

Atochem Inc./Polymers Div.
266 Harristown Rd., POB 607
Glen Rock, NJ 07452

Atochem
4, cours Michelet, La Défense 10-Cedex 42
92091 Paris la Défense France

Rilsan Corp./Subsid. of Atochem, France
139 Harristown Rd.
Glen Rock, NJ 07452

Badische Corp.
PO Drawer D
Williamsburg, VA 23187

Bamberger Polymers, Inc.
3003 New Hyde Park Rd.
New Hyde Park, NY 11042

BASF Corp.
100 Cherry Hill Rd.
Parsippany, NJ 07054

BASF Canada Ltd.
PO Box 430
Montreal, Quebec H4L 4V8, Canada

BASF (UK) Ltd.
PO Box 4, Earl Rd., Cheadle Hulme
Cheadle, Cheshire 5K8 60QG, UK

BASF Belgium S.A.
avenue Hamoir-Iaan 14
B-1180 Bruxelles/Brussel
Belgium

BASF AG
ESA/WA-H 201
D-6700 Ludwigshafen, West Germany

BASF Espanola S.A.
Apartado 762
Barcelona 8, Spain

BASF S.A., Compagnie Francaise
MC-NT, 140, Rue Jules Guesde
92303 Levallois-Perret, France

BASF India, Ltd.
Maybaker House, S.K. Ahire Marg., PO
Box 19108
Bombay 400 025 India

BASF Japan Ltd.
C.P.O. Box 1757
Toyko 100-91
Japan

Bayer AG
Sitz der Gesellschaft, Leverkusen
Eintragung, Amtsgericht
Leverkusen HRB 1122, Germany

Bayer UK Ltd.
Bayer House, Strawberry Hill
Newbury, Berkshire RG13 1JA, UK

Belding Chemical Industries
1430 Broadway
New York, NY 10018

Calgon Corp.
P.O. Box 1346
Pittsburgh, PA 15230

Cal Polymers, Inc.
2115 Gaylord St.
Long Beach, CA 90813

R.H. Carlson Co.
41 Chestnut St.
Greenwich, CT 06830

Chemplast, Inc.
150 Dey Road
Wayne, NJ 07470

Chemplex Co.
Rolling Meadows, IL 60008

Chevron Chemical Co.
575 Market St.
San Francisco, CA 94105

Chromex Chemical Corp.
19 Clay St.
Brooklyn, NY 11222

Ciba-Geigy Corp.
PO Box 18300
Greensboro, NC 27419

Ciba-Geigy Corp./
Plastics & Additives Div.
Three Skyline Dr.
Hawthorne, NY 10532

Ren Plastics/Ciba-Geigy Corp.
4917 Dawn Ave.
Lansing, MI 48823

Ciba-Geigy Corp.
CH-4002
Basle, Switzerland

Ciba-Geigy PLC
30 Buckingham Gate
London SW1E 6LH, UK

Ciba-Geigy Dyestuffs & Chemicals
Ashton New Road, Clayton
Manchester M11 4AR, UK

Colloids Inc.
394 Frelinghuysen Ave.
Newark, NJ 07114

Compounding Technology, Inc.
3140 Pullman St.
Costa Mesa, CA 92626

Conap Inc.
1405 Buffalo St.
Olean, NY 14760

**Conoco Chemicals/Continental Oil
Co.**—*See Vista*

Crucible Chemical Co.
POB 6786, Donaldson Center
Greenville, SC 29606

Cyanamid—*See American
Cyanamid*

Dow Chemical Co.
1703 S. Saginaw Rd.
Midland, MI 48640

Dow Chemical Co. Ltd.
Stana Place, Fairfield Ave.
Staines, Middlesex TW18 4SX, UK

Dow Chemical Europe S.A.
Bachtobelstrasse 3, CH-8810
Horgen, Switzerland

Dow Chemical Pacific Limited
P.O. Box 711
Hong Kong

Dow Corning Corp.
Box 0994
Midland, MI 48640

Drew Chemical Corp./Drew Ind. Div., Ashland Chem. Co.
One Drew Plaza
Boonton, NJ 07005

E.I. DuPont de Nemours & Co.
Nemours Bldg.
Wilmington, DE 19898

Du Pont Co./
Polymer Prods. Dept.
15990 N. Barkers Landing Rd., POB 19029
Houston, TX 77224

DuPont Canada
PO Box 26, Toronto Dominion Center
Toronto, Ont. M5K 1B6 Canada

DuPont (UK) Ltd.
Wedgwood Way, Stevenage
Herts SG1 4QN, UK

DuPont (UK) Ltd./
Polymer Prods. Dept.
Maylands Ave., Hemel Hempstead
Herts HP2 7DP, UK

DuPont France
0137, rue de l'Universite
F-75334 Paris Cedex 07
France

DuPont de Nemours International S.A.
Case Postale CH-1211
Geneva 24, Switzerland

DuPont Japan Ltd.
P.O. Box 37, Akasaka
Tokyo, 107-91 Japan

Eastman Chemical Products, Inc./ Subsid. of Eastman Kodak Co.
PO Box 431
Kingsport, TN 37662

Eastman Chemical International A.G.
Hemel Hempstead
P.O. Box 66
Kodak House, Station Road
Herts, HP1 1JU England

Eastman Japan Ltd.
Nishi-Shinbashi Mitsui Bldg.
1-24-14 Nishi-Shinbashi
Minato-Ku, Tokyo 105
Japan

El Paso Products Co.
PO Box 3986
Odessa, TX 79760

Emerson & Cuming—*See W.R. Grace & Co.*

Emser Industries Inc.
PO Box 1717
Industrial Park and Corporate Way
Sumter, SC 29151-1717

Fiberite Corp.
501-559 W. Third St.
Winona, MN 55987

Firestone Tire & Rubber Co.
381 W. Wilbeth Rd.
Akron, OH 44301

Firestone Plastics Co.
Box 699, Firestone Blvd.
Pottstown, PA 19464

Firestone Synthetic Rubber & Latex Co.
PO Box 2786, Firestone Park Sta.
Akron, OH 44301

Freeman Chemical Corp./Subsid. of H.H. Robertson Co.
PO Box 247
Port Washington, WI 53074

GAF Corp./Chemical Products
1361 Alps Rd.
Wayne, NJ 07470

GAF Europe
40 Alan Turing Rd., Surrey Research Park
Guildford, Surrey, UK

General Electric Co.
General Electric Co./Plastics Div.
One Plastics Ave.
Pittsfield, MA 01201

General Electric Co./Silicone Prod. Div.
Mechanicsville Rd.
Waterford, NY 12188

General Electric Plastics B.V.
Plasticslaan 1, P.O. Box 117
4600 AC Bergen op Zoom
The Netherlands

GE Plastics Ltd.
Birchwood Park
Risley, Warrington
Cheshire WA3 6DA
England

General Electric (USA) Plastics Japan Ltd.
No. 35 Kowa Building 5F
14-14 Akasaka 1-chome
Minato-ku, Tokyo 107
Japan

Goldschmidt Chemical Corp.
914 Randolph Rd., Box 1299
Hopewell, VA 23860

Goldschmidt AG, Th.
Goldschmidtstr. 100, Postfach 101461
D4300 Essen-1, West Germany

B.F. Goodrich Co.
6100 Oak Tree Blvd.
Cleveland, OH 44131

BFGoodrich Chemical (U.K.) Ltd.
The Lawn
100 Lampton Road
Hounslow, Middlesex TW3 4EB
England

BFGoodrich Chemical (Deutschland)
GmbH
Goerlitzer Str. 1
4040 Neuss 1
Federal Republic of Germany

Goodyear Tire & Rubber Co.
1144 East Market St.
Akron, OH 44316

Goodyear Canada Inc.
45 Raynes Ave.
Bowmanville, Ontario, Canada

W.R. Grace & Co,
W.R. Grace & Co./Organic Chem. Div.
55 Hayden Ave.
Lexington, MA 02173

Emerson & Cuming/W.R. Grace & Co.
869 Washington St.
Canton, MA 02021

W.R. Grace Ltd.
Northdale House, North Circular Rd.
London NW10 7UH, UK

Graden Chemical Co., Inc.
426 Bryan St.
Havertown, PA 19083

Hamblet & Hayes Co.
PO Box 730, Colonial Rd.
Salem, MA 01970

Harcros Chemicals Inc.
5200 Speaker Rd., PO Box 2383
Kansas City, MO 66110

Harcros Chemicals UK Ltd.
Lankro House
PO Box 1, Eccles
Manchester, M3O 0BH, UK

Hart Chemicals Ltd.
256 Victoria Rd. South
Guelph, Ontario N1H 6K8 Canada

Hefti Ltd. Chemical Products
PO Box 1623, CH-8048
Zurich, Switzerland

Henkel Corp.
480 Alfred Ave.
Teaneck, NJ 07666
Henkel Corp./Process Chemicals
350 Mt. Kemble Ave.
Morristown, NJ 07960

Henkel Chem. (Canada) Ltd.
9550 Ray Lawson Blvd.
Ville d'Anjou, Quebec H1J 1L3, Canada

Henkel Chemicals Ltd.
Organic Products Division
Merit House, The Hyde
Edgeware Rd., London NW9 5AB, UK

Henkel Argentina S.A.
Avda. E. Madero Piso 14
1106 Capital Federal
Argentina

Henkel KGaA
Postfach 1100 D-4000
Dusseldorf 1, West Germany

Henkel-Nopco SA, Process Chem. Div.
185 Ave. de Fontainebleau
77310 St. Fargeau
Pontheirry, France

Herbert Lushan Plastics Corp.
189 Wells Ave.
Newton, MA 02159

Hercules Inc.
Hercules Plaza
Wilmington, DE 19894

Hercules B.V.
8 Veraartlaan, PO Box 5822
2280 HV Rijswijk, The Netherlands

Hercules Ltd., Hercules European Hdqts.
20 Red Lion St.
London WC1R 4PB UK

Hexcel Corp./Chemical Products Div.
215 N. Centennial St.
Zeeland, MI 49464

Hexcel Corp./Rezolin Div.
20701 Nordhoff St.
Chatsworth, CA 91311

Hoechst-Celanese Corp.
Route 202-206 North
Somerville, NJ 08876

Hoechst AG
Verhaufkanststoffe, D-6230
Frankfurt (M) 80, West Germany

Hoechst Celanese Plastics Ltd.
78-80 St. Albans Rd.
Watford, Herts
England WD2-4AP

Hoechst Japan
10-33, 4-Chome-Akasaka, Minato-ku
Tokyo, Japan

Huls America Inc.
10 Link Dr.
Rockleigh, NJ 07647

Hüls AG, Chemische Werke
Postfach 1320 D-4370
Marl 1, West Germany

Huls (UK) Ltd.
Cedars House, Farnborough Common
Orpington, Kent BR6 7TE, UK

Hysol Div./The Dexter Corp.
15051 E. Don Julian Road
Industry, CA 91749

ICI Americas Inc.
New Murphy Rd. & Concord Pike
Wilmington, DE 19897

ICI Australia Operations Pty, Ltd.
ICI House, 1 Nicholson St.
Melbourne 300, Australia

ICI Europe Ltd.
Everslaan 45
B-3078 Kortenberg, Belgium

ICI Ltd./Organics Div., Chemical
Auxiliaries Business Group
Smith's Rd., Bolton
Lancs, BL3 2QJ, UK

ICI Plant Protection Div.
Fernhurst, Haslemere
Surrey GU27 3JE, UK

ICI PLC/Petrochemicals & Plastics Div.,
England
PO Box 90, Wilton Middlesbrough
Cleveland TS6 8JE, UK

ICI Specialty Chemicals
Everslann 45 B-3078
Kortenberg, Belgium

ICI Japan Ltd.
Osaka Green Building
1, 3-Chome Kitahama
Higashi-Ku
J-Osaka 541
Japan

S.C. Johnson & Son, Inc.
3225 North Verdugo Rd.
Glendale, CA 91208

Kempen, Elektrochemische Fabrik Kempen GmbH
Postfach 100 260
D-4152 Kempen 1, West Germany

LNP Corp.
412 King St.
Malvern, PA 19355

LNP Plastics U.K.
Unit 25 Monkspath Business Park
Solihulll, West Midlands
England

M.A. Industries Inc., Polymer Div.
PO Box 2322
Peachtree City, GA 30269

Millmaster-Onyx International
PO Box 1045
Fairfield, NJ 07007

Millmaster-Onyx UK
Marlborough House, 30-2 Yarm Rd.
Stockton-on-Tees, Cleveland, UK

Miranol Chemical Co., Inc.
68 Culver Rd., PO Box 411
Dayton, NJ 08810

Represented by:
Alfa Chemicals Ltd.
Broadway House, 7 Shute End
Wokingham Berkshire RG11 1BH, UK

Mitsubishi International Corp./ Chemicals Dept.
277 Park Ave.
New York, NY 10017

Mitsubishi Chemical Industries Ltd.
6-3, Marunouchi 2-chome, Chiyoda-ku
Tokyo, Japan

Mobay Chemical Corp./Organic & Rubber Chem. Div., Plastics Div.
Penn Lincoln Parkway West
Pittsburgh, PA 15205

Mobil Oil Corp./Special Products Dept.
3225 Gallows Rd.
Fairfax, VA 22037

Mobil Polymers U.S. Inc.
591 W. Putnam Ave.
Greenwich, CT 06836

Monmouth Plastics, Inc.
814 Asbury Ave.
Asbury Park, NJ 07712

Monsanto Co.
800 N. Lindbergh Blvd.
St. Louis, MO 63167

Monsanto PLC
Monsanto House
Chineham Court, Chineham
Basingstoke, Hants RG24 0UL, UK

Monsanto Europe S.A.
Avenue de Tervuren 270-272
P.O. Box 1
B-1150 Brussels, Belgium

TRADENAME PRODUCT MANUFACTURERS

NL Industries, Inc.
NL Chemicals, Inc./
Industrial Chemicals Div.
PO Box 700
Hightstown, NJ 08520

NL Treating Chemicals
PO Box 60020
Houston, TX 77205

NL Chem Canada, Inc.
2140 Edifice Sul Life Building
Montreal, Quebec H3B 2X8, Canada

Norchem Inc.—*See Northern Petrochemical*

Northern Petrochemical Co.
2350 East Devon Ave.
Des Plaines, IL 60018

Norchem Inc.
Two Central Park Plaza
Omaha, NE 68102

Nortech
830 Main St.
Clinton, MA 01510

Occidental Chemical Corp.
360 Rainbow Blvd, S., POB 728
Niagara Falls, NY 14302

Occidental Chemical Corp./
Durez Resins & Molding Materials
Walck Rd., Box 535
North Tonawanda, NY 14120

Olin Chemicals
120 Long Ridge Rd.
Stamford, CT 06904

Olin U.K. Ltd.
Site 7, Kidderminster Road
Cutnall Green
Worcestershire
England WR9 0NS

Olin Japan, Inc.
Shiozaki Building
7-1 Hirakawa-Cho 2-Chome
Chiyoda-ku
Tokyo 102, Japan

Pantasote Inc.
26 Jefferson St.
Passaic, NJ 07055

Perma Flex Mold Co.
1919 East Livingston Ave.
Columbus, OH 43209

Phillips Chemical Co./Subsid. of Phillips Petroleum Co.
14 Phillips Building
Bartlesville, OK 74004

Plastics Engineering Co.
3518 Lakeshore Rd.
Sheboygan, WI 53081

Polymer Applications
3445 River Road
Tonawanda, NY 14150

The Polymer Corp.
2120 Fairmont Ave., PO Box 422
Reading, PA 19603

Polymer Industries
Viaduct Rd.
Stamford, CT 06907

Polymer Research Corp. of America
2186 Mill Ave.
Brooklyn, NY 11234

Polymer Systems Corp.
16 Edgeboro Rd.
E. Brunswick, NJ 08816

Polysar Inc.
2603 West Market St.
Akron, OH 44313

Polysar Ltd.
201 Front St.
Sarnia, Ontario N7T 7M2 Canada

Polysar International S.A.
Route de Beaumont 10
P.O. Box 1063
Ch-1701-Fribourg, Switzerland

379

Polysar Technical Centre NV
P.O. Box 354
B-2000 Antwerp
Belgium

Polyvinyl Chemical Industries
730 Main St.
Wilmington, MA 01887

PPG-Mazer
3938 Porett Dr.
Gurnee, IL 60031

Presto Mfg. Co., Inc.
2 Franklin Ave.
Brooklyn, NY 11211

Proctor Chemical Co.
PO Box 399
Salisbury, NC 28144

Quantum Chemical Corp.
26250 Euclid Ave.
Euclid, OH 44132

Quantum Chemical Corp., USI Div.
Lange Bunder 7
4854 MB Bavel, The Netherlands

Quantum Chemical Ltd.
365 Evans Avenue, Suite 601
Toronto, Ontario M8Z 1K2

K.J. Quinn & Co., Inc.
195 Canal St.
Malden, MA 02148

Reichhold Chemicals Inc.
RCI Building
White Plains, NY 10602

Ren Plastics—See Ciba-Geigy Corp.

Rexene Products—See El Paso
Products Co.

The Richardson Co.
2400 Devon Ave.
Des Plaines, IL 60018

Rilsan Corp.—See Atochem

RITA Corp.
PO Box 556
Crystal Lake, IL 60014

Represented by:
Maprecos
4, Rue des Passe-Loups
7770 Fontaine
Le Port, France

Rogers Corp.
One Technology Dr.
Rogers, CT 06263

Rohm & Haas Co.
Independence Mall West
Philadelphia, PA 19105

Rohm & Haas (Australia) Pty. Ltd.
969 Burke Rd., PO Box 115 Camberwell
Victoria 3124 Australia

Rohm and Haas (UK) Ltd.
Lennig House
2 Mason's Avenue
Croydon CR9-3NB
England

Rohm & Haas Co. European Operations
Chesterfield House, Bloomsbury Way
London WC1A 2TP, UK

Rohm and Haas Asia, Ltd.
Kaisei Building
8-10 Azabudai 1-Chome
Minato-ku, Tokyo 106 Japan

Ruco Polymer Corp.
New South Rd.
Hicksville, NY 11802

Sandoz Chemicals Corp.
4000 Monroe Rd.
Charlotte, NC 28205

Sandoz Chemicals Corporation
Dorva, Quebec H9R 4P5
Canada

Sandoz Ltd./Chemicals Div.
Lichtstrasse 35
CH-4002 Basel, Switzerland

Sandoz Ltd.
Calverley Lane, Horsforth
Leeds LS18 4RP, UK

Sartomer Co./Subsid. of The Atlantic Richfield Co.
PO Box 799
West Chester, PA 19380

Sartomer International, Inc.
Kingswick House
Sunninghill
Berkshire SL5 7BH, England

Sartomer International Deutschland GmbH
Konigsallee 60F
4000 Dusseldorf 1
West Germany

Schenectady Chemicals, Inc.
PO Box 1046
Schenectady, NY 12301

Scott Bader Co. Ltd.
Wollaston, Wellingborough
Northamptonshire NN9 7RL, UK

Servo Chemische Fabriek B.V.
PO Box 1, 7490 AA
Delden, Holland

Shell Chemical Co.
One Shell Plaza
Houston, TX 77002

Shell Chemicals UK Ltd.
1 Northumberland Av.
London WC2N 5LA, UK

Shell Chimie France
27 Rue de Berri
7539, Paris Cedex 08 France

Shell International Chemical Co., Ltd./
Agrochemicals Div. (CSAA)
Shell Centre
London SE1 7PG, UK

Shell Nederland Chemie B.V.
PO Box 187, 2501 CD
The Hague, Netherlands

Shenango Phenolics Corp.
Clover Lane
New Castle, PA 16105

Solidur Plastics Co.
200 Plum Industrial Ct.
Pittsburgh, PA 15239

Soltex Polymer Corp.—*See Solvay & Cie*

Solvay & Cie
Rue du Prince Albert 44-
1050 Brussels, Belgium

Soltex Polymer Corp
3333 Richmond Ave., POB 27328
Houston, TX 77098

Stauffer Chemical Co.
Nyala Farm Rd.
Westport, CT 06880

Stepan Co.
Edens & Winnetka Roads
Northfield, IL 60093

Stepan Europe
BP127
38340 Voreppe, France

Texaco Chemical Co./Div. Texaco Inc.
PO Box 15730
Austin, TX 78761

Texaco Canada Inc.
1210 Shepherd Ave. East, Willowdale
Ontario, Canada M2K 2S8

Texaco, Ltd.
Petrochemical Department
195 Knightsbridge
London SW7 1RU England

Texaco France S.A.
Division Chimie
Tour Arago
5, rue Bellini
F-92806 Puteaux Cedex, France

S.A. Texaco Petroleum N.V.
Ghent Chemical Plant
J.F. Kennedylaan 29
B-9020 Ghent, Belgium

Sanseki-Texaco Chemicals Co. Ltd.
Kasumigaseki Building 23F
No. 2-5, 3-chome
Kasumigaseki Chiyoda-Ku
Tokyo 100, Japan

Texapol Corp.
177 Mikron Rd.
Bethlehem, PA 18017

Thermoset Plastics, Inc.
5101 East 65th St.
Indianapolis, IN 46220

TSE Industries
5260 113 Ave. North, POB 17225
Clearwater, FL 33520-0225

Union Carbide Corp.
39 Old Ridgebury Rd.
Danbury, CT 06817

Union Carbide U.K. Limited
Rickmansworth WD 3 1RB Herts
England

Union Carbide Europe S.A.
15, Chemin Louis-Dunant, 1211 Geneve 20
Geneva, Switzerland

Union Carbide Brazil
Rua Dr. Eduardo De Souza Aranha
153, Sao Paulo 04530
Brazil

Union Carbide Japan KK
Toranomon 45 Mori Bldg.
1-5 Toranomon
5-Chome Minato-Ku
Tokyo 105, Japan

Uniroyal Inc./Chemical Group
World Headquarters
Middlebury, CT 06749

Uniroyal Chemical
Brooklands Farm, Cheltenham Road
Evansham, Worcestershire WR11 6LW
England

United States Polymers Inc.
19012 Clymer St.
Northridge, CA 91326

R.T. Vanderbilt & Co., Inc.
30 Winfield St.
Norwalk, CT 06855

Vista Chemical Co.
15990 N. Barkers Landing Rd.
PO Box 19029
Houston, TX 77224

Vista Chemical Europe
Hilton Tower
Boulevard de Waterloo, #39
81000 Brussels, Belgium

Vista Chemical Far East, Incorporated
Kasumigaseki Building, 25th Floor
Post Office Box 110
Tokyo, Japan 100

Wacker Silicones Corp.
3301 Sutton Rd.
Adrian, MI 49221

Washington Penn Plastic Co., Inc.
2080 North Main St.
Washington, PA 15301

Wellman Inc./Plastics Div.
Hwy. 41
Johnsonville, SC 29555

Witco Chemical Corp./
Organics Div. & Sonnneborn Div.
277 Park Ave.
New York, NY 10017

Witco Canada Ltd.
2 Lansing Square/Suite 1200
Willowdale, Ontario M2J4Z4

Witco Chemical Ltd. (UK)
Union Lane, Droitwich
Worcester WR9 9BB, UK

Witco S.A.
Rue Gravetel
Saint Pierre les Elbeuf 76320, France

Witco B.V.
P.O. Box 5
Koogaan de Zaan, the Netherlands

www.ingramcontent.com/pod-product-compliance
Lightning Source LLC
Chambersburg PA
CBHW060754220326
41598CB00022B/2431